Y0-CBW-957

The Historical Development
of
Quantum Theory

Jagdish Mehra
Helmut Rechenberg

The Historical Development
of
Quantum Theory

VOLUME 1

Part 1

The Quantum Theory of
Planck, Einstein, Bohr and Sommerfeld:
Its Foundation and the Rise of Its Difficulties
1900–1925

Springer-Verlag

New York   Heidelberg   Berlin

Library of Congress Cataloging in Publication Data
Mehra, Jagdish.
The quantum theory of Planck, Einstein, Bohr and Sommerfeld: Its foundation
and the rise of its difficulties, 1900–1925.
(The historical development of quantum theory/Jagdish Mehra and Helmut
Rechenberg; v. 1, pt. 1)
1. Quantum theory—History. I. Rechenberg, Helmut, joint author
II. Title. III. Series: Mehra, Jagdish. The historical development of quantum
theory; v. 1, pt. 1.
QC173.98.M44   vol. 1, pt. 1   530.1'2'09s   81-18451
                 [530.1'2'09]   AACR2

All rights reserved. No part of this book may be translated or reproduced in any
form without written permission from Springer-Verlag, 175 Fifth Avenue,
New York, New York 10010, USA.

© 1982 by Springer-Verlag New York Inc.
Printed in the United States of America.

9 8 7 6 5 4 3 2 1

ISBN 0-387-90642-8   Springer-Verlag   New York   Heidelberg   Berlin
ISBN 3-540-90642-8   Springer-Verlag   Berlin   Heidelberg   New York

Ô récompense après une pensée
Qu'un long regard sur le calme des dieux!
PAUL VALÉRY
Le Cimetière marin

I took pains to determine the flight of
crook-taloned birds, marking which were
of the right by nature, and which of the
left, and what were their ways of living,
each after his kind, and the enmities and
affections that were between them, and
how they consorted together.
AESCHYLUS
Prometheus Vinctus

Contents

Preface

Bertrand Russell quotes Callimachus, the Alexandrian poet, to say that 'A big book is a big evil!'[1] Russell himself wrote some big books which had a large influence. As evils go, this book is a minor one. Nevertheless it calls for an explanation. The discovery and development of quantum theory in the twentieth century is an epic story and demands appropriate telling. This story cannot be told in the fullness of its glory without analyzing in some detail the multitude of problems which together came to constitute the fabric of quantum theory. Much more than the relativity theories, both special and general, which completed the edifice of classical mechanics, the quantum theory is unique in the history of science and the intellectual history of man: in its conceptions it made a complete break with the past and fashioned a new worldview about the structure of matter and radiation and many of the fundamental forces of nature.

My own intellectual development and choice of occupation, how I began to pursue the historical and conceptual development of modern physics, and how this work came to be written, are bound up with the story of my encounters with major quantum physicists and some literary figures. I shall briefly narrate it here. Since my early youth I had a number of heroes among physicists, mathematicians, philosophers and literary personalities, about whose lives and achievements I wanted to find out more. In my studies I had been greatly intrigued by the theories of modern physics, especially relativity and quantum theory, and had encountered the names of Planck, Einstein, Bohr, Born, Pauli, Heisenberg, Dirac and Schrödinger. Among all these personalities the name of Albert Einstein had a powerful attraction, and I read all about Einstein and his work—including his own scientific and general writings—that I could lay my hands on. In spring 1952 I wrote an essay with the pretentious title of 'Albert Einstein's Philosophy of Science and Life,' which only youthful exuberance could allow. This essay won a small prize in a competition, but the real reward came from Einstein himself. He sent me a one-sentence letter which said:

112 Mercer Street
Princeton, N. J.
July 2nd, 1952

Dear Sir:
    Apart from too unwarranted praise I find your characterization of my convictions and personal traits quite veracious and showing psychological understanding.
With kind greetings and wishes,
Sincerely yours,
Albert Einstein

[1] Bertrand Russell: *Wisdom of the West*, Crescent Books, Crown Publishers, Inc., New York, 1978, p. 5.

This kind letter inspired me for a long time and I hoped that one day I shall have occasion to meet some of the great quantum physicists in person, and that I shall pursue with them the question of how they had come upon their discoveries. It happened sooner than I had expected.

In the following I will mention the names of many famous physicists. I should at once state that I do not do so to express any easy familiarity with them. It was an enormous privilege and good fortune that I encountered them in the course of my work on the historical development of physics in the twentieth century, a privilege and fortune for which I am deeply grateful. Many of them became my 'sources.' I came to have great respect and, in a number of cases, affection for them; some of them began by being my heroes and became my good friends. Oscar Wilde says somewhere that: 'There are two tragedies in life: one is not to meet any of your heroes, the other is to meet them all.' In my case, I met most of them, and I have had the greatest pleasure.

Encounters with Quantum Physicists

In fall 1952 I received a fellowship, awarded by the German industrialist Oskar Henschel, with the help of which I could pursue advanced studies in almost any university of Western Europe that would accept me. My ambition was to go to Göttingen to study under Werner Heisenberg, whose name and work, after Einstein's, captivated me the most. But in mid-November 1952 I first went to Zurich to see Wolfgang Pauli. At the *Eidgenösische Technische Hochschule* (E.T.H.), at 35 *Gloriastrasse*, on Tuesday, 11 November, my first encounter was with Walter Thirring, who had just come down from Göttingen after a period of stay at the *Max-Planck-Institut für Physik und Astrophysik*, where Heisenberg was. Thirring told me that not much active work in physics was going on at Göttingen, and that Zurich was a better place to be. That afternoon I attended the theoretical physics seminar at the E.T.H., at which Robert Schafroth spoke about some work on superconductivity. In the lecture hall I sat in the back row next to a much older person and struck a conversation with him before the seminar began. He told me he was Otto Stern. I knew about the Stern–Gerlach experiment and was delighted to make the acquaintance of a prominent quantum physicist right away. Stern told me he was visiting Pauli. I told him that I had also come to see Pauli, but that I was rather awed by the prospect because I had heard about his terrible temper. Stern reassured me that Pauli was actually very kind. We agreed to meet the next day for lunch. Next morning, before meeting Pauli I saw Paul Scherrer, who was kind and charming. He also assured me that behind his gruff manner Pauli was really a friendly person, and he offered to take me to Pauli and make the introduction. Thus braced, I went in to see Pauli. He looked like the owner of a delicatessen, who consumed his own wares more than was good for him, but his eyes had a spiritual radiance. He received me kindly enough and talked with me about my background and education, about

Zurich and the E.T.H., and told me what the various theoreticians were doing at the *Physikalisches Institut der E.T.H.* As he did so he kept bobbing his head up and down and rocking his body from side to side. He asked me what I wanted to do. I expressed my wish to work on some aspect of quantum theory, perhaps on a problem of quantum electrodynamics. I also mentioned that I wanted to learn about the development of quantum theory from various angles and that some day I hoped to write about it. Pauli laughed, at what probably seemed to him to be my audacity. He said the 'creators' of modern quantum theory were all still around and I could find things from them, 'but you will have to learn an awful lot to be able to write about the development of quantum theory; it may seem easy but it is not.' He said that since I had a fellowship and was provided for, I could work at his Institute if I so desired. However, he said, I should meet Heisenberg before deciding on Zurich as a place to study, but if I chose Göttingen he would still be glad to talk to me and 'you can use me as a source.' I told Pauli that I was going to Göttingen to see Heisenberg and would inform him how things developed, and took his leave.

Afterwards I had lunch with Stern in a bistro at *Limmat-Quai* near the railway station in Zurich. I reported to him about my meeting with Pauli. Stern told me that Pauli had himself witnessed the entire development of quantum theory and participated in it since 1920, that Pauli alone was really qualified to write about it, and that of all people he had the 'moral right to do so.' But Stern was not sure whether Pauli would actually ever do it.

That same evening, Wednesday, 12 November, I took the night train from Zurich to Göttingen; it was the express which travelled from Rome to Hamburg, Copenhagen and Stockholm. The train arrived in Göttingen at 4:30 A.M. It was still very dark, and I was the only passenger to alight on the platform. The porter took me to the waiting room with my luggage. I anxiously waited in the coffee shop, then at 8:30 I took a taxicab to the *Max-Planck-Institut für Physik*, *Böttingerstrasse* 4. I did not speak any German, but somehow I made it clear enough to the concierge that I wanted to see Professor Heisenberg. The concierge called Fräulein Giese, Heisenberg's secretary, a kind and courteous young lady, who took me to meet Harry Lehmann, then Heisenberg's assistant. He sat in a small office in which there was barely room for another chair. I had already informed Heisenberg about my arrival in Göttingen, and Lehmann knew I was coming. Lehmann spoke good English and he discussed my plans with me. He promised that he would soon arrange my meeting with Heisenberg. At about 9:30 A.M. a middle-aged, well-dressed gentleman looking like a prosperous haberdasher walked into Lehmann's office. He and Lehmann talked to each other in German; the visitor gave a nod to me with a smile, shook hands with me and soon left, leaving behind a stack of papers for Lehmann. Lehmann told me that this was Professor Heisenberg and my appointment to see him was fixed for 10 o'clock the next morning, Friday, 14 November. I felt very happy that I had already been introduced to the famous Heisenberg without realizing it. Lehmann introduced me to several members of the Institute, and I spent the day with them: Reimar Lüst, Sebastian von Hörner, Kurt Symanizik, the American physicist Richard Farrell, the Italian physicist Paolo Budini, and a few others. They were all much senior to me; I was over ten years younger than Symanzik, who was then the youngest member of the Institute. I found the German and foreign physicists I met at the Institute very congenial. I was invited to spend that night at the apartment of a member of the Institute, but I was full of anticipation about my meeting with Werner Heisenberg the next morning.

At the appointed time I presented myself at Heisenberg's office. Fräulein Giese immediately ushered me into his inner office, past a double door, the inner door made of foam rubber and cork as in a sound studio. It was an elegantly furnished, comfortable office, done in soft blue, with a large uncluttered executive desk and chair which Heisenberg used, a light beige sofa and easy chairs for visitors, and a huge vase of fresh cut flowers at the coffee table. Heisenberg received me warmly. He wore a medium-gray worsted suit and a striped blue-and-white tie, with a gold pin bearing the letter $\hbar$ (which I later learned was a present from the members of the Theoretical Physics Seminar at Leipzig) stuck into the knot of his tie. He was gentle and soft-spoken. He spoke excellent English with a faint touch of German–Danish accent, a reminder of the fact that he had learned his English in Copenhagen, but it had been perfected over the years.

Heisenberg made me feel at home and welcome and we talked about many things: about science, literature and theatre, and about Göttingen, Munich and Copenhagen. He told me about the great tradition of Göttingen in mathematics and physics and about the cultural life of the city. Then, gently but inevitably, he asked me about my background, education, and the subjects I had studied and liked (though he had already received the transcripts of my academic record earlier). He also asked me about my plans, but in a tone that he earnestly wished to encourage me in any worthwhile pursuit. I told him what I had told Pauli: that I wanted to learn all about quantum theory and its historical and conceptual development, that I wanted to have the opportunity of talking to all the living pioneers, and that hopefully I would write about it some day. I could not fail to notice the contrast with Pauli: Pauli, corpulent, overbearing and forbidding; Heisenberg, gentle, urbane, civilized and, at fifty-two, in the best of health, with a rather round face and balding head. Heisenberg thought that this was a 'wonderful idea,' that quantum theory had 'introduced a new way of thinking about physical problems' and had 'produced new insights about the workings of nature.' 'It would be wonderful,' he said, 'to write about the historical development of quantum theory in a rigorous and detailed manner.' He explained to me that probably the best way to proceed was to learn about quantum physics by doing research problems. 'In this way,' he said, 'you would learn the theory from the inside. And you can build on your knowledge by studying the original papers and finding out about their scientific and human background, about how they came to be written. Work on actual research problems of today will help you master the subject more easily.' At the institute, he said, the main fields of research were nuclear physics, elementary particle physics and quantum field theory. He told me that numerous theoreticians at the Institute were able to work with the S-matrix and were well versed in quantum field theory. He suggested that I should study Walter Heitler's book on quantum theory of radiation, Gregor Wentzel's book on quantum theory of fields, and Freeman Dyson's notes of lectures (delivered at Cornell University) on 'quantum electrodynamics,' a few copics of which were in circulation among the theoreticians at the *Max-Planck-Institut für Physik*. He advised me to study carefully the papers of Tomonaga, Schwinger, Feynman and Dyson on quantum electrodynamics. Heisenberg also assigned to me a research problem about which Paul Scherrer had written to him from Zurich: this was to construct a theory of neutron–deuteron scattering at 3.24 meV. Heisenberg said that I was welcome to work at the Institute, and he invited me to stay in the guest quarters at the Institute itself until I had found lodgings of my own. This was a most rewarding meeting and I felt full of admiration for Heisenberg and hopeful about my possibilities.

A desk was assigned to me in an office and I immediately became engaged in the tasks given to me. In addition, I attended courses given by Heisenberg, Carl Friedrich von Weizsäcker and Richard Becker at the University of Göttingen, as well as other courses in the institutes of mathematics and philosophy. I carried a full load of activities and enjoyed the challenging assignments.

There were regular seminars at the *Max-Planck-Institut für Physik*: Wednesday mornings on nuclear physics and Thursday afternoons on quantum field theory. On Thursdays there would be tea and cookies after the seminar. After the first Thursday seminar on field theory and tea, Heisenberg invited me to go for a walk with him. The Institute was situated at the edge of the city, and soon we were walking on a meadow close to the woods in the falling darkness. I took advantage of this opportunity to ask Heisenberg about the old days of quantum theory in the early 1920s when he became Sommerfeld's disciple. Heisenberg first told me about himself, about his growing up and schooling in Munich, about his interests as a youngster, and about the difficult economic and political times during and following the First World War. He then told me about his first encounter with Arnold Sommerfeld and how at the very beginning of his university studies he got into research. During the walk I interrupted him with many questions and he responded to them candidly. On my return to my room I wrote down an almost verbatim account of our conversation and gave it to Heisenberg the following Monday to check, add or subtract. The following Thursday he again asked me to take a walk, and then it became our custom to take a short or a long walk, depending on the time available to Heisenberg, on most Thursdays when he was in town and weather permitted. As I learned more about quantum theory from original papers, I was able to ask him more searching questions about the context in which the problems had arisen, the manner in which they were approached, and the human interactions at the time the work was done; I regularly wrote down my notes and had them edited by him. For me this process was wonderful and Heisenberg's company inspiring.

But I noticed that Heisenberg enjoyed it too. He was young enough to remember still all the details of everything that had happened and old enough to enjoy looking back at the exciting moments of the past and telling about them; I was the aspiring chronicler of these ideas and events.

I pursued the story of the period of the creation of quantum mechanics during at least a dozen encounters with Heisenberg. It had been the most thrilling period of his life, and he was full of memories, feelings and sentiment about it. He told me about his visit to Berlin to give a colloquium on quantum mechanics (on 26 April 1926) at the invitation of Planck and Einstein. It was a great day for Heisenberg to address the famous Berlin Colloquium with many illustrious physicists in the audience. After his talk and discussion in the colloquium he walked with Einstein to his house and they had their first conversation in which Einstein wanted to know how Heisenberg had come upon his fundamental idea leading to the discovery of quantum mechanics. That evening Heisenberg went to a party in his honour at the *Studentenhaus*. 'I was very happy,' he said. 'I played a Beethoven sonata on the piano for the students. I had come of age as a physicist!'[2]

My discussions with Heisenberg continued even after I left Göttingen in April 1955. I saw him over the years in Göttingen, Munich, Varenna, Lindau, Geneva, Trieste and Brussels, and our conversations were resumed without suffering from the intervening lapses of time. When we did not go for walks and talked in his office, I often recorded our conversations on tape. My last conversation with him —dealing with his latest views about the nature of elementary particles, and some questions I still wanted to discuss with him about Einstein, Born and Pauli —took place in his office on 25 February 1975. During the last year of his life he was very ill and died on 1 February 1976. The record of my edited notes and transcripts of taped conversations with him covered several hundred typed pages. He had read the first draft of Volume 2 of this work, dealing with the discovery of quantum mechanics, and approved it.

During my first semester at Göttingen there came three physicists whose visits were especially significant for me; Sin-Itiro Tomonaga from Japan, Aage Bohr from Copenhagen, and Max Born from Edinburgh. Tomonaga had been with Heisenberg in Leipzig many years previously. He was lean of build, delicate of health and features, friendly, charming and simple, and I found immediate contact with him. It had now become my custom to keep notes and tapes of conversations with major physicists.[2a] I had the opportunity of renewing my

[2] The mathematician Salomon Bochner attended Heisenberg's colloquium in Berlin and was also present at his piano recital. Bochner confirmed Heisenberg's recollection to me.

[2a] The records of my conversations and interviews with major quantum physicists will be deposited with the archives of the *Max-Planck-Gesellschaft zur Förderung des Wissenschaften* in West Berlin, Federal Republic of Germany. All first person accounts and references to conversations in the text and footnotes refer to my encounters with the quantum physicists in question, unless otherwise identified. Complete references to all quotations indicated as 'Conversations' and 'Interviews' will be provided when all references to them are collected together in the last volume of this work.

contact with Tomonaga many years later; in June 1960 I spent several days with him in Tokyo and pursued further the history of quantum electrodynamics. On the latter occasion I also had conversations with Hideki Yukawa in Kyoto. Aage Bohr was young and robust and full of confidence, but polite and friendly, and he gave a talk about the moments of inertia of nuclei; this work led to the line of investigations with Ben Mottelson that would win them the Nobel Prize in Physics in 1975. At that time, however, I found personal contact with Aage Bohr and, through him and Heisenberg, with Niels Bohr. These contacts were maintained in the following years.

Early in 1953 Max Born came for a visit to Göttingen. He was about to retire from the Tait Professorship of Natural Philosophy at Edinburgh, Scotland, and was looking for a place to live (close to Göttingen) in Germany, where he was to receive full pension as a former university professor. He stayed in a well-appointed guestroom in the *Akademische Burse*, where I also lived. I walked with Born every morning after breakfast to the Institute, and during two weekends we went on walking tours of Göttingen and its environs. Born showed me numerous houses and places connected with the names of great mathematicians and physicists and historic events, such as the house on *Wilhelm-Weber-Strasse* 29, where David Hilbert had lived, and the house in *Merkel Strasse* in which Max Planck had lived after the war. We made a tour of the *Hainberg*, went to *Café Rohn*, and walked to *Nikolausberg* and the *Bismarck Turm*. I had explored all these places before, but coloured with the many memories of an old-time resident of Göttingen, this sightseeing with Max Born had a special charm. During these two weeks with Born I had many discussions with him about the early days of quantum theory and relativity, his work on specific heats with von Kármán, his work on matrix mechanics, and his associations with Einstein, Hilbert, Minkowski, Felix Klein, Bohr, Franck, Pauli, Heisenberg, Hund, Jordan and Dirac, and numerous others. Born disliked Dirac's notation of *bra* and *ket* vehemently; I found it strange for a man who was such a keen mathematical formalist himself. He discussed almost everything quite calmly; he had a genuine feeling of happiness about his associations with Einstein, Hilbert and Minkowski, and much pride in the school of theoretical physics he had founded in Göttingen. He expressed a sense of intimidation in talking about Pauli, and a touch of resentment about the glory of the discovery of quantum mechanics which, in a large measure, had gone to Heisenberg. His spirit was soothed with regard to the latter in 1954, when he was awarded the Nobel Prize in Physics (which he shared with Walther Bothe) for his discovery of the statistical interpretation of the wave function. Born settled in Bad Pyrmont, where I visited him from time to time until 1968 and pursued with him the discussion about the historical development of modern physics, especially quantum theory.

In spring 1953 I went on a long trip to meet several quantum physicists and traditional philosophers with introductions from Heisenberg and von Weizsäcker: Niels Bohr in Copenhagen, Pascual Jordan in Hamburg, Friedrich Hund in Frankfurt, Wolfgang Pauli in Zurich, Louis de Broglie in Paris, and the philosophers Karl Jaspers in Basle, Gabriel Marcel in Paris, Martin Heidegger in Freiburg-im-Breisgau, and the Catholic philosopher and theologian Romano Guardini in Munich. In Paris I also met the Indian physicists Homi Bhabha and S. N. Bose.[3] I started a series of discussions with Jordan, Hund, Pauli and de Broglie that were to continue for many years to come.

Pascual Jordan treated me most kindly. He had the disadvantage of stuttering badly, but he was a warm-hearted person. He told me how he had pursued many threads of knowledge as a 'natural philospher' and not just as a theoretical physicist. He had lively recollections of his work on matrix mechanics and quantum field theory, and he cherished pleasant memories of his associations with Heisenberg, Born, Pauli, Dirac, Wigner and Oskar Klein, although all of them had overshadowed him: in the fundamental papers on quantum mechanics and quantum field theory his name was always that of the second or the third author and the other partners had made bigger names for themselves for other, independent contributions, hence Jordan did not receive much credit even for what he did with them. In talking to him about matrix mechanics (the first work with Born, then with Born and Heisenberg) it soon became evident that Jordan had played a central role in its development; I also learned from others that as a formalist Jordan had been an equal of Pauli. I did not meet Jordan again until 1964, when I spent several days with him in Hamburg. From then on I kept fairly regular contact with him and learned many things from him about the development of quantum theory and the various personalities connected with it. In September 1972 he attended the Symposium on the Development of the Physicist's Conception of Nature in the Twentieth Century, which I had organized in honour of P. A. M. Dirac's seventieth birthday.[4] In 1976 he read the first draft of Volume 3 of this work, dealing with the formulation of matrix mechanics and its modifications, and recommended me most graciously for the award of the *Humboldt-Preis* of the *Alexander von Humboldt-Stiftung* and the research grant of the *Alfred Krupp von Bohlen und Halbach-Stiftung*. From then until his death on 31 July 1980 we exchanged a number of letters; he had read the final version of Volume 3 and approved it.

From 1953 to 1976 I met Louis de Broglie in Paris on seven different occasions. Always punctual, kind and considerate, he told me many things about the background of his family, about his brother Maurice, about his work on wave mechanics and its reception by Langevin and Einstein, and about the first and fifth Solvay Conferences in Brussels that had meant so much to him; about

[3] I visited S. N. Bose many years later in Calcutta, India, on 30 August 1970. In 1974, after his death on 4 February, the Council of the Royal Society invited me to write his biographical memoir. (See J. Mehra: *Satyendra Nath Bose, Biographical Memoirs of Fellows of the Royal Society* **21**, pp. 117–154, 1975.)

[4] See J. Mehra (Ed.): *The Physicist's Conception of Nature*, D. Reidel Publishing Company, Dordrecht, Holland/Boston, 1973.

the first Conference he had learned from his brother and its proceedings,[5] while he had attended the fifth himself.[6]

In Göttingen I had the great privilege of knowing Otto Hahn. He was a likable, charming and kind man. Very often we walked together from the Institute complex (where he had his office as President of the *Max-Planck-Gesellschaft*) to the bus stop in *Weenderstrasse* opposite the *Rathaus*. Occasionally I had lunch with him in the Institute's cafeteria. He told me many stories about Rutherford, about his work with him, and about the years in Berlin. He recalled the months during which he was interned at Farm Hall near Cambridge, England, with a number of other German nuclear scientists after the war, and he gave me impressions of his various co-internees. He was one of the most unpretentious of men, warm and friendly, and oblivious of his great reputation as the man who had discovered nuclear fission.

Another quantum physicist whose acquaintance I made in Göttingen was Max von Laue. He came to the Institute from time to time, always riding his bicycle, with clips to hold the cuffs of his trousers from getting into the chain. He was very accessible, kind and rather old-worldly. I learned a lot from him about the early development of relativity theory, about his discovery of X-ray interference patterns in collaborations with Walter Friedrich and Paul Knipping, about his relations with Planck, Einstein, Pauli and Heisenberg, about the scientific community in Europe during the Second World War, and the internment of German nuclear scientists after the war at Farm Hall. Von Laue was always courteous and candid.

On numerous occasions I had discussions with C. F. von Weizsäcker about science, literature and philosophy. He was learned, brilliant and civilized. He spoke German and English eloquently and was a master of the speculative argument. His presence at the Thursday afternoon teas at the *Max-Planck-Institut für Physik* added much lustre to the spontaneous conversations that took place there. Once at tea he built a whole case on how much science (including navigation) and geography Columbus knew on the basis of his reported voyages. He had a scintillating mind, and his monologues could be captivating. He had a controversial reputation though: among the physicists he was regarded as a philosopher and among the philosophers as an atomic physicist. Often he was referred to as a 'quantum theologian,' indicating at once his capability as a physicist, philosopher and public speaker with a religious zeal. I had many conversations with him about his ideas, about Heisenberg, about Leipzig and Copenhagen (where he had studied with Heisenberg and Bohr, respectively), and about Aldous Huxley, Bertrand Russell and the Indian mystic and philosopher

---

[5] *La Théorie du Rayonnement et les Quanta, Rapports et Discussions de la Réunion tenue a Bruxelles, du 30 Octobre au 3 November* 1911, *Publiés par MM. P. Langevin et M. de Broglie,* Gauthier-Villars, Paris, 1912.

[6] *Electrons et Photons, Rapports et Discussions du Cinquième Conseil de Physique tenu à Bruxelles du 24 au 29 Octobre* 1927, Gauthier-Villars, Paris, 1928.

Sri Aurobindo. In the 1960s I visited him in Hamburg, and in 1972 he attended the Dirac Symposium at Trieste at my invitation, where he gave an extemporaneous lecture on physics and philosophy.[7] Early in 1975 I had some very serious conversations with him about the activities of certain prominent German scientists during World War II (in Geneva, where my friend Charles Enz and I had invited him to give the Joseph Jauch Memorial Lecture). In recent years I have met him occasionally in Munich, a conversation with him always being a memorable experience.

In Göttingen I also made the acquaintance of the mathematicians Theodor Franz Eduard Kaluza and Carl Ludwig Siegel. Kaluza had made important contributions to geometry and general relativity, and Siegel had later become one of the world's leading mathematicians. Siegel told me many things of interest about the Göttingen tradition in mathematics, as did Richard Courant whom I met many years later in New York.

In my meetings with Niels Bohr in Copenhagen in 1953 and 1954 I explored the story of his relations with J. J. Thomson and Ernest Rutherford, his early work on the Bohr atom, but especially the period from 1922 to early 1930s, beginning with Pauli's arrival in Copenhagen up to the Bohr–Rosenfeld analysis of the field quantities in quantum electrodynamics. Bohr was patient and loved to reminisce; though it was not easy to comprehend him, it was always interesting to listen to him. One had to be very attentive, what with the soft Danish accent and his habit of swallowing not only words but entire phrases. Some of the most fascinating things were his remarks about the contributions of Pauli, Kramers, and Heisenberg. The young Heisenberg had been for Bohr like a messiah who had brought forth the solution of the quantum riddle and dispelled all doubts and gloom. I also discussed with Bohr his visits to Berlin in April 1920, when he first met Einstein in person, and to Leyden in December 1925 to attend the golden jubilee of Lorentz' doctorate, when he and Einstein gave their blessing to the hypothesis of electron spin put forward by Uhlenbeck and Goudsmit (and the following furor with Pauli). In 1954 I also met Oskar Klein in Copenhagen, but I did not see him again until 1968 in Trieste and again in 1974 at Leyden. I did not see Bohr again until summer 1959 when on Thursday, 25 June, he came to inaugurate the John Jay Hopkins Laboratory of Pure and Applied Science in San Diego, California.[8] On that occasion I spent some time with him and he invited me to visit him in Copenhagen to talk further about the development of quantum theory from his perspective. 'And perhaps you will tell me what you have found out yourself in this very fascinating matter,' he said. It did not happen until June 1962. Then I was invited to attend the symposium of the

---

[7] See C. F. von Weizsäcker, 'Physics and Philosophy,' in footnote 4, Chap. 40, pp. 736–746.

[8] Niels Bohr gave an address on 'Science and Technology.' On this occasion I also made the acquaintance of Lothar Nordheim, who had worked with David Hilbert at Göttingen in the 1920s on the mathematical foundations of quantum mechanics. He told me a good deal about Hilbert and the young John von Neumann.

Nobel Prize Winners in Physics in Lindau at Lake Constance as an observer. Many laureates were present on that occasion: Edward Appleton, John Bardeen, Niels Bohr, Max Born, Walter Brattain, John Cockcroft, James Franck, Werner Heisenberg, George de Hevesy, Robert Hofstadter and Harold C. Urey. On this occasion I interviewed James Franck and had several discussions with Niels Bohr. I was able to cover numerous points that had interested me about Bohr's role in the development of quantum theory from 1922 to 1927: the formulation of the dispersion-theoretic approach, Bohr's discussions with Schrödinger in summer 1926 about the interpretation of the wave function, his discussions with Heisenberg in February 1927 about the uncertainty principle, the Como Conference in September 1927 and the Solvay Conference in Brussels in October 1927, and finally the rise of the interpretation of quantum mechanics and his discussions with Einstein at the fifth and sixth Solvay Conferences that were continued afterwards. Bohr talked with me kindly and patiently enough, but he was rather unhappy on this occasion in Lindau, for some tension had developed between him and Heisenberg. Somehow the subject had come up again of Heisenberg's meeting with Bohr in Copenhagen at the end of October 1941 when the question of the atomic bomb had been broached and the misunderstanding that had persisted between Bohr and Heisenberg ever since.[8a] Even on this occasion over twenty years later Heisenberg was not able to dispel Bohr's doubts about his intentions in that conversation during World War II, and Bohr was quite aggravated. I had looked forward to seeing Bohr again in Copenhagen at Christmas that year, but he died on 18 November 1962.

In spring 1954 I had written to Hermann Weyl about the possibility of meeting him, especially if he would be in the vicinity of Göttingen sometime. He proposed that I come to the 7th International Congress of Mathematicians that was taking place that year in Amsterdam from 2 to 9 September. I thought that was a marvellous opportunity, and I had three meetings of two hours each with him. He told me about the old days in Göttingen and Zurich, about his relations with Hilbert, Klein, Born, Pauli and Schrödinger, about the work on matrix mechanics in Göttingen and wave mechanics in Zurich (where he was at the time), and his own work on general relativity, unified field theory, and quantum mechanics, as well as his lectures on group theory and quantum mechanics at Zurich. At one point he mentioned that there used to be a feeling that the *Eidgenössische Technische Hochschule* (E.T.H.) in Zurich was considered a first class waiting room ('*Wartesaal erster Klasse*'), where one received calls to go to Göttingen or Berlin; this had happened to Minkowski and then to Weyl himself, both of whom were called to Göttingen, and to Schrödinger, who went to Berlin,

---

[8a] For an account of Heisenberg's visit to Copenhagen in October 1941 and the controversy surrounding it, see: Robert Jungk, *Brighter Than a Thousand Suns*, Harcourt Brace Jovanovich Inc., New York, 1958, Chap. 6, pp. 99–102; Elisabeth Heisenberg; *Das politische Leben eines Unpolitischen: Erinnerungen an Werner Heisenberg*, R. Piper & Co., Verlag, Munich, 1980, V. Kapitel, pp. 89–103.

but Hitler changed all that by dismissing Jewish professors and destroying the schools of mathematics and physics in Göttingen and Berlin.[9]

At the end of my fellowship in April 1955 I left Göttingen. I had learned much about the development of quantum physics, as well as about the relations among the quantum physicists through direct contacts with them. However, my stay at Göttingen had not been as successful as I could have wished. I had lived there as a gentleman-at-large with a handsome fellowship which permitted me a convenient life-style and travel. I had enjoyed the closeness to Heisenberg and the accessibility to other major physicists and mathematicians. I was bent upon finding all I could about the development of quantum physics in a hurry. However, I was too young, and not yet practiced enough in advanced mathematical techniques to take the fullest advantage of the possibilities of research at the frontiers of quantum field theory which the *Max-Planck-Institut für Physik* and several of its members offered.

In May 1955 I took the open competition for the United Kingdom Scientific Civil Service. I was selected as a Scientific Officer with the Department of Scientific and Industrial Research, now the Science Research Council, to begin my first job in August: research on problems of theoretical hydrodynamics and magnetohydrodynamics at a laboratory of D.S.I.R. Several wonderful encounters occurred that summer. C. P. Snow, as a Civil Service Commissioner, had interviewed me in the competition; soon we became very good friends, indeed friends for life. Snow had a profound admiration for the founders of quantum theory, especially Dirac, and he always continued to encourage me in my work on the historical development of quantum theory. In spring and summer 1955 I also met and had discussions with Aldous Huxley, Bertrand Russell, T. S. Eliot, E. M. Forster, Arthur Koestler, J. B. Priestley, Arnold Toynbee, and Hesketh Pearson (biographer of George Bernard Shaw, Oscar Wilde, and Sir Walter Scott). Of these the most important were meetings in summer 1955 and the following years with Bertrand Russell and Arnold Toynbee, and the meeting and continued friendship with Aldous Huxley. When Huxley learned about my interest in quantum theory and quantum physicists, he told me that quantum theory was the greatest revolution intellectually and scientifically that had happened in the twentieth century and the story of its development was worth writing about. Huxley himself was in contact with Pauli at that time about the influence of archetypal conceptions and the role of the subconscious in intellectual and scientific creativity, as well as the nature of mystical experience common to highest creativity in arts, sciences and religion. (Pauli had written an essay on this subject in a book published jointly with C. G. Jung.[10]) Huxley and I

---

[9] Hermann Weyl told me the following story. In the mid-1930s Hilbert was once seated next to the Prussian Minister of Education at a meeting. The Minister asked Hilbert: 'Herr Geheimrat, I hope the departure of the Jewish mathematicians has not seriously affected the activities of your institute?' Hilbert replied: 'No, Herr Minister, not at all. [Pause] It just doesn't exist anymore.'

[10] C. G. Jung and W. Pauli: *Naturerklärung und Psyche. Herausgegeben von C. A. Meier*, Rascher-Verlag, Zurich, 1952; C. G. Jung; *Synchronizität als ein Prinzip akausaler Zusammenhänge*; W. Pauli; *Der Einfluss archetypischer Vorstellungen auf die Bildung naturwissenschaftlicher Theorien bei Kepler*.

became friends and we had many things to talk about. He was one of my literary heroes. His knowledge was so vast, his sensibilities so refined, his personal and intellectual elegance so acute, that it was a transforming experience to be in his company. A conversation with him often began with his question to me: 'And why are precious stones precious?' The answer to this question would always lead us into new perspectives about human possibilities and potentialities. The answer in a variety of ways was not the monetary worth of precious stones, but their value as symbols of intellectual and spiritual experience, of light, beauty and truth, of eternity and the divine spark in the human being—symbols of transcendental experience. That's why the cathedrals of Chartres and Rheims celebrate scriptural, spiritual and pastoral scenes—scenes of human glory and suffering, of aspirations, hopes and dreams—in light and colour through stained-glass windows. Huxley encouraged me greatly to write about the historical and conceptual development of quantum theory. We continued to meet regularly until spring 1962. His last visit to my house in Orange, California, was on 6 May 1962. We talked about precious stones, English and French poetry, perennial philosophy and quantum theory, and I drove him back to where he was then staying at Mulholland Drive in Hollywood. This was the last I saw of him. I left for Europe in June 1962 for over two years. Huxley died in Hollywood on 22 November 1963.

In spring 1955 I visited Cambridge, where at St. John's College I met Paul Adrien Maurice Dirac. I saw him two or three times during 1955–1956, and then again only ten years later at the Symposium of the Nobel Prize Winners in Physics at Lindau in June 1965, to which I was invited as an observer. Among the Nobel laureates attending this symposium were Max Born, John D. Cockcroft, P. A. M. Dirac, Werner Heisenberg, Gustav Hertz, J. H. D. Jensen, Rudolf Mössbauer, George P. Thomson and Hideki Yukawa, most of them quantum physicists. On this occasion I had much contact with Born, Heisenberg, Hertz, Thomson and Yukawa, but most of all with Dirac. It was only then that my contact and friendship with Dirac began to flourish. From then on we met from time to time at various places: New York, Miami, Austin, Trieste, Cambridge, Lindau, Geneva and Chicago. From the first encounter in May 1955, which I described at the Dirac Symposium in Trieste in September 1972,[11] through his visit to me in Geneva in the beginning of July 1973,[12] until most recently a few months ago, my encounters with Paul Dirac have been some of the most encouraging and heartwarming of any I have had with the great quantum physicists, and having known him as a friend during many years has been one of the greatest privileges of my life.

[11] See my remarks in footnote 4, Chap. 44 ('The Banquet of the Symposium'), pp. 817–818.

[12] After the Nobel Prize Winners' Symposium at Lindau in the beginning of July 1973, Paul Dirac and his wife Mancie visited me in Geneva for several days. He had recently received the Order of Merit and told me that he had brought along the insignia to show me; I was most touched. Dirac, Mancie, and I went on an extensive sightseeing trip of French-speaking Switzerland by car, and I had the opportunity of having some of the most personal conversations with Dirac.

On 3 July 1973, at the Reformation Wall at the University of Geneva, Dirac and I had a long conversation. Dirac suddenly remarked to me; 'When I talk with you I almost become loquacious.'

In May 1955 I also met P. M. S. Blackett at the Imperial College in London. I did not see him again until fall 1968 at the University of Chicago, and then again several times while he was President of the Royal Society. Behind a gruff exterior he hid a heart of gold. I learned many things from him about Rutherford and his own work at Cambridge, about the Kapitza Club, and about his stay with James Franck at Göttingen.

Early in June 1955 I went to meet Erwin Schrödinger in Dublin. I had an appointment to see him on the morning of Tuesday, 7 June, at 10 o'clock at the Dublin Institute for Advanced Studies. When I arrived he was not in his office. His secretary told me that Professor Schrödinger was in the workshop and I could join him there. I found him sitting on a stool in front of a workbench contemplating a transformer and some circuitry. After I greeted him he asked me if I knew why the transformer hummed. I said something about faulty laminations and mechanical resonance and we discussed it a bit, but soon we were talking about his recollections of the old days: first his student days with Franz Exner and Fritz Hasenöhrl in Vienna, then the years of his professorship at the University of Zurich, the exciting months in which wave mechanics was created, and his relations with Einstein, Planck, Bohr, Born, Pauli, Weyl, Wilhelm Wien, Heisenberg and others. I discussed with him at length about the year 1925 when wave mechanics was conceived and 1926 when he wrote his great papers.[13] We talked about de Broglie's papers of 1923 and 1924, and his (Schrödinger's) quick grasp of the significance of de Broglie's thesis; he gave me to understand that he had thought about some similar things in 1922.[14] I stayed with Schrödinger for three days on that occasion. I had fortunately struck a resonance with him. I visited him later on several occasions after his return to Austria in spring 1956, and I maintained contact with him until he died on 4 June 1961.

In a visit to Dublin in December 1956 I also met Cornelius Lanczos and J. L. Synge. Lanczos had done some work in 1925 on a field-theoretical formulation of quantum mechanics, a kind of modification of matrix mechanics, but his work was quickly superseded by the wave mechanics of Schrödinger. Lanczos' admiration for Einstein amounted to hero-worship, which I shared, and I maintained my contacts with him over the years. In September 1972 he attended the Dirac

---

[13] Schrödinger told me that in spring and summer 1925 he had been quite unhappy in Zurich. He did not find the atmosphere very congenial and he longed to return to Austria. He also did not earn enough money at the University of Zurich and was negotiating for a position with the University of Innsbruck. Innsbruck did not make him a good offer and he let the matter drop. In September, when the Born–Jordan paper came out, he disliked its mathematics very much, and looked forward to doing 'something better and more *anschaulich*,' which he did. He also left Zurich soon after his work on wave mechanics was completed: in 1927 he became Max Planck's successor in the chair of theoretical physics at the University of Berlin.

[14] Early in October 1922, Schrödinger had completed a paper, '*Über eine bemerkenswerte Eigenschaft der Quantenbahnen eines einzelnen Elektrons*' ('On a Remarkable Property of the Quantum Orbits of a Single Electron,' Schrödinger, 1922c), in which he derived Bohr's quantum condition by applying Hermann Weyl's extension of general relativity to an electron orbit. There was an interesting parallel between the ideas of Schrödinger and de Broglie, though Schrödinger had not considered a wave being associated with a material particle in his paper.

Symposium at Trieste. On that occasion B. L. van der Waerden gave a talk on Lanczos' early work on quantum mechanics and Pauli's response to it. Van der Waerden did not know that Lanczos was in the audience; when it was pointed out to him that this was so, it provided one of several thrilling moments at the Symposium.[15]

Though I had met Pauli several times between 1952 and 1956, it was only in 1956 that I got a bit closer to him. Then I learned much from him about the development of quantum theory, about his own work on its various problems, about his relationships with various physicists, about his deep interest in the theatre and his love for wine and ice cream. These meetings with Pauli took place in his office at the E.T.H. in *Gloriastrasse* in Zurich and his house at 35 *Bergstrasse* in Zollikon, a suburb of Zurich. In his small study in the house on *Bergstrasse* he had hung framed reproductions of *apsaras*, images of goddesses from the Ajanta caves, which he had brought back from a visit to India (to the Tata Institute for Fundamental Research, Bombay) in December 1952. These images used to reflect his own moods. He would say, 'The goddesses look happy today.' On other days they looked sad or unhappy. During my visit to Pauli in spring 1956 I again met Scherrer and made the acquaintance of Res Jost.

In fall 1957 I went to the United States with teaching appointments in Southern California. Then on 26 February 1958 Pauli arrived in Berkeley, having travelled by boat via Gibraltar to New York, and from there to California. He stayed in Berkeley until the end of May. He gave a series of lectures on the CPT theorem, continuous groups in quantum mechanics, and related matters. I went often to Berkeley during this period, and once spent three weeks there in order to be close to Pauli. In these meetings I discussed with him his work on the exclusion principle, the roles of the various quantum physicists in the development of the theory, the discovery of quantum mechanics, his work with Heisenberg on quantum field theory, his conception of the neutrino and its later discovery, his work on spin and statistics, violation of parity, and his recent work in collaboration with Heisenberg on nonlinear spinor theory (which he was about to abandon), etc. Among various personalities we discussed the roles of Bohr, Heisenberg, Kramers, Oppenheimer,[16] Rosenfeld,[17] Stückelberg, Wentzel, Weyl,[18] Wigner and several others in the development of recent physics. He gave his candid judgments, friendly or acerbic. Before he left Berkeley for his return to

---

[15] See footnote 4, pp. 285–287.

[16] Pauli told me several things about the young Oppenheimer and also about his contacts with him in Princeton after the war. About Oppenheimer's gestures and ponderous remarks he said: '*Er ist eine Karikatur des schaffenden Gottes.*' ('He is a caricature of God in action.')

[17] As a malicious pleasantry, Pauli referred to Rosenfeld as '*der Chorknabe des Papstes*' ('the choir boy of the Pope [i.e., Niels Bohr]').

[18] Pauli liked and appreciated Hermann Weyl both personally and for his important contributions to mathematics and mathematical physics. But he could not resist taking a dig at Weyl with the remark: '*Man muss erst die Schminke von seinem Gesicht abkratzen, um seine Gedanken zu verstehen.*' ('One must first penetrate his façade in order to understand his thoughts.' Literally, 'remove the make-up from his face.')

Europe, I went to a bar with him after dinner one evening. There he reminisced about the various exciting developments of quantum physics. Then he remarked: '*Als ich jung war, glaubte ich der beste "Formalist" meiner Zeit zu sein. Ich glaube, ich wäre ein Revolutionär. Wenn die grossen Probleme kämen, würde ich sie lösen und darüber schreiben. Die grossen Probleme kamen und gingen vorüber, andere lösten sie und schrieben darüber. Ich war doch ein Klassiker und kein Revolutionär.*' ('When I was young I thought I was the best formalist of the day. I thought I was a revolutionary. When the great problems would come I shall be the one to solve them and to write about them. The great problems came and went. Others solved them and wrote upon them. I was of course a classicist rather than a revolutionary.') And then, as an afterthought, he added: '*Ich war so dumm als ich jung war.*' ('I was so stupid when I was young.') Well, he was not to be taken literally. I travelled at the same time as Pauli to Europe. He arrived in Zurich on 2 June 1958, and took part in the International Conference on High Energy Physics (the 8th Rochester Conference) in Geneva, 30 June–5 July. Pauli presided on Session 4 on 1 July in which Heisenberg spoke on 'Research on the Non-Linear Spinor Theory with Indefinite Metric in Hilbert Space.' He took Heisenberg severely to task publicly for the imperfections of the nonlinear theory of elementary particles and Heisenberg's 'excessive optimism.' This was his last outburst on a question concerning physics, public or private. He fell very sick and died on 15 December 1958.

This then was how my initial encounters with the principal architects of quantum mechanics—Bohr, Born, Pauli, Heisenberg, Jordan, Dirac and Schrödinger—came about and then flourished for many years. My great regret was that I did not have the opportunity of meeting and talking with Einstein, Sommerfeld, Kramers and Fermi about their work on quantum theory. However, I did the next best thing: I studied their works and talked to numerous physicists who had known them well and worked with them.

From 1958 to 1975 I developed further resources for my work on the historical development of quantum theory by having discussions with a number of quantum physicists not mentioned above. In the early 1960s I learned about the project on *Sources for History of Quantum Physics*, sponsored by the American Physical Society and the American Philosophical Society, under the leadership of Thomas S. Kuhn.[18a] In almost all cases this archive concentrated its efforts up to the period leading to the formulation of quantum mechanics and just following it. I had enough resources of my own on the background of the formulation of quantum mechanics, which could be handsomely supplemented by the large collections of the *Sources*; the latter included systematic interviews with surviving quantum physicists, unpublished notebooks and manuscripts, and unpublished scientific correspondence of the various physicists. I then decided to restrict my own efforts to those interviews and conversations in which I could pursue the

---

[18a] An inventory of the *Sources* is given in: Thomas S. Kuhn, John L. Heilbron, Paul Forman and Lini Allen, *Sources for History of Quantum Physics*, The American Philosophical Society, Philadelphia, 1967.

development of certain specific problems of quantum mechanics and its applications to nuclear physics, elementary particle physics, solid state physics and, of course, to the whole range of problems of quantum electrodynamics and quantum field theory.

The opportunities for new encounters took place in seminars, symposia and conferences, and the contacts were then developed in private meetings on a more personal basis. The Solvay Conferences on Physics at Brussels (especially, the 1961, 1967, 1970, 1973 and 1978 Conferences) were attended by many of the more eminent quantum physicists; for me these Conferences were memorable events, enriched by encounters with physicists who helped to enhance my knowledge and perspective of the development of quantum theory and its applications to various fields.[19] Other such occasions were provided by: the 11th International Conference on High Energy Physics Geneva, Switzerland, 1962[20]; the International Conference on Particles and Fields, Rochester, New York,

---

[19] See J. Mehra: *The Solvay Conferences on Physics: Aspects of the Development of Physics since 1911*, D. Reidel Publishing Company, Dordrecht, Holland/Boston, 1975.

[20] On this occasion I made the acquaintance of H. B. G. Casimir and spent much time with J. Robert Oppenheimer between the sessions. I had made Oppenheimer's acquaintance in a visit to the Institute for Advanced Study, Princeton, in Spring 1958, then seen him again at Caltech in fall 1959. At Caltech Oppenheimer had addressed the Physics Research Conference (physics colloquium) on Thursday, 22 October 1959. He spoke on 'Why is gravitation related to geometry?,' dealing with some work of Bruno Bertotti's (on 'Uniform Electromagnetic Field in the Theory of General Relativity,' *Phys. Rev.* **116**, No. 5, 1 December 1959) which had been done at the Institute in Princeton. Oppenheimer's lecture drew a large crowd of attentive listeners. The lecture itself was quite poor and I was surprised by the thunderous ovation accorded to it; however, I soon realized that the ovation was not for the lecture but for the hero who was visiting his old haunt. After the lecture I had the opportunity of visiting Oppenheimer at the Athanaeum, Caltech's Faculty Club, where Oppenheimer was staying and we talked about poetry and my interest in the historical development of quantum theory. In Geneva, in July 1962, we had long conversations about Göttingen and poetry. I determined that Oppenheimer did not read Sanskrit, though he was quite familiar with certain Sanskrit scriptures in English translations and liked the sound of Sanskrit poetry. I had a pocket *Bhagavad Gita* with me, and he asked me to recite to him passages from the second and eleventh chapters ('The Path of Knowledge' and 'The Vision of the Universal Form,' respectively). I saw Oppenheimer again at the January 1966 meeting of the American Physical Society at which he gave an impressive lecture on 'Thirty Years of Mesons' (see *Physics Today*, November 1966, pp. 51–58); it bore no comparison to his performance at Caltech; at the APS meeting he looked like a prophet and spoke like one. (See footnote 16.) I met him again in Princeton on 17 November 1966. He was quite sick at that time. We talked about French symbolist poetry and I recited to him the last four stanzas of Paul Valéry's *Le Cimetière marin*, the last of which being such a glowing affirmation of life:

'*Le vent se lève!* . . . *il faut tenter de vivre*!
L'air immense ouvre et referme mon livre,
La vague en poudre ose jaillir des rocs!
Envolez-vous, pages tout éblouies!
Rompez, vagues! Rompez d'eaux réjouies
Ce toit tranquille où picoraient des focs!'

('The wind is rising! . . . We must try to live! The huge air opens and closes my book, the powdered wave dares to spout from the rocks! Fly away, bedazzled pages! Break, waves! Break with your joyous water this calm roof where sails packed liked doves!')

These lines seemed to cheer him. I did not see Oppenheimer again; he died on 18 February 1967. On 1 July 1973, I. I. Rabi and I travelled together from Lindau to Zurich on the train, and I told him about Oppenheimer's more than passing interest in the Hindu scriptures and their Sanskrit poetry. Rabi laughed and said: 'That was his tragedy. He should have studied Hebrew and read the *Torah*. He would have been a happier man.'

1967[21]; the International Symposium on Contemporary Physics, Trieste, Italy, 1968 (organized by Abdus Salam)[22]; the Symposium on the Past Decade (1960–1970) in Particle Theory, Austin, Texas, 1970[23]; the Symposium on the Development of the Physicist's Conception of Nature in the Twentieth Century, Trieste, Italy, 1972 (organized by me, this Symposium was attended by participants from all over the world, including a number of physicists with whom I had had discussions since 1952 about the historical and conceptual development of modern physics)[24]; the Symposium on the Present and Future Goals of Science, Los Angeles, California, 1973 (again, organized by me, this Symposium brought together many eminent quantum physicists)[25]; the Symposia of the Nobel Prize Winners in Physics at Lindau, West Germany (1962, 1965, 1973[26]), which I attended as an observer; and the organizational meetings and symposia of the *Institut de la Vie* on theoretical physics and biology, which always included a number of senior quantum physicists.[27] The number of physicists I met at these symposia and conferences and those whom I met otherwise and with whom I sought personal interviews is too long to describe each encounter individually,

[21] At the symposium in Rochester I again met Richard Feynman. I had known Feynman since fall 1958. I first interviewed him about his work on quantum electrodynamics in his office at Caltech on 10 June 1962, but had not seen him since. On this occasion I talked to him about the systematic development of quantum field theory; I also discussed the same subject with Gunnar Källén and Hideki Yukawa.

[22] At the International Symposium on Contemporary Physics, Trieste, Italy, 1968, several old masters—Heisenberg, Dirac, Wigner, Bethe, and Oskar Klein—gave evening lectures on their lives in physics, and E. M. Lifschitz spoke on Lev Landau's scientific work. These lectures were collected together, in a slightly condensed form, in a pamphlet entitled *From a Life of Physics*, a special supplement of the International Atomic Energy Agency Bulletin, 1969. At this symposium I had the opportunity of having long discussions with many of my 'sources,' including Vladimir Fock, with whom I discussed the development of quantum theory and relativity.

[23] For me the highlight of this occasion was the presence of Dirac, Feynman, Rosenfeld, Salam and Wigner, with whom I had discussions on many questions of the development of physics in the twentieth century. Also, C. P. Snow inaugurated the Program on Public Understanding of Science which I had developed at the University of Texas at Austin, 1970–1973.

On this occasion, Abdus Salam and I had the first discussion about my idea to hold a symposium on the development of the physicist's conception of nature in the twentieth century and his idea to hold a symposium at Trieste, Italy, in 1972 in honour of Paul Dirac's seventieth birthday. Salam proposed that we combine the two symposia into one and that I organize and direct it.

[24] See footnote 4.

[25] Among the participants and speakers at this symposium were Felix Bloch, Owen Chamberlain, Leon N. Cooper, Sir John Eccles, Murray Gell-Mann, Alfred Kastler, Willis E. Lamb, Edwin McMillan, Emilio Segrè, Allan Sandage, Robert Sinsheimer, and Julian Schwinger.

[26] The laureates who participated in the 1973 Symposium of Nobel Prize Winners in Physics at Lindau included Hannes Alfvén, Walter H. Brattain, Leon N. Cooper, P. A. M. Dirac, Dennis Gabor, Robert Hofstadter, Alfred Kastler, Willis E. Lamb, Isidor I. Rabi, J. Robert Schrieffer and Chen Ning Yang.

Dennis Gabor sat in on several of my discussions with Heisenberg and Dirac and was fascinated by the story of the development of quantum mechanics. (See D. Gabor: '*Eindrücke von der Lindauer Tagung der Nobelpreisträger (Physik) 1973*,' in *Nobel führte sie zusammen, Begegnungen in Lindau*, Belser A. G., Stuttgart/Zurich, 1975).

[27] The meetings of the *Institut de la Vie* at Versailles, France, are well known for their relaxed, elegant ambiance. I had some wonderful encounters over the years at these meetings, and there arose marvellous opportunities of discussions with Pierre Auger, Felix Bloch, Leon Cooper, Max Delbrück, Sir John Eccles, Manfred Eigen, Herbert Fröhlich, Kurt Mendelsohn, Lars Onsager and I. I. Rabi.

but the most important interactions took place with: Edoardo Amaldi, Pierre Auger, John D. Bernal, Hans Bethe, Felix Bloch, Sir Lawrence Bragg, H. B. G. Casimir, S. Chandrasekhar, Leon N. Cooper, Peter Debye, Max Delbrück, Freeman Dyson, Paul S. Epstein, P. P. Ewald, Richard Feynman, Marcus Fierz, Vladimir Fock, James Franck, Herbert Fröhlich, Dennis Gabor, George Gamow, Murray Gell-Mann, Walther Gerlach, Samuel Goudsmit, Walter Heitler, G. Herzberg, Res Jost, Gunnar Källén, Peter Kapitza, Alfred Kastler, Edwin Kemble, Oskar Klein, Ralph Kronig, Polykarp Kusch, Willis E. Lamb, Alfred Landé, T. D. Lee, Maria Goeppert Mayer, Philip Morrison, N. F. Mott, Yuval Ne'eman, Lars Onsager, J. Robert Oppenheimer, Rudolf Peierls, Francis Perrin, Ilya Prigogine, I. I. Rabi, C. V. Raman, Léon Rosenfeld, Abdus Salam, Leonard Schiff, Julian Schwinger, Manne and Kai Siegbahn, E. C. G. Stückelberg, Edward Teller, Sin-Itiro Tomonaga, G. P. Thomson, Charles Townes, George Uhlenbeck, Léon Van Hove, J. H. Van Vleck, B. L. van der Waerden, Ivar Waller, Victor F. Weisskopf, Gregor Wentzel, John A. Wheeler, Norbert Wiener, Arthur Wightman, Eugene P. Wigner, Chen Ning Yang, and Hideki Yukawa. Among these I was fortunate in being able to develop close personal relations with Casimir, Lamb, Ne'eman, Prigogine, Rosenfeld, Salam, Schwinger, Wentzel, Wheeler and Wigner. Over the years I learned much from Eugene Wigner about the development of physics in the twentieth century, and my friendship with him turned out to be just as heartwarming and personally inspiring as with his 'famous brother-in-law' Paul Dirac.

Initial Use of Historical Sources

During this long period my collection of notes and transcripts of tape recordings of conversations, discussions and interviews had become quite large. It was supplemented by copies of all the relevant original papers, unpublished manuscripts and notebooks, and letters exchanged between the principal quantum physicists. In addition, Aage Bohr had given me access to study the archives of his father, Niels Bohr, and of the *Sources for History of Quantum Physics* that were deposited at the Niels Bohr Institute in Copenhagen. Thus, there resulted vast materials related to the historical development of quantum theory (old quantum theory, quantum mechanics and quantum field theory).

My aim, however, was not to collect materials for documentary purposes alone. I first used these sources to prepare lectures, which I gave at different places, covering many themes from the quantum theory of Planck, Einstein, Bohr and Sommerfeld (its foundation and the rise of its difficulties), from the discovery and development of quantum mechanics, from wave mechanics, from quantum field theory, from the history of the theory of nuclear structure, and from the history of the theory of weak and strong interactions. All these topics were treated in a deeply human context. Then, at the University of Texas from 1969 to 1973, I had the opportunity of organizing my historical materials for a regular course and seminars on the development of physics in the twentieth century and, in addition to quantum theory, other themes—such as the origins of special and general relativity theories, and the history of statistical mechanics— were added. Later, opportunities arose to give lectures on the history of quantum theory at many universities in the United States and Western Europe, including the Planck Lecture (on 'Max Planck and the Discovery of the Quantum of Action' at the *Akademie der Wissenschaften*, East Berlin, 15 December 1975, on the occasion of the 75th anniversary of quantum theory), the Heisenberg Memorial Lecture (on 'The Birth of Quantum Mechanics,' at CERN, Geneva, on 30 March 1976),[28] the *Fondation Louis de Broglie* Lecture (on 'The Ideas which Led Schrödinger to Wave Mechanics,' University of Paris, 24 May 1976, on the occasion of the 50th anniversary of wave mechanics), and numerous Albert Einstein Centennial Lectures during 1979 at various universities. In these lectures I drew on the historical source materials I had collected.

Early in 1972 I was invited by Abdus Salam and Eugene Wigner to contribute an article to the *Dirac Festschrift*, which they were editing and which was to be presented to Dirac at the Symposium in September at Trieste. The response to my article on 'The Golden Age of Theoretical Physics: P. A. M. Dirac's Scientific Work from 1924 to 1933'[29] was quite favourable. Many of my friends and former 'sources'—Dirac, Heisenberg, Lamb, Wigner and, in particular, Joseph Maria Jauch, Ilya Prigogine, Abdus Salam and C. P. Snow—advised me that there was no time to waste and that I should undertake to write a full-scale account of the historical development of quantum theory.

Joseph Jauch and Ilya Prigogine invited me to do the writing in Geneva and Brussels, respectively. In spring 1973 I was invited as a visiting professor in the *Troisième Cycle* of the French–Swiss universities. For three months I gave a series of lectures on the conceptual development of physics in the twentieth century, primarily at the University of Geneva and *École Polytechnique* of Lausanne, but a few colloquia at Zurich as well. In fall 1973 I was appointed to a professorship of physics and the history of physics at *Instituts Internationaux de Physique et de Chimie* (*Solvay*), Brussels, at the initiative of Ilya Prigogone. My initial assignment at Brussels was to make use of my own resources and archives of the Solvay Institutes to write a book on the history of the Solvay Conferences on Physics. I completed the writing of this book from October 1973 to November 1974.[30] I also gave a series of lectures in Brussels on the history of the quantum theory and on 'the quantum principle: its interpretation and epistemology.' In the meantime Joseph Jauch made efforts to invite me to Geneva with the specific goal of writing my work on the historical development of quantum theory. At Jauch's urging a proposal was submitted by B. L. van der Waerden (University of Zurich) and Martin Peter (University of Geneva) to *Fonds National Suisse*

---

[28] See J. Mehra, 'The Birth of Quantum Mechanics,' the Heisenberg Memorial Lecture, 30 March 1976, CERN Yellow Report 76-10, 14 May 1976, Geneva; also *Uspekhi* **22**, pp. 719–744 (1977) in Russian, and *Postepy Fizyki* **29**, pp. 275–304 (1978) in Polish.

[29] See J. Mehra: 'The Golden Age of Theoretical Physics: P. A. M. Dirac's Scientific Work from 1924 to 1933,' in *Aspects of Quantum Theory*, A. Salam and E. P. Wigner (Eds.), Cambridge University Press, London, 1972.

[30] See footnote 19.

pour la Recherche Scientifique, Berne, for the support of my project. Jauch, van
der Waerden and Peter argued that in view of the early roots of quantum theory
in Switzerland—based on the work of Einstein, von Laue, Debye, Schrödinger,
Pauli and Wentzel—the writing of my work should be done there. They received
the support of a large number of distinguished quantum physicists in my behalf,
and a handsome grant for the carrying out of my project at the University of
Geneva was awarded by *FNSRS* for two years. I was also elected an invited
professor of the University of Geneva for the duration of my project there.

Organization of Materials

I began by organizing my materials for writing a detailed account of the
historical development of quantum theory, which would deal at the same time
with the mutual relations of the quantum physicists and their personal, educational and scientific backgrounds, about which I had learned so much during my
conversations, discussions and interviews with them. Thus, the vision of a book
emerged, which would satisfy the following requirements: first, it should describe
as accurately as possible the main ideas and events in the history of quantum
theory consistent with the written sources (the original scientific papers and
correspondence) and recollections of the leading physicists; second, it should
bring in the necessary technical discussion and mathematical formalism; third, it
should describe the personalities and scientific backgrounds—i.e., formation as
scientists up to the point of their decisive discoveries—of the physicists, mathematicians and chemists involved in the crucial work that led to the major
discoveries in quantum theory; fourth, it should illuminate the background of
these people, their institutes and universities and places where they lived and
worked; fifth, it should present clearly the fundamental concepts which quantum
theory brought into science.

My collection of materials had become enormous and to tackle it and fulfill
all these requirements seemed to be an awesome task. I therefore looked around
for a collaborator to work with me. Werner Heisenberg, who knew about my
plans for many years, helped me in selecting and persuading Helmut Rechenberg, who had been his last doctoral student (1968) in Munich and was a
member of the research group of Hans-Peter Dürr at the *Max-Planck-Institut für
Physik* in Munich, to work as my collaborator in my challenging enterprise.
Rechenberg had received excellent training as a theoretical physicist (quantum
field theory) and he shared my strong cultural interests in the history of twentieth
century physics. Peter Dürr was helpful and willing to allow Rechenberg to
pursue a different field of research in collaboration with me. Dürr and Heisenberg submitted a proposal to the *Deutsche Forschungsgemeinschaft* for Rechenberg's financial support for two years to devote full time working with me in
Geneva. This proposal was duly approved early in 1975.

From April 1970 to June 1972 Helmut Rechenberg had been at the University
of Texas at Austin. Part of his duties had included working with me. During that

period he had already assisted me in the organization of my materials in folders according to specific problem areas. The folders were divided into 39 parts, starting from 1. The Beginnings of Quantum Theory (1900) up to 39. Quantum Theory and Philosophy (Causality, Determinism and Free Will). The organization thus obtained suggested a division of the materials into five major parts: I. The Old Quantum Theory (1900–1922); II. The Creation of Quantum Mechanics (1922–1927); III. The Conceptual Development of Quantum Theory through Its Applications (1925–1960); IV. The Development of Quantum Electrodynamics and Quantum Field Theory (1927 to Present); V. The Development of the Interpretation of Quantum Theory (1925 to Present). During 1970–1972 Rechenberg collaborated with me in several projects on the history of physics and assisted me in planning my lectures in this field.

In late spring 1974 my project was funded by the *Fonds National Suisse pour la Recherche Scientifique*, and Rechenberg's appointment to work with me was expected later that year. Early in 1974 my friend Charles P. Enz, Professor of Theoretical Physics at the University of Geneva, and I edited a book—*Physical Reality and Mathematical Description*—as a *Festschrift* for Joseph Jauch, the publication of which was intended to coincide with Jauch's sixtieth birthday and my arrival in Geneva in August. Unfortunately Jauch died of a sudden heart attack three weeks before the book was to be presented to him on his birthday, though he knew it was coming. Jauch's death was a great shock to me and his many friends. I was robbed of the opportunity of close association with him for an extended period of time; he had also been looking forward to deepen his already profound cultural interest in the development of modern physics by our frequent personal discussions. It was very sad. I postponed my departure from Brussels to Geneva for a few months.

I arrived in Geneva early in January 1975 and Rechenberg joined me in March. We set up offices and archives in the quarters assigned to us in the *École de Physique*. When work began in Geneva I tried hard, with Rechenberg's help, to organize as logical an outline of the whole work as possible, to which we could adhere closely in the time to come, one that would suffer only minimum changes and reflect the chronological development of quantum theory adequately. The whole writing was divided into ten major parts: *Parts One* and *Two*, The Quantum Theory of Planck, Einstein, Bohr and Sommerfeld: Its Foundation and the Rise of Its Difficulties, 1900–1925; *Part Three*, The Discovery of Quantum Mechanics (1925); *Part Four*, The Formulation of Matrix Mechanics and Its Modifications (1925–1926); *Part Five*, The Fundamental Equations of Quantum Mechanics (1925–1926); *Part Six*, The Reception of the New Quantum Mechanics (1925–1926); *Part Seven*, The Formulation of Wave Mechnics (1926); *Part Eight*, The Rise of the Mathematical Formulation of Unified Quantum Mechanics and Its Fundamental Applications (1926 to Present); *Part Nine*, The History of Relativistic Quantum Mechanics and Quantum Field Theory (1927–1975); *Part Ten*, The Physical Interpretation of Quantum Mechanics (1927 to Present). These parts were divided into chapters and the chapters into sections and subsections. The organization of the first six parts was almost the same as it now is in the contents of this work; original plans had to be modified only minimally.

We began by writing *Part Three*, the Heisenberg chapter on The Discovery of Quantum Mechanics and continued on to *Parts Four*, *Five* and *Six*. Our procedure for work turned out to be the following. Rechenberg and I discussed the details of what the chapters should contain, what the contents of each section should be, and how the different sections should relate to each other. We would thrash out the content of each section before beginning work on it; naturally it would be modified to some extent in the process of writing. Then Rechenberg, who had mastered the organization of my materials, would prepare a folder for a given section, containing the original papers, interviews (appropriate parts duplicated and cut out), secondary literature, as well as the letters and biographies pertaining to that section. We would discuss all these materials and analyze all the problems involved in the historical context, then Rechenberg would prepare a set of notes covering the information contained in these materials and our analysis of it. He would then give me the folder and the set of notes, and I would prepare the penultimate draft. In the process of writing this draft I would raise any number of questions—about points that were obscure, about more information that was necessary, about what should be added or deleted, etc.—and give my draft and questions to Rechenberg. He would then look for the missing elements, search the answers to my questions (we actually divided this task among ourselves), and we would complete the footnotes. I would then write the final version. Rechenberg and I held discussions several times a day to make sure that our information and analysis pertaining to each problem were as complete as possible. The materials which were not in my folders, or which we somehow lacked, were obtained for us by the librarian of the *École de Physique* from the libraries of the University of Geneva, the E.T.H., the *Staatsbibliothek* in Berne, or wherever else they could be found.

In January 1977 I returned to Brussels and Rechenberg joined me there for three months. We continued our work there on *Part Five*, dealing with Dirac's formulation of the fundamental equations of quantum mechanics. At the end of March, Rechenberg returned to Munich. Meanwhile I had received the U.S. Senior Scientist Award of the *Alexander von Humboldt-Stiftung*, which made it possible for me to go to Munich periodically for several weeks at a time. Our work thus made progress and *Part Five* was completed by the end of 1977. *Part Five* fully met the criteria we had established regarding contents, chronology, analysis, and the arrangement of text and footnotes.

Rechenberg and I then sought to revise *Parts Three* and *Four* in the light of what we had learned from *Part Five*. Rechenberg diligently helped me in preparing a new set of notes, rearranging our previous version and adding or redistributing information. I then wrote the final version. By September 1979 we had fully completed the revision of all parts from *Three* to *Six*, and they were typed. Now the old quantum theory, *Parts One* and *Two*, had to be worked on and was to form the background of the discovery of quantum mechanics. We had the choice of giving a summary treatment or match the detailed style of the rest. We opted for something in between: we covered the story of the old quantum theory in detail, without giving it quite the epic treatment of quantum mechanics. We began this work in October 1979, again with an agreed upon outline of chapters and sections, but by then the pattern of our work was well organized and everything fell into place in a suitable arrangement. The writing of *Parts One* and *Two*, with the same division of labour—with the responsibility of writing the final version resting upon me—was completed by December 1980. Beginning in January 1979 I had commitments to give lectures in Brussels, Houston and Irvine, and I carried on my work on the writing of *Parts One* and *Two* at all these places, but Rechenberg and I maintained regular contact with each other by my periodic visits to Munich, otherwise by telephone and express mail.

The Sources and Their Use

In writing about the old quantum theory as well as the discovery and development of quantum mechanics, Rechenberg and I have tried to present the historical events in detail with the best accuracy—we believe—that can be achieved at the present time, i.e., more than five to eight decades after the events have taken place. For this purpose we have followed certain rules:

(i) We have attempted to use all possible sources that may throw light on the historical development of quantum theory: first, the published papers in various journals, including the original research papers addressed to an audience of specialists as well as the broader reviews on quantum physics and related topics addressed to the more general community of scientists; second, the available scientific correspondence exchanged between the scientists involved or interested in quantum phenomena; third, the testimonies, recollections and interviews, published or unpublished, given by the principal contributors to the old quantum theory and quantum mechanics; fourth, the biographies of the physicists in question, throwing light on their education and upbringing, their research goals, their special aptitudes and turns of mind and temperament, as well as their relations with other physicists; fifth, the intellectual tradition and cultural and scientific environment of the places, universities and institutions where quantum theory was primarily developed.

(ii) From all these sources, and in special cases from other relevant information, we have sought to reconstruct a picture of ideas and events seen as much as possible from the point of view of a contemporary observer. From the rich material of the available sources, we have used for a given idea or event only those documents and statements that were written or made at the time in question. This was found to be especially necessary for the period in which quantum mechanics was created; in this period the pace of development was very rapid, and occasionally the active physicists changed their attitude towards certain problems from week to week. Therefore, one can hardly describe adequately—in case one wants to give an accurate microscopic presentation—an event occurring, say, in the beginning of 1925 by quoting a letter of fall 1925. For example, we have put together the puzzle of scattered information to present a nearly daily description of Werner Heisenberg's crucial steps in May and June 1925; or of his visits to Leyden (the date of which was confirmed by the signatures of the houseguests on the wall of Ehrenfest's living room in his old house in Leyden[31]) and to Cambridge, England, in July (the title and date of Heisenberg's talk at the Kapitza Club were recorded in the *Minute-Book of the Club*); or of Paul Dirac's discovery of the fundamental equations of quantum mechanics in September and October of the same year.

(iii) For the purpose of reconstructing events we have tried to match the information available from various sources to clarify possible disagreements. So the contents of the letters exchanged among physicists had to be related to the results given in the papers; and their recollections, expressed in interviews, discussions, or published autobiographical accounts—written long after the occurrence of certain events—had to be checked against the story emerging from the original sources, i.e., published papers, unpublished manuscripts and letters of the olden days. In performing the task of matching, comparing and checking the various sources of information, we noticed a most remarkable fact: in all the discussions and interviews none of the quantum physicists had made false claims for himself. At times the recollections revealed mistakes in dating events accurately, or a rather peculiar interpretation of certain aspects, or even peevishness at lost opportunities for discoveries that were made by others at the time or with people who beat them at the game—which, Heisenberg, Dirac and Schrödinger, for instance, evidently did—but no dishonesty. This fact speaks very well for the attitude of the scientific community which came to maturity in the 1920s and early 1930s, and of course before that, and it may be attributed to the impeccable leadership of such men as Planck, Rutherford, Einstein, Bohr and Sommerfeld.

(iv) Quite often I have been assisted in analyzing tricky developments by personal knowledge of the physicists active in this dramatic story—that is, by knowing their habits of mind and peculiar ways of thinking and solving problems. This knowledge was achieved through personal contacts described earlier, which in several cases extended over twenty years, through prolonged discussions, and careful reading and analysis of the styles and contents of the original papers and scientific correspondence. In the analysis of Heisenberg's papers, Rechenberg also had the advantage of prolonged exposure to Heisenberg's thinking.

General Goal of the Present Work

Our aim in the present work is to describe the discovery of quantum mechanics and its immediate consequences (the formulation of matrix mechanics and its modifications and of the fundamental equations of quantum mechanics) with the background of the quantum-physical problems of the first quarter of the twenti-

---

[31] At my request Peter Mazur, Director, *Instituut-Lorentz*, had these photographs made and sent to me.

eth century. The interest at that time concentrated on the constitution of matter (the microscopic composition of solids, liquids and gases; the structure of atoms and molecules, and even the atomic nucleus); and on the nature of radiation and its interaction with matter and its constituents. The solution of these problems, should it be achieved, promised greater insight in the fundamental problem of science, which had concerned man since the times of the ancient Greeks: the microscopic structure of matter and its most elementary building blocks. In telling the story of the historical development of quantum mechanics, we not only want to show the revolution brought about by the quantum-theoretical concepts in various fields of physics, but we also seek to throw light on the external conditions in which the physicists then worked. What, at that time, were the scientific and personal relations between professors and their students, between senior and junior physicists, between experimentalists and theoreticians? How did the institutes of physics function in the 1920s, and how were they funded in those economically difficult times? What were the connections between the research institutes and universities of different countries as far as quantum physics was concerned? How did the individual scientists interact with each other after World War I, when the universal international relations of science had suffered serious blows? Where did the physicists meet and discuss with each other, when travelling was expensive and restricted? While ours is primarily an 'internalist' history of the development of quantum theory, such questions get answered in the process of pursuing the scientific careers of the major quantum physicists; these answers then supply some of the essential background of the life of the flourishing community of quantum physicists, a community which in the early 1920s consisted of a number of senior members, such as Planck, Einstein, and Rutherford, Sommerfeld, Bohr, Born and Franck, and their associates and students. Without going into the details of the answers, an important result can be stated: under the peculiar conditions of that time a new class of physicists emerged, doing research in a manner different from the one used earlier. Papers were written and collaborations established in just about the same fashion as it happens today; however, in the 1920s the period of the isolated scientists working on their private problems, the *Privatgelehrten*, came to an end. A new period started when people worked, alone or together, on widely acknowledged problems, many arising from quantum mechanics, its extensions and applications.

Diverse Ways of Approaching Quantum Theory

In spite of the change in the methods of doing physics, and in spite of the large number of scientists concerned with quantum problems around 1925, still comparatively few individuals played a dominant role in the creation of quantum mechanics. These individuals belonged to different age groups—including the fifty-six-year-old Sommerfeld, forty-three-year-old Niels Bohr, thirty-eight-year-old Erwin Schrödinger, thirty-year-old Hendrik Kramers, twenty-five-year-old

Wolfgang Pauli, twenty-three-year-old Werner Heisenberg and twenty-two-year-old Paul Dirac. By no means did all these physicists approach the problems of quantum theory from the same point of view. The confrontation of goals and attitudes, which had occurred earlier in the history of quantum theory and had been represented by such pairs of scientific opponents as Max Planck and Albert Einstein, or Johannes Stark and Arnold Sommerfeld, was continued in the 1920s by numerous others. Thus, Albert Einstein and Niels Bohr began their arguments on the foundation of quantum and theoretical physics, in which they were to assume totally incompatible views on the fundamental aspects of physical theories. Moreover, people who worked on the same problems of quantum theory disagreed deeply in their approaches. Thus, Bohr and Sommerfeld derived very different consequences from the same spectroscopic data and, even when they arrived at the same formula—as in the case of hydrogen—they did not give it the same interpretation. Max Born, trained in the mathematical school of Göttingen like Arnold Sommerfeld, used certain methods which Sommerfeld had never considered as being adequate for treating physical problems. And Wolfgang Pauli quarrelled for several years constantly with Werner Heisenberg about the correct theory of the anomalous Zeeman effects. Separated from those mentioned in the foregoing, the quantum physicists, who—with the exception of Albert Einstein—belonged to the same great school of atomic theory, Louis de Broglie and Erwin Schrödinger treated quantum problems by following their own, quite different lines of thought.

Looking at the widely different approaches to treat the same phenomena, which ranged from highly mathematical treatments—as preferred by David Hilbert, his disciple Max Born, and Born's student Pascual Jordan or by Paul Dirac—to a very physical treatment—as preferred by Niels Bohr or Paul Ehrenfest—one must be quite surprised that after all the various efforts the same theoretical description finally emerged. Quantum mechanics, as it was finally completed in 1927, provides a most remarkable example of the universality of science, i.e., the fact that the rational description of nature converges to one and the same universal scheme no matter what the starting point.

Some Decisive Personalities

Just as the old quantum theory is associated with the names of Planck, Einstein, Bohr and Sommerfeld, the discovery of quantum mechanics is associated with the names of Werner Heisenberg, Max Born, Pascual Jordan, Paul Dirac, Louis de Broglie and Erwin Schrödinger. These physicists were the principal actors and heroes of the history of quantum mechanics; they developed the equations and formulae for describing the known phenomena correctly and for predicting many that were to be discovered later; they performed the hard and laborious task of transforming the old atomic dynamics into a genuine quantum theory, while the others stayed aside and watched their endeavours. Still the story of the discovery of quantum mechanics cannot be told without taking into account the behind-the-scene contributions of certain physicists who did not publish papers themselves on the burning questions of the day. In this context we think less of Niels Bohr, who was the acknowledged leader of most quantum physicists in the 1920s and who influenced all the main actors on the stage, but of the much junior Wolfgang Pauli, who was to become one of the greatest physicists of the twentieth century.

In early 1920s Pauli performed the critical analysis of some of the most difficult problems of quantum theory. Nobody, not Louis de Broglie, nor Werner Heisenberg, nor Erwin Schrödinger, was prepared as much as Pauli was in demanding a total, revolutionary change of the old concepts. This critical and radical attitude was present not only in his papers, but more so in his letters which he wrote between 1922 and 1925 to Bohr, Heisenberg and Sommerfeld. In his letters Pauli dropped the cautious, diplomatic manner, in which he phrased the difficulties of the theory in his publications; in his letters he wrote in clear and outspoken words about the insurmountable difficulties. He rejected sharply, if not cynically, most of the proposals that were made to cure the difficulties; and often he expressed despair about the state of the entire atomic theory. And yet, the man who saw most profoundly at the very basis of the problems and who was most familiar with the sickness which afflicted atomic theory remained strangely inactive in finding the proper cure. He demanded the cure in strong words, but he did nothing himself until somebody else suggested a possible medicine.

There is still another aspect to Pauli's story during the decisive years of the discovery of quantum mechanics. Pauli was so clearly aware of the depths of the difficulties that he wanted to prevent the acceptance of any solutions that appeared to him too cheap. So he, recognized as he was as a great authority on spectroscopic problems, rejected several proposals made by younger or older colleagues: he turned against Heisenberg's theories of the anomalous Zeeman effects as well as against the Bohr–Kramers–Slater theory of radiation; he suppressed Ralph Kronig's idea of an intrinsic angular momentum or spin of the electron, and wanted to do the same when this idea reappeared in the proposal of Uhlenbeck and Goudsmit; he initially despised the matrix method of Born and Jordan as 'Göttingen's deluge of formal learning' ('*Göttinger formaler Gelehrsamkeitsschwall*'), claiming that it would ruin the future of quantum theory. There was hardly any step in the establishment of the new atomic theory, which was not criticized at least once by Pauli, except Heisenberg's work in May, June and July 1925 and Schrödinger's papers in spring 1926. Pauli behaved as if he represented 'the conscience of quantum mechanics' and later on as 'the conscience of physics' itself, but in many ways it was a flawed conscience[32]; more

---

[32] Pauli never called himself 'the conscience of physics'; he just acted as if he was. This epithet was applied to him by his former students and collaborators. (See, e.g., the remarks of I. I. Rabi and V. F. Weisskopf in 'Wolfgang Pauli: Tributes by M. E. Fierz and V. F. Weisskopf,' *Physics Today*, July 1959, pp. 18–19; also N. Bohr, 'he [Pauli] more and more became the very conscience of the community of theoretical physicists,' in *Theoretical Physics in the Twentieth Century, A Memorial Volume to Wolfgang Pauli*, M. Fierz and V. Weisskopf (Eds.), Interscience Publishers, New York, 1960, p. 4.)

often he hit his less conscious contemporaries like—in Ehrenfest's phrase—
'God's whip' (*'Geißel Gottes'*), rather than as the divine light of new and
revolutionary reason. In fact, through most of his life Pauli remained the prophet
of gloom and doom regarding the new innovations in physics. This was behind
his own remark to me in spring 1958: *'Ich war doch ein Klassiker.'*

Pauli turned out to be wrong in many, if not most, of his negative opinions.
He harmed and delayed the publication of the work of some persons who
followed his advice. But the strongest personality, on whom he lavished most of
his criticism, did not suffer at all: Heisenberg took Pauli's judgment seriously and
tried to turn Pauli's arguments around until he found an answer, which allowed
him (Heisenberg) to progress along his own lines.[33]

Paul Ehrenfest also possessed a critical mind, though not as sharp and biting
as Pauli's, of whom he was always a bit afraid. Through his analysis of statistical
mechanics early in his scientific career, Ehrenfest had isolated many problematic
points in the understanding of radiation theory and kinetic theory. In the 1920s
he often checked the consistency of the new results—e.g., of the Bose–Einstein
statistics—with the accepted principles of physical theory. Ehrenfest applied his
criticism in a friendly way to young people, whom he encouraged even to follow
ideas that implied certain difficulties, as was the case initially with the electron-
spin hypothesis. Thus, he combined a critical and analytical mind with the
virtues of a wise teacher. Though Ehrenfest did not write any decisive paper on
quantum mechanics in 1925–1926, he certainly assisted in its birth.

Relation to Philosophy

As someone has said, with a sleepy nod to his Greek predecessor—Dionysius of
Halicarnassus, I think—'History is Philosophy teaching by examples.' This is the
attitude Rechenberg and I have taken in treating the historical development of
quantum theory. We have not used any *a priori* philosophy to guide our analysis.
That is, we have not assumed the framework of a traditional philosophical
doctrine—like idealism, positivism or materialism—in dealing with the historical

---

[33] Pauli's first wife, Kate Goldfinger, *née* Deppner—(after she left Pauli, she married the chemist
Paul Goldfinger from Berlin; they later moved to Brussels, where Goldfinger became Professor of
Chemistry at *Université libre de Bruxelles*)—told me: 'When Pauli and I got married in Zurich he
always told me that he was someone really and profoundly important in the world of physics. He
used to get letters from many physicists, especially Heisenberg. He used to walk around like a caged
lion in our apartment, formulating his answers in the most biting and witty manner possible. This
gave him great satisfaction' (Conversation with Kate Goldfinger, 12 March 1974). Pauli's widow,
Franca, gave me a glimpse of his later feelings toward Heisenberg: 'Pauli had a great respect for
Heisenberg's intuition. He often used to tell me: "Heisenberg's intuition triumphs over all the reasons
against it. The combination of his powerful intuition and childlike optimism is invincible"' (Conversation with Franca Pauli, 23 December 1974). In relation to the nonlinear spinor theory Pauli was
angry at Heisenberg's optimism in spring and summer 1958, but he told Heisenberg that fall in
Varenna: 'You have to go forward in this task. You have always had the right intuition. My powers
have failed; I don't have the strength. You should go on' (Conversation with Heisenberg, 25 February
1975).

process and events in physics. Nor have we used any philosophical framework for discussing how the scientific development leading to quantum mechanics occurred. We believe that the presupposition of a philosophy involves certain prejudices in assembling and interpreting the material at hand. It appears more adequate to present first the details of the historical development and their rigorous scientific analysis—the examples, that is—and only afterwards to derive any philosophical moral from the sequence of ideas and events. In other words, we suggest that philosophy of science should be based on the rigorous analysis of the given facts and their logical relations, rather than serve as an *a priori* guide to represent the facts.

We have also not adopted the point of view of considering only those events which led to the results accepted later, but have often presented and discussed also those physical ideas and results which were falsified later. The reason is that in any period when a major development occurs, the physicists will necessarily be occupied for a longer or shorter period with wrong ideas; indeed truth and error will then be close neighbours. Moreover, among the various premises and arguments upon which a new and eventually successful set of physical relationships rests, about half of them may later on be proved to be unjustified. As a rule, valid nearly without any exception, we may take for granted the statement that a consistent and complete foundation of a new theory can be given only *a posteriori*, i.e., after the theory has been established and tested by many phenomena and physical facts. The same is true, even more so, when we consider the problem of a physical or philosophical understanding of the new theory: the correct interpretation will occur only months, if not years, after the discovery of the physical theory. This does not mean, however, that physicists, while analyzing experimental data and obtaining new relations between them, have no physical picture in mind; but it does mean that this physical picture may be ephemeral and subject to change.

Contents of This Work

Having mentioned several important aspects that have determined this presentation of the story of the historical development of quantum theory, I shall now give an outline of the present work. It is divided into four volumes:

Volume 1:   The Quantum Theory of Planck, Einstein, Bohr and Sommerfeld: Its Foundation and the Rise of Its Difficulties, 1900–1925;

Volume 2:   The Discovery of Quantum Mechanics, 1925;

Volume 3:   The Formulation of Matrix Mechanics and Its Modifications, 1925–1926;

Volume 4:   The Fundamental Equations of Quantum Mechanics, 1925–1926; The Reception of the New Quantum Mechanics, 1925–1926.

Each part consists of several chapters. The material covered includes the discussion of the crucial happenings in quantum theory from 1900 to the middle of 1926, but not including the events connected with the immediate discovery of wave mechanics. Thus it deals with: the origin of the quantum theory in Planck's formula for blackbody radiation in 1900; Einstein's light-quantum hypothesis of 1905; the quantum theories of specific heats from 1907 to 1913; Bohr's quantum theory of atomic structure of 1913 and its extension by Sommerfeld, Ehrenfest and Bohr himself in the following decade; Einstein's statistical treatment of radiation, 1916–1917; the triumphs and failures of the description of many-electron atoms in the early 1920s; the Stern–Gerlach and Compton effects, 1922; the Bohr–Kramers–Slater theory of radiation and the dispersion-theoretic approach, 1924–1925; the Bose–Einstein statistics, 1924–1925; Louis de Broglie's phase waves associated with matter, 1924–1925; the exclusion principle, 1924–1925; the discovery of electron spin, 1925–1926; Heisenberg's discovery of the quantum-theoretical reformulation of the classical equations of atomic theory, summer 1925 (including the steps leading to it, 1922–1925); Born and Jordan's matrix scheme of quantum mechanics, summer and early fall 1925; Born, Heisenberg and Jordan's completion of the matrix scheme, fall 1925; Pauli's success in calculating the energy states of the hydrogen atom, October 1925; Dirac's formulation of quantum mechanics with q-numbers, fall 1925; Born and Wiener's formulation of quantum mechanics with operators, November–December 1925; application of the Göttingen–Cambridge quantum mechanics to one and many-electron atoms including spin (hydrogen fine structure, anomalous Zeeman effects: Heisenberg and Jordan, March 1926) and to diatomic molecules (Pauli and Mensing, spring 1926).

The primary aim of Volume 1 is to provide the prehistory of the story of quantum mechanics, i.e., to give a general outline of the development of quantum theory from 1900 up to the eve of the discovery of quantum mechanics. During that time a large number of quantum phenomena was discovered and the main problems of the theory arose. In the various chapters of Volume 1 we define and explain these problems, the reasons why they came about and how they could be approached by modifications of the existing theories. The remaining difficulties of the theoretical description then provided the starting point for the endeavours to develop quantum mechanics. Besides the immediate purpose of setting the stage for the following parts, Volume 1 also serves as an outline of the history of quantum theory in the first quarter of the twentieth century, an abbreviated treatment of the events in much the same spirit in which we treat the later parts dealing with the main subject of our work: the discovery and development of quantum mechanics.

Volume 2 deals with the events connected with and leading to Heisenberg's discovery of the quantum-theoretical reformulation of the mechanical equations of atomic theory. In particular, we present an account of young Werner Heisenberg's growing up, education and scientific career, the influences upon him and the problems he grappled with on his way to the discovery of quantum mechanics. The Heisenberg story throws a strong light on the situation in quantum theory after 1920, for in these years he came into contact with all the three main schools of quantum theory: Munich, Göttingen and Copenhagen.

In Volume 3, the development of the first theoretical scheme of quantum mechanics is discussed. Max Born and Pascual Jordan in Göttingen used the mathematical theory of quadratic forms of infinitely many variables or infinite quadratic matrices, which David Hilbert had developed—also in Göttingen— about twenty years earlier, to formulate Heisenberg's ideas in a mathematically consistent manner. Thus, they arrived at a new mechanical theory, whose equations showed great similarity with those of classical mechanics, except that Hermitian matrices now described the dynamical variables of quantum systems. The edifice of this theory was completed by the joint collaboration of Born, Heisenberg and Jordan in fall 1925. This so-called matrix mechanics could be applied to specific problems of atomic theory, especially to determine the energy states of the hydrogen atom in complete agreement with Bohr's theory of 1913. In order to solve more complicated atomic problems, the matrix scheme had to be extended, for the matrices seemed to be unable to describe infinitely increasing position variables such as angles of the orbiting electrons. Born and Wiener recognized that a suitable generalization of the matrix scheme could be achieved by using mathematical operators, and Pauli, Heisenberg, Jordan and Gregor Wentzel succeeded in solving problems like the relativistic hydrogen atom or the anomalous Zeeman effects with the help of the extended matrix theory.

The most abstract scheme of quantum mechanics was proposed, independently of all others, by Paul Dirac in Cambridge, and we describe this approach in Volume 4. Dirac, six months younger than Heisenberg, studied electrical engineering and applied mathematics in Bristol and theoretical physics in Cambridge, England. Upon becoming acquainted with Heisenberg's first paper on quantum mechanics, he obtained by geometrical reasoning a calculus of q-numbers, which turned out to be equivalent to Born and Jordan's matrix mechanics but more powerful than it, for it allowed him to deal with action-angle variables as well.

The theory thus completed did not remain the only available quantum mechanics. Actually, quantum mechanics emerged from a twin birth, of which wave mechanics, discovered by Erwin Schrödinger in the second half of 1925 and published in early 1926, was the other twin. However, before Schrödinger's papers became known, the Göttingen–Cambridge version of quantum mechanics became widespread and was recognized as a decisive step forward in the solution of quantum-theoretical problems. Many quantum physicists in Europe and America applied the new theory with striking success. We report on the reception of the new theory also in Volume 4.

One may ask how our work relates to or compares with the other accounts of the history of quantum theory.[34] The depth and scope of our work are different

[34]There exist two considerable treatments of the history of quantum theory: Max Jammer: *The Conceptual Development of Quantum Mechanics*, McGraw-Hill Book Company, New York, 1966; and Friedrich Hund: *Geschichte der Quantentheorie*, Bibliographisches Institut AG, Mannheim, 1967,

from any attempted thus far in this field: we bring in all the physical, mathematical and human details to provide the reader a complete account of the old quantum theory and the discovery and development of quantum mechanics. We believe that the detailed and meticulous treatment we have employed serves to illuminate the intricate path which the development of quantum theory took in the crucial years after 1920; this approach can be justified on the grounds that this period belongs to one of the most creative and brilliant in the whole history of science.

We are aware of the fact that several accounts dealing with certain parts of the story we cover already exist in print; they deal with the development of certain specific quantum problems and the work of certain physicists (especially, Planck, Einstein and Bohr). Our aim, however, goes much beyond such works; we want to give the full story of all significant problems and their interplay in leading to the discovery and completion of quantum mechanics and to discuss the roles and contributions of individual quantum physicists properly and adequately. In this spirit we expect to continue our project of writing the remaining parts of the history of quantum theory, dealing with the discovery of wave mechanics, its merging with the Göttingen–Cambridge theory to form a unified quantum mechanics, the fundamental applications of quantum mechanics, the historical development of quantum field theory, and the physical interpretation and epistemology of quantum theory.

We are hopeful that this work will help physicists in understanding the rise of one of the most fundamental theories they work with. We hope, however, that this work might be useful to chemists and biologists as well, for quantum theory plays an important role in their fields. Mathematicians might be interested in how certain powerful mathematical methods were made use of in the description of nature. Those who pursue the development of the human mind, of new conceptions and ideas, and of new intellectual communities—such as the historians, philosophers, psychologists and sociologists of science—might find a valuable field in the lives and growth of conceptions of the quantum physicists treated here, and we hope that they will not be put off by at times difficult physical and mathematical concepts employed here.

Support for the Project

Joseph Maria Jauch actually set me on the road to writing this work. It had just been a dream before, a project on the drawing board. Jauch had the gift of encouraging people to do their best. By dint of his vision and efforts he had established the *École de Physique* at the University of Geneva as a preeminent

---

second edition, 1975. Both of these books deal with the development of quantum theory from 1900 to 1928, including the old quantum theory and quantum mechanics. They serve as useful outlines of the history of quantum theory.

institution for research in physics, a place where constant concern with the ideas and culture of physics, its history and philosophy, were most important. It was his insistence that my writing should not be postponed, his wish that the work be written and find fruition at his _École_, and his successful efforts to obtain the necessary means for doing it from _Fonds National Suisse pour la Recherche Scientifique_, Berne, that got me really started. My gratitude to Jauch's memory is great.

Since it was begun in Geneva in early 1975, the writing of this work has been continued in Brussels (Belgium), Munich (West Germany), Ithaca (New York), Irvine (California), and Houston (Texas). It has been supported by many people and institutions. Many of the persons listed here among my _Encounters with Quantum Physicists_ have encouraged me at one time or another in carrying on with this task. A number of them offered advice, counsel, help, encouragement and friendship, for which I am grateful. The most continuous, warm and friendly support came from H. B. G. Casimir, P. A. M. Dirac, Werner Heisenberg, Pascual Jordan, Willis E. Lamb, Yuval Ne'eman, Ilya Prigogine, Abdus Salam, C. P. Snow, Gregor Wentzel and Eugene P. Wigner, and I wish to thank them sincerely. Dirac, Lamb, Prigogine, Salam and Wigner have helped and comforted me in difficult moments and my debt to them is great. The philosopher–novelist Raja Rao has always encouraged me with his warmth and solicitude and I feel special gratitude towards him. I also owe much to the joy and inspiration I have received from the perspicaciousness and luminosity of the writings of two historians of modern science, Gerald Holton and Martin Klein, and I wish to thank them for giving me this pleasure.

For several years before the writing of this work was undertaken, Helmut Rechenberg shared with me the excitement of pursuing the historical and conceptual development of quantum theory. I wish to acknowledge the loyal friendship and devotion, intelligence and hard work, with which he has supported me in the work on this project. Without his help and active collaboration since 1975 this work would not have been successfully completed up to the point it has. Because of our close association, this work has taken on the character of a joint collaborative effort, and I have invited him to associate himself formally as its co-author. I look back on our collaboration with satisfaction, and look forward to our continuing association on the remaining parts of this project with anticipation. This acknowledgement of Rechenberg's help is combined with my gratitude to Werner Heisenberg and Hans-Peter Dürr for making his services available to me, and to the _Deutsche Forschungsgemeinschaft_ for making Rechenberg's collaboration with me in Geneva and Brussels possible by granting him a fellowship (1975–1977).

The initial outline of this work was carefully examined by John Archibald Wheeler, and his criticism and guidance helped to improve the organization. Volume 2 was read by Werner Heisenberg and Volume 3 by Pascual Jordan. Their comments were invaluable. Both Heisenberg and Jordan assured me that this work represented a faithful reconstruction and interpretation of the glorious period of the discovery of quantum mechanics, and that through it they had relived some wonderful moments of their lives. I am grateful that this work afforded them some pleasure before their death. Paul Dirac has read portions of Volume 4, and I have greatly benefited from his remarks. Large parts of the first draft of this work were read by B. L. van der Waerden, and his extensive comments and criticism were valuable. Some other readers have also helped to improve this work by their comments. However, faults and errors do remain, for which I alone am responsible.

JAGDISH MEHRA

Acknowledgements

During the writing of this work I have incurred many debts and obligations which I have not yet acknowledged and shall do so here.

Since fall 1973 I have enjoyed a close relationship with the *Instituts Internationaux de Physique et de Chimie* (*Solvay*), Brussels, Belgium. I wish to express my gratitude to Ilya Prigogine, Director of the International Solvay Institutes, for his continuous support, encouragement and sponsorship of my work.

I thank B. L. van der Waerden (University of Zurich) and Martin Peter (University of Geneva) for making the proposal to the *Fonds National Suisse pour la Recherche Scientifique*, Berne, Switzerland for the funding of my project, and I am grateful to the *Fonds National Suisse* for its support of my work during 1975 and 1976.

I am thankful to the faculty of the *École de Physique* and the Rector of the University of Geneva for appointing me as *Professor Invité* during 1974–1976. I wish to thank my colleagues in the *École de Physique* for their friendship and support, in particular my old friend Charles P. Enz. After Joseph Maria Jauch's death, it was especially Enz' presence in the *École de Physique*, University of Geneva, that meant much to me. In long walks on Sundays in the *Jura-Genèvois* or on top of *Le Salève* and frequently at dinners in our favourite restaurants and bistros in the greater Geneva region, I expounded to him interminably about my work with Helmut Rechenberg on the history of quantum theory. He was always a patient listener, his comments and criticism always intelligent and helpful, his enthusiasm for the success of my project great. The association and friendship with Charles Enz were very dear to me and I thank him.

My sincere thanks are due to the *Alexander von Humboldt-Stiftung*, Bonn-Bad Godesberg, West Germany, and its Secretary General, Heinrich Pfeiffer, for giving me the U.S. Senior Scientist Award (*Humboldt-Preis*), 1976, in support of my work on the historical development of quantum theory. I also wish to thank Hans-Peter Dürr, Director of the *Max-Planck-Institut für Physik*, Munich, for the hospitality of the Institute during several periods of my stay there in 1977.

In spring 1978 I was invited as a Senior Scholar of the Society for the Humanities at Cornell University, Ithaca, New York. I wish to thank Michael Kammen, the Society's Director, for his warm hospitality and support of my work during my stay. I also wish to thank Hans Bethe, Paul Ewald, Michael Fisher, Thomas Gold and Martin Harwit for encouraging me in my work on quantum theory during my stay at Cornell.

The *Alfred Krupp von Bohlen und Halbach-Stiftung*, Essen-Bredeney, West Germany, gave me a generous grant in support of my work from June 1978 to May 1980. I am grateful for this support to the *Stiftung* and its President, Berthold Beitz. Two members of the *Kuratorium* of the *Stiftung* took special interest in my work and encouraged me continuously: Hans Leussink, former Minister for Science and Technology in the administration of Chancellor Willy Brandt in West Germany, and Reimar Lüst, President of the *Max-Planck-Gesellschaft zur Förderung der Wissenschaften*. I deeply appreciate the support they have given me.

I am thankful to the Regents and President of the University of California for appointing me to the Regents' Professorship at the University of California at Irvine in the spring quarter of 1980. My work moved smoothly forward during my stay at Irvine and I received much encouragement and support from colleagues in the Department of Physics and Astronomy, especially from Frederick Reines and A. A. Maradudin. During my stay at Irvine, I gave lectures on the history of quantum theory at several campuses of the University of California and at Caltech, and I received much encouragement from Richard Feynman and Julian Schwinger. My grateful thanks to them.

I wish to thank the Board of Regents and Chancellor of the University of Houston for appointment as Distinguished University Professor in the spring semesters of 1979 and 1980. It was in Houston that the last portion of the present work (on Volume 1, Parts 1 and 2) was completed.

I also wish to acknowledge with thanks and gratitude the following: all the many quantum physicists whom I interviewed for my work and who (or their heirs) have allowed me to make use of conversations, discussions and interviews with them (as well as their interviews with others, especially the *Archives for the History of Quantum Physics*), their published or unpublished autobiographical accounts, unpublished manuscripts and notebooks, and, of course, their published scientific papers; Aage Bohr for giving me full access to the *Bohr Archives* and *Archives for the History of Quantum Physics* held at the Niels Bohr Institute in Copenhagen; to Franca Pauli and Victor F. Weisskopf for giving me permission to make use of Pauli's scientific correspondence for this work; to the Einstein Estate, the holders of the copyright on Einstein's writings, for making use of Einstein's writings and correspondence, which are so essential for any account of the development of quantum theory. I also wish to acknowledge with thanks two conversations in Brussels in March 1974 with Kate Goldfinger, *née* Deppner, Pauli's first wife, about Pauli's early years as a professor at the E.T.H. in Zurich, and a conversation in Zollikon (Zurich) on 23 December 1974 with Franca Pauli about a number of questions dealing with Pauli's life and career.

I wish to thank the librarians of the University of Geneva, the *Staatsbibliothek* in Berne, the *Eidgenössische Technische Hochschule* in Zurich, the Mathematics–Physics Library of the *Université libre de Bruxelles*, the *Niedersächsischen Staats-und Universitätsbibliothek* in Göttingen, the *Staatsbibliothek Preussische Kulturbesitz* in Berlin, the *Max-Planck-Institut für Physik*, and the *Handschriften*

Abteilung of *Deutsches Museum* in Munich, for all the help they have given me and Helmut Rechenberg in obtaining copies of relevant materials for our use.

Last but not least I would like to thank the *Max-Planck-Institut für Physik* for a handsome subsidy towards the costs of publication of this work and Springer-Verlag New York Inc. for meeting my wishes with regard to its publication.

JAGDISH MEHRA

The Quantum Theory of
Planck, Einstein, Bohr and Sommerfeld:
Its Foundation and the Rise of Its Difficulties
1900–1925

Prologue

Like many other activities of the human mind, the discoveries in science depend on past developments. Thus quantum theory, which was initiated in the first year of the twentieth century, emerged logically and systematically—though with certain ingenious innovations—from the major discoveries and principal branches of physics of the nineteenth century. The scientific heritage, which we wish to summarize in this prologue, consists of three parts: first, the great theoretical schemes which reached their perfection in the nineteenth century; second, the revival of the atomistic picture of matter; and third, the recognition of new, unexpected phenomena and the genesis of new theories before and around 1900. Before dealing with this heritage, let us mention another tradition in science that was established during the nineteenth century, namely, the creation of the profession of a specialized scientist, notably a physicist or a chemist. Naturally, we do not imply that in earlier times there were no people who occupied themselves primarily or even exclusively with physical or chemical problems. At many universities appropriate chairs had been established before 1800. However, it happened in the nineteenth century that the main scientific activity moved to universities and university-like institutions. The professors and the teaching and research staffs at those places then constituted the main part of the scientific community. They founded their professional societies, such as the *Physikalische Gesellschaft zu Berlin* in 1845, the *Deutsche chemische Gesellschaft* in 1867, the British Institute of Physics in 1874, the American Chemical Society in 1876, or the American Physical Society in 1899, which supplemented the already existing interdisciplinary societies like the *Gesellschaft deutscher Naturforscher und Ärzte* (founded in 1822), the British Association for the Advancement of Science (founded in 1831) or the American Association for the Advancement of Science (founded in 1848). The members of these societies included an increasing number of physicists and chemists who did not work in academic institutions, for —due to growing industrialization—many scientists were employed in chemical and electrical plants and laboratories. During the second half of the nineteenth century an especially fruitful collaboration developed between industry and science, leading not only to numerous technical applications of scientific results and methods in the production of new machines and materials, but also to many challenging tasks in pure research. It was in this atmosphere and under these general conditions that the various branches of science, especially physics and chemistry, grew and assumed their dominant position.

The Great Theoretical Schemes of Physics

Mechanics, the oldest among the principal schemes to describe the behaviour of inanimate objects, had been formulated in 1687, when Isaac Newton (1642–1727) published his '*Philosophiae Naturalis Principia Mathematica*,' containing the principles of dynamics and the law of gravitation. During the eighteenth century the mathematicians, such as Leonhard Euler (1707–1783), worked out the details of Newtonian mechanics and its applications, say to extended rigid bodies and fluids. Joseph Louis Lagrange (1736–1813) finally formulated the fundamental dynamical equations in a particularly suitable, elegant form as analytical dynamics. In this scheme he incorporated, for instance, the 'principle of virtual work,' by means of which Jean le Rond d'Alembert (1717–1783) had described the equations of motion. Thus mechanics could be taken over as a fully elaborated scheme into the nineteenth century; all that seemed to be left to do was to calculate with its help the problems pertaining to more and more complicated systems in physics and astronomy. However, certain mathematicians still sought to add to it. Thus, in 1829, Carl Friedrich Gauss (1777–1855) proposed a new principle, applicable to all mechanical problems, be they statical or dynamical in nature: this was the 'principle of least stress,' which stated that the motion of an arbitrary system of material points under whatever constraints takes place as close to a free motion as possible, i.e., it takes place under minimum stress. With it Gauss tried to generalize the earlier principle of d'Alembert; however, since the concept of stress was a rather complex one, the principle of minimum stress did not in practice replace the use of Lagrange's equations. In this respect, a scheme of mechanics devised by William Rowan Hamilton (1805–1865) and further developed by Carl Gustav Jacob Jacobi (1804–1851) turned out to be more successful.[1] Hamilton started out from the 'principle of least action,' which had been introduced by Pierre Louis Moreau de Maupertuis (1698–1759) in 1747, stating that the motions of objects in nature always occur with minimum action. Maupertuis had chosen as the quantity describing action the path-integral over the product $m \cdot v$, m being the mass of

---

[1]William Rowan Hamilton was born on 4 August 1805 in Dublin. He studied at Trinity College, Dublin, from 1823 to 1827. In 1827 he was appointed to the Andrews Chair of Astronomy in Trinity College as the successor of John Brinkley. He spent the rest of his life at the Observatory at Dunsink and died on 2 September 1865. Hamilton started out by working on optical problems (systems of rays: Hamilton, 1828). From these he obtained certain ideas which he employed in his great memoirs on the methods of dynamics (Hamilton, 1834, 1835). He discovered and propagated the calculus of quaternions. Hamilton received many honours: he was the Royal Astronomer for Ireland and was knighted in 1835.

Carl Gustav Jacob Jacobi was born at Potsdam on 10 October 1804. He studied at the University of Berlin and received his doctorate in 1825 with a thesis on partial fractions. Two years later he become extraordinary, and finally ordinary, professor of mathematics at the University of Königsberg (1829–1842). Then he went to Berlin, living on a pension from the King. He died there on 18 February 1851. Jacobi contributed important papers to several mathematical fields, especially the theory of numbers, elliptic functions (a subject which, with Niels Abel, he founded), partial differential equations, analytical dynamics and the theory of determinants. He also studied the equilibrium shapes of rotating liquid masses (Jacobi ellipsoids).

the moving object and v its velocity; his principle thus stated that $\int mv\,ds$ should be a minimum over the actual path of the object. Hamilton defined the action in a different, more general way as the time-integral over the so-called Lagrange function L, where L was defined as the difference of kinetic and potential energies; hence the integral, $\int L\,dt$, extending from time t_1 to the time t_2, had to be a minimum for the actual motion considered (Hamilton, 1834, 1835). Hamilton showed that the principle led to a new form of the equations of motion, later called Hamilton's equations, which were equivalent to Lagrange's. He also suggested a method of integrating these equations, involving the partial differential equation for the characteristic or action function. His method was perfected by Jacobi and applied to solve in particular the dynamical many-body problem in astronomy (Jacobi, 1866). In Jacobi's and later treatments of mechanical systems the role of the constants of motion was recognized. These constants of motion corresponded to certain physical properties, for which conservation laws held. The most important such conserved quantity represented the energy. The discovery of its conservation was connected with another branch of physics, i.e., thermodynamics.

The development of the description of thermal phenomena in nature had begun early in the eighteenth century with the invention of a reliable gas thermometer by Gabriel Daniel Fahrenheit (1686–1736); with it one could determine quantitatively the temperature of a heated body. It had been assumed that heat represented a substance, which was added to or subtracted from matter, thus giving it a higher or lower temperature, respectively. In 1798 Benjamin Thompson, Count Rumford (1753–1814), showed, however, that heat could be produced mechanically (i.e., by the motion of bodies) without changing the properties of the bodies involved, in contrast to the prediction based on the hypothesis of a heat substance. More than forty years later the true nature of heat was recognized by the physician Julius Robert Mayer (1814–1878) and the physicist James Prescott Joule (1818–1889): they claimed that heat corresponded fully to energy and determined a number for the mechanical equivalent of a unit quantity of heat. Hermann von Helmholtz then developed in full generality what he called the 'principle of conservation of force' or the principle of conservation of energy (Helmholtz, 1847).[2] This principle constituted the first law of thermodynamics; the second law originated from the investigations of Sadi Carnot (1796–1832) early in the century, reported in his *Réflexions sur la Puissance Motrice du Feu* (Carnot, 1824). Both laws were first listed explicitly in two memoirs of Rudolf Emanuel Clausius (1822–1888), entitled '*Über die bewegende*

---

[2] Among the different forms of energy, Helmholtz included the potential energy, the energy of gravitation, the energy of the static electric and magnetic fields, and the electrodynamic energy.

The word 'energy' occurred already in Johannes Kepler's *De Harmonice Mundi* of 1619, meaning the source of all motion in the world originating from God. The word 'force' was derived from Gottfried Wilhelm Leibniz' *vis viva*. In the nineteenth century Thomas Young and William John Macquorn Rankine (1820–1872), for instance, used the word 'energy.' Then William Thomson, later Lord Kelvin, introduced it into thermodynamics (Kelvin, 1853).

Kraft der Wärme' (Clausius, 1850).[3] Clausius showed in particular that Carnot's result concerning the production of mechanical work in a process, in which heat passes from a hotter to a colder body, could be upheld, even when the validity of the first law—the transformation of heat into work—in contrast to Carnot's original assumption was taken into account. William Thomson, Lord Kelvin (1824–1907), who had earlier foreseen difficulties in adapting Carnot's principle to the new foundation of heat theory, as provided by energy conservation, agreed with Clausius' formulation of the laws of what he now called the 'dynamical theory of heat' (Kelvin, 1853). The second law of thermodynamics could also be expressed in physical terms as: it is impossible to obtain work by cooling matter below the temperature of the coldest of the surrounding objects without any other change. In analogy to expressing energy conservation as the nonexistence of a *perpetuum mobile* (i.e., a machine which produces work without input of energy), the second law then meant the impossibility of the existence of another *perpetuum mobile* ('of the second kind'): that no energy can be produced by just establishing a larger temperature difference between two objects. Clausius gave the second law a more precise mathematical formulation. He used the absolute scale of temperature, T, introduced by William Thomson in 1848 (a temperature of $-273°C$ defines absolute zero in this scale), and showed that the integral, $\int dQ/T$, where dQ denotes the differential of heat, assumed the value zero if extended over a reversible cycle (i.e., a thermodynamic process whose initial and final state was the same), and a positive value if extended over an irreversible thermodynamic cycle. Thus he showed that the quotient, Q/T, where Q denotes the heat content of a body, determined the maximum energy which could be derived from it. He further found that this quantity represented a variable characterizing the thermodynamic state of a body and called it 'entropy' (Clausius, 1865, p. 390). Then the second law stated: the entropy of a system either remains constant or increases. Or, if the entire universe is considered: the entropy of the universe tends to a maximum ('*Die Entropie der Welt strebt einem Maximum zu*,' Clausius, 1865, p. 400). Based on the first and second law, the thermodynamic properties of matter were investigated in detail: Clausius developed the theory of equilibrium between the different phases of the same substance, while Jacobus Henricus van't Hoff (1852–1911) and Josiah Willard Gibbs (1839–1903) studied the equilibrium between different substances. Thus, for instance, van't Hoff succeeded in reducing the chemical concept of affinity to the physical concepts of energy and entropy differences. In discussing the concept of affinity further, Walther Nernst (1865–1941) was led to his heat theorem, the so-called third law of thermodynamics (Nernst, 1906), with the help

---

[3] Clausius was born in Köslin on 2 January 1822 and studied at the University of Halle, receiving his doctorate in 1848. Two years later he was appointed professor of physics at the Royal Artillery and Engineering School, Berlin; at the same time he lectured as *Privatdozent* at the University of Berlin. In 1855 he became ordinary (i.e., full) professor at the Zurich Polytechnic (*Eidgenössische Technische Hochschule* or E.T.H.), and in 1867 he moved to Würzburg and two years later to Bonn. Clausius died in Bonn on 24 August 1888.

of which he could determine the entropy of systems at the absolute zero of temperature.

The third great theoretical scheme of the nineteenth century was electrodynamics. A large variety of electrical phenomena had been studied already in the previous century: not only had experiments been performed with natural electricity (like the one occurring in thunderstorms), but also with artificial electricity, produced, for example, with the help of the Leyden jar—invented by Ewald Georg von Kleist (1700–1748) and Pieter van Musschenbroek (1692–1761)—or the cells of the physician Luigi Galvani (1737–1798). Thus the attraction and repulsion of electrically charged bodies, the phenomena of conduction of electric charge by certain materials and the creation of electric sparks had provided private and public amusement to many people. The first quantitative description of an electric effect was provided by the law which Charles Auguste de Coulomb (1736–1806) had obtained from measurements in 1785; it stated that the electrical force (attractive or repulsive) between two small charged spheres—similar to the gravitational attraction—was inversely proportional to the distance between their centres. At the turn of the nineteenth century, research in the science of electricity was eagerly promoted. Men like Alessandro Volta (1745–1827), Alexander von Humboldt (1769–1859), Johann Wilhelm Ritter (1776–1810) and Humphrey Davy (1778–1829) investigated the detailed mechanism of galvanic cells; Humboldt (1797) and Ritter (1799) discovered the fundamental process of electrolysis and Davy (from 1800) investigated it further. Others concerned themselves with the effects of electric currents. Hans Christian Oersted (1777–1851) discovered in 1820 that a magnetic needle was deflected if a current passed through a conducting wire nearby. André Marie Ampère (1775–1836) observed in the same year that electric currents flowing in the same direction give rise to a repulsive force (bending the neighbouring conducting wires away from each other), while currents flowing in opposite directions create an attractive force; moreover, a solenoid of conducting wire behaved like a magnet if a current passed through it. Two years later Ampère called the theory, which described the magnetic and mechanical effects of electric currents, 'electrodynamics' (Ampère, 1822). At about the same time Jean Baptiste Biot (1774–1862) and Félix Savart (1791–1841) derived from their experiments the law determining the force exerted by electric currents on magnetic poles (Biot, 1820). They found that the force, or the magnetic field $\mathbf{H}$, exerted by a line element ds of a conducting wire carrying the current $\mathbf{j}$ on a magnetic pole at a distance r, was proportional to $(j\,ds\sin\alpha)/r^2$, where α was the angle between the direction of the current and the line joining the line element to the magnetic pole and j denoted the absolute value of the current.

The new theory of electrodynamics established a unified description of electric and magnetic phenomena, which had previously been considered as having different origins. Further analysis of the properties of electric currents led to important scientific and technical results. In 1826 Georg Simon Ohm (1787–1854) formulated a quantitative law describing the relation between the electric

current passing through a given conductor and the potential difference created
by a galvanic cell between the ends of a conductor. Just seven years later Carl
Friedrich Gauss and Wilhelm Weber (1804–1891) constructed their telegraph,
which demonstrated one of the early, important applications of electric currents.
These results, brilliant as they were, were just the beginning of a period of great
experimental and theoretical development of electrodynamics, in which Michael
Faraday (1791–1867) and James Clerk Maxwell (1831–1879) played the dominant role. Faraday, born on 22 September 1791, at Newington Butts, Surrey,
began his scientific career in 1813 as secretary to Humphrey Davy, then
professor at the Royal Institution in London. He began to publish scientific
papers in 1816, announcing important results in quick succession: in 1821 the
discovery of electromagnetic rotation (showing that a circular magnetic force
exists around a current-carrying wire); in 1823 the liquefaction of chlorine; in
1824 the discovery of benzene. In 1824 he was elected a Fellow of the Royal
Society, and in 1825 he became director of the laboratory at the Royal Institution. While his reputation grew steadily—he was appointed Fullerian Professor of
Chemistry in 1833, became scientific advisor to Trinity House in 1836, and was
awarded an annual pension of £300 in 1835—Faraday entered upon a series of
most fundamental researches on electricity and magnetism. In 1831 he carried
out experiments leading to the discovery of electromagnetic induction, i.e., of the
effect that a changing magnetic field (provided by a suddenly interrupted current
in a conducting coil) creates an electric field (i.e., a potential difference in a
neighbouring coil). This effect would enable one to construct motors driven by
electricity and generators producing strong electric currents. Shortly afterwards
Faraday showed that all known kinds of electricity (frictional, thermal, galvanic,
voltaic, and magnetic, i.e., by induction) were fundamentally the same. He
obtained the quantitative laws of electrolysis in 1834. Eleven years later he again
made a fundamental contribution by demonstrating the influence of a magnetic
field on the plane of polarization of light. However, he tried unsuccessfully to
establish a relation between gravity and electricity (in 1850) and to discover the
effect of a magnetic field on the frequency of spectral lines (in 1865). When he
died on 25 August 1867 near Hampton Court, Surrey, where he had retired to a
comfortable house provided by Queen Victoria, Faraday left behind not only a
wealth of experimental discoveries, but also conceptions which changed the
theory of electrodynamics. For instance, he introduced the notion of lines of
electric force, in analogy with the lines of magnetic force; and he explained the
mechanism of dielectrics. Based on Faraday's work and ideas Maxwell could
proceed to build electrodynamic theory; indeed, he was introduced to electrodynamics by reading Faraday's books on the experimental researches in Cambridge, and he began his own research on this subject by translating Faraday's
electric and magnetic lines of force into a proper mathematical language.[4] Later

[4] James Clerk Maxwell was born in Edinburgh on 13 November 1831. He was educated at
Edinburgh Academy (1841–1847), the University of Edinburgh (1847–1850), and at Trinity College,
Cambridge (1850–1854). After his graduation from Cambridge (where he was Second Wrangler) he
spent two more years at Trinity College. He was appointed to the chair of natural philosophy at

Maxwell presented a detailed theory of electrodynamics in a series of papers, entitled 'On Physical Lines of Force' (Maxwell, 1861a, b; 1862a, b) and 'A Dynamical Theory of the Electromagnetic Field' (Maxwell, 1865). In them he expounded what later came to be called Maxwell's equations of electrodynamics. They included, in particular, the 'displacement current,' $\partial \mathbf{D}/\partial t$, obtained from the displacement of electric charges in dielectric materials when electric fields are present; thus the current effective in the creation of a magnetic field $\mathbf{H}$ consisted of two parts, the so-called convection current $\mathbf{j}$—arising from the motion of charged particles—and the displacement current, i.e., curl $\mathbf{H} = 4\pi(\mathbf{j} + \partial \mathbf{D}/\partial t)$. During the following years Maxwell derived numerous results from his system of electrodynamic equations and wrote the *Treatise on Electricity and Magnetism* (Maxwell, 1873).

Maxwell's theory of electrodynamics stood in some opposition to other theories, especially to the ones developed by Franz Ernst Neumann (1798–1895), Wilhelm Weber and Rudolf Clausius. In these (the so-called potential theories) the emphasis was on establishing a law describing the force between two charges; it was assumed, in particular, that this force depended not only on the distance between the charges but also on their velocities and accelerations. Both schemes, Maxwell's and the potential theory, provided a consistent description of the known electrodynamic phenomena. But Maxwell's went a step further. Already in 1861 he had drawn an important conclusion upon studying a mechanical model for the medium, which—in his view—the electric and magnetic lines of force could penetrate; from this model of an elastic medium he calculated the velocity of transverse waves in it and found it to be equal to the velocity of light. 'We can scarcely avoid the inference,' he wrote, 'that *light consists in the transverse undulations of the same medium which is the cause of electric and magnetic phenomena*' (Maxwell, 1862a, p. 22; *The Scientific Papers I*, 1890, p. 500). Although Maxwell was the first to claim the full identity of light as an electromagnetic wave phenomenon, the possibility of such a connection had already been indicated in earlier investigations. Thus Wilhelm Weber, who had concerned himself since 1852 with a detailed comparison of the electromagnetic and the electrostatic units of the electric charge, had found that their ratio had the dimension of a velocity, and he determined this velocity (from experiments) to be equal to the velocity of light *c in vacuo* (Weber and Kohlrausch, 1856). Gustav Kirchhoff (1824–1887) had then studied the propagation of electrical

---

Marischal College, Aberdeen (1856–1860), and then as professor of physics and astronomy at King's College, London (1860–1865). After retiring from regular academic life to write his *Treatise on Electricity and Magnetism*—during which time he also served as examiner or moderator for the Mathematical Tripos in Cambridge—Maxwell was appointed in 1871 as the first professor of experimental physics at Cambridge University and director of the Cavendish Laboratory. He died on 5 November 1879 in Cambridge.

Maxwell's main contributions to physics were in electrodynamics and the kinetic theory of gases. But he also worked on several other experimental and theoretical topics: on colour vision, the theory of Saturn's rings, geometrical optics, thermodynamics and viscoelasticity. He wrote four books and edited the papers of Henry Cavendish (1731–1810). In 1861 he was elected a Fellow of the Royal Society.

disturbances along a wire and had observed their velocity to be again equal to c (Kirchhoff, 1857).

The possibility that light might be of electromagnetic origin had received some general support from the development of optics in the nineteenth century. At the beginning of the century Thomas Young (1773–1829) interpreted the well-known interference effect as proving the wave nature of light (Young, 1800). In 1809 Etienne Louis Malus (1775–1812) discovered the polarization of light, upon which Young claimed—in 1817—that light consisted of transverse waves; this was confirmed by the investigations of Augustin Jean Fresnel (1788–1827) and Dominique François Arago (1786–1853). The outstanding theoretical problem then became to understand the process of propagation of light. It was known that light travelled through spaces that were free of matter; it was assumed, therefore, that the undulations of light were transported through a special medium, the ether.[5] The ether was imagined to be an elastic solid, infinitely penetrable for all ponderable matter, and had such properties as to suppress all longitudinal wave motions in it. Many detailed models of the ether were developed after the work of Fresnel, e.g., by Claude-Louis Marie Navier (1785–1836), Augustin Louis Cauchy (1789–1857), George Green (1793–1841) and George Gabriel Stokes (1819–1903). Maxwell's ether model of 1861 also fitted into these endeavours (Maxwell, 1861a, b; 1862a, b). However, an actual proof of light as an electromagnetic phenomenon was not provided until Heinrich Hertz (1857–1894) demonstrated in 1888 the existence of electromagnetic waves, created by purely electromagnetic processes, which possessed all the properties of light waves (Hertz, 1888a, b). Only after 1888 did optical theory become an integral part of electrodynamics. At the same time those electrodynamic theories, which were not able to account for electromagnetic waves and light, were discarded and Maxwell's theory emerged as one of the great theoretical schemes of physics.

The Atomism of the Nineteenth Century

The concept of the atom is supposed to be as old as science. It had been introduced by the ancient Greek philosophers, especially by Democritus of Abdera (born 470–460 B.C.?, died around 370 B.C.), to account for the complexity

---

[5]The concept of the ether is quite old. Meaning originally the 'blue sky' (in Greek) it had been used in ancient cosmology to represent the celestial regions. René Descartes (1596–1650) then gave the ether a place in his system of nature: he stated in particular that space is filled by a medium which, though imperceptible to the senses, is capable of transmitting forces and exerting effects on material bodies immersed in it. This view was taken over to some extent by Isaac Newton in his memoirs on optics: all space, he assumed, is filled by an elastic medium or ether, which is capable of propagating vibrations (in the same way as the air propagates the vibrations of sound, just with much greater velocity); further, the ether fills all material bodies, thus causing their cohesion. But according to Newton the vibrations of the ether did not constitute light. Christiaan Huygens (1629–1695), however, claimed that light consisted of disturbances propagated with a large velocity in a highly elastic medium. For a detailed account of the development of the ether problem, see Edmund Whittaker's *A History of the Theories of Aether and Electricity* (Whittaker, 1951).

of natural phenomena and their changes. According to this view there should exist innumerably many, unchangeable atoms having different shapes and sizes, which move about constantly. Atoms were the smallest constituents of all matter; the different kinds of matter consisted of different atoms, having different positions with respect to each other. In ancient times the atomic view of nature had been propagated by Epicurus of Samos (341–270 B.C.) and Lucretius Carus (96–55 B.C.), but in the philosophy of the middle ages it had not played a major role. It was rediscovered in the seventeenth century, notably by Pierre Gassendi (1592–1655), and then incorporated into the new science. For example, Robert Boyle (1627–1691) connected it with chemistry, and Isaac Newton used it in mechanics and optics. In spite of the great heuristic value of the concept of smallest particles or atoms in explaining the behaviour of matter, the atomic hypothesis did not contribute essentially to the progress of science in the following hundred years. This may be ascribed to the fact that practically nothing was known about the specific properties of atoms. However, the situation changed towards the end of the eighteenth century because of the fundamental work of the chemist Antoine Laurent Lavoisier.[6] He studied in detail the processes of combustion, which he recognized as being connected with the binding of oxygen by certain substances. From these studies he derived the law of conservation of mass in chemical reactions. Further, he defined a number of substances as elementary ones or elements, because they could not be separated by chemical methods; they included some of the best known metals like silver, gold, mercury, antimony and zinc.[7] Lavoisier's new chemical nomenclature, together with his law of mass action, started a period of great prosperity in chemistry. The research of Joseph Louis Proust (1754–1826) established another fundamental theorem, the law of definite proportions: it claimed that the elements in a given compound were present in a fixed proportion by weight. This law was first stated explicitly by John Dalton (1766–1844), who is held responsible for having established atomism in chemistry through his publications between

[6] Lavoisier was born on 26 August 1743 at Paris. He studied astronomy, chemistry and botany at the *Collège Mazarin*, Paris (1754–1761). Then he quickly made a name for himself by writing papers on the analysis of gypsum, on thunder, on the aurora and other topics. He also contributed to the mineralogical atlas of France. Having received in 1766 a gold medal of the Academy of Sciences, Paris, for an essay on the best method for lighting a large town, Lavoisier was nominated *adjoint chimiste* of the Academy in 1768; later, in 1785, he served as director and, in 1791, as treasurer of the Academy. In 1775 he became director of the governmental gunpowder works; he also got into agricultural research. Through this and other work on public commissions, he became involved in politics. After a long and distinguished career in various offices, he was sent to the guillotine on 8 May 1794 in the *Place de la Révolution* during the French Revolution.

Lavoisier's work on the combustion of sulphur and phosphorus and on the reduction of metallic oxides disproved the phlogiston hypothesis of combustion. It further led to a new nomenclature in chemistry based on elementary substances. He also carried out some of the earliest thermochemical investigations and he explored the processes of fermentation, respiration and animal heat, recognizing their chemical nature. Thus he became one of the founders of modern chemistry.

[7] As elementary substances, Lavoisier also listed the gases oxygen, nitrogen and hydrogen, and the solid materials sulphur, phosphorus, carbon, baryta, magnesia and silica. However, he did not add to this list the alkali metals, for he had doubts about their elementary nature.

1803 and 1808.[8] Dalton assumed, in particular, that the ultimate particles or atoms in a chemically homogeneous substance had the same weight and the same shape. He then set forth a table of relative weights of the atoms for a number of elementary substances, derived by the chemists of his time from the analysis of water, ammonia, carbon dioxide and other compounds. Moreover, he added to Proust's law of definite proportions the law of multiple proportions stating: if a given amount of weight of a chemical substance combines with different amounts of weight, $m_1, m_2, \ldots$, of another element to form compounds, the ratios m_1/m_2, etc., can be represented by the ratios of small integers. For example, 100 g of tin combine with 13.5 g of oxygen as well as with twice that amount, i.e., 27 g. From this fact Dalton concluded that in the first case one atom of tin associates itself with one atom of oxygen, and in the second case with two atoms of oxygen. The laws of Proust and Dalton were substantiated by an observation of Joseph Louis Gay-Lussac (1778–1850), subsequently called the law of Gay-Lussac. It stated: when gases combine with each other, they do so in simple proportions of their volumes, and the volume of the compound, if gaseous, bears a simple relation to the volumes of the reacting gases. From it Amadeo Avogadro (1776–1856) in 1811 drew a far-reaching conclusion: under identical conditions of temperature and pressure equal amounts of gases contain the same number of molecules (Avogadro, 1811).[9] In order to formulate this hypothesis, Avogadro distinguished between the ultimate particles of an element, the atoms, and the ultimate particles of a chemical compound, which he called molecules. The validity and importance of Avogadro's hypothesis was not acknowledged right away, and this neglect caused considerable confusion in chemistry during the following decades. Only when Stanislao Cannizzaro (1826–1910) accepted it fully in 1858 was he able to establish a consistent nomenclature for chemical substances (Cannizzaro, 1858).[10] Nevertheless, some opposition still

[8] John Dalton was born on 6 September 1766 at Eaglesfield near Cockermouth, Cumberland. He acquired his education by private studies; then from 1785 to 1793, together with his brother, he was manager of the school at Kendal. In 1793 he moved to Manchester and took the position of a teacher of mathematics and natural philosophy at New College, until it was moved in 1799 to York. Dalton remained in Manchester and earned his living as a public and private teacher of mathematics and chemistry. He died on 27 July 1844.

Dalton worked on meteorology, colour vision (colour blindness), the absorption of gases by liquids and on a system of chemistry. In 1822 he was elected a Fellow of the Royal Society and in 1830 he succeeded Humphrey Davy as Foreign Associate of the Academy of Sciences, Paris.

[9] Amadeo Avogadro, *Conte di Quaregna e Ceretto*, was born on 9 August 1776 at Turin. He studied law and later (1800–1805) science at the University of Turin. In 1809 he became professor of mathematics and physics at the Royal College of Vercelli, and in 1820 he was appointed to the first Italian chair of mathematical physics at the University of Turin. He occupied this chair until 1822, when it was suppressed due to political reasons; after its reestablishment in 1834, Avogadro remained in it until 1850. He died at Turin on 9 July 1856.

[10] Cannizzaro was born on 13 July 1826 at Palermo. He was educated at Palermo, Naples, Pisa, Turin and Paris. In 1851 he became professor of physics and chemistry at the Technical Institute of Alessandria, Piedmont, in 1855 professor of chemistry at the University of Genoa, in 1861 professor of inorganic and organic chemistry at Palermo and in 1871 professor of chemistry at the University of Rome. Cannizzaro, an ardent Italian patriot, participated in the Sicilian Revolution of 1848 and joined Garibaldi in 1860. He worked mainly on problems of organic chemistry. He died on 10 May 1910 in Rome.

arose against Cannizzaro's proposals, because there seemed to be exceptions from Avogadro's hypothesis; however, in the following years they were recognized as originating from partial or complete thermal dissociation of the substances involved. As a consequence, especially the scientists who worked on dissociation processes, like Jacobus Henricus van't Hoff and Walther Nernst, advocated the atomic hypothesis. To the title of his book on theoretical chemistry, Nernst added: 'from the point of view of Avogadro's rule' (Nernst, 1893). This was an important statement, for the majority of the chemists—since Dalton's time—had given up the belief in atoms. The chemists more or less tended to interpret the results of Lavoisier, Proust and Dalton as expressing simple ratios not necessarily connected with the existence of ultimate particles.

At about the same time, as the atomic hypothesis lost credit among the chemists, it gained partisans among the physicists. In principle, the development of physics during the first half of the nineteenth century did not favour the atomic constitution of matter. Rather, the results in optics and electrodynamics pointed to the necessity of continuum theories. However, there was one field which brought about a change: the kinetic theory of gases. The basic ideas contained in this theory had been present already in the views of Democritus, but only two thousand years later did the physicists return to them. After Daniel Bernoulli (1700–1782) had taken the first step in kinetic gas theory in 1738, further development began only about a hundred years later. By that time there were available several laws describing the properties of gases in some approximations: Boyle's law (1662) stating that as a gas is compressed without change of its temperature, its pressure increases directly as its density; Gay-Lussac's law (1802) stating when a gas is heated at constant volume, its pressure increases at a rate which is the same for all gases and proportional to (what was later defined as) the absolute temperature; Dalton's law (1802) stating that if a gas consists of a mixture of different kinds of gases, then the total pressure is just the sum of the partial pressures (i.e., of the pressures of the individual gas components). Based on these laws and on Avogadro's hypothesis of 1811 (Avogadro, 1811), the physicists developed shortly after 1855 a systematic kinetic theory of gases.[11] Then detailed accounts were published by August Karl Krönig (1822–1879), a *Gymnasium* teacher in Berlin, and by Rudolf Clausius (Krönig, 1856; Clausius, 1857). The main assumptions involved in the theory were the following: a gas consists of molecules, which may be represented roughly by elastic spheres; the molecules move with a certain velocity in straight lines until they collide with each other or the wall. Then the gas pressure could be explained immediately by

---

[11] Concerning the origin of the kinetic theory of gases we refer to the relevant literature, e.g., to the articles of Stephen G. Brush (1957a, b; 1965; 1966). Surely Daniel Bernoulli must be considered as the first author to have treated this subject. In the nineteenth century John Herapath (1790–1868) apparently published the first paper on kinetic theory in 1821 (Herapath, 1821). James Prescott Joule (1818–1889) knew about this paper and quoted it when he discussed the heat motion of gas molecules thirty years later (Joule, 1851). In December 1845 John James Waterston (1811–1883) submitted a paper containing ideas similar to Herapath's to the Royal Society of London (Waterston, 1846); it was not published then because of the unfavourable reports of the referees; Lord Rayleigh retrieved it over forty-five years later and had it published in the *Philosophical Transactions* (Waterston, 1893).

the elastic impacts of the molecules on the wall. Further, the kinetic energy of the molecules was assumed to be proportional to the absolute temperature, which provided a relation for calculating the velocity of the molecules. (Such a calculation had been made already in 1851 by James Prescott Joule.) In a later paper Clausius introduced the concept of the mean free path of molecules, denoting the average distance a molecule can travel in a gas before colliding with another (Clausius, 1858); evidently, this distance depended on the size of the molecules and their number in a given volume. The kinetic theory of gases was perfected by Clausius and Maxwell in the following years. Maxwell removed the assumption in Krönig's and Clausius' treatments that all molecules possess the same velocity by showing that the velocities of the molecules were distributed according to a certain law (Maxwell, 1860). James Clerk Maxwell, Ludwig Boltzmann (1844–1906) and others—e.g., Oskar Emil Meyer (1834–1909)—then succeeded in deriving the detailed properties of gases, such as internal friction (viscosity), diffusion and heat conduction from the kinetic theory. The theory also provided the first determination of the size of molecules: Joseph Loschmidt (1821–1895) calculated from the mean free path (obtained from measurements of air viscosity) the results that the diameter of air molecules is of the order of 10^{-8} cm and about 10^{23} molecules are contained in one gram-atom (Loschmidt, 1865).[12]

The combined efforts of chemists and physicists gave the hypothesis of atomic and molecular constitution of matter important support. The question arose whether chemical and physical atoms were the same. From the point of view of the atomic hypothesis the question had to be answered in the affirmative, and several facts hinted in this direction. For example, certain properties—such as the specific heats of solids, via the rule of Pierre Louis Dulong (1785–1838) and Alexis Thérèse Petit (1791–1820), and the vapour density—had been used to determine details of the chemical composition of molecules. But the closest connection between the chemical and the physical atoms was established when one studied the organization and classification of the chemical elements of Lavoisier and Dalton. Already in the early decades of the nineteenth century several chemists, notably Johann Wolfgang Döbereiner (1780–1849) and Jean André Baptiste Dumas (1800–1884) had attempted to find relations between the atomic weights of elements and their properties. For instance, it had been observed in the case of homologous elements, like the triple, chlorine, bromine and iodine, that the middle element, bromine, showed physical and chemical

---

[12] Joseph Loschmidt was born on 15 March 1821 in Putschirn, close to Karlsbad in Bohemia. After studying philosophy, mathematics, physics and chemistry at the University of Prague and (after 1842) at the University of Vienna and the Vienna Polytechnic, he founded—together with a colleague—a chemical plant in the vicinity of Vienna. The enterprise went bankrupt in 1849, after which Loschmidt spent several more years in the chemical industry until he returned to Vienna. In 1856 he became a schoolteacher, and a year later he obtained his *Habilitation* at the University of Vienna. In 1860 he was appointed extraordinary, and in 1872 ordinary, professor of physical chemistry at the University of Vienna. He retired in 1891, and died on 8 July 1895 at Vienna. Loschmidt worked on problems of chemical constitution, atomic theory and diffusion of gases.

properties between those of chlorine and iodine. As the century progressed, more data on the properties of elements were collected, and towards the end of the 1860s two chemists, Dmitri Ivanovich Mendeleev (1834–1907) in St. Petersburg and Julius Lothar Meyer (1830–1895) in Karlsruhe, felt safe in proposing a complete organization, the periodic system of elements (Mendeleev, 1869; L. Meyer, 1870).[13] In this system the elements were basically arranged according to rising atomic weights; because of the periodic change of their properties, homologous elements appeared in the same vertical columns. Altogether Mendeleev and Meyer proposed eight columns, numbered with Roman numerals from I to VIII; they especially suggested the splitting of the so-called long periods consisting of seventeen elements (the noble gases were not yet known) into a period of seven elements, a group of three elements and another period of seven elements. Thus in the case of the fourth period, containing the elements from potassium to bromine, they put the first seven elements in columns I to VII, the three central elements—iron, cobalt and nickel—in column VIII, and then the following elements from copper to bromine again into columns I to VII. The arrangement of elements in the periodic system immediately led to certain consequences: first, several small errors in the previously accepted atomic weights could be corrected; second, Mendeleev predicted the existence and properties of several as yet undiscovered elements, e.g., eka-boron, eka-aluminum and eka-silicon. These elements were discovered in the following decades and were called scandium, gallium and germanium, respectively. In addition, Lord Rayleigh (1842–1919) and Sir William Ramsay (1852–1916) added, beginning in 1894, a group 0 to the periodic system, containing the inert gases helium, neon, argon, krypton and xenon. Since then the number of newly discovered elements has increased considerably, and they can all be fitted into the periodic system.

---

[13] Mendeleev was born on 7 February 1834 at Tobolsk, Siberia. After attending the *Gymnasium* in his hometown, he went to St. Petersburg to study science at the Pedagogical Institute and (from 1850) at the University of St. Petersburg. There he graduated in 1856 (he received his doctorate in 1865). Then he pursued further studies abroad, in Heidelberg (with Robert Bunsen) and Paris (with Henri Regnault). He returned to St. Petersburg in 1861 and became professor of chemistry at the Institute of Technology two years later. In 1865 he obtained a professorship at the University of St. Petersburg, which he held until his retirement in 1890. Three years later he was appointed director of the Bureau of Weights and Measures in St. Petersburg. He died on 2 February 1907 at St. Petersburg. Mendeleev was a member of many learned societies, including the Royal Society and the Academy of Sciences, Paris. Besides the systematic organization of chemical elements, he worked on the physical properties of gases and liquids (he pointed to the existence of the critical temperature) and on the character of petroleum.

Julius Lothar Meyer was born on 19 August 1830 at Varel, Oldenburg. He studied medicine at Zurich and Würzburg, where he obtained his medical doctor's degree. Then he turned to physiological chemistry at Heidelberg, and finally to mathematical physics at Königsberg and Breslau. He received his doctorate from Breslau in 1858. A year later he became director of the chemical laboratory at the Physiological Institute of the University of Breslau; in 1866 he was named professor of natural history at Eberswalde; from 1868 to 1876 he served as professor of chemistry at the *Technische Hochschule*, Karlsruhe; finally, he became the first occupant of the chair of chemistry at the University of Tübingen (1876–1895). Meyer worked on the chemistry of blood and the physiology of respiration before he attacked the problem of organizing the elements in the periodic system. He died in Tübingen on 11 April 1895.

The important conclusion obtained from the organization of Mendeleev and Meyer was that the periodicity extended both to chemical and physical properties. For example, Lothar Meyer showed that if the atomic volumes were plotted against the atomic weights of elements, the curve obtained exhibited a series of maxima and minima, the most electropositive elements always appearing at the peaks of the curve (L. Meyer, 1870). Mendeleev then used this curve to predict the density of the unknown elements eka-boron and eka-aluminum, in excellent agreement with the later findings. In spite of the unchallenged successes of the periodic system of elements, strong reservations were expressed by some scientists against the atomic hypothesis, the principal spokesmen being Ernst Mach (1838–1916), Wilhelm Ostwald (1853–1932) and Pierre Duhem (1861–1916). They argued in particular that all physical phenomena could be described solely in terms of energy considerations and that it was erroneous to attribute reality to atoms and molecules as constituents of matter. At the Lübeck meeting of the Society of German Scientists and Physicians (*Gesellschaft deutscher Naturforscher und Ärzte*) in September 1895, Ostwald and his supporter Georg Helm (1851–1923), then professor at the *Technische Hochschule* of Dresden, presented the results of what they called 'energetics' ('*Energetik*'). At the same meeting Ludwig Boltzmann defended the atomic hypothesis; he claimed that none of the existing physical theories could be developed purely on the basis of Ostwald's assumptions (Boltzmann, 1896). Boltzmann provided a proper description of the whole situation concerning the atomic hypothesis in the foreword to the second volume of his lectures on kinetic gas theory. He stated there:

As the first part of *Gas Theory* [*Vorlesungen über Gastheorie*, Boltzmann, 1895c] was being printed, I had already almost completed the present second and last part, in which the more difficult parts were not to have been treated. It was just at this time that attacks on the theory of gases began to increase. I am convinced that these attacks are merely based on a misunderstanding, and that the role of gas theory in science has not yet been played out. (Boltzmann, 1898, p. V)

Boltzmann then went on to quote the results of Johannes Diderik van der Waals on the theory of real gases, William Ramsay on the atomic weight of argon leading to the subsequent discovery of neon, and of Marian von Smoluchowski (1872–1917) on the heat conduction in rarefied gases as strongly supporting the kinetic theory of matter. And he continued:

In my opinion it would be a great tragedy for science if the theory of gases were temporarily thrown into oblivion because of a momentary hostile attitude toward it, as was for example the wave theory because of Newton's authority. I am conscious of being only an individual struggling weakly against the stream of time. But it still remains in my power to contribute in such a way that, when the theory of gases is again revived, not too much will have to be rediscovered. Thus in this book [the Second Part] I will now include the parts that are the most difficult and most subject to misunderstanding, and give (at least in outline) the most easily understood exposition of them. (Boltzmann, 1898, pp. V–VI)

With the most difficult part of gas theory Boltzmann had in mind what later came to be called statistical mechanics, including the H-theorem and the statistical interpretation of entropy. Boltzmann was rather disturbed by the opposition to kinetic theory of such respected scientists as Mach and Ostwald. His unhappiness about this opposition and the fact that there seemed to be no direct proof of the atomic constitution of matter may have been one of the causes of his suicide in 1906; at that time, however, such a proof had already been provided by the work of Marian von Smoluchowski and Albert Einstein (1879–1955) on Brownian motion.

New Experimental Discoveries and New Theories

While the debate on the physical reality of atoms went on, experimental discoveries were made that had the greatest relevance to this question. Thus the investigations concerning the nature and properties of cathode rays by Joseph John Thomson (1856 1940), Wilhelm Wien (1864–1928), Emil Wiechert (1861–1928) and others established the existence of the electron, the atom of electricity. According to the theory of Hendrik Antoon Lorentz (1853–1928), the same electrons could be made responsible for the magnetic separation of spectral lines into triplets as observed by Pieter Zeeman (1865–1943). Shortly before this, two other important discoveries had become known. In 1895 Wilhelm Conrad Röntgen (1845–1923) demonstrated the existence of what he called X-rays, a penetrating radiation created by the impact of cathode rays on metals.[14] The discovery of X-rays not only stimulated active research devoted to understanding their nature and applications, it also encouraged Antoine-Henri Becquerel (1852–1908) to investigate whether phosphorescent materials emitted similar rays; Becquerel selected uranium salts as being particularly suitable for his purpose

---

[14]Röntgen was born in Lennep, Rhineland, on 27 March 1845. He studied physics at the University of Utrecht, and mechanical engineering and physics at the Zurich Polytechnic (E.T.H.). In Zurich he attended the lectures of Clausius and worked in the laboratory of August Kundt. After obtaining his doctorate in 1869 (at the E.T.H) he became assistant to Kundt and moved with his teacher to Würzburg the same year; three years later he went to Strasbourg, where he became *Privatdozent* in 1875. Soon thereafter he was appointed professor at the Academy of Agriculture in Hohenheim near Stuttgart and returned to Strasbourg the following year as professor of physics. Three years later, in 1879, he went to Giessen. In 1888 he succeeded Friedrich Kohlrausch in Würzburg, and in 1900 he finally accepted the chair of physics at the University of Munich. He died on 10 February 1923 in Munich.

Röntgen worked in many fields of physics including elasticity, capillarity, specific heats of gases, conduction of heat in crystals, the rotation of the plane of polarization of light by a magnetic field and piezoelectricity. A fundamental experiment was his demonstration, in 1888, of the magnetic effects caused by a moving, electrically polarized dielectric (it corresponded to what one later called the 'Röntgen current'); this experiment confirmed Faraday's concepts. Röntgen was awarded the first Nobel Prize in Physics in 1901 for his discovery of X-rays.

and discovered a new kind of radiation—radioactive rays—which blackened photographic plates.[15] His findings inspired Marie and Pierre Curie (1867–1934, 1859–1906, respectively) to study radioactive phenomena in detail; by analyzing large amonts of uranium minerals chemically, they isolated two highly radioactive elements, which they called polonium (having properties similar to bismuth) and radium (having properties similar to barium).[16] These results stimulated a lively research in the new field of radioactivity. Among the scientists who worked on it were Ernest Rutherford (1871–1937) in Cambridge and Montreal, André Louis Debièrne (1874–1949) in Paris, Friedrich Giesel (1852–1927) in Braunschweig, and Egon von Schweidler (1873–1948) in Vienna. They not only extended the list of known elements by discovering further radioactive ones between radium and uranium, but also initiated a new field of physics dealing with the structure of atoms.

The new discoveries definitely spoke in favour of the atomic hypothesis. However, the final proof of the physical reality of atoms resulted from a purely theoretical investigation in the theory of statistical mechanics. Statistical mechanics grew out of kinetic gas theory in the last decades of the nineteenth century and the early years of the twentieth century, its creators being Ludwig Boltzmann and Josiah Willard Gibbs.[17] The problem which led to this theory—to which

---

[15] Becquerel came from a family of scientists: his father, Alexandre-Edmond Becquerel (1820–1891), and his grandfather, Antoine-César Becquerel (1788–1878), had distinguished themselves as physicists. He was born on 15 December 1852 in Paris. He entered the *École Polytechnique* in 1872; two years later he joined the Government Department of Ponts-et-Chausées. There he became an engineer in 1877 and chief engineer in 1894. In 1888 he obtained his doctorate. He also worked (since 1878) at the Museum of Natural History, Paris, where he occupied (from 1892) the chair of applied physics. In 1895 he was appointed professor of physics at the *École Polytechnique*. He died at Le Croisic in Brittany on 25 August 1908.

Becquerel performed research on magnetism, polarization of light, and absorption of light in crystals and phosphorescence. For his discovery of radioactivity he shared the 1903 Nobel Prize in Physics with Pierre and Marie Curie.

[16] Marie Curie (*née* Marja Sklodowska) was born on 7 November 1867 at Warsaw. After studying in Warsaw and Cracow, she went to Paris to continue her studies at the Sorbonne. There she obtained in 1893 a *Licence-ès-science physique* and her doctorate in 1903. She married Pierre Curie in 1895; they had two daughers, Irène, who became a famous physicist in her own right, and Eve. In 1906 Marie Curie succeeded her husband as professor of general physics in the science faculty of the Sorbonne (the first woman to obtain such a position), and in 1914 she was appointed director of the Radium Institute of the University of Paris. She won a second Nobel Prize, the 1911 Chemistry Prize. She died on 4 July 1934 in the sanatorium of Sancellemont, French Alps.

Pierre Curie was born in Paris on 15 May 1859. He studied at the Sorbonne and obtained his *Licence* in physics in 1878. He worked as a demonstrator in the physics laboratory of the Sorbonne until 1882, when he became director of the laboratories at the municipal school for physics and industrial chemistry. In 1895 he obtained his doctorate and became professor of physics at the Sorbonne. He was killed in a street accident in Paris on 19 April 1906. Pierre Curie did research in crystallography, discovered piezoelectricity and the critical point of ferromagnetic substances—beyond which they lose their ferromagnetism (Curie temperature)—before he embarked on research in radioactivity.

[17] Ludwig Eduard Boltzmann was born on 20 February 1844 in Vienna. He studied physics and mathematics at the University of Vienna (1861–1865) under Joseph Petzval, Andreas von Ettinghausen and Joseph Stefan; he received his doctorate in 1865 and his *Habilitation* a year later. Then he continued as Stefan's assistant until he was appointed professor of mathematical physics at the

Gibbs gave the name statistical mechanics—consisted in providing a suitable mechanical foundation for the second law of thermodynamics. In the late 1860s Clausius and Boltzmann believed that they had demonstrated a proof of the second law on the basis of purely mechanical arguments; however, a decade later Boltzmann discovered that one had to invoke probabilistic arguments, such as had been introduced earlier in kinetic gas theory by Maxwell in order to obtain the velocity distribution of molecules (Maxwell, 1860). Boltzmann succeeded in establishing a connection between the entropy of a gas and the probability associated with the particular state of the gas (Boltzmann, 1877). The clarification of the meaning of this relation and of the origin of irreversibility occupied him for the rest of this life. Gibbs, in his *Elementary Principles of Statistical Mechanics* (Gibbs, 1902), dealt with the same problem. He discussed very general systems described by dynamical variables and investigated their thermodynamic behaviour; under special assumptions—the systems had to belong to certain ensembles, like the canonical, the microcanonical, etc.—he was able to prove that the systems did indeed approach thermodynamic equilibrium. The theory of Boltzmann and Gibbs was characterized by the fact that the actual thermodynamic properties of the systems showed deviations from the mean, equilibrium values. These fluctuations, which seemed to be very small in gases (because of the large number of molecules contained in any volume under investigation), were further explored by Marian von Smoluchowski and Albert Einstein.[18] Their investigations, especially on the so-called Brownian motion, led, in the first decade of the twentieth century, to a convincing confirmation of the existence of molecules.

Just as statistical mechanics emerged by unifying the theoretical schemes of mechanics and thermodynamics, another theory grew by bringing together mechanics and Maxwell's electrodynamics, namely, the relativity theory. In 1864 Maxwell had already considered the extension of his electrodynamic equations to

---

University of Graz in 1869. From 1873 to 1876 he served as professor of mathematical physics at the University of Vienna, then went back to Graz as professor of experimental physics. In 1890 he accepted a call to Munich as the first professor of theoretical physics at the University. Four years later he moved to the University of Vienna in the chair of theoretical physics, which he occupied—with an interruption of two years (1900–1902) at the University of Leipzig—until his death. On 5 September 1906 he committed suicide at Duino, near Trieste, Italy. Boltzmann devoted his life to the mechanical foundation of the second law of thermodynamics and the theory of irreversible processes. He also concerned himself with many topics of experimental and theoretical physics and mathematics, among them problems of electrodynamics and radiation theory.

See footnote 69 for the scientific background of Josiah Willard Gibbs.

[18] Marian von Smoluchowski was born on 28 May 1872 in Vorder-Brühl near Vienna. He studied at the University of Vienna under Josef Stefan and Franz Exner from 1890 to 1894 and obtained his doctorate in 1895. Then he worked in Paris (1895–1896, with Gabriel Lippmann), Glasgow (1896–1897, with Lord Kelvin) and Berlin (with Emil Warburg). On his return to Vienna, he became *Privatdozent* at the University in 1898, after which he went to Lemberg, being promoted to an extraordinary professorship in 1900 and to a full professorship in 1903. In 1913 he accepted a call to the University of Cracow. He died in Cracow on 5 September 1917. Smoluchowski worked on the heat conduction of gases, the kinetic theory of gases, fluctuation phenomena (Brownian motion and critical opalescence), and later on the physics of colloids.

describe the behaviour of electrified moving bodies as well. In 1890 Heinrich
Hertz developed a detailed formalism for the electrodynamics of moving bodies
by extending Maxwell's equations (Hertz, 1890b). However, a special difficulty
arose in this theory because of the underlying concept of the ether. The
experiment of Albert Abraham Michelson (1852–1931) and Edward Williams
Morley (1838–1923), performed in 1887, had failed to show any effect of the
motion of the ether relative to the earth, as had been expected on the basis of the
ether models (Michelson and Morley, 1887). In the analysis of this situation
Hendrik Antoon Lorentz and later the French mathematician Jules Henri
Poincaré (1854–1912) became deeply involved.[18a] Lorentz was especially interested in constructing a consistent theory of the electron—prior work on this
subject had been performed by John Henry Poynting (1852–1914) and Joseph
John Thomson—which he considered to be of primary importance for the
constitution of matter and the explanation of its optical and electrical properties.
So in 1892 he suggested a hypothesis, which agreed with the one given earlier by
George FitzGerald (1851–1901) and allowed him to overcome the difficulties
presented by the outcome of the Michelson–Morley experiment: the dimensions
of all material bodies, especially of the electrons, should be slightly altered when
they are in motion relative to the ether, i.e., they should be contracted in the
direction of the motion as compared to their extension at rest (Lorentz, 1892b;
FitzGerald, 1889). The contraction of moving bodies could be formulated
mathematically with the help of a transformation involving the space coordinates
and the time, by which different frames of reference were related. These transformations, which appeared first in a paper of Woldemar Voigt (1850–1919), did
not agree with the ones connecting the admitted frames of reference (the inertial
frames) in mechanics (Voigt, 1887). That is, the Newtonian equations of motion
remained valid for all frames of reference which move with uniform velocity with
respect to the standard one; the equations of motion of electrodynamics of
moving bodies, on the other hand, were valid in a different system of reference
frames connected by the above-mentioned transformations (which Poincaré later
called 'Lorentz transformations'). In spite of this difficulty, Lorentz pursued his
investigations further, arriving in 1904 at a complete and consistent system of
equations (Lorentz, 1904). Poincaré, after examining in detail the clash between

---

[18a] Jules Henri Poincaré was born at Nancy on 29 April 1854. In 1873 he entered the *École
Polytechnique* and two years later the *École Nationale Supérieure des Mines*, Paris, receiving his
doctor's degree in mathematical sciences from the University of Paris (Sorbonne) in 1879. He began
his career as an instructor in mathematical analysis at the University of Caen, from where he moved
in 1881 to the chair of physical and experimental mechanics at the Sorbonne. He stayed on in Paris,
and later occupied the chair of mathematical physics and finally the chair of mathematical
astronomy at the Sorbonne. Poincaré worked in many fields of pure and applied mathematics. He
established his reputation with his work on automorphic functions, which he did partly in competition with Felix Klein. Later he contributed greatly to celestial mechanics (the three-body problem),
the theory of rotating fluid masses (which C. G. J. Jacobi had worked on earlier), and the dynamics
of the electron. He received numerous honours, including memberships in the Academy of Sciences,
Paris (1887), and the Royal Society of London (1894). Because of his literary writings on science he
was elected to the *Académie Française* in 1908. Poincaré died in Paris on 17 July 1912.

See footnote 121 for the scientific background of H. A. Lorentz.

the electrodynamic and the mechanical inertial frames, developed a dynamics of the electron in agreement with Lorentz' results; it implied, however, a changed, a new mechanics (Poincaré, 1906). Finally, Einstein completed the unification of mechanics and electrodynamics in his theory of (special) relativity: from a fundamental analysis of space and time measurements he showed that Newton's concepts of absolute space and absolute time had to be abandoned, and the concept of an ether had to be replaced by the new combined concepts of space and time (Einstein, 1905d). These concepts were developed by Hermann Minkowski (1864–1909) in detail (Minkowski, 1909). The theory of relativity could be further generalized to include the theory of gravitation as well; to that end, the old Newtonian theory had again to be modified, but as in the case of special relativity the experiments clearly confirmed the predictions of the new theory (Einstein, 1916c).

With the above sketch of the new theories of statistical mechanics and relativity we have left the nineteenth century and proceeded far into the twentieth. In this volume we shall have to mention occasionally some aspects of these theories in more detail when developing the story of quantum theory between 1900 and 1925, i.e., those parts of quantum theory which preceded the discovery of quantum mechanics. This development may be separated into two periods. In the first, which will be discussed in Chapters I, II and III, fall the origin of quantum theory and its successful applications to atomic phenomena, especially to the theory of atomic structure. Then, in the years between 1922 and 1925, the failures of this 'old' quantum theory showed up clearly; their recognition and the new points of view that arose in the discussion of specific problems, such as the complex structure of atomic spectra and radiation theory, will be treated in Chapters IV, V and VI.

Synopsis of Chapters I to VI

Any historical account of quantum theory must begin with Max Planck's theory of blackbody radiation and Albert Einstein's subsequent analysis of the structure of radiation leading to the light-quantum hypothesis. The investigations of these two pioneers, besides providing a proof of the atomic structure of matter, invoked also the unexpected new concept which Planck called the 'quantum of action.' In the years after 1905 an increasing number of quantum phenomena were discovered. At the same time it was recognized that the quantum concepts did not fit into the known classical theories. The various aspects of these developments are discussed in Chapter I.

In Chapter II we turn to a specific application of the quantum theory, namely, an attempt to explain the structure of atoms and the discrete radiation emitted by them. Niels Bohr and Arnold Sommerfeld succeeded in developing a detailed theory of atoms and molecules: the quantum theory of multiply periodic systems. It enabled the physicists to view the atom as a mechanical system, made up of electrons and a positively charged nucleus (as had been shown by experiments), and to organize the great wealth of spectroscopic data that had been collected over the past hundred years. The difficulties that arose because of the use of classical theories in atomic problems could be minimized by applying certain fundamental principles, i.e., Ehrenfest's adiabatic hypothesis, Einstein's statistical interpretation of the radiation processes, and Bohr's correspondence principle.

In Chapter III we report the triumph of the theory of atomic structure, especially the explanation of the periodic system of elements by Bohr in the early twenties, which led to the prediction of the correct properties of the element hafnium discovered in late 1922. This successful period was illuminated by two events: Bohr's lectures of June 1922 at Göttingen, the so-called 'Bohr Festival,' and by the award of the 1922 Nobel Prize in Physics to Niels Bohr in December of that year.

However, at the very moment when the adherents of quantum theory expected to have in hand adequate methods to describe the properties of arbitrary atomic systems, they were confronted by serious difficulties. These difficulties showed up (in the treatment of atoms having more than one electron) immediately after the triumphant year of 1922. Thus the calculation of the energy states of the helium atom turned out to yield results in obvious disagreement with experiments. And the explanation of the multiplet structure of spectral lines and their Zeeman effects failed completely, as will be described in Chapter IV.

A further fundamental difficulty arose in connection with the investigations on the nature of radiation, which we treat in Chapter V. The effect discovered by Arthur Holly Compton in December 1922 seemed to prove the light-quantum hypothesis beyond doubt. However, this hypothesis seemed to be at variance with Bohr's correspondence principle, the most reliable tool in atomic theory at that time. An attempt to rescue a consistent description of atoms and radiation by Niels Bohr, Hendrik Kramers and John Slater led to consequences contradicted by experiment. While these difficulties plagued Bohr and his collaborators, other scientists worked out steps towards a more satisfactory theory of the light-quantum. The new statistical ideas, developed in this field by Satyendra Nath Bose, were applied by Albert Einstein to gas theory. And his results seemed to be confirmed by Louis de Broglie's unified treatment of radiation and matter and the prediction of matter waves.

In the last chapter we shall review some of the ideas which came up in atomic theory after the failure of the Bohr–Sommerfeld theory was recognized, notably the disperson-theoretic approach, the introduction of the exclusion principle and the discovery of electron spin. Although these ideas did not solve the fundamental problems and create a consistent quantum theory, they helped to overcome some of the temporary difficulties and ultimately found their proper place in the new quantum mechanics.

Chapter I

Quanta and Molecules: The Quantum Theory of Planck, Einstein and Nernst

In the Friday evening lecture at the Royal Institution of Great Britain, London, on 27 April 1900, Lord Kelvin gave a talk entitled 'Nineteenth Century Clouds over the Dynamical Theory of Heat and Light,' in which he addressed himself to two outstanding difficulties in the otherwise well-established theory.[19] The first cloud concerned the relative motion of the ether and ponderable bodies; the second cloud had to do with a failure of the Maxwell–Boltzmann theorem on the equipartition of energy among the degrees of freedom of molecular systems. The speaker, himself a pioneer of the mechanical theory of heat, spent most of his time on the second problem which manifested itself in the fact that from the observed specific heats of molecules fewer degrees of freedom resulted than expected on the basis of kinetic theory. Lord Kelvin discussed some details in a paper which he submitted nearly a year later to *Philosophical Magazine*, in which he extended his Royal Institution lecture by new calculations and demonstrations (Kelvin, 1901). At the end of his paper he referred to an article of Lord Rayleigh's on the equipartition theorem, where Rayleigh had also noticed the above-mentioned difficulties and stated that a way out must be found. Lord Kelvin picked up this suggestion and concluded his paper by remarking:

> The simplest way of arriving at the desired result [i.e., the agreement between experiment and theory of specific heats] is to deny the conclusion [derived from the equipartition theorem]; and so, in the beginning of the twentieth century, to lose sight of a cloud which has obscured the brilliance of the molecular theory of heat and light during the last quarter of the nineteenth century. (Kelvin, 1901, p. 40)

On 14 December 1900, less than eight months after Kelvin's lecture, Max Planck gave a talk at the German Physical Society in Berlin, in which he outlined a theoretical derivation of his law—proposed some weeks earlier on 19 October

[19] William Thomson (Lord Kelvin) was born on 26 June 1824 in Belfast, Ireland. He entered the University of Glasgow at the age of eleven and continued his studies at Cambridge University, where he took his degree as Second Wrangler in 1845. He then worked in the laboratory of Henri Victor Regnault (1810–1878) in Paris, and in 1846 became Professor of Natural Philosophy at the University of Glasgow. Being strongly influenced by James Prescott Joule, he pursued research in thermodynamics and formulated, in particular, the second law of thermodynamics. Later, he also worked on electromagnetic theory and its applications, especially to telegraphy. He was knighted in 1866 and raised to the peerage in 1892 with the title of Baron Kelvin of Largs. In 1899 Lord Kelvin retired from his chair in Glasgow. Five years later he was elected Chancellor of the University of Glasgow. He died on 17 December 1907 at his residence, Netherhall, near Largs, Scotland, and was buried in Westminster Abbey.

—describing the energy distribution in blackbody radiation. Planck made use of a fundamental constant, h, which had the dimension of an action. The same constant then played a decisive role in the light-quantum hypothesis proposed by Albert Einstein in spring 1905 in order to explain several peculiar phenomena connected with the emission and absorption of light that could not be understood on the basis of the undulatory theory of radiation. Two years later Einstein —who had already dispersed Lord Kelvin's first cloud by formulating the theory of special relativity (Einstein, 1905d)—again showed that Planck's quantum theory enabled one to obtain a more satisfactory theory of specific heats which avoided the earlier difficulties with the equipartition theorem (Einstein, 1906g). This new application of the quantum concept was soon followed by others. As a result, quantum theory, a theory which exhibited features clearly distinct from the previous classical theories of mechanics and electrodynamics (including relativity theory), turned into one of the most revolutionary fields of physics in the early twentieth century.

I.1   The Law of Blackbody Radiation[20]

Towards the end of 1859 Gustav Robert Kirchhoff, Professor of Physics at the University of Heidelberg, submitted two papers to the Prussian Academy concerning questions of radiation.[21] One of them dealt with the explanation of the so-called Fraunhofer lines, i.e., the dark lines observed in the solar spectrum (Kirchhoff, 1859a); it contributed, together with the experimental investigations carried out by Kirchhoff and his colleague Robert Bunsen, to establishing the methods of chemical spectral analysis (Kirchhoff and Bunsen, 1860). The other did not lead immediately to any practical application; in it a rather specific property of bodies emitting light and invisible heat radiation was expounded, namely, the fact that the ratio of emissivity to absorptivity must be the same for all bodies, provided a given wavelength of the radiation is observed and the

---

[20] For this section we have obtained useful information from Hans Kangro's book on the prehistory of Planck's law (Kangro, 1970).

[21] Gustav Robert Kirchhoff was born on 12 March 1824 in Königsberg. From 1842 he studied at the University of Königsberg, especially under Franz Neumann. After receiving his doctorate in 1847 he went to Berlin and became *Privatdozent* at the University of Berlin a year later. In 1850 he was called to Breslau as an *Extraordinarius*; there he met the chemist Robert Wilhelm Bunsen (1811–1899). Bunsen, who moved to Heidelberg the following year, proposed (in 1854) his friend Kirchhoff for the Professorship of Physics at the University of Heidelberg, which had become vacant when Philipp von Jolly left for Munich. Finally, in 1875 Kirchhoff was appointed to the chair of theoretical physics at the University of Berlin, where he died on 17 October 1887.

Kirchhoff made numerous contributions to experimental and theoretical physics. He worked on electricity, especially on the connection between electrostatic and electrodynamic concepts, and derived a theorem which gives the distribution of the currents in a network (Kirchhoff's rules). He worked on various other problems, such as the thermal conductivity of iron, reflection and refraction from crystals, and the thermodynamics of solutions. However, his most important researches concerned the emission and absorption of light; these were important not only for physics but also for chemistry and astrophysics.

bodies have the same temperature (Kirchhoff, 1859b). Kirchhoff concluded further:

> The ratio of the power of emission to the power of absorption, e/a, common to all bodies, is a function depending on the wavelength [of the radiation emitted or absorbed] and the temperature. At low temperatures this function assumes the value zero in the case of the wavelengths of visible radiation, and values different from zero for larger wavelengths; at higher temperatures the function $[e/a]$ takes on a finite value also for the wavelengths of visible rays. At the temperature for which this function (considered for the wavelength of a given visible ray) ceases to be zero, all bodies begin to emit light having the colour of this ray, except those [bodies] which possess a negligibly small power of absorption for [rays of] this colour at this temperature; the larger the power of absorption, the more light does a body emit. (Kirchhoff, 1859b, p. 726)[22]

In a subsequent paper in *Annalen der Physik* Kirchhoff discussed the relation between emissivity and absorptivity in great detail; in particular, he introduced there the concept of a completely black body, defining it as a body which absorbs *all* radiation falling upon it (Kirchhoff, 1860).

Although Kirchhoff's discussion of the emission and absorption of visible and (invisible) heat radiation appeared to deal with an abstract problem of only theoretical interest, many physicists concerned themselves during the following decades with the properties of the function $\Phi(\lambda, T)$,

$$\Phi(\lambda, T) = \left(\frac{e}{a} \right)_\lambda, \qquad (1)$$

which expressed the ratio e/a, emissivity to absorptivity, for a given wavelength of radiation, λ, on the temperature T. (The absorptive power of a body, a_λ, is defined as the ratio of the intensity of the absorbed to the incident radiation for a given wavelength λ.) The reason for that interest may be easily understood if one takes into account the general situation towards the end of the nineteenth century. At that time, for instance, the illumination of the cities at night, both by gas and electricity, was introduced; evidently, for practical and economic reasons

---

[22] Kirchhoff was not the first to discuss the relation between emission and absorption of (heated) bodies. After the experimental work of John Leslie (1766–1832), establishing the proportionality between the total emission and absorption of two bodies (Leslie, 1804), especially the Scottish physicist and meteorologist Balfour Stewart (born on 1 November 1828 at Edinburgh, died on 19 December 1887 at Manchester) concerned himself with the problem. In paper, entitled 'An Account of Some Experiments on Radiant Heat' and read in March 1858 before the Royal Society of Edinburgh, Stewart presented his empirical results claiming that the absorption of a heated plate equals its emission for all kinds of radiation (Stewart, 1858). In this paper Stewart also gave a theoretical proof for his finding; he argued, from the assumption that particles in the interior of bodies emit equal amounts of radiation independently when at uniform temperature, that emission and absorption of any type of heat radiation must be equal for any part of a body. His proof did not, however, allow him to compare different materials and to postulate, as Kirchhoff did, a universal function of temperature and wavelengths. A historical discussion of the respective achievements of Stewart and Kirchhoff in radiation theory has been given by Daniel M. Siegel (1976).

it became necessary to find the most efficient means of producing visible light. Second, since the exploration of the earth and the continents seemed to move towards the state of completion, man turned his attention increasingly to the study of celestial objects and stars. The investigation of the spectra of stars, in particular, seemed to provide important hints concerning their properties, such as their surface temperatures. The latter could be derived, in principle, by observing the distribution of intensity of the radiation among the wavelengths and comparing the result with Kirchhoff's function $\Phi(\lambda, T)$. Third, after the recognition of light as electromagnetic waves one turned to explore systematically the full spectrum of these waves, discovering quite new forms of radiation. Thus, Samuel Pierpont Langley extended the range of observable heat radiation to include wavelengths of one-millionth of a metre by his bolometer method (Langley, 1886) and Heinrich Hertz demonstrated the existence of electromagnetic waves emitted by oscillating currents in open circuits, having wavelengths of the order of a metre (Hertz, 1888a). On the other hand, the electromagnetic spectrum was also extended on the ultraviolet side to shorter and shorter wavelengths, especially through the discovery of X-rays and of γ-radiation from radioactive substances, which took place shortly before the end of the nineteenth century and in the beginning of the twentieth century, respectively.[23]

The discoveries mentioned above evidently required an immense improvement of the experimental methods; completely new techniques had to be developed to register, say, the long electromagnetic waves, and the precision and reliability of known techniques had to be increased. In this development the physicists received help from outside; because of the needs of growing industrial production in Europe and the United States a strong pressure was exerted on governments and educational institutions to provide the means for such improvements of techniques and methods. On one hand, laboratories were installed at universities and technical institutions were established, equipped with modern instruments and apparatus; thus, for example, the establishment of the Cavendish Chair and Laboratory at Cambridge University was strongly motivated by the idea that the work performed in the Laboratory might result in obtaining better and more accurate standards for industrial purposes. On the other hand, for the same reason, special national institutes and laboratories were also created, which had not only to control the standard weights and measures but also to develop methods and standards for new technologies. The first of these institutes (or bureau of standards), the *Physikalisch-Technische Reichsanstalt* in Berlin, came into being in 1887, with substantial help from the industrialist and inventor Werner von Siemens, the founder of the electrical firm of *Siemens und Halske*.[24] The various precision methods developed at the *Physikalisch-Technische Reich-*

[23] The electromagnetic nature of X- and γ-rays was confirmed only much later, i.e., in connection with the discovery of their interference effects in crystal lattices (Friedrich, Knipping and Laue, 1912).

[24] For a history of the *Physikalisch-Technische Reichsanstalt* see the article of Karl Scheel (1913).

sanstalt were combined towards the end of the nineteenth century to provide fundamental data concerning the energy distribution in blackbody radiation— data on which a theoretical understanding of Kirchhoff's function could be achieved.

Ever since Friedrich Wilhelm (William) Herschel had demonstrated the heat effects of infrared radiation in the very beginning of the nineteenth century, physicists had been interested in the study of the radiation emitted by hot bodies. In order to investigate the visible part of the spectrum, optical and, later on, photographic methods were used. The latter were also found useful for the ultraviolet part. The infrared region, the heat radiation in the original sense of the word, could be observed by its temperature-increasing effect, which was objectively registered, say, by a thermocouple. The experimental observation of infrared radiation was substantially improved when Samuel Pierpont Langley invented the so-called 'bolometer,' an instrument using the temperature-dependent change of the resistivity of platinum wire blackened by carbon (Langley, 1881).[25] With it Langley observed the radiation emitted from heated copper, including wavelengths up to 5.3 μ; he demonstrated, in particular, a definite displacement of the maximum of intensity with increasing temperature of the copper—the temperature of the probe was varied between 330°C and 815°C—towards smaller wavelengths (Langley, 1886). Langley's measurements were extended by Friedrich Paschen, then at the Technical University of Hanover.[26] Paschen entered the field in 1892 and concerned himself for several years with various techniques of observing and measuring the intensity of heat radiation, employing many different substances such as incandescent gases and

[25] S. P. Langley was born on 22 August 1834 in Roxbury, Boston, Massachusetts. He completed his formal education on graduating from Boston High School in 1851. After travelling in Europe (1864–1865), he became an assistant at the Harvard Observatory in 1865. In 1866 he was appointed assistant professor of mathematics at the U.S. Naval Academy and placed in charge of the observatory. In 1867 Langley was made Director of the Allegheny Observatory and Professor of Physics and Astronomy at Western University, Pennsylvania. From 1887 onwards Langley was Secretary of the Smithsonian Institution in Washington, D.C. He died in Aiken, South Carolina, on 27 February 1906.

Besides astronomical researches, especially on stellar spectra, Langley made aerodynamical studies and designed, for instance, engine-driven airplanes. He received many honours and prizes, including memberships in the U.S. National Academy of Sciences, and the Royal Societies of London and Edinburgh. He was awarded the Rumford and Henry Draper Medals.

[26] Louis Carl Heinrich Friedrich Paschen was born in Schwerin, Mecklenburg, on 22 January 1865. He studied mathematics, natural science and physics at the Universities of Strasbourg and Berlin from 1884 to 1888, receiving his doctorate from Strasbourg in 1888. In 1888 he became an assistant to Johann Wilhelm Hittorf (1824–1914) at the University of Münster; in 1891 he became an assistant of Heinrich Kayser (1853–1940) in Hanover, where he received his *Habilitation* two years later and held the position of a *Dozent* for 'Physics and Photography' from 1895. In 1901 Paschen was appointed *Ordinarius* (full professor) of physics at the University of Tübingen. He stayed there, apart from a short period at the University of Bonn (1919–1920), until 1924. then he was called to Berlin to succeed Emil Warburg as President of the *Physikalisch-Technische Reichsanstalt*, where he remained from 1924 to 1933. He died in Potsdam on 25 February 1947.

Paschen worked primarily in spectroscopy, both on heat radiation and optical spectroscopy. Under his leadership Tübingen became the 'mecca of spectroscopy.'

solid matter. Finally he obtained detailed results from several solid emitters including carbon, copper oxide and platinum; these results converged to yield an empirical formula for Kirchhoff's function, that is,

$$\Phi(\lambda, T) = c_1 \lambda^{-\alpha} \exp\left(- \frac{c_2}{\lambda T}\right), \tag{2}$$

where c_1 and c_2 were constants and the negative exponent $-\alpha$ assumed the value of about -5.5 (Paschen, 1896). Shortly before Paschen's publication of his empirical law, Wilhelm Wien had arrived at the same result on the basis of theoretical arguments (Wien, 1896).

Equation (2) describes the energy density distribution, ρ_λ, among the various wavelengths of the radiation emitted by an idealized body, which Gustav Kirchhoff had called a 'completely black' or just a 'black' body, i.e., a body which possesses the property of absorbing all the radiation falling upon it (Kirchhoff, 1860, Section 1). The first attempts to derive theoretically the energy distribution of blackbody radiation, which is identical with Kirchhoff's function, i.e., $\Phi(\lambda, T) = \rho_\lambda$, were made by Eugen Lommel (1837–1899), then Professor of Physics at the University of Munich, by Vladimir Alexandrovich Michelson (1860–1927), then at the University of Berlin, and by Ludwig Boltzmann, then professor of physics at the University of Graz. While Lommel had based his approach on a mechanical model describing the vibrations in a solid body (Lommel, 1878), Michelson had used arguments from kinetic gas theory, including Maxwell's velocity distribution of molecules, to arrive at a formula for $\Phi(\lambda, T)$ (Michelson, 1888).[27] On the other hand, Ludwig Boltzmann had based his treatment right away on Maxwell's electromagnetic theory of light when he derived the law obtained empirically by Joseph Stefan, which accounted for the total intensity of the radiation (both visible and invisible) emitted by a heated blackbody (Stefan, 1879). Boltzmann had also used a connection between radiation and the second law of thermodynamics that had been put forward several years earlier by the Italian physicist Adolfo Bartoli.[28] Bartoli had considered, in particular, the possibility of increasing (by an adiabatic process) the temperature of a body situated in an enclosure through which heat radiation passed; this adiabatic process consisted in the reduction of the volume of the enclosure, and he had found that, in order not to violate the second law of

---

[27] For a detailed discussion of Michelson's result, see Kangro, 1970, Section 2. Let us just mention here that from his function followed the constancy of the product, $\lambda_{max}^2 \cdot T$, where λ_{max} is the wavelength having the maximum intensity and T the absolute temperature.

[28] Adolfo Bartoli, born on 19 March 1851 in Florence, studied physics in Pisa and Bologna, and became Professor of Physics at the University of Sassari (1878) and the Technical Institute of Florence (1879–1886), then Professor and Director of the Observatory at the University of Catania, and finally Professor at the University of Pisa. He died on 18 July 1896 in Pavia. Bartoli worked on various problems, including the specific heat and the dissociation of water and on electrolysis. He demonstrated the existence of radiation pressure.

thermodynamics, one had to take into account the pressure exerted by the heat radiation on the body (Bartoli, 1876). In his first note analyzing Bartoli's result, Boltzmann had arrived at the conclusion that the existence of a radiation pressure of magnitude $p = \frac{1}{3}\rho$, where ρ is the radiation density, was consistent with Stefan's law (Boltzmann, 1884a). In a second note Boltzmann had then sharpened his conclusion to state the following: if the second law was valid and if heat radiation—like any radiation—possessed a pressure of the above magnitude, Stefan's law, i.e.,

$$\rho = AT^4, \tag{3}$$

with A a constant, would follow immediately (Boltzmann, 1884b).[29]

Bartoli and Boltzmann's idea of bringing both electrodynamics and thermodynamics into the treatment of heat radiation was picked up and followed further by Wilhelm (Willy) Wien, who had served as an assistant at the *Physikalisch-Technische Reichsanstalt* since 1890.[30] In his first publication on this subject, entitled '*Über eine neue Beziehung der Strahlung schwarzer Körper zum zweiten Hauptsatz*' (Wien, 1893a), he immediately extended the earlier treatments: instead of talking about the energy of the full spectrum—as Boltzmann had

[29]To find this result, Boltzmann used the equation, $T\,dp - p\,dT = \rho\,dT$, which can be derived from the second law of thermodynamics. (It follows from the fact that dQ/T is a total differential, where dQ is the sum of the changes in the internal energy and the external work.) In this equation he substituted the relation $p = \frac{1}{3}\rho$ and integrated the resulting equation, putting the constant of integration equal to zero, from which Eq. (3) followed.

[30]Wilhelm Carl Werner Otto Fritz Franz Wien was born on 13 January 1864 in Gaffken near Fischhausen, East Prussia. After attending the *Gymnasium* in Rastenburg and Königsberg, he studied mathematics and physics in Göttingen (1882), Berlin (1882 to the winter semester 1883–1884), Heidelberg (summer 1884) and again Berlin (winter semester 1884–1885 to winter semester 1885–1886), receiving his doctorate in 1886 under Helmholtz with a thesis on a problem of the diffraction of light by sharp edges. Wien then worked for several years on the agricultural estate (*Landgut*) of his father until he received the appointment as Helmholtz' assistant at the *Physikalisch-Technische Reichsanstalt* in 1890. He obtained his *Habilitation* two years later with a theoretical work on the 'localization of energy.' (Wien considered the concept of 'localization of energy' in connection with Poynting's theory of the propagation of energy in the electromagnetic field and the subsequent work of Heinrich Hertz.) Wien discussed the flux of energy in hydrodynamics, in elastic bodies and finally electrodynamics, including radiation theory. (See Wien, '*Über den Begriff der Lokalisierung der Energie*' ('On the Conception of Localization of Energy'), 1892.) In 1896 he was called to the *Technische Hochschule*, Aachen, as *Extraordinarius* for physics (succeeding Philipp Lenard); in 1899 he succeeded Otto Wiener as *Ordinarius* in Giessen, and a year later he moved as *Ordinarius* to Würzburg (this time succeeding Wilhelm Conrad Röntgen). Finally in 1920 he obtained the Chair of Experimental Physics at the University of Munich, again succeeding Röntgen. He died in Munich on 30 August 1928.

Wilhelm Wien was one of those rare twentieth-century physicists who worked as a specialist both in experimental and theoretical physics. His researches on blackbody radiation won him the Nobel Prize in Physics in 1911. Wien worked in thermodynamics and hydrodynamics and made pioneering experimental studies of the electric and magnetic deflection of canal and cathode rays (the latter contributed to the discovery of the electron); he also worked on X-rays and on the recombination of ions.

done—he was interested in the energy distributed among given wavelengths.[31] Wien studied, especially, the enlargement of the energy density of radiation caused by the following two processes: the increase of the temperature and the adiabatic decrease of the volume of the enclosure containing the radiation; and he demanded that both processes should lead to the same energy distribution among the wavelengths if the same final temperature were reached. He achieved this goal after taking into account Doppler's principle, that is, the fact that λ, the wavelength of the emitted radiation, depends on the velocity of the source. In particular, he obtained the result that the densities, ρ_{λ_0} and ρ_λ, associated with λ_0 and λ, the wavelengths before and after the volume change, respectively, were related as $\rho_\lambda/\rho_{\lambda_0} = (\lambda_0/\lambda)^4$; hence, due to the Stefan–Boltzmann law, Eq. (3), there followed the equation

$$T\lambda = T_0\lambda_0 = \text{const.} \tag{4}$$

Equation (4) expresses the fact, which later on came to be called 'Wien's displacement law,' that for blackbody radiation the product of temperature and the related (through the adiabatic process) wavelength remains a constant; a special example of this law is provided by taking for λ the wavelength, λ_{max}, having the maximum intensity or energy density at the given temperature.[32]

Wien continued to investigate the properties of blackbody radiation during the following years. Especially, he used the concept of entropy to rederive Eq. (4) and to obtain an inequality, that is,

$$\rho_\lambda \leqslant \frac{\text{const.}}{\lambda^5}, \tag{5}$$

for ρ_λ, the energy density of blackbody radiation in the interval of wavelength between λ and $\lambda + d\lambda$ (Wien, 1894). A couple of years later he arrived at an

[31] Wien became interested in the problem of heat radiation on his own. Thus he spoke in his autobiography (written in 1927, and published posthumously in 1930) about the freedom which he had at the *Reichsanstalt* to deal, besides his official task of establishing a standard of light intensity, also with theoretical topics of his choice, such as water waves and cyclones. He wrote:

The independence, which I soon achieved in scientific work, immediately bore fruits. I turned to the field of heat radiation and succeeded in discovering in it—without really great effort— new laws, which received the approval of the scientific world. Even today it gives me great satisfaction that my first paper on the displacement law [Wien, 1893a] was presented by Helmholtz—though after some resistance—to the Berlin Academy. He [Helmholtz] told me that he had thought earlier that the radiation could not be treated thermodynamically, but that he had been convinced [by my work] that I was right. I did not understand Helmholtz' reservations, for in my opinion radiant heat was an integral part of heat itself and had to satisfy the same laws. Only later did I come to know that Lord Kelvin had spoken much more definitely against my theories, for he said: "Thermodynamics are going mad." I gave the general formulation of my ideas on the properties of the heat radiation in my paper on '*Temperatur und Entropie der Strahlung*' ('Temperature and Entropy of Radiation' [Wien, 1894]). Apart from a few minor points, it has been fully accepted into theoretical physics. (Wien, 1930, pp. 16–17)

[32] In the latter sense, Wien's law was used in future. The name 'displacement law' ('*Verschiebungsgesetz*') first occurred in a paper by Otto Lummer and Ernst Pringsheim six years later (Lummer and Pringsheim, 1899b, p. 219).

explicit law for ρ_λ by applying the 'molecular hypothesis' in addition to thermodynamic arguments (Wien, 1896). In particular, Wien started by assuming that blackbody radiation was emitted by molecules obeying Maxwell's velocity distribution; that is, the number of molecules having the velocity v was proportional to the quantity $v^2 \exp(-\text{const.}v^2/T)$, where T denotes the absolute temperature. Further, he assumed that λ, the wavelength of radiation emitted by a molecule, was a function only of its velocity v and vice versa. In this way he obtained the result

$$\rho_\lambda = F(\lambda)\exp\left(-\frac{f(\lambda)}{T}\right), \tag{6}$$

where $F(\lambda)$ and $f(\lambda)$ are functions of the wavelength λ. By applying the displacement law, Eq. (4), Wien determined the function in the exponent to be

$$f(\lambda) = \frac{c_2}{\lambda T} \; ; \tag{6a}$$

and from Stefan–Boltzmann's law, Eq. (3), he derived that

$$F(\lambda) = \frac{c_1}{\lambda^5} . \tag{6b}$$

Hence Wien's Eq. (6) agreed with Friedrich Paschen's empirical law, Eq. (2), provided one could identify ρ_λ with Kirchhoff's function $\Phi(\lambda, T)$, Eq. (1), and take the value 5 for Paschen's power exponent; thus it reproduced the observed data perfectly.

In June 1896 Willy Wien left Berlin to take a professorship at the *Technische Hochschule* of Aachen, at a time when the *Physikalisch-Technische Reichsanstalt* became increasingly involved in the preparations for absolute measurements of the blackbody radiation law. It was very fortunate that Max Planck, who had succeeded Gustav Kirchhoff as Professor of Theoretical Physics at the University of Berlin, at that time became a *Haustheoretiker* (resident theoretician) of the experimentalists working on blackbody radiation. Max Karl Ernst Ludwig Planck was born in Kiel on 23 April 1858, the son of a law professor at the university of that city; he had received his early education in Kiel and Munich and had studied physics and mathematics at the Universities of Munich (1874–1877) and Berlin (1877–1878).[33] Among his professors were Philipp von Jolly (1809–1894) in Munich and Hermann von Helmholtz (1821–1894) and Gustav

---

[33] Planck's father, Johann Julius Wilhelm von Planck, came from an academic family (both his grandfather, Gottlieb Jakob Planck, and his father, Heinrich Ludwig Planck, were professors of theology at the University of Göttingen). His second wife, Emma Patzig, came from Greifswald. The Planck's had seven children, Hugo and Emma (from Johann Julius Planck's first wife), Hermann, Hildegard, Adalbert, Max and Otto (from his second wife); thus Max Planck was the fourth son of Johann Planck.

Kirchhoff (1824–1887) in Berlin. Planck obtained his doctorate, *summa cum laude*, from the University of Munich in 1879 with a thesis entitled '*Über den zweiten Hauptsatz der Wärmelehre*' (Planck, 1879), and became *Privatdozent* in Munich the following year. In 1885 he accepted a call to an *Extraordinariat* (associate professorship) at the University of Kiel, and four years later he moved to the University of Berlin, where he was promoted to a full professorship of theoretical physics in 1892. Planck devoted his early scientific career to the investigation of one general topic, the second law of thermodynamics, especially the concept of entropy and its application to problems of physical and chemical equilibrium, such as phase transitions and electrolytic dissociation.[34] His first published papers already exhibited the characteristic features of his later work: on one hand, he carefully worked out the details of his theories and calculated results that could be compared immediately with the available experimental data (see, e.g., Planck, 1881, 1890); on the other, he put great emphasis on clear definitions of the fundamental concepts. Having been deeply influenced by the writings of Rudolf Clausius, he had sought in particular to establish the 'principle of the increase of entropy' in thermodynamics on the same level as the law of conservation of energy. In this endeavour he had pointed out the crucial role of irreversible processes, i.e., the fact that in nature processes occur, such as the conduction of heat, which cannot be reversed completely. Planck's analysis of mechanical and thermodynamic concepts had brought him into disagreement with Wilhelm Ostwald and the partisans of 'energetics' ('*Energetik*'). For example, at the Lübeck Assembly of Natural Scientists (*Naturforscherversammlung*) in 1895, where Ostwald and Georg Helm had defended the views of the energeticists and Boltzmann had attacked them, Planck had been on Boltzmann's side.[35] However, unlike Boltzmann, who had argued against *Energetik* on the basis of the molecular hypothesis and the kinetic theory of matter, Planck had used thermodynamic reasoning in order to criticize Ostwald's concept of 'volume energy.' Indeed, Planck had adopted a very cautious attitude towards the molecular hypothesis and, in a paper entitled '*Gegen die neuere Energetik*' ('Against the New Energetics'), had declared concerning this point: 'I do not intend, at this point, to enter the arena [on behalf of] the mechanistic view of nature; for that purpose, one has to carry out far-reaching and, to some extent, very difficult investigations' (Planck, 1895b, p. 73).

At the time of the controversy about *Energetik* Planck had turned his

---

[34] Planck did fundamental work on these fields, i.e., both on phase transitions and electrolytic dissociation. However, his contributions were not the first: on phase transitions, Josiah Willard Gibbs had published his pioneering papers several years earlier, while on dissociation the Swedish chemist Svante Arrhenius was slightly ahead of Planck.

[35] Later on Planck recalled the discussions about *Energetik* at the Lübeck *Naturforscherversammlung* in the following words:

> It is evident that this fight, in which especially Boltzmann and Ostwald opposed each other, was carried out rather spiritedly; it also led to some drastic effects since both opponents were well-matched in quickness and wit. In it, according to what I said earlier [concerning my position], I could only play the role of a second to Boltzmann, whose services were not only not recognized but even not liked by him [Boltzmann]. (Planck, 1948a, p. 20)

attention to what was a new field of investigation for him: heat radiation. Several reasons may be cited why he had become interested in this field. First, there was the general interest of many physicists in the phenomena of electromagnetic waves after Heinrich Hertz' successful experiments.[36] A second reason was Planck's concern with the importance of thermodynamic arguments in electromagnetism. Thus, for example, in his inaugural address (*Antrittsrede*) to the Prussian Academy on 28 June 1894, he had expressed the hope 'that we can obtain a closer understanding also of those electrodynamic processes, which are directly caused by the [action of] temperature and which show up especially in heat radiation, without having to follow the laborious detour through the mechanical interpretation of electricity' (Planck, 1894b, p. 643). Evidently, Planck had had the occasion to participate in many discussions between his Berlin colleagues, such as Willy Wien and Heinrich Rubens, who worked actively on the problems of heat radiation. And, before starting his theoretical investigations, he knew that detailed experiments to explore the properties of blackbody radiation were being made ready at the *Physikalisch-Technische Reichsanstalt*. Thus he reported later in his scientific autobiography:

> By the measurements of [Otto] Lummer and [Ernst] Pringsheim at the *Physikalisch-Technische Reichsanstalt*, performed to investigate the spectrum of heat radiation, my attention was directed to the theorem of Kirchhoff, stating that in an evacuated cavity—bound by totally reflecting walls and containing emitting and absorbing bodies that are completely arbitrary [in shape, material and temperture]—a state will be established in course of time, in which all bodies assume the same temperature and in which all properties of the radiation [contained in the cavity]— even its spectral energy distribution—do not depend on the structure and composition of the bodies, but solely on the temperature. This so-called normal energy distribution [of the radiation thus obtained in the cavity], therefore, represents an absolute quantity; and, since the search for the absolute always appeared to me to be the most beautiful ("*schönste*") task of research, I eagerly started to deal with it. (Planck, 1948a, pp. 23–24)

Max Planck was aware, of course, of Wien's law, Eq. (6), which represented the existing observations extremely well. However, the derivation given by Wien, though it employed thermodynamic arguments, did not satisfy Planck's taste completely. He hoped to arrive at the same or a similar equation in a more systematic way, using fewer hypotheses than his predecessor.

---

[36] The general interest in Maxwell's electrodynamics and its consquences may be discerned, for example, by the great number of lecture courses on it at the University of Berlin during the early 1890s. Among others, courses were given by Willy Wien (in the summer semester 1892 and winter semester 1892–1893), Max Bernhard Weinstein (1852–1918) who had translated Maxwell's *Treatise on Electricity and Magnetism* into German (in the winter semester 1891–1892), and Heinrich Rubens (on the experimental foundations of the subject in the winter semester 1894–1895). Max Planck also lectured in Berlin regularly on the theory of electricity and magnetism (e.g., winter semesters 1889–1890, 1892–1893, and 1895–1896) (see Kangro, 1970, Sections 6.1.2 and 6.1.3). In this connection, it is worth noting that Planck gave the memorial address on Heinrich Hertz on 16 February 1894 at the Berlin Physical Society, in which he emphasized Hertz' role in the development of Maxwell's theory and electromagnetic waves (Planck, 1894a).

Planck presented the first paper, in which he concerned himself with the properties of electromagnetic radiation, to the Prussian Academy in its session of 21 March 1895. In this work he studied the processes of absorption and emission of radiation by an electrically charged system having the same eigenfrequency as the radiation or, as he called it, a 'resonator' (Planck, 1895a, p. 296).[37] He continued the investigation of resonating systems during the following years, including, for example, the treatment of the damping effect arising from the emission of radiation (Planck, 1896). He explained the goal of his endeavours in the following words:

> The study of conservative damping [i.e., of radiation damping, which Planck called "conservative," because it did not violate energy conservation] appears to me to be of fundamental importance due to the fact that through it one's view is opened towards the possibility of a general explanation of irreversible processes with the help of conservative forces—a problem which confronts the theoretical research in physics more urgently every day. (Planck, 1896, p. 154)

By the 'pressing problem' Planck referred to the question of whether the thermodynamic concept of entropy could be defined in a rational way either in mechanics or in electrodynamics. Ludwig Boltzmann had proposed an expression for the entropy derived from kinetic gas theory (Boltzmann, 1872). But in 1896 Planck's assistant Ernst Zermelo suggested that Boltzmann's result was not correct (Zermelo, 1896). He argued, especially, that according to a mechanical theorem of Henri Poincaré, a conservative system of gas molecules would always return to its initial state after a finite period of time; hence, a mechanical

---

[37] Planck referred in his paper first to the 'secondary conductor, whose characteristic period almost coincides with the period of the primary wave' and, therefore, will be 'excited by resonance to perform electric oscillations, the more so the less the periods differ from each other' (Planck, 1895a, p. 289). With this statement he had in mind Heinrich Hertz' experimental setup consisting of 'primary and secondary conducting systems' ('*primäre und secundäre Leiter*,' Hertz, 1888a, p. 552), which Hertz had used to determine the propagation of electromagnetic waves; that is, the waves were created in the primary system and directed in the secondary system. In a later paper Hertz had also given the theory of his systems; he represented both the primary and the secondary systems by a conductor of length *l*, in which an electric charge of magnitude *e* performed harmonic oscillations, and he investigated the properties of the electromagnetic waves that were emitted by the primary system and absorbed and reemitted by the secondary system (Hertz, 1888b). Planck quoted these results of Hertz when establishing his theory of absorption and emission of electric waves by resonance (see Planck, 1895a, p. 290, footnote 1).

It should be noted that Planck later on used the words 'resonator' and 'oscillator' synonymously for the elementary systems which absorb and emit electromagnetic radiation (see Planck, 1900a, p. 69). The idea that oscillating ions were responsible for the emission and absorption of electromagnetic radiation by matter was emphasized, in particular, by Hendrik Antoon Lorentz. In his book *Versuch einer Theorie der elektrischen und optischen Erscheinungen in bewegten Körpern* (*Attempt at a Theory of the Electric and Optical Phenomena in Moving Bodies*, Lorentz, 1895), dealing with electrical and optical phenomena in moving bodies, in which Lorentz presented this idea, he also drew attention to the fact that his formulae for the components of the electric and magnetic fields emitted by the oscillating ions agreed with the expressions 'by which Hertz [Hertz, 1888b] has described the oscillations in the neighbourhood of his vibrator' (Lorentz, 1895, p. 54). Thus, by the end of the century, Planck took into account the identity of Lorentz' molecular oscillators and his elementary resonators.

definition of entropy which implied an evolution of the system in time towards a more probable state in the sense of Boltzmann was not at all possible. Against this objection of Zermelo, Boltzmann argued that his conclusion could be avoided by taking into account the statistical nature of the H-theorem.[38] That is, the entropy could be defined only by a probability consideration, and it was this definition which Planck did not like at all. He preferred to argue that: 'A rigorous theory of friction on the basis of the kinetic theory [of matter] will be achieved only with the help of an additional hypothesis' (Planck, 1897a, p. 58). In contrast to this situation in gas theory, however, he hoped that he would be able to prove (without reference to a probability assumption) the existence of irreversible processes in a cavity, i.e., a volume surrounded by completely reflecting walls and filled with (heat) radiation and one resonator *à la* Hertz, which absorbs and emits radiation. In particular, he claimed:

> Such a resonator will be excited by absorbing energy from the [electromagnetic] radiation incident upon it from the outside, and it will be damped by emitting energy. Now the emitted energy will not, in general, be of the same type [especially, it will not have the same energy distribution] as the absorbed energy; hence the resonator will change by its vibration the nature of the electromagnetic waves propagating through its vicinity to some extent. It can be shown that these changes possess, in several respects, a certain direction, i.e., the tendency to homogenize [some properties, especially the temperature, of the incident radiation]. (Planck, 1897a, p. 59)

In five contributions, entitled '*Über irreversible Strahlungsvorgänge*' ('On Irreversible Radiation Processes') and submitted to the Prussian Academy between February 1897 and May 1899, Planck tried to prove his above statement (Planck, 1897a, b, c, 1898, 1899). These investigations resulted in establishing Wien's law, Eq. (6), as *the* law defining the energy distribution of blackbody radiation.

Planck proceeded in the following manner. He first calculated the effect of a resonator (having a large wavelength and small damping constant) on incident electric waves (Planck, 1897a). After rejecting a criticism of Boltzmann, who claimed that all processes used by Planck were reversible, Planck went on to prove the existence of irreversibility in a system, in which a resonator situated in the centre of a spherically shaped totally reflecting cavity absorbs and emits radiation (Planck, 1897c).[39] However, in carrying out the proof he had to make an assumption concerning the nature of radiation. He examined this assumption

[38] A discussion about the significance of $-H$, Boltzmann's mechanical quantity for the entropy of a system, had started already in late 1894, when several British authors, including George Hartley Bryan (1864–1924) and Samuel Hawksley Burbury (1831–1901), had written critical notes in *Nature* on that question. In his reply Boltzmann had pointed out that his H-theorem, i.e., the statement about the decrease of the expression H (and the consequent increase of entropy), could not be proved by purely mechanical means (Boltzmann, 1895b).

[39] In particular, Boltzmann argued that Planck had only used outgoing, but not incoming, spherical waves in his derivation (Boltzmann, 1897). Planck replied, however, that his assumption about the incoming wave (that it should always have finite intensity) forbade the use of incoming spherical waves (Planck, 1897b).

in more detail in his fourth communication, where he introduced the concept of 'natural radiation' (*natürliche Strahlung*) and claimed: 'It will be shown, in particular, that each radiation process, exhibiting the properties of "natural" radiation, necessarily proceeds in an irreversible manner, such that the waves after passing the resonator have radiation intensities with smaller fluctuations than before' (Planck, 1898, p. 450).[40] For example, if natural radiation belonging to two different temperatures was absorbed and reemitted by the resonator, the emitted radiation would correspond to a more uniform temperature. By invoking the hypothesis of natural radiation Planck not only succeeded in obtaining a relation between the energy of the resonator and the intensity of radiation for a given wavelength or frequency, but also in defining the entropy of radiation by a proper expression such that the change of the total entropy (of the resonator plus radiation) was always a positive quantity. In the last communication Planck finally generalized his proof of irreversibility to a cavity containing heat radiation and many resonators; he further identified the 'electromagnetic' entropy with the thermodynamic entropy and derived from it the law of blackbody radiation (Planck, 1899). He summarized the results of the irreversible radiation processes in a long paper, which was received by *Annalen der Physik* on 7 November 1899 (Planck, 1900a).

Within less than three years Planck had achieved his goal of connecting the thermodynamic and electrodynamic theories. He had been able, especially, to introduce the concept of irreversible processes in the treatment of systems consisting of radiation and resonators absorbing and emitting radiation. He had to pay a price for this success; this was the assumption of what he called 'natural radiation,' which could be interpreted physically as:

> If it is stated that an electromagnetic ray possesses the properties of natural radiation, then in short it should mean the following: The energy of radiation is distributed in a completely *irregular* manner among the individual partial vibrations, of which the ray can be thought to be composed. (Planck, 1900a, p. 73)

In 1899 Planck was aware of the fact that

> The hypothesis of natural radiation, if assumed to be valid for all space points at all times, implies at its core the second law of thermodynamics when applied to the processes of radiation; that is, it [the hypothesis of natural radiation] is another expression of the same law. (Planck, 1900a, p. 74)[41]

Hence the programme of finding a system in which the second law of thermodynamics followed from electrodynamics without further assumptions had failed.

---

[40] Planck defined 'natural radiation' by certain averaging procedures over the phases of the radiation. (See Planck, 1898, p. 468, Eqs. (69).)

[41] In this connection Planck referred to Boltzmann's hypothesis of 'molecular chaos' ('*molecular Unordnung*'), which Boltzmann had introduced in his lectures on gas theory in order to account for irreversibility in the kinetic theory of matter (Boltzmann, 1895c, p. 21).

The hypothesis of natural radiation enabled Planck, however, to define the entropy as a function of time. Even more:

> By identifying the electromagnetic entropy with the thermodydnamic entropy, there follows immediately the electromagnetic definition of the temperature of a ray of heat radiation; and the maximum of the entropy defines thermal equilibrium, i.e., the state in which the heat radiation becomes stationary. The law of spectral energy distribution in this state turns out to be identical with the law which Wien has derived in a different manner, a law which, in some approximation, has been substantiated recently—as is known—by the experimental investigations of F. Paschen, as well as of O. Lummer and E. Pringsheim. (Planck, 1900a, p. 74)

Planck's procedure in deriving the blackbody radiation law involved the following steps. At first, he established a relation between U_ν, the average energy of a resonator having frequency ν, and ρ_ν, the energy of incident radiation having the same frequency. Thus he found from electrodynamics the equation

$$U_\nu = \frac{c^3}{8\pi\nu^2}\, \rho_\nu. \tag{7}$$

Then he defined the entropy of the resonator as

$$S = -\frac{U_\nu}{a\nu}\ln\frac{U_\nu}{b\nu} \tag{8a}$$

and the entropy of a ray of radiation of frequency ν in a given direction as

$$S = -\left(\frac{K_\nu}{a\nu}\ln\frac{c^2 K_\nu}{b\nu} + \frac{K'_\nu}{a\nu}\ln\frac{c^2 K'_\nu}{b\nu} \right), \tag{8b}$$

where a and b are constants and K_ν and K'_ν (with $\rho_\nu = (4\pi/c)(K_\nu + K'_\nu)$) denote the intensities of the ray in the two main directions of polarization perpendicular to each other. The sum of the two entropies, given by Eqs. (8a) and (8b), could be shown to increase in the course of time; hence Planck believed his definitions to make sense. Finally, by taking the maximum of the total entropy, he obtained a relation between the resonator energy and the ray intensity in the state of equilibrium, namely,

$$U_\nu = \frac{c}{\nu^2} K_\nu = \frac{c}{\nu^2} K'_\nu. \tag{9}$$

In this equilibrium state the temperature could be defined by the relations

$$\frac{1}{T} = \frac{dS}{dU_\nu} = -\frac{1}{a\nu}\ln\frac{U_\nu}{b\nu} = \frac{1}{a\nu}\ln\frac{b\nu^3}{c^2 K_\nu}. \tag{10}$$

Equation (10) implied a relation between U_ν and T. In the case of equilibrium, Planck derived with the help of Eq. (7) a relation for ρ_ν, the energy density of radiation of frequency ν (incident on the resonator), and T, that is,

$$\rho_\nu = \frac{8\pi b \nu^3}{c^3} \exp\left(-\frac{a\nu}{T}\right). \tag{11}$$

Equation (11) thus represented the energy distribution of blackbody radiation and agreed with Wien's law, given by Eqs. (6), (6a) and (6b), if the quantities ν and ρ_ν were rewritten in terms of the wavelength λ and the corresponding energy density distribution ρ_λ.[42] From Paschen's recent measurements (Paschen, 1899a), Planck derived the values for his constants a and b, i.e., $a = 0.4818 \times 10^{-10}$ (seconds $\times$ degree Centigrade) and $b = 6.885 \times 10^{-27}$ (erg $\times$ seconds).

In his paper of November 1899 in *Annalen der Physik* Planck claimed that the definition of entropy, Eqs. (8a) and (8b), and therefore also the radiation law, Eq. (11), were a *necessary* consequence of the second law of thermodynamics. 'If, on the other hand, one tries,' he argued, 'to start from any distribution law different from the one given by Eq. [(11)], and computes from it in turn the entropy, then one arrives again and again at contradictions with the law of the increase of entropy' (Planck, 1900a, p. 118). Several months later, in March 1900, he submitted a further paper on this subject (Planck, 1900b). He had meanwhile convinced himself that 'the law of the increase of entropy by itself does not suffice to determine the expression of the entropy as a function of the energy, but that a closer investigation of the physical significance of the entropy function is necessary for this purpose' (Planck, 1900b, pp. 730–731). Hence he had to look for another method of defining the entropies and he did so by presenting a 'method for the direct calculation of the radiation entropy' ('*Weg zur direkten Berechnung der Strahlungsentropie*', Planck, 1900b, p. 720). For this purpose Planck wrote down an expansion for the quantity dS, expressing the change of the total entropy of the system (cavity radiation plus resonator), in the vicinity of the equilibrium situation (denoted by the suffix zero), that is,

$$dS = dU \cdot \Delta U \cdot \frac{3}{5}\left(\frac{d^2S}{dU^2}\right)_0, \tag{12}$$

where dU denotes the change of energy of the resonator associated with the entropy change dS, and $\Delta U = U - U_0$ is the difference of the average energy of the resonator, U, and the equilibrium energy, U_0. (We have suppressed the suffix ν here.[43]) Due to the equation of motion of the resonator (damped by the emission of radiation), ΔU and dU have opposite signs.[44] Since the entropy

---

[42] To see the equality, one has to make use of the equation, $\int \rho_\nu \, d\nu = \int \rho_\lambda \, d\lambda$, with both integrations from zero to infinity, and the fact that $d\nu$ can be written as $-(c/\lambda^2)\,d\lambda$ (c being the velocity of light *in vacuo*).

[43] Equation (12) was obtained by expanding both the change of the entropy of the resonator and of the entropy of the radiation, in a Taylor series and observing that the first-order differential quotients cancel at the equilibrium position. The subscript ν, referring to the frequency of the oscillator (in U_ν), has been dropped in the following derivation.

[44] The time change of the (average) energy of a damped resonator is given by the equation $dU/dt + 2\sigma\nu\,\Delta U = 0$, where the positive damping constant σ is related to the emitted intensity of radiation of frequency ν.

change, dS, must be a positive quantity, the factor $\frac{3}{5}(d^2S/dU^2)_0$ on the right-hand side of Eq. (12) has to represent a negative quantity, say $-f(U)$, where $f(U)$ is a positive function of the resonator energy $U = U_\nu$. In order to determine $f(U)$, Planck considered n resonators in the cavity and postulated the additivity of the resonator entropies; thus he obtained from Eq. (12) the result

$$n\,dS = -n^2\,dU\,\Delta U\,f(nU). \tag{13}$$

The obvious means of satisfying this equation was to take $f(U)$ as proportional to U^{-1}. Then the differential quotient, $(d^2S/dU^2)_0$, could be written as

$$\left(\frac{d^2S}{dU^2}\right)_0 = -\frac{\text{const.}}{U}, \tag{14}$$

with a positive constant, which depended only on the frequency ν of the oscillator. By integrating Eq. (14), Planck obtained an expression for dS/dU, namely, $-\alpha U \ln(\beta U)$, with factors α and β that might depend on the frequency. By comparing this expression with Eq. (10), Planck found that α and β corresponded to $(a\nu)^{-1}$ and $(b\nu)^{-1}$, respectively. Then he immediately derived the validity of Wien's law, Eq. (11), for the energy distribution of blackbody radiation.

At the same time as Planck presented his, as he claimed, rigorous derivation of Wien's law, the experimentalists reported deviations from the latter. As we have already mentioned, some members of the *Physikalisch-Technische Reichsanstalt* had become increasingly involved in studying the properties of blackbody radiation during the 1890s. Interestingly enough, no plan had existed at the *Reichsanstalt* to perform measurements to establish the form of Kirchhoff's function. It just so happened that several scientists working in different departments got into this problem. Some came to it while investigating certain problems of thermometry, e.g., the question of determining the temperature in regions where the usual methods (say, the mercury and gas thermometers) failed. But also a more technical problem, the search for a suitable standard unit of luminosity, led scientists to develop techniques which could later be used for the study of blackbody radiation. In particular Otto Lummer, who had just joined the *Reichsanstalt* in 1887 and had become the leading person responsible for optical measurements, turned his attention after 1890 to the problem of measuring the temperatures of light sources employing incandescent platinum.[45] To that

---

[45] Otto Richard Lummer was born in Gera, Thüringen, on 17 July 1860. He studied at the University of Berlin, where he obtained his doctorate in 1884 with a thesis on a new interference effect observed in plane parallel glass plates. He continued to work as an assistant to Hermann von Helmholtz at the University of Berlin; when Helmholtz became President of the *Reichsanstalt*, he took Lummer along with him. Lummer served at the *Reichsanstalt* first as an assistant, then (from 1889) as a scientific member, and received the title of 'professor' at the same institution in 1894. In 1901 he became *Privatdozent* at the University of Berlin, and three years later he was appointed *Ordinarius* of physics at the University of Breslau. He died in Breslau on 5 July 1925.

Lummer made important contributions to optics and photometry (Lummer's interferometer, Lummer–Brodhun photometer). He also participated in the discovery of the liquefaction of carbon.

end he developed both photometric and bolometric methods in collaboration with his colleague Ferdinand Kurlbaum.[46] In 1895 Lummer published a short note with Willy Wien, entitled '*Methode zur Prüfung des Strahlungsgesetzes absolut schwarzer Körper*' ('Method for Testing the Radiation Law of Perfectly Black Bodies'); in it the authors proposed to represent a blackbody by a cavity which was 'brought to as uniform a temperature as possible, and radiation allowed to pass outwards through an opening' (Wien and Lummer, 1895, p. 453).[47] A year later Lummer described the method in full detail; he pointed out that all the investigations carried out hitherto on the properties of blackbody radiation suffered from errors, because the radiators employed were not completely black, and recommended the use of cavities to determine Kirchhoff's function (Lummer, 1896).

In the first paper on cavity radiation, Wien and Lummer had already suggested the testing of Stefan–Boltzmann's law and Wien's displacement law by the new method (Wien and Lummer, 1895). The authors had then started to make preliminary experiments to achieve this goal. After Wien's departure for Aachen, Lummer looked for collaborators, and he received the help of Ferdinand Kurlbaum and Ernst Pringsheim for his work on blackbody radiation.[48] In less than three years these three experimentalists improved the techniques of observation to such an extent that the problem of the measurement of Kirchhoff's function could be attacked in earnest.[49] In performing their investigation, however, Lummer and Pringsheim made use of another experimental development which concerned the analysis of very long wavelengths. The principal

[46] Ferdinand Kurlbaum was born in Burg near Magdeburg on 4 October 1857. He studied at the Universities of Heidelberg and Berlin and received his doctorate from the latter university in 1887 with a thesis on a new determination of wavelengths of absorption lines in the solar spectrum. He first went to the *Technische Hochschule* in Hanover as an assistant with his thesis supervisor Heinrich Kayser for four years. In 1891 he returned to Berlin and joined the optical laboratory of the *Reichsanstalt*, which was directed by Lummer. Ten years later he had his own laboratory at the *Reichsanstalt*, and in 1904 he accepted a call to a professorship of physics at the *Technische Hochschule* of Berlin, from where he retired in 1925. Kurlbaum died on 29 July 1927 in Berlin.

[47] The idea of using a cavity for representing a blackbody goes back to Kirchhoff (1860). However, Wien and Lummer were not the first to think of using Kirchhoff's proposal for experimental purposes. Already the Danish physician Christian Christiansen (1843–1917), one of the discoverers of the anomalous dispersion of light, had experimented with the absorption of (visible) radiation by cavities having a small hole and observed that they could be taken as a good approximation of a blackbody (Christiansen, 1884). A few months later, independently of Wien and Lummer, the American physicist Edward St. John, then a student at the University of Berlin, also noticed the blackbody properties of a cavity radiator (St. John, 1895).

[48] Ernst Pringsheim was born in Breslau on 11 July 1859. He studied physics and mathematics in Heidelberg, Breslau, and Berlin, from 1877 to 1882. He received his doctorate at the University of Berlin in 1882, where he also became *Privatdozent* four years later and Titular Professor in 1896. In 1905 he accepted a call to a full professorship in theoretical physics at the University of Breslau, where he joined Otto Lummer, his collaborator for many years. Pringsheim worked on problems of radiation including luminescence, on optics (interferometry) and on physics of the sun. He died on 28 June 1917 in Breslau.

[49] Special progress in that direction was achieved through the construction of a suitable cavity, namely, an electrically heated plantinum cylinder which was blackened inside with iron oxide and divided by diaphragms and enclosed in a larger asbestos cylinder (Lummer and Kurlbaum, 1898).

contribution in that field had been made by Heinrich Rubens, since 1892 *Privatdozent* at the University of Berlin and from 1895 Professor of Physics at the *Technische Hochschule* in Berlin.[50] In 1889 he had already begun to measure the wavelengths of invisible infrared radiation with the help of a Rowland grating and the bolometer method. During the 1890s he penetrated further into the infrared spectral region by using various techniques. Finally, in late 1896 he published, together with a visitor from the United States, Ernest Fox Nichols (1869–1924), a new method for measuring long wavelengths (Rubens and Nichols, 1896). This new method, which was later called the method of 'residual rays' ('*Reststrahlenmethode*'), made use of the fact that all substances reflect radiation especially strongly in the region of strong absorption, hence it was possible to isolate certain wavelengths by multiple reflection; thus, by 1898, wavelengths of 61.1 μ were reached by using a sylvine crystal as the reflecting substance (Rubens and Aschkinass, 1898). The tools were thus made ready for a fresh attack on the empirical determination of the radiation law.

On 3 February 1899, at a session of the German Physical Society in Berlin, Lummer presented the first results of his and Pringsheim's investigations on the energy distribution in the spectrum of blackbody radiation (Lummer and Pringsheim, 1899a). They had observed wavelengths between 0.2 μ and 6 μ emitted from cavities heated to temperatures between 800 and 1400 degrees absolute; they found that the energy distribution agreed in general with Paschen's law, Eq. (2), although the constant c_2 seemed to increase systematically with decreasing temperatures. In order to eliminate experimental errors Lummer and Pringsheim continued to improve their methods in the following months. Later in the year, on 3 November 1899, Lummer again spoke at a session of the German Physical Society; now he was certain that there were 'discrepancies of a systematic nature between theory and experiment' ('*Abweichungen zwischen Theorie und Experiment systematischer Natur*,' Lummer and Pringsheim, 1899b, p. 222) which could not be attributed to experimental errors. In contrast to Lummer and Pringsheim, Friedrich Paschen—in two contributions to the Prussian Academy, which were presented by Planck at the sessions of 27 April and 7 December 1899, respectively did not find any deviations from Wien's law (Paschen, 1899a, b), and the situation remained undecided in the beginning of 1900.

[50] Heinrich Leopold Rubens was born in Wiesbaden on 30 March 1865. He first studied electrical engineering at the Technical Universities (*Technische Hochschulen*) of Darmstadt and Berlin (1884–1885), then physics at the Universities of Strasbourg and Berlin. He obtained his doctorate under August Kundt from the University of Berlin in 1889; he stayed on there as an assistant and *Privatdozent* until he was appointed in 1896 an '*außerplanmäßiger*' professor of physics at the *Technische Hochschule* of Berlin. He became a full professor in 1900 and moved six years later to the University of Berlin, succeeding Paul Drude. He devoted nearly his entire scientific career to the study of electromagnetic waves, especially the infrared region which he extended immensely by his researches. He received many honours, including the Rumford Medal of the Royal Society and an honorary D.Sc. from Cambridge University; he was a member of the scientific academies of Berlin and Göttingen. Rubens died in Berlin on 17 July 1922. The importance of his role in quantum theory was later described by Max Planck as follows: 'Without the intervention of Rubens the formulation of the radiation law and thereby the foundation of quantum theory would perhaps have [arisen] in quite a different manner, or perhaps not have developed in Germany at all' (Planck, 1923a, p. cxi).

During 1900 the meetings of the German Physical Society in Berlin became the main forum where the problems connected with the law of blackbody radiation were discussed most frequently because most of the leading experts were resident members; besides Max Planck, Otto Lummer, Ernst Pringsheim, Ferdinand Kurlbaum and Heinrich Rubens, all of whom we have already mentioned, there were scientists like Ludwig Holborn, Eugen Jahnke, Friedrich Kohlrausch (the Acting President of the *Reichsanstalt*), Max Ferdinand Thiesen and Emil Warburg, who became interested in the subject.[51] The first session on the problems of blackbody radiation took place on 2 February. In it Max Thiesen presented a theoretical discussion of blackbody radiation (which he called 'black radiation'), Ernst Pringsheim reported about new measurements by Lummer and himself, and finally Planck discussed the assumptions which had entered into his derivation of Wien's law (Planck, 1900b).[52] Thiesen analyzed the structure of the existing laws of blackbody radiation—i.e., the Stefan–Boltzmann law as well as Wien's displacement law and the radiation law, Eq. (6)—and suggested a more complicated formula, generalizing Wien's Eq. (6) in order to account for the deviations observed especially by Lummer and Pringsheim (Thiesen, 1900a). Pringsheim showed that these deviations became more significant if one studied—for the same temperature of the emitter—the long wavelengths of the spectrum. (The new experiments went up to 18 μ: Lummer and Pringsheim, 1900.)[53]

The February session of the German Physical Society started a series of considerations to examine the validity of Wien's radiation law. Thus Lummer, together with Eugen Jahnke, a high school teacher, submitted a paper to *Annalen der Physik* in which they made proposals to generalize Eq. (6) (Lummer and Jahnke, 1900). Such attacks on his equation encouraged Willy Wien to reply to the criticisms; he did so in his talk at the International Congresss of Physics

---

[51] The German Physical Society or, more accurately, the *Deutsche Physikalische Gesellschaft zu Berlin*, grew out of the *Berliner Physikalische Gesellschaft* which was founded on 14 January 1845 after a colloquium by Gustav Magnus at the University of Berlin. Among the founders one finds the names of the chemist Wilhelm Heintz, the physiologists Emil du Bois Reymond and Ernst Brücke, and the physicists Hermann Knoblauch, Wilhelm Beetz and Gustav Karsten; they were soon joined by Hermann von Helmholtz and Werner Siemens (later 'von' Siemens). The sessions of the *Gesellschaft* took place regularly in the Institute of Physics of the University of Berlin; towards the end of the century there took place about one session every two weeks, except during the semester vacations. At the end of 1898 the members of the Berlin Physical Society decided to change the name to German Physical Society; the members were partly Berlin residents (about 50%) and partly nonresidents. (For historical details, see Schwalbe, 1900.)

[52] The report of this session, after mentioning the talks of Thiesen and Pringsheim, stated:
Following the two talks, Mr. Planck discusses in detail the assumptions on which the radiation theory developed by him is founded, and reports on a direct '*Deduktion der Strahlungs-Entropie aus dem zweiten Hauptsatz der Thermodynamik*' ('Deduction of the Radiation Entropy from the Second Law of Thermodynamics'), which he found recently and will publish soon. (*Verh. d. Deutsch. Phys. Ges.* (2) 2, Nr. 3, 1900, p. 37)

[53] Pringsheim's report was not published in spring 1900 but only in the fall. The publication in the *Verhandlungen* represented essentially the content of a talk given on 18 September 1900 before the *Versammlung Deutscher Naturforscher under Ärzte* in Aachen (Lummer and Pringsheim, 1900, p. 163).

(*Congrès International de Physique*), held at Paris on the occasion of the World's Fair from 6 to 12 August 1900, and in a paper submitted to *Annalen der Physik* in October 1900 (Wien, 1900a and 1900b, respectively). He came to the conclusion that on the basis of thermodynamic and kinetic-theoretic arguments it followed that Eq. (6) must describe the energy distribution among short but not among long wavelengths.

Wien's conclusion was confirmed by the results obtained by Heinrich Rubens and Ferdinand Kurlbaum, who investigated the spectrum of cavity radiation up to the longest wavelengths obtained until then, namely, the residual rays of rock-salt having a wavelength of 51.2 μ; actually, the crucial quantity was the product of wavelength and temperature, which reached values up to 90,000 in the measurements of Rubens and Kurlbaum, in contrast to the value of 32,000 obtained earlier by Lummer and Pringsheim (Rubens and Kurlbaum, 1900). The main outcome of the observations of Rubens and Kurlbaum could be stated briefly as follows: for very long wavelengths λ and very high temperatures T, the intensity of radiation increases in proportion to T, i.e.,

$$\rho_\lambda = \text{const.}\, T, \tag{15}$$

in agreement with the formulae derived earlier by Lord Rayleigh (1900b) and Lummer and Jahnke (1900). (See Rubens and Kurlbaum, 1900, p. 941.)

Planck received the news about the results of Rubens and Kurlbaum before their public announcement. In a report about these events Planck's student Gerhard Hettner recalled later: 'When on Sunday, 7 October 1900, Rubens together with his wife visited Planck, the discussion turned to the measurements with which Rubens was occupied. He [Rubens] said that for the longest wavelengths [which he could achieve], the law recently proposed by Lord Rayleigh was valid' (Hettner, 1922, p. 1036).[54] On receiving this information from Rubens, Planck set down and studied the theoretical implications for the equilibrium entropy; he noticed especially that if Rubens' results were correct, then for long wavelengths—or, more accurately, for large λT—the entropy had to satisfy the equation[55]

$$\left(\frac{d^2S}{dU^2} \right)_0 = - \frac{\text{const.}}{U^2}. \tag{16}$$

Then he combined Eq. (16) with Eq. (14), which was valid for short wavelengths

[54] Hettner obtained the information from Planck himself. (See Kangro, 1970, Section 8.8; English edition, pp. 198–200.)

[55] Equation (16) follows from the definition of the entropy, $(dS/dU)_0 = 1/T$, and the proportionality of the energy U and the temperature T. (This gives the relation $(dS/dU)_0 \sim 1/U$, from which one derives the expression for $(d^2S/dU^2)_0$. The index zero on the left-hand side of Eq. (16) is added to emphasize the fact that this equation describes only the equilibrium situation. Again, the subscripts referring to frequency and wavelength have been dropped in the following derivations.)

(more accurately, small λT) and obtained the following equation for the equilibrium entropy:

$$\left(\frac{d^2S}{dU^2}\right)_0 = \frac{\alpha}{U(\beta + U)}, \tag{17}$$

where α and β were constants depending on the wavelengths of the radiation. Planck then integrated Eq. (17) to obtain the expression for $(dS/dU)_0$, which he put equal to the inverse of the absolute temperature. By again integrating the relation between entropy and temperature, he found an equation for the energy of the resonator, from which he derived—with the help of Eq. (7)—a new radiation formula, that is,

$$\Phi(\lambda, T) = \rho_\lambda = \frac{c_1}{\lambda^5} \frac{1}{\exp(c_2/\lambda T) - 1}, \tag{18}$$

where c_1 and c_2 were constants similar to the ones entering Paschen's law, Eq. (2).[56] As Hettner reported later on these events: 'The same evening still he [Planck] reported this formula to Rubens on a postcard, which the latter received the following morning . . . One or two days later Rubens again went to Planck, and was able to bring him the news that the new formula agreed perfectly with his [Rubens'] observations' (Hettner, 1922, p. 1036). At the following meeting of the German Physical Society, on Friday, 19 October 1900, Kurlbaum talked about his and Rubens' experiments on 'the emission of long wavelengths from black bodies' ('*Über die Emission Langer Wellen durch den Schwarzen Körper*'). The report on this meeting, given in the *Verhandlungen* then stated: 'In the lively discussion following this talk, Mr. M. Planck spoke on "An Improvement of Wien's Spectral Law".'[57]

With Planck's formula, whose validity for all wavelengths and all temperatures became established again and again in the following years, the correct law for the energy distribution of blackbody radiation had been given.[58] For Planck,

---

[56] Equation (18) was obtained as follows. From the integration of Eq. (17), one found an expression for $(dS/dU)_0$, which could be put equal to the inverse of the absolute temperature, i.e.,

$$\left(\frac{dS}{dU}\right)_0 = \frac{\alpha}{\beta} \ln \frac{U}{\beta + U} = \frac{1}{T}.$$

By integrating this equation, Planck found the result

$$U = \beta\left[\exp\left(-\frac{\beta}{kT}\right) - 1\right]^{-1}$$

for the equilibrium energy of the resonator of a given frequency, $U = U_\nu(T)$. The result agreed with Eq. (18) if he took into account Eq. (7) between U_ν and ρ_ν and rewrote it as an equation for ρ_λ, the energy distribution over the wavelengths. The constants α and β of Eq. (17) were now replaced by the constants c_1 and c_2, that is, $\alpha = -c_1/c^2c_2$ and $\beta = c_1/c^2\lambda$.

[57] See *Verh. d. Deutsch. Phys. Ges. 2* (1900), p. 181.

[58] In November 1900 Paschen submitted a paper to *Annalen der Physik* in which he reported that his own recent measurements agreed with those of Rubens and Kurlbaum (Paschen, 1901).

however, the immediate and more fundamental task was to provide the theoretical foundation and physical motivation for the basic Eq. (17). This task led him to the discovery of a new constant of nature: the quantum of action.

I.2   The Significance of the Constants in Planck's Law

In October 1900 it had become certain on experimental grounds that Wien's law, Eq. (6), did not describe the entire energy distribution in blackbody radiation, and Max Planck—so far a firm believer in Wien's law—had proposed a new law, Eq. (18), which agreed quantitatively with the most recent results (Planck, 1900c).[59] The new situation implied that some of the assumptions, which Planck had made earlier in his theory of blackbody radiation, were definitely wrong. Planck had immediately identified the crucial point on which he had to improve: while he had previously assumed that $(d^2S/dU^2)_0$, the second derivative of the resonator entropy at equilibrium, was proportional to U^{-1}, the inverse of the average resonator energy, he now arrived at the conclusion that Eq. (14) should be given up. In particular, he had remarked:

> I believe, on the contrary, that it is possible—though still not easy to grasp and in any case difficult to prove—that the expression on the left-hand side [of Eq. (14] does not possess in general the significance I had attributed to it. In other words: the values assumed by U_n, dU_n and ΔU_n [denoting, respectively, the total energy of n resonators in a cavity and its changes] do not suffice at all for determining the change of entropy [of the n resonators] considered; in addition, U [i.e., the energies of the individual resonators] must be known. By following this idea I have finally been able to construct totally arbitrary expressions for the entropy, which—though more complicated than Wien's expression [i.e., the entropy derived from Wien's law and given by Eq. (8a)]—still seem to satisfy, as completely as the latter, all requirements of thermodynamics and electrodynamics. (Planck, 1900c, p. 203)

Thus he also studied the entropy function S satisfying Eq. (17), that is,

$$S_0 = -\alpha\left[\left(\frac{U}{\beta'\nu} + 1\right)\ln\left(\frac{U}{\beta'\nu} + 1\right) - \frac{U}{\beta'\nu}\ln\frac{U}{\beta'\nu}\right], \tag{19}$$

with the constant $\beta' = \beta/\nu$ and the suffix zero denoting the equilibrium function.[60] This entropy had then yielded the radiation law, Eq. (18).

---

[59] Not only did the measurements of the long wavelength distribution of Rubens and Kurlbaum (1900), quoted above, contribute to Planck's conversion, but Friedrich Paschen, whose measurements had always confirmed Eq. (6) (e.g., Paschen, 1899a, b), had also written a letter to Planck at about that time pointing to 'appreciable departures from Wien's law.' (See Planck, 1900c, p. 202, footnote 1.)

[60] In the integration of Eq. (17) we have suppressed all terms not depending on the (average) energy of the resonator U. Note that in Eq. (19), U is a function of the frequency ν and the temperature T.

Now, after he had proposed the successful formula, Planck was faced with the task of obtaining a theoretical derivation of the expression for the equilibrium entropy S_0. He had already observed earlier—in February 1900—that thermodynamic arguments alone did not suffice to define the entropy function of the resonator and, therefore, also of the blackbody radiation (Planck, 1900b). The additional argument which had enabled him to arrive at, as he believed, a unique definition of the entropy via Eq. (14), had now been discovered to fail. On what basis could he, in the light of these negative experiences, claim that Eq. (18) yielded, up to additional constants, the correct entropy of the resonator in a cavity? How could he justify, in particular, the existence of two constants, α and β', and why did he not have to introduce any other constants? Would he have to give up essential parts of his theory of irreversible radiation processes which he had developed over so many years? After all, Wien, in his paper in *Annalen der Physik*, had just recently raised two arguments against this theory, namely:

> First, one misses the demonstration that the hypothesis of natural radiation which has been introduced [by Planck] is the only one that might lead to irreversibility. Hence it remains doubtful whether the processes considered [i.e., the absorption and reemission of radiation by the resonators in the cavity] are related in any way to the heat radiation. Secondly, in the derivation of the expression for the entropy the radiating resonators are assumed to be independent of each other. But the derivation [of the expression for the entropy, Eq. (14)] depends on the assumption that several resonators exist. In my opinion, this seems to imply a contradiction. (Wien, 1900b, pp. 538–539)[61]

Though Planck accepted the opinion that his theory was, so far, incomplete, he did not for a moment think of giving up his original plan of research, namely, the principal idea that the condition of irreversibility of radiation processes would determine the law of blackbody radiation. He also disagreed strongly with Wien on another point of principle, which the latter had stated in his paper in *Annalen der Physik* as follows:

> I must emphasize, first, that I still stick, in contrast to Mr. Planck [see Planck, 1900a, p. 725], to my previously stated opinion [see Wien, 1893b, p. 633] that short and long electromagnetic waves—as far as their relations to heat radiation are concerned—differ more than just quantitatively from each other. With respect to the process of absorption it is assumed generally that longer waves can be described by a single vector or, what amounts to the same, that matter can be regarded in that case as being continuous; however, for shorter wavelengths the influence of the molecular constitution of bodies comes into play. Exactly the same must be true for emission processes. Therefore, I believe it to be improbable from the very beginning to assume that a radiation law, which rests on the molecular hypothesis, should also

---

[61] Planck had listened to Wien's arguments already at the Paris Conference (Wien, 1900a). He read them again in Wien's paper for *Annalen der Physik* before its publication because he was assisting the editor of the journal (Wien, 1900b). Planck had seen the paper certainly soon after it was received on 12 October, for he replied to some of the other questions raised in Wien's paper in November (Planck, 1900d, e).

be valid for very long waves. The agreement [of the radiation law] with the empirical data in the case of short wavelengths obviously shows that the assumptions made [in the derivation of the law] are approximately satisfied for waves whose wavelengths are not too long. (Wien, 1900b, pp. 537–538)

If a qualitative difference existed, as Wien claimed, between the heat radiations of long and short wavelengths, then the unity of electromagnetic radiation had to be given up, and the great advance brought about by Maxwell's electromagnetic theory would be lost again. Such an attitude contradicted the scientific goals of Max Planck, who had always sought to establish relations and connections between different phenomena rather than to separate them. He was convinced that a unified theory had to describe all blackbody radiation and that this theory was based on electrodynamics and thermodynamics. On the other hand, he recognized the necessity of taking into account even other physical ideas in order to derive a radiation formula applicable to all wavelengths.

As to the direction in which to look for these supplementary ideas, Planck could obtain some indication from his previous experience in the theory of cavity radiation as well as from Wien's critical remarks quoted above. In order to explain Eq. (17), the *Ansatz* for the entropy function, it was evidently necessary to consider *several* resonators in the cavity and to assume that the entropy would not only depend on the total energy of all resonators but also on the distribution of energy among the individual resonators. Still, the question remained open as to how one might proceed in detail. The most obvious method to apply seemed to follow Wien's example of 1896, that is, to treat the objects emitting heat radiation in a way similar to the molecules in kinetic theory. However, if one wanted to arrive at a theory including the long waves as well, more features of kinetic theory had to be used than Wien had employed earlier. Now the most detailed and complete scheme of kinetic theory had been developed by Ludwig Boltzmann. Planck had been aware of Boltzmann's work for quite some time. He had even had the opportunity of discussing with Boltzmann a specific question of kinetic theory, namely, the derivation of the Maxwellian velocity distribution for molecules in a gas.[62] However, Planck—though he had joined Boltzmann in fighting against the views of the energeticists—had not developed closer relations with him. As he explained later: 'In particular he [Boltzmann] was unhappy about the fact that I was not only indifferent towards the atomistic theory, which constituted the foundation of his entire research work, but even opposed it' (Planck, 1948a, p. 21). In particular, he had criticized the probabilistic interpretation of entropy. However, in the course of his treatment of the theory of heat radiation, Planck had withdrawn his earlier criticism and increasingly approached Boltzmann's views on the origin of irreversibility, especially by intro-

---

[62] Planck had edited Gustav Kirchhoff's lectures on the theory of heat (Kirchhoff, 1894), and Boltzmann had criticized a passage in the book dealing with a proof of Maxwell's velocity distribution formula (Boltzmann, 1894). Planck had then replied to Boltzmann's objections and formulated the proof more accurately (Planck, 1894c), with which Boltzmann agreed (Boltzmann, 1895a).

ducing the hypothesis of 'natural radiation' and recognizing the similarity of this concept with that of 'molecular chaos,' which Boltzmann had introduced in kinetic theory. Thus in fall 1900 Planck was prepared well enough and willing to search in Boltzmann's statistical theory of matter for the physical interpretation of his radiation formula, Eq. (18).

> This problem [he recalled later in his Nobel lecture] led me automatically to a consideration of the connection between entropy and probability, that is, to Boltzmann's trends of ideas, until after some weeks of the most strenuous work of my life, light came into the darkness and a new, hitherto undreamt of perspective began to open up for me (Planck, 1920, p. 5)

The reconstruction of the step which led Planck to the interpretation of his radiation formula, Eq. (18), must begin by analyzing the expression for the equilibrium entropy of the resonator on the right-hand side of Eq. (19). For some time Planck tried to use only thermodynamic and electrodynamic arguments, but without any success. Then, 'since no other path [seemed to be] open,' he turned to the 'method of Boltzmann' (Planck, 1943, p. 157). This method employed the statistical interpretation of the entropy, in particular the fact that the entropy of any molecular system in a given state could be identified with the 'permutation measure' ('*Permutationsmaß*'), that is, the logarithm of the number of 'complexions' or possibilities of permuting the molecules without changing the state of the system (Boltzmann, 1877). Planck now formulated Boltzmann's relation as

$$S = k \ln \mathscr{P},\tag{20}$$

where $\mathscr{P}$ denotes the number of complexions and k is a natural constant. With Eq. (20) in mind he then returned to the interpretation of S_0, the equilibrium entropy of the resonator given by Eq. (19). The expression within the square brackets on the right-hand side of Eq. (19) should denote the permutation measure of the resonators, but it was not immediately obvious how. Surely, a permutation measure made sense only in the case of a system of, say, n resonators, in which case it described the distribution of energy among the individual resonators. Would Planck be able to find in Boltzmann's work an expression for the permutation measure of a system of n objects having a structure similar to the expression within the square brackets on the right-hand side of Eq. (19)? For this purpose Planck studied in detail Boltzmann's great memoir, and he discovered right away the desired expression in its first section (Boltzmann, 1877). There Boltzmann had given an equation for J, the total number of complexions or possibilities of distributing λ discrete, equal energy values ϵ among n molecules, as

$$J = \frac{(\lambda + n - 1)!}{\lambda!\,(n - 1)!}\,.\tag{21}$$

He had rewritten this equation for large values of n and λ with the help of Stirling's formula as

$$J = \frac{1}{\sqrt{2\pi}} \frac{(\lambda + n - 1)^{\lambda + n - 1/2}}{(n-1)^{n-1/2}\lambda^{\lambda + 1/2}} . \tag{21a}$$

Then the logarithm of J, after neglecting the terms -1 and $\pm \frac{1}{2}$ in comparison to n and λ, would take the form

$$\ln J = n\left[\left(\frac{\lambda}{n} + 1 \right)\ln\left(\frac{\lambda}{n} + 1 \right) - \frac{\lambda}{n}\ln\frac{\lambda}{n} \right]. \tag{22}$$

(Note that on the right-hand side of this equation a subtractive term of magnitude $\ln\sqrt{2\pi}$ has been dropped.) Planck immediately noticed the similarity between $\ln J$ and his expression, Eq. (19), for the equilibrium entropy of the resonator in the following way: if he considered n resonators in a cavity having the equilibrium entropy nS_0, then the expressions within the square brackets of Eqs. (19) and (22) were identical provided he put λ/n equal to $U/\beta'\nu$. Now U, the average energy of one resonator, assumed the value $(\lambda/n)\epsilon$; hence he concluded that the discrete energy value in blackbody radiation was given by the equation

$$\epsilon = \beta'\nu. \tag{23}$$

Thus the constant β', which Planck would later denote by the letter h, possessed the dimension of an action. It determined, when multiplied by the frequency ν, the size of the energy elements. Finally, in order to make the agreement between S_0 and $k \ln J$ complete, Planck identified the constant $-\alpha$ with the natural constant k occurring in Eq. (20).

The question arises whether Planck was really allowed to take over Boltzmann's expression for J to describe the equilibrium situation of his cavity resonators. Boltzmann, in his memoir of 1877, had given a different prescription for calculating the equilibrium entropy of molecular systems. He had first talked about the distribution of λ energy values ϵ among n molecules and had presented the method of obtaining the number of complexions; then he had claimed that the equilibrium state of the system was defined by taking the maximum of all possible numbers of complexions. Planck did not follow this procedure at all, but just referred to the expression J, which entered into Boltzmann's calculations as the factor normalizing $\mathcal{P}$, the number of complexions for any given state, such that the quotient $(\mathcal{P}/J)$ could be interpreted as a probability. More than thirty years later Planck tried to recall the psychological reasons that motivated his step:

In short [he wrote to Robert Williams Wood], I can characterize the whole procedure as an act of despair, since, by nature I am peaceable and opposed to doubtful adventures. However, I had already fought for 6 years (since 1894) with the problem of equilibrium between radiation and matter without arriving at any successful result. I was aware that this problem was of fundamental importance in physics, and I knew the formula describing the energy distribution in the normal spectrum [i.e., the spectrum of a blackbody]; hence a theoretical interpretation *had* to be found at any price, however high it might be. (Planck to Wood, 7 October 1931)

In that desperate situation Planck would have turned to any suitable formula of Boltzmann's which had to do with the number of complexions. Actually, the choice of taking J really agreed with Planck's goals; only one did not have to stick to its interpretation in Boltzmann's original memoir. Planck was interested in calculating the entropy of the equilibrium state; thus he interpreted J, Eq. (21), right away as the number of complexions for the equilibrium state and, consistent with it, he identified the expression $(\lambda/n)\epsilon$ with the average energy of the resonators in the equilibrium state. At that moment he was not interested in Boltzmann's procedure for obtaining the equilibrium distribution by selecting from different distributions the one associated with the maximum number of complexions; he rather assumed that he *knew* the equilibrium state already. And, after all, that was indeed the case in his considerations, which were aimed at obtaining a physical interpretation of the entropy S_0 derived from the radiation law, Eq. (18), i.e., from the equilibrium situation.[63]

Planck had the physical interpretation of his radiation law in hand before the middle of November 1900, but he presented his results first at the meeting of the German Physical Society in Berlin on 14 December 1900 in a contribution entitled '*Zur Theorie des Gesetzes der Energieverteilung im Normalspektrum*' ('On the Theory of the Law of Energy Distribution in Normal Spectrum,' Planck, 1900f).[64] In this contribution he outlined in short 'the most essential point of the whole theory as clearly as possible' (Planck, 1900f, p. 238). Thus, after emphasizing the necessity of introducing Boltzmann's probability arguments in the theory of blackbody radiation, he started right away by considering the distribution of a given amount of energy, E, among N cavity resonators having the frequency ν. He noted:

If E is considered as an infinitely divisible quantity, the distribution can be made in an infinite number of ways. However, we consider—and this is the most important point of the entire calculation—E as being composed of a completely definite number of finite, equal parts, and make use for that purpose of the natural constant $h = 6.55 \times 10^{-27}$ [erg · sec]. This constant, when multiplied by the common frequency ν of the resonators, yields the energy element ϵ in ergs; and by dividing E

[63] For the reasons given above, we disagree with Thomas Kuhn's conclusion (Kuhn, 1978, pp. 282–283, footnote 20) that Planck had to take his formula for the number of complexions in radiation theory from sources different from Boltzmann's 1877 memoir.

[64] Concerning the status of his theoretical derivation of the radiation law, Planck wrote to Wien on 13 November 1900: 'My new formula [Eq. (18)] is well satisfied; I now have also obtained a theory for it, which I shall present in four weeks at the Physical Society here [in Berlin].'

by ϵ we obtain P, the number of energy elements which have to be distributed among the N resonators. If the quotient $[E/\epsilon]$ thus calculated does not happen to be an integral number, then one has to take for P an integer close to it [the quotient]. (Planck, 1900f, pp. 239–240)

He then calculated $\mathcal{R}$, the number of complexions for the distribution of the energy E among the N resonators of frequency ν, according to the combinatorial equation (20). Further, he considered the cavity to contain also resonators of different frequencies, say N' of frequency ν', N'' of frequency ν'', etc., among which the energy amounts of E', E'', etc., were distributed in $\mathcal{R}'$, $\mathcal{R}''$, etc., ways. Evidently, the number of complexions denoting the possibilities of the distribution of discrete energy elements $\epsilon \ (= h\nu)$, $\epsilon' \ (= h\nu')$, $\epsilon'' \ (= h\nu'')$, etc., among the resonators was identical with the product $\mathcal{R} \cdot \mathcal{R}' \cdot \mathcal{R}'' \cdots$. In case of thermal equilibrium, a temperature could be defined through the equation

$$\frac{1}{T} = k \frac{d \ln \mathcal{R}_0}{dE_0} , \tag{24}$$

where $\mathcal{R}_0$ is the maximum value of the number of complexions $\mathcal{R} \cdot \mathcal{R}' \cdot \mathcal{R}'' \cdots$ (or the number of complexions in equilibrium) and $E_0 \ (= E + E' + E'' + \cdots)$ denotes the sum of all energies distributed among the frequencies $(\nu, \nu', \nu'', \dots)$ of the cavity resonators. Planck did not carry out any maximalization procedure with the total number of complexions; he rather assumed that the equilibrium distributions were already established. Then it was sufficient to study the (equilibrium) energy distribution among the resonators of a single frequency, say ν, and to obtain from it the expression for U_ν, the average energy of the resonator (at temperature T), that is,

$$U_\nu = U_\nu(T) = h\nu \left[\exp\left(\frac{h\nu}{kT} \right) - 1 \right]^{-1}. \tag{25}$$

When Planck combined Eq. (25) with Eq. (7) he obtained an equation for the energy distribution among the frequencies in blackbody radiation, which he could easily transform into Eq. (18) for the distribution among the wavelengths.

In his *Atombau und Spektrallinien*, Arnold Sommerfeld called 14 December 1900 the 'birthday of quantum theory' ('*Geburtstag der Quantentheorie*,' Sommerfeld, 1919, p. 213). He referred in particular to the fact which Planck had considered to be the 'most essential point' of his derivation of the blackbody radiation law: namely, the assumption that the energy is distributed among the cavity resonators only in integral multiples of finite energy elements. The existence of these finite energy elements followed from the application of Boltzmann's relation between entropy and number of complexions also to the system of resonators in a cavity, i.e., from the equation

$$S = k \ln \mathcal{R}. \tag{26}$$

Moreover, the application of Eq. (26) to the theory of blackbody radiation led to an important consequence, which Planck also mentioned in his talk at the German Physical Society in Berlin on 14 December 1900. If one considered a system of gas molecules and of radiating resonators in equilibrium, the total entropy was given by the sum of the entropies of the molecules and the resonators, that is,

$$S_0 = f \ln(\mathcal{P}_0 \mathcal{R}_0) = f \ln \mathcal{P}_0 + f \ln \mathcal{R}_0, \tag{27}$$

where $\mathcal{P}_0$ and $\mathcal{R}_0$ denote the respective numbers of complexions. Now Planck observed that the constant factor f had to be proportional to the universal gas constant R, with a proportionality factor ω denoting the ratio of the mass of the molecule to the mass of a gram-atom (or gram-molecule) of the gas (Boltzmann, 1877). Evidently, the inverse of ω then represented the number of molecules in one gram-atom, the so-called Avogadro number (or, in German scientific literature, Loschmidt number) N_0. Since, due to Eqs. (20) and (27), f must be equal to the constant k of radiation theory—k may be expressed in terms of the constants c_1 and c_2 of the blackbody radiation formula, Eq. (18), as $k = c_1/c^2 c_2$—Planck obtained for N_0 the equation

$$N_0^{-1} = \omega = \frac{k}{R}. \tag{28}$$

Upon substituting the observed values for k and R (i.e., $k = 1.346 \times 10^{-16}$ and $R = 8.31 \times 10^7$ erg/°C): 'A real molecule is 1.62×10^{-24}-th fraction of a gram-molecule, or: an atom of hydrogen weighs 1.64×10^{-24} gm, since H [i.e., the weight of one gram-atom of hydrogen in units of grams] $= 1.01$; or, a gram-molecule of any substance consists of $1/\omega = 6.175 \times 10^{23}$ real molecules' (Planck, 1900f, p. 244). He noted that this value of ω^{-1} or N_0 agreed well with the best estimate known from kinetic gas theory, namely, $N_0 = 6.40 \times 10^{23}$ (Meyer, 1899, p. 337). With the help of the ratio k/R he also calculated the 'elementary quantum of electricity e, that is, the electric charge of a positive single-valued ion or electron' according to the equation

$$e = F \cdot \frac{k}{R}, \tag{29}$$

where F (Faraday's constant) is the charge of a gram-atom of single-valued ions. Thus he found that the unit charge assumed the value 4.69×10^{-10} electrostatic units, adding that 'F. Richarz [1894] found 1.29×10^{-10}, while J. J. Thomson [1898b] has recently found 6.5×10^{-10}' (Planck, 1900f, p. 245).

Planck claimed repeatedly that the relations between the constants of the radiation law, c_1 $(= c^3 h)$ and c_2 $(= ch/k)$, and the constants ω and e which determine the 'elementary quanta of matter and electricity' had to hold as accurately as the given experimental data.

Hence the accuracy of the numbers calculated [i.e., of 6.175×10^{23} for ω^{-1} and of 4.69×10^{-10} for the elementary quantum of charge e] is essentially the same as the accuracy of the most uncertain constant of radiation, k, and it surpasses by far [in accuracy] all determinations of these quantitites that have been made until now. [Planck argued, and added:] An examination of these numbers by more direct measurements will constitute a task for future research, which is as important as it is necessary. (Planck, 1900f, p. 245)

About eight years later Ernest Rutherford and Hans Geiger succeeded in measuring the electric charge of α-particles as 9.3×10^{-10} electrostatic units; hence they concluded the value of the elementary charge to be 4.65×10^{-10} electrostatic units, because α-particles carry two elementary charges (Rutherford and Geiger, 1908). 'The agreement of this figure with the number calculated by me, 4.69×10^{-10}, could be taken as decisive confirmation of the usefulness of my theory,' remarked Max Planck in his Nobel lecture (Planck, 1920, p. 15). Towards the end of the first decade of the twentieth century several different methods for determining experimentally the number of molecules per gram-atom were available, as for instance the observation of Brownian motion or critical opalescence. The results obtained from them agreed fairly well with the theoretical value for ω^{-1} ($= N_0$) which Planck had derived in 1900 from his radiation law.

While it took several years to improve the experimental techniques to obtain reliable determinations of the values of Avogadro's number N_0 and the elementary charge e, other than the ones derived by Planck from the law of blackbody radiation, Planck himself continued with his attempts to deepen the theoretical foundations and to elucidate the physical interpretation of his theory of blackbody radiation. In a paper, which he submitted to *Annalen der Physik* in January 1901, Planck again presented the theoretical derivation of his radiation formula, Eq. (18), introducing certain changes in his earlier treatment (Planck, 1901a). The main difference with the earlier derivation consisted in the fact that Planck did not obtain the magnitude of the discrete energy elements, ϵ ($= h\nu$), by comparing the entropy expression, Eq. (19), derived from the radiation law, with the one derived from the combinatorial treatment, Eq. (22). He rather began with the statistical entropy arising from the distribution of P_ν energy elements of magnitude ϵ_ν among N_ν resonators of frequency ν, that is,

$$S_\nu = kN_\nu \left[\left(1 + \frac{U_\nu}{\epsilon_\nu}\right) \ln\left(1 + \frac{U_\nu}{\epsilon_\nu}\right) - \frac{U_\nu}{\epsilon_\nu} \ln \frac{U_\nu}{\epsilon_\nu} \right], \tag{30}$$

where $P_\nu \epsilon_\nu = N_\nu U_\nu$, and then determined the value of ϵ_ν applying two well-known and well-established laws of the theory of heat radiation: the Stefan–Boltzmann law, Eq. (3), and Wien's displacement law, Eq. (4). From those laws Max Thiesen had obtained in February 1900 the equation

$$E_\lambda = T^5 \psi(\lambda T), \tag{31}$$

where $\psi(\lambda T)$ is a function of the product of wavelength and temperature alone (Thiesen, 1900a, p. 66). By investigating the consequences of this 'simplest formulation of Wien's displacement law' (Planck, 1901a, p. 561), Planck found that the entropy of a resonator in a cavity would depend only on the ratio of its energy and its frequency, that is,[65]

$$S_\nu = f\left(\frac{U_\nu}{\nu}\right). \tag{32}$$

Because of Eq. (32), then, the energy element in Eq. (30) had to be proportional to the frequency, or

$$\epsilon_\nu = h\nu. \tag{23'}$$

And, therefore, the entropy per resonator, S_ν, took the form

$$S_\nu = k\left[\left(1 + \frac{U_\nu}{h\nu}\right)\ln\left(1 + \frac{U_\nu}{h\nu}\right) - \frac{U_\nu}{h\nu}\ln\frac{U_\nu}{h\nu}\right], \tag{30'}$$

where h and k denote two universal constants. The expression for the (equilibrium) energy of a resonator having frequency ν, Eq. (25), and the radiation law, Eq. (18), then followed in the same way as Planck had outlined already in December 1900 and which we have described above.[66]

Planck returned to questions connected with the radiation spectrum emitted by a blackbody, which he also called at that time the 'normal spectrum,' twice again in 1901.[67] In the first communication, presented to the Prussian Academy on 9 May, he discussed the consistency of the definition of entropy, Eq. (30'), with the second law of thermodynamics (Planck, 1901c). For this purpose he had to consider the entropy of the system (cavity plus resonators) and to prove that, starting from any distribution of the energy density, say $\hat{\rho}_\nu$, among the frequencies, the total entropy never decreases. Planck defined the entropy of a monochromatic polarized ray of radiation of frequency ν by the expression

$$\begin{aligned}
\mathcal{L}_\nu = \frac{k\nu^2}{c^2}&\left[\left(1 + \frac{\hat{\rho}_\nu}{8\pi h\nu^3/c^3}\right)\ln\left(1 + \frac{\hat{\rho}_\nu}{8\pi h\nu^3/c^3}\right)\right. \\
&\left. - \frac{\hat{\rho}_\nu}{8\pi h\nu^3/c^3}\ln\left(\frac{\hat{\rho}\nu}{8\pi h\nu^3/c^3}\right)\right]
\end{aligned} \tag{33}$$

[65] Planck had noted this consequence already when he proposed the new radiation formula, Eq. (18), in October 1900 (Planck, 1900c, p. 201, footnote 1).

[66] Planck had, of course, noticed in December 1900 that the resonator entropy, Eq. (19), was of the form given by Eq. (32).

[67] The notion of 'normal spectrum' had occurred in Planck's publications since 1899, for the first time in his fifth communication to the Prussian Academy on irreversible radiation processes (Planck, 1899, Section 23). The normal spectrum was defined as the radiation spectrum obtained in a perfectly reflecting cavity with resonators; that is, it was identical with the equilibrium spectrum of a blackbody.

and studied the change of $\mathfrak{L}_\nu$ during the process of absorption and reemission of the ray by a resonator having the same frequency ν. He found that the change of the total entropy of the system (radiation plus resonator) was indeed positive until the stationary normal density distribution was obtained, when $\hat{\rho}_\nu$ assumed the equilibrium value ρ_ν (at temperature T).[68]

In the other paper, dedicated to the Dutch physicist Johannes Bosscha on the occasion of his seventieth birthday, Planck made some further comments on the relation between entropy and probability (Planck, 1901d). In particular, he compared the ways of obtaining the number of complexions in kinetic gas theory and radiation theory. Thus he outlined first Boltzmann's derivation of the number of possibilities, $\mathcal{P}$, of distributing a given amount of energy, E, among N molecules of mass m, that is,

$$\mathcal{P} = \frac{N!}{\Pi\left[\,f(x,\,y,z,\xi,\eta,\zeta)\,\right]d\sigma!}\,, \qquad (34)$$

where $f(x,\,y,z,\xi,\eta,\zeta)$ denotes the distribution function depending on the coordinates $(x,\,y,z)$ and the velocities (ξ,η,ζ) of the molecules, and $d\sigma$ represents six-dimensional 'elementary regions.' (Evidently, $N = \int f\,d\sigma$ and $E = (m/2)$ $\times \int(\xi^2 + \eta^2 + \zeta^2)\,d\sigma$.) The important point in Eq. (34) is that all elementary regions have to have the same magnitude, which is arbitrary but chosen in such a way that in each region a sufficiently large number of molecules is contained (Boltzmann, 1877, section II). In the case of Planck's resonators in radiation theory the number of complexions, $\mathcal{R}$, could also be obtained by counting the possibilities of a certain distribution of energy; especially, one had to distribute P_ν energy elements of magnitude ϵ_ν ($= h\nu$) among N_ν resonators of frequency ν, $P_{\nu'}$ energy elements of magnitude $\epsilon_{\nu'}$ ($= h\nu'$) among $N_{\nu'}$ resonators of frequency ν', etc. The corresponding number of complexions was given by the product

$$\mathcal{R} = \frac{(P_\nu + N_\nu - 1)!}{P_\nu!\,(N_\nu - 1)!} \cdot \frac{(P_{\nu'} + N_{\nu'} - 1)}{P_{\nu'}!\,(N_{\nu'} - 1)!}\,\cdots\,. \qquad (35)$$

If one assumes with Boltzmann and Planck that $\mathcal{P}$ and $\mathcal{R}$ determine the probability for the respective states of the systems under consideration (i.e., for the energy distribution among the parts of the systems considered), then one may conclude: 'The entropy of a system in any state depends only on the probability of that state' (Planck, 1901d, p. 638).

These additional remarks completed the theory of blackbody radiation, a theory which had occupied Planck for the previous six years. He then shifted his field of interest, partly back to problems of the theory of dissociation and solutions and partly to other problems of radiation theory, such as the dispersion and absorption of radiation by nonconducting and conducting materials. Planck considered the theory of heat radiation as a completed subject, and in the winter

---

[68] By means of irreversible absorption and reemission processes the polarization of the original radiation, if any was present before, will also be removed.

semester of 1905–1906 he delivered lectures on it which were published as a book (Planck, 1906). Planck's *Lectures on the Theory of Heat Radiation* provided a systematic and detailed presentation of all aspects of the subject. He started from the elementary optical phenomena and developed the theory by adding the electrodynamic and thermodynamic descriptions. He devoted special attention to the statistical definition of entropy (in Part 4) and to the demonstration of the irreversibility of radiation phenomena (Part 5) because they seemed to throw new light on the physical interpretation of the constant h. The *Lectures* show that Planck had continued to think about radiation theory since his last paper on it in 1901; he even made use of some recent progress in the kinetic theory of matter for that purpose.

As we have emphasized above, the central role in establishing the blackbody radiation formula, Eq. (18), was played by the relation between entropy and probability of the stationary state of the system consisting of radiation and resonators. In 1900 Planck had based his approach completely on the procedure which Boltzmann had developed in 1877. Then, two years later, he received a new book entitled *Elementary Principles in Statistical Mechanics*; it was sent to him by the author, Josiah Willard Gibbs, himself, then professor at Yale University, New Haven, Connecticut (Gibbs, 1902).[69] Planck esteemed Gibbs highly because of his important contributions to the thermodynamic description of phase transitions; hence he was very interested to read what Gibbs had to say about kinetic theory, especially about those parts which constituted the foundation of his radiation theory.[70] Gibbs had based his treatise, to which he gave the qualifying subtitle 'with especial reference to the rational foundation of thermodynamics,' on considerations of the phase space; that is, he had studied the description of dynamical systems of n degrees of freedom in a $2n$-dimensional

---

[69] Josiah Willard Gibbs was born on 11 February 1839 in New Haven, Connecticut. Beginning in 1854 he studied at Yale University, graduating in 1858. Upon receiving his doctorate at Yale in 1863 he was appointed a tutor for three years. He continued his studies abroad: Paris (1866–1867), Berlin (1867–1868), and Heidelberg (1868–1869); at Heidelberg he studied with Kirchhoff and Helmholtz. He returned to New Haven in 1869 and two years later was appointed Professor of Mathematical Physics at Yale University, a position which he occupied until he died on 28 April 1903 in New Haven.

Gibbs wrote his first paper on the thermodynamics of fluids in 1873. He continued to work on thermodynamical problems, contributing in 1876 and 1878 three memoirs on the equilibrium of heterogeneous substances, which made him famous among the specialists in America and Europe. Besides thermodynamics, he worked on electromagnetic theory of light and developed vector analysis. The book on kinetic theory, *Elementary Principles in Statistical Mechanics* (Gibbs, 1902), represented his contribution to the bicentennial celebration of Yale College.

Gibbs was a member of many learned societies including the U.S. National Academy of Sciences, the Royal Society of London, the Royal Institution of Great Britian, the Prussian Academy of Sciences and the *Gesellschaft der Wissenschaften zu Göttingen*. He received honorary degrees from Erlangen, Christiana and Princeton Universities and was awarded the Rumford Medal (1881) of the American Academy of Arts and Sciences, Boston, and the Copley Medal (1901) of the Royal Society.

[70] Gibbs sent copies of his book on *Statistical Mechanics* to many of the well-known physicists in the United States and Europe, including Lord Rayleigh, H. A. Lorentz and Max Planck, who acknowledged receipts of the book. It was translated into German by Planck's former assistant Ernst Zermelo and published in 1905.

space of position and momentum variables, $q_1, \ldots, q_n$ and $p_1, \ldots, p_n$. All thermodynamic properties of the systems could be obtained from the distribution of the variables in phase space. Thus Gibbs arrived, for example, at a definition of the entropy of a system having the kinetic energy E_{kin} as (Gibbs, 1902, Chapter XIV),

$$S = k \ln V, \tag{36}$$

where V represents the volume of phase space,

$$V = \int \cdots \int dq_1 \ldots dp_n, \tag{37}$$

occupied by all systems of the same or smaller kinetic energy (i.e., $p_1^2/2m + p_2^2/2m + \cdots + p_n^2/2m \leqslant E_{\text{kin}}$, if m is the mass of the particles in the system).

Planck, though he admired the generality and elegance of Gibbs' presentation, found that for practical purposes his definition of entropy was not as useful as it pretended to be. In a paper on the mechanical interpretation of temperature and entropy, in which he compared Boltzmann's definition of entropy via the complexions (1877) and Gibbs' definitions (1902), Planck complained that Gibbs —in contrast to Boltzmann—had not given any prescription for calculating the entropy of a nonequilibrium state.

> Therefore I believe [I am] allowed to state the result of this investigation as follows [he concluded in July 1903:] For the overwhelming generality with which, as it strikes one on first inspection, Gibbs has formulated his various definitions of entropy—i.e., without referring to the nature of the systems considered—he has to pay by restricting their physical significance. For all reversible processes, Gibbs' definitions will serve as well as several other already existing definitions of a more formal nature. However, for irreversible processes, which alone endow entropy with its full significance and which provide the key to a complete understanding of thermal equilibrium, Boltzmann's definition of the entropy turns out to be the most appropriate and productive among all those known hitherto. (Planck, 1904b, pp. 121–122)

About a year later Planck discovered that Gibbs' entropy defintion, Eq. (36), was very useful in radiation theory, for it provided a physical interpretation of the constant h occurring in his radiation law. He immediately presented this interpretation in the lectures on heat radiation (Planck, 1906, Section 150, pp. 154–156).

The method of the phase space allowed one, as Planck recognized, to derive the entropy of the resonator directly from its dynamical variables in the following way. The probability that the energy of the resonator assumes values between E and ΔE is given according to Eq. (37) 'by the magnitude of that area of the state variables f and g, which is bounded by the curves $E = \text{const.}$ and $E + \Delta E = \text{const.}$' (Planck, 1906, p. 155). Thus Planck represented the resonator by a system of two charges of opposite sign which oscillate towards each other along a fixed line. The Hamiltonian H of the system is given by

$$H = E = \tfrac{1}{2}Kf^2 + \tfrac{1}{2}L\dot{f}^2, \tag{38}$$

where f is the product of the positive charge times the distance from the centre of the dipole and $\dot{f}$ its time derivative; K and L denote the constants of the system. Now f and $g = L\dot{f}\,(= \partial H/\partial \dot{f})$ constitute a pair of canonically conjugate dynamical variables of the linear oscillator or resonator. According to Gibbs the state of the latter may be determined by points in the two-dimensional (f, g)-phase space. The curves $E = \text{const.}$ and $E = \text{const.} + \Delta E$ are ellipses having the area E/ν and $(E + \Delta E)/\nu$, respectively; hence the difference of the two areas in phase space becomes $\Delta E/\nu$.[71] Planck now noted:

> If we assume that the entire phase-plane [*Zustandsebene*', i.e., the phase space of the one-dimensional resonator] is divided by a large number of these ellipses into separate sections, such that the ring-shaped areas bounded by two successive ellipses become equal in magnitude to one another, such that $\Delta E/\nu = \text{const.}$, then we obtain a method of determining those sections ΔE of the energy, which correspond to equal probability and which we may therefore call energy elements. [And he continued:] If we put the magnitude of an energy element ΔE equal to ϵ, and put the constant in the previous equation [$\Delta E/\nu = \text{const.}$] equal to h, then we arrive exactly at the former equation [(23')] without having made use of Wien's displacement law. At the same time the elementary quantum of action h acquires a new meaning, namely it gives the area of an elementary region in the phase-plane of a resonator, no matter what its frequency. (Planck, 1906, pp. 155–156)

In other words, Gibbs' entropy definition clarified the concept of the energy elements in radiation theory by identifying the constant h in Eq. (23') with the elementary region in the two-dimensional phase space of the resonator. The same h appears in the equation for the energy elements of arbitrary frequency ν. Moreover, if the resonator is replaced by any other electrodynamic system (of one degree of freedom), which is capable of absorbing and emitting radiation such that an equilibrium can be established with radiation in a cavity, then the same constant h determines the elementary regions in its phase space. Hence Planck took this constant as representing a universal feature of nature and rightly named h as the 'elementary quantum of action.'[72]

While the relation between entropy and the probability of the resonator's state together with the hypothesis of the elementary quantum of action provided the correct description of the stationary (i.e., equilibrium) state in a cavity, it did not yet guarantee that the final state was actually approached. To some extent the

---

[71] The semiaxes of the ellipse, $E = \text{const.}$, are $\sqrt{2E/K}$ and $\sqrt{2EL}$; the area becomes $2\pi E\sqrt{L/K}$ or E/ν, since the frequency is given by $\nu = (1/2\pi)\sqrt{K/L}$.

[72] Planck introduced the notion of the 'elementary quantum of action' with the following words: 'I want to designate this [i.e., the constant h] as the "elementary quantum of action" ["*elementares Wirkungsquantum*"] or as "element of action" ["*Wirkungselement*"], because it has the same dimension as the quantity which owes its name to the Principle of Least Action' (Planck, 1906, p. 154).

hypothesis of natural radiation—i.e., the assumption that radiation in the original state consists of a disordered mixture of plane waves having arbitrary phases and directions (which form a cone of finite aperture)—made sure that by the interaction of this radiation with resonators the entropy is actually increased; especially, the intensity and polarization of the radiation of a given frequency will be changed to complete uniformity.[73] Still, the available resonator mechanisms of absorption and emission of radiation (having a given frequency) could not establish an exchange of energy with radiation of different frequencies, a process which is necessary in order to obtain the stable final equilibrium distribution. Planck suggested that the collisions of resonators might imply frequency-changing interactions, and called for a better theoretical description in the future.

> From such a theory [he concluded], one must then surely expect to obtain far-reaching information about the constitution of oscillators that exist in nature; this is motivated by the fact that the theory must, in any case, also yield a deeper explanation of the physical significance of the universal quantum of action h, an explanation of the same importance as that of the elementary quantum of electricity. (Planck, 1906, pp. 127–128)

Planck's *Lectures on the Theory of Heat Radiation* represented the rational formulation of his vision of a unified theory of all radiation phenomena. The basis for this unification was provided by thermodynamics, which accounts for both the reversible and irreversible process in nature. In radiation theory the phenomena of propagation, reflection and diffraction of light could be described as reversible processes, while the phenomena of absorption and emission of light and the transformation of frequencies seemed to require a description as irreversible processes. As soon as Planck had written down his unified conception of radiation, it was attacked severely from two sides: On the one hand, Albert Einstein destroyed the unity of heat radiation by emphasizing the qualitatively different nature of short and long wavelength radiations; on the other, closer investigations showed that the radiation formula, Eq. (18), contradicted the fundamental concepts of existing electrodynamic theory. We shall treat these developments in the following sections.

I.3 Fluctuations and Light-Quanta

On 13 December 1900, one day before Max Planck presented the derivation of his radiation law in Berlin, Albert Einstein in Zurich signed a paper on '*Folgerungen aus den Capillaritätserscheinungen*' ('Inferences Drawn from Capillary Phenomena') and submitted it to *Annalen der Physik* (Einstein, 1901). In the following years Einstein regularly submitted further contributions to the *Annalen*,

---

[73] One cannot speak of the entropy of a plane electromagnetic wave.

dealing with problems of the foundations of thermodynamics and kinetic theory. Finally, in 1905 he derived two important consequences from the latter. One, he claimed that particles suspended in a fluid must execute an irregular motion even if the fluid is at rest (Einstein, 1905c). The other consequence concerned the nature of radiation; it led him to expound a new, as he said, 'heuristic point of view' (Einstein, 1905b). Einstein described it in the introduction of his paper with the following words:

> According to the assumption considered here, when a light ray starting from a point is propagated, the energy is not continuously distributed over an ever increasing volume, but it consists of a finite number of energy quanta, localized in space, which move without being divided and which can be absorbed or emitted only as a whole. (Einstein, 1905b, p. 133)

With this hypothesis of the existence of localized energy quanta in radiation Einstein wanted to explain certain phenomena connected with the creation and conversion of light, especially photoluminescence, the production of cathode rays by ultraviolet light and the observed blackbody radiation. He thus placed himself in contrast to Max Planck, who had based his derivation of the energy distribution in blackbody radiation entirely on Maxwell's electromagnetic theory of light. How did he come to such a conclusion? In order to answer this question we shall briefly outline the background of the young Einstein.

Albert Einstein was born on 14 March 1879 in Ulm, Württemberg, Southern Germany, the son of the Jewish merchant Hermann Einstein and his wife Pauline, née Koch.[74] The family moved to Munich the following year, where Hermann Einstein opened a small electrotechnical workshop together with his brother Jakob. In Munich Einstein received his early education; in 1885 he became a pupil in a Catholic elementary school and, from 1889, he attended the *Luitpold-Gymnasium*.[75] Several months after the family had moved to Milan in 1894, Einstein left the *Gymnasium*; he first joined the family in Italy and was then sent to Switzerland to complete his education. He finished high school in Aarau and began to study mathematics and physics at the *Eidgenössische Technische Hochschule* (E.T.H. or the Swiss Federal Institute of Technology) in Zurich. Following his examination for the *Diplom* (actually the physics teacher's certificate) in summer 1900, Einstein did not succeed in obtaining the position of an assistant in a university in order to be able to pursue scientific research. Instead of embarking upon a scientific career he had to accept various short-term teaching jobs until he joined the *Eidgenössisches Amt für geistiges Eigentum* (Swiss

---

[74] The Einsteins had been settled for centuries in the Swabian region between the Danube and the Lake of Constance, especially in the town of Buchau where Hermann Einstein was born in 1847. Hermann married Pauline Koch of Cannstadt (close to Stuttgart) in 1876; she was the daughter of a grain merchant and court purveyor of Stuttgart. In 1877 the Einstein couple moved to Ulm.

[75] Albert Einstein's sister, Maria (Maja), was born in Munich on 18 November 1881. She would later on join her brother in Switzerland and marry one of his friends, Paul Winteler. She also joined Einstein in the United States and died there on 25 June 1951.

Patent Office) in Berne in June 1902.[76] This secure position allowed him to marry Mileva Marič, a former fellow student, and to found a family.[77] It also offered him enough spare time to work on problems of theoretical physics in which he had become interested.

Already at the *Gymnasium*, Einstein had been attracted by certain questions of mathematics and physics. Thus, for example, he had learned Euclidean geometry from a book and the general methods of science from a series of popular books; he had also become acquainted with the principles of differential and integral calculus.[78] 'I had also already studied some theoretical physics when, at the age of 17, I entered the Polytechnic Institute of Zurich as a student of mathematics and physics,' he recalled later (Einstein, 1949, p. 15). At the Zurich Polytechnic, Einstein had not been a particularly good student. As he confessed in his Autobiographical Notes:

There I had excellent teachers (for example, [Adolf] Hurwitz, [Hermann] Minkowski), so that I really could have gotten a sound mathematical education. However, I worked most of the time in the physical laboratory, fascinated by the direct contact with experiment. The balance of the time I used in the main in order to study at home the works of Kirchhoff, Helmholtz, Hertz, etc. (Einstein, 1949, p. 15)[79]

As a result, he obtained most of his education in theoretical physics from books written by the foremost scientists; these included, besides the ones already mentioned, Ludwig Boltzmann, Hendrik Lorentz, James Clerk Maxwell and Henri Poincaré.[80] However, there was one exception. In his last semester Einstein

[76] Einstein obtained the position at the Patent Office on the recommendation of the father of Marcel Grossmann, his friend and former fellow student. His provisory initial appointment as 'Expert III. Class,' made on 23 June 1902, was confirmed in 1904; in 1906 he was promoted to 'Expert II. Class.' Einstein became a Swiss citizen (citizen of Zurich) on 21 February 1901; he had already renounced his Württemberg citizenship in early 1896.

[77] Mileva Marič, born in 1875 near Novy Sad, Serbia, had also studied mathematics and physics at the Zurich Polytechnic (E.T.H.) and had graduated in 1901. The Einsteins had two sons, Hans Albert (1904–1973) and Eduard (1910–1965). Later, when Albert Einstein moved to Berlin in spring 1914, Mileva remained in Zurich with the children. They were divorced in 1919; Einstein then married Elsa Löwenthal, the niece of his mother. For further details of Einstein's life see one of his biographies, such as Carl Seelig (1954), Ronald W. Clark (1971), or Banesh Hoffmann (1972).

[78] In his old age Einstein still remembered the details of his early reading (see Einstein, 1949, p. 14). In some of his reading—e.g., to study Aaron Bernstein's *Naturwissenschaftliche Volksbücher*—Einstein was stimulated by Max Talmey, a young medical student at the University of Munich and a friend of the Einstein family (see Clark, 1972, pp. 33–34).

[79] In his later scientific career Einstein continued to take an interest in direct contact with experimental observations. Among his published papers quite a few dealt with experimental questions, notably the joint papers with Wander Johannes de Haas on the experimental proof of Ampère's molecular currents (Einstein and de Haas, 1915; de Haas, 1916; Einstein, 1916b). Moreover, experimental verification always played an important role in his theoretical work.

[80] The professor of physics at E.T.H., Heinrich Friedrich Weber, did not lecture, for instance, on Maxwell's electrodynamics. Since Weber did not have a particularly high esteem for Einstein, he did not offer him, after the examination, a position as an assistant. (See Kollros, 1956, pp. 21–22.)

attended Hermann Minkowski's lectures on capillarity. 'That,' he is reported to
have remarked, 'is the first lecture [course] on mathematical physics which we
have heard at the Poly[technic]' (Kollross, 1956, p. 21).[81] The lectures stimulated
Einstein to work on capillarity himself and he made it the subject of his first
scientific paper (Einstein, 1901).

By capillarity one denotes a special form of energy connected with the shape
and position of the surface of fluids; thus, for instance, it determines the increase
in the level of a fluid in a thin capillary tube. During the nineteenth century
many scientists had concerned themselves with the problem of capillarity, among
them Thomas Young, Pierre-Simon de Laplace, Carl Friedrich Gauss, James
Clerk Maxwell, Josiah Willard Gibbs, Johannes Diderik van der Waals and
Henri Poincaré.[82] The existence of capillarity had been derived from the hypothesis of cohesive forces exerted by the molecules of a fluid on each other (Laplace,
1806). As a consequence one can obtain, from an experimental determination of
the capillarity of a fluid, information about the intermolecular forces. Exactly
this possibility interested Einstein in his first paper, which he entitled 'Inferences
Drawn from Capillary Phenomena' (Einstein, 1901, p. 513). He assumed, 'guided
by the analogy to the gravitational force' (Einstein, 1901, p. 515), that a universal
function of the intermolecular distance determined the attraction between any
two molecules, and he suggested the possibilities of deriving the parameters
entering into this universal function in the case of given molecules from the data
on fluids. He established, in particular, a relation between the energy necessary
to create a unit area of surface of a fluid and the parameters characterizing the
cohesive forces. By analyzing the data from various fluids, Einstein found fair
agreement between his theory and experiment. He concluded: 'In summing up
we can say that our fundamental assumption has stood the test: To each atom
there corresponds a molecular field of attraction, which is independent of the
temperature and independent of the type of chemical binding between atoms'
(Einstein, 1901, p. 523).[83]

The study of intermolecular forces continued to occupy Einstein. 'I am now
almost sure,' he wrote in April 1901 to Marcel Grossman, his friend from E.T.H.,
'that my theory of the power of attraction can be extended to gases and that the

---

[81] It is of interest to note that Einstein did not learn much mathematics from Minkowski. Instead,
he enjoyed Karl Friedrich Geiser's lectures on infinitesimal geometry (see Einstein, 1956, p. 11).

   Otherwise, during my student years I was not much interested in higher mathematics [he
   recalled in his *Autobiographische Skizze*, and added:] Mistakenly it seemed to me that it was a
   field with so many branches that one could easily squander one's entire energy in one of its
   remote parts. Furthermore, I thought in my innocence that it sufficed for a physicist to
   understand elementary mathematical concepts clearly in order to be able to apply them, and
   that the remainder consisted of subtleties that were fruitless for a physicist. (Einstein, 1956, p.
   11)

[82] See Hermann Minkowski's review article on capillarity in the *Encyklopädie der mathematischen
Wissenschaften* (Minkowski, 1907).

[83] It should be noted that the cohesive forces were assumed to be different from gravitational
forces, having identical structure for all molecules; that is, they were taken to be forces depending
only on (the same) inverse power of the distance between molecules with different coefficients for
different types of molecules.

characteristic constants for nearly all elements could be specified without undue difficulty.' And he added:

> It is a magnificent feeling to recognize the unity of a complex of phenomena which appear to be things quite apart from the direct visible truth. (Einstein to Grossmann, April 1901; quoted in Seelig, 1956, p. 53)

His next paper, however, did not deal with the intermolecular forces in gases, but with forces existing at the boundary between solid and fluid matter. Einstein submitted it in April 1902, again to *Annalen der Physik*; he sought to explain in it the difference of the electric potential between metals and dissociated solutions of their salts (Einstein, 1902a). He calculated the potential difference by using the same assumption concerning the nature of the intermolecular forces as in his previous paper. Then he suggested a procedure for determining the parameters of the cohesive forces from observations of the potential difference. But the procedure seemed to offer little promise for a practical application, and Einstein concluded:

> Finally I wish to apologize for the fact that I have sketched here only a poor plan for a laborious investigation, without contributing personally to the experimental solution; however, I am not in a position to do so. Still this paper will achieve its goal, if it stimulates a scientist to attack the problem of molecular forces from the aspect indicated above. (Einstein, 1902a, p. 814)

Einstein recognized that the theories he had attempted to construct so far did not bring him closer to his 'major aim to find facts which would guarantee as much as possible the existence of atoms of finite size' (Einstein, 1949, p. 47). He, therefore, extended his theoretical tools: thus far he had based his approach to the problems of physics only on mechanics and thermodynamics, now he would also apply the methods of the kinetic theory of matter. Within less than two months after the paper on the thermodynamic theory of potential difference (Einstein, 1902a), he submitted a new paper to *Annalen der Physik* dealing with the kinetic theory of heat equilibrium and the second law of thermodynamics (Einstein, 1902b). Two further papers, in which he developed the kinetic theory of heat, followed in the next two years (Einstein, 1903, 1904). They not only closed the gaps, which according to Einstein existed in the previous theories of Maxwell and Boltzmann, but also prepared the ground for the proof of 'the existence of atoms of finite size' (see Mehra, 1975a).[84]

---

[84] Einstein mentioned the goal of his investigation in the introduction of his first paper on statistical thermodynamics in the following words:

> However great may have been the successes of the kinetic theory of heat in the field of gas theory, until now mechanics has not been able to provide an adequate foundation for the general theory of heat. The reason is that it has not yet been possible to derive the theorems concerning thermal equilibrium and the second law by using the mechanical equations and the probability calculus alone, though the theories of Maxwell and Boltzmann have come quite close to this goal. The aim of the following considerations is to fill this gap. (Einstein, 1902b, p. 417)

Statistical considerations had been introduced into the kinetic theory of gases by James Clerk Maxwell (1860) and Ludwig Boltzmann (1868, 1872, 1877). The main reason for their introduction was the recognition of the fact that purely dynamical arguments did not suffice for establishing the thermodynamic behaviour of gases. Boltzmann and Gibbs then wrote the first monographs on the new kinetic theory of matter (Boltzmann, 1895c, 1898; Gibbs, 1902), which Gibbs called 'statistical mechanics.' Thus the theory was incorporated into classical theoretical physics as a standard subject at the beginning of the twentieth century. Still it caused many controversial discussions and its value was doubted by many scientists, for it lacked a physical basis; in particular, there existed no definite proof of the molecular structure of matter, although a large variety of phenomena could be explained consistently on the basis of kinetic theory. The fundamental difficulty lay in the fact that the number of molecules in any given volume of matter appeared to be so inconceivably large that it appeared hopeless to find any direct evidence for the existence of molecules. Gradually the methods for determining that number began to be clarified; thus, for example, the number of molecules per gram-atom derived from the cross sections of gas molecules, on the one hand, and from Planck's radiation theory, on the other, agreed well, as both methods yielded the value of about 6×10^{23}. However, all methods so far involved intermediate steps or additional assumptions, hence they could not provide immediate proof of the existence of molecules. It was in the midst of this situation that Einstein started his work on kinetic theory. A peculiar point concerning Einstein's endeavours was the fact that, as he confessed later, he was 'not acquainted with the investigations of Boltzmann and Gibbs, which had appeared earlier and actually exhausted the subject' (Einstein, 1949, p. 47).[85]

In his first paper on statistical thermodynamics Einstein presented what he called a 'kinetic theory' of two fundamental laws of thermodynamics: (i) the law that between any systems which are brought into contact a thermal equilibrium will be established, and (ii) the second law of thermodynamics (Einstein, 1902b). Einstein considered two mechanical systems described by momentum and position variables, and he assumed that on the masses of each system two kinds of forces may act: forces which can be derived from a potential (such as gravitation) and forces which cannot (such as the ones exerted by the fixed walls enclosing the system); he assumed that the latter changed fast in time and were responsible for the transfer of heat to the system. Then he considered N identical systems having energy values between E and $E + \delta E$, and studied their distribution among possible states, that is, among the regions of the multidimensional phase space. By applying Liouville's theorem Einstein obtained

---

[85] Since, already in his first paper on the subject, Einstein quoted only Boltzmann's lectures on gas theory (Einstein, 1902b, p. 420, footnote 1), it may well be that he did not know the details of the foundations of kinetic theory which Boltzmann had discussed in his original papers. On the other hand, Gibbs' book on statistical mechanics, which appeared in 1902, became available to Einstein only later in the German translation of 1905. (The young Einstein read publications in German and French, but not very well in English.)

the equation

$$dN = A \int_g dp_1 \cdots dq_n, \tag{39}$$

for the number dN of systems whose variables, $p_1, \ldots, q_n$ (where the p's and q's represent momentum and position variables of the system, respectively), lie in that infinitesimal region of phase space g which is consistent with the allowed energy values. (A denotes a quantity independent of the p's and q's.) By regarding each of the N systems as being in a (stationary) interaction with a large system—the so-called heat reservoir—Einstein derived from Eq. (39) an expression for the probability dW of a state (which is characterized by the fact that the $2n$ dynamical variables assume values between p_1 and $p_1 + dp_1$, p_2 and $p_2 + dp_2, \ldots, p_n$ and $p_n + dp_n$, q_1 and $q_1 + dq_n$, q_2 and $q + dq_2, \ldots, q_n$ and $q_n + dq_n$, respectively), i.e.,

$$dW = A \exp(-2hE) dp_1 \cdots dq_n, \tag{40}$$

where A denotes a constant. The factor h in the exponent represented a positive quantity depending only on the state of the heat reservoir, hence Einstein called it the 'temperature function' (Einstein, 1902b, p. 425).[86] He also showed that two systems, each having the temperature function h, may be joined into one system having the same h; that is, if two isolated systems possess the same temperature, the combined system will also do so. Further, he established between h and the absolute temperature T the equation

$$\frac{1}{4h} = \kappa T, \tag{41}$$

where κ denotes a universal constant. Having prepared this ground, Einstein approached the second task, namely, to establish a mechanical definition of the entropy of a system at temperature T. He found, in particular, that the quotient dQ/T, where dQ denotes the heat transferred to a system at temperature T, represented a total differential whose integration yielded the expression

$$S = \frac{E}{T} + 2\kappa \ln \int \exp(-2hE) dp_1 \cdots dq_n \tag{42}$$

for the entropy of a system with energy E. (The phase-space integration of the right-hand side of Eq. (42) extends over all possible values of the dynamical variables.)

Einstein concluded his paper by drawing attention to two specific points. First, in the derivation of the expression for dQ/T he did not have to require

---

[86] Evidently, the value of the constant A entering Eqs. (39) and (40) can be found by integrating dW over all states consistent with an energy between E and $E + \delta E$. The integration must yield the result $A \cdot \int \exp(-2hE) dp_1 \cdots dq_n = 1$.

that the forces, which he had introduced in order to carry out the adiabatic changes of the systems, really exist in nature; thus, for example, one was allowed to use the so-called semipermeable membranes separating two types of gas molecules without being concerned about their experimental verification. This fact confirmed a generalization of the second law of thermodynamics, which Einstein had already noted earlier: 'One remains in agreement with experience if one applies the second law of thermodynamics to physical mixtures, on the individual components of which conservative forces are acting' (Einstein, 1902a, p. 799).[87] The second point consisted in the observation that in the definition of the entropy, Eq. (42), only the total energy E of the system entered and not its separation into potential and kinetic parts. 'This fact leads us to suspect,' argued Einstein, 'that our results are much more general than the mechanical representation employed here' (Einstein, 1902b, p. 433). He returned to this point in detail in his next paper on the subject, where he answered in the affirmative the question 'whether the kinetic theory is really necessary for the derivation of the fundamental laws of thermodynamics [i.e., the law of temperature equilibrium and the second law], or whether perhaps assumptions of a general type may be sufficient for this purpose' (Einstein, 1903, p. 170).

  Einstein gave the new work the title '*Eine Theorie der Grundlagen der Thermodynamik*' ('A Theory of the Foundations of Thermodynamics') in order to emphasize the independence of his procedure from mechanical models. The systems he now considered were most general ones; the only restriction was that they could be described by n dynamical variables (including both position and momentum variables) whose equations of motion led to just one constant of motion, the energy E. He assumed further that each system should, after a sufficient period of time, be in a stationary state which is defined by the fact that the dynamical variables, say $p_1, \ldots, p_n$, always assume the same values with the same frequency. In other words, in a stationary state the variables of the system do change; however, they stay in a given volume of the phase space for a given, constant fraction of time. This concept of a stationary state yielded a new expression for the probability of a system to assume a particular state, say the one in which the dynamical variables have the values between p_1 and $p_1 + dp_1, \ldots$, and p_n and $p_n + dp_n$, respectively. Einstein found

$$\ln W = \text{const.} - \int \epsilon \ln \epsilon \, dp_1 \cdots dp_n, \tag{43}$$

where the quantity ϵ (which may depend on the variables, $p_1, \ldots, p_n$, and on time t) represented the distribution of the systems—i.e., $dN = \epsilon \, dp_1 \cdots dp_n$. By assuming that the evolution of the N systems proceeds in such a way that the probability W, Eq. (43), increases in the course of time, Einstein proved the second law of thermodynamics (Einstein, 1903, Sections 8 and 9).

---

[87]The extended version of the second law of thermodynamics allowed Einstein to replace real processes by idealized ones involving auxiliary forces; the results thus obtained remained valid in the real cases provided they did not depend on the auxiliary forces.

Einstein completed his foundation of the molecular theory of heat in a third paper, submitted to *Annalen der Physik* at the end of March 1904 (Einstein, 1904). It served two purposes. One was to show that the definition of entropy, Eq. (42), agreed with the ones given by previous authors, notably Boltzmann and Planck; the other to provide a new interpretation of the universal constant κ entering into the relation between the temperature function h and the absolute temperature T, Eq. (41). The first goal was easily achieved by rewriting the right-hand side of Eq. (42) in the form

$$S = 2\kappa \ln\left[\omega(E)\right]. \tag{44}$$

The energy function $\omega(E)$ was defined with the help of the equation

$$\omega(E)\,\delta E = \int_E^{E+\delta E} dp_1 \cdots dp_n, \tag{44a}$$

whose right-hand side denotes the volume of phase space assumed by the variables of the systems under consideration, which have energies between E and $E + \delta E$. The definition of entropy, Eq. (44), was now identical with what Planck had called 'Gibbs' third entropy definition' in his article for the *Boltzmann Festschrift* (Planck, 1904, p. 119).[88] One possibility of obtaining a physical interpretation of the universal constant κ could be achieved very easily by introducing systems consisting of mass-particles; then the average kinetic energy of each particle was computed to be $3\kappa T$, and the quantity 2κ had to be taken equal to the ratio R/N_0, where R is the universal gas constant and N_0 the number of molecules per gram-atom; in other words, the relation

$$2\kappa = \frac{R}{N_0} = k \tag{45}$$

existed between Einstein's κ and Planck's universal constant k (called the Boltzmann constant). However, Einstein did not content himself with this inter-

---

[88] Gibbs' expression, apart from a multiplicative factor (which was Boltzmann's constant), was $d\ln V/dE$, representing the derivative of the phase-space volume with respect to the energy E of the systems. Evidently, $\omega(E)$ could be identified with $d\ln V/dE$.

The *Festschrift Ludwig Boltzmann gewidmet zum sechzigsten Geburtstag, 20. Februar 1904* was published as a supplementary volume of *Annalen der Physik* early in 1904. In contained contributions by many authors: Max Planck's paper on the mechanical significance of temperature and entropy (Planck, 1904), Max Abraham's paper on radiation pressure and blackbody radiation (Abraham, 1904), Marian von Smoluchowski's paper on the influence of the irregularities of the distribution of molecules on entropy (Smoluchowski, 1904), and Walther Nernst's paper on chemical equilibrium and temperature gradient (Nernst, 1904). These and other articles did interest Einstein, and it is likely that he obtained valuable information from the *Festschrift*. Indeed, he reviewed several articles of the *Festschrift* for the *Beiblätter zu den Annalen der Physik*; his reviews appeared in its issue No. 5 of 1 March 1905 (see, e.g., Einstein, 1905a). During 1905 Einstein wrote further reviews of papers on thermodynamics and statistical mechanics for the *Beiblätter*. However, he stopped writing these reviews in the following year, when he contributed only the review of Planck's book on heat radiation (Einstein, 1906e).

pretation, but continued to search for a 'general significance' ('*allgemeine Bedeutung,*' Einstein, 1904, p. 359) of κ, that is, for an interpretation which did not depend on the representation of the system as a mechanical system of gas molecules. He, therefore, started from the equation for dW, the probability that a stationary system of temperature T has an energy value between E and $E + dE$, i.e.,

$$dW = C \exp\left(-\frac{E}{2\kappa T}\right) \cdot \omega(E) dE, \tag{46}$$

where the constant C may be determined by integrating over all energies ($\int_0^\infty C \exp(-E/2\kappa T) \cdot \omega(E) dE = 1$). With the help of dW he calculated the average energy $\bar{E}$ ($= \int_0^\infty CE \exp(-E/2\kappa T) \cdot \omega(E) dE$). Evidently, the equation

$$\int_0^\infty (\bar{E} - E) \exp\left(-\frac{E}{2\kappa T}\right) \cdot \omega(E) dE = 0 \tag{47}$$

was also satisfied. From it, by differentiating with respect to the temperature T, Einstein obtained the result

$$2\kappa T^2 \frac{d\bar{E}}{dT} = \overline{E^2} - \left(\bar{E}\right)^2 = \overline{\epsilon^2}. \tag{48}$$

He commented on Eq. (48) as follows:

> In general the actual value E of the energy deviates from [the average value] E by a certain quantity, which we call "energy fluctuation"; we put $E = \bar{E} + \epsilon \cdots$. The quantity $\overline{\epsilon^2}$ is then a measure of the thermal stability of the system; the larger $\overline{\epsilon^2}$ is, the smaller is its stability. The absolute constant κ therefore determines the thermal stability of the systems. The last relation [Eq. (48)] is interesting, because it contains none of the quantities which recall the assumptions on which the [kinetic] theory is based. (Einstein, 1904, p. 360)

Einstein immediately sought to confirm his new interpretation of the constant κ. 'The equation just derived [Eq. (48)] would allow an exact determination of the universal constant κ,' Einstein noted, 'if it were possible to determine the average value of the square of the energy fluctuation of a given system; this determination, however, cannot be performed with the present state of our knowledge' (Einstein, 1904, p. 360). The difficulty was, of course, that in the available molecular systems, say gases, the energy fluctuations were extremely small because of the large numbers of molecules present in any volume; hence they could not be observed.[89] On the other hand, Eq. (48) was applicable to any

---

[89]Marian von Smoluchowski, in his article for the *Boltzmann Festschrift*, studied another kind of fluctuations in gases: the fluctuations of density. He found that in the smallest microscopically observable volumes one would obtain deviations from the average density of one *per mille* (Smoluchowski, 1904, p. 628).

system which exhibited thermodynamic behavior, especially to heat radiation. And Einstein saw the opportunity of checking the applicability of Eq. (48) to this system. 'From our experience we may surmise that only in one type of physical systems—if at all—an [observable] energy fluctuation will exist,' he argued. 'This is the empty space filled with heat radiation' (Einstein, 1904, p. 361). He knew, of course, that the linear dimension of any given volume was very large in comparison with the wavelengths of the observed heat radiation; for example, the wavelength emitted by a body having a temperature equal to that of the sun's surface (i.e., 6000°K) is of the order of a fraction of one-millionth of a metre. As a consequence, the amount of the energy fluctuation would be small in comparison with the average energy of radiation. 'However, if the space volume containing the radiation has the linear dimensions of a wavelength,' Einstein pointed out, 'then the energy fluctuation will have the same order of magnitude as the radiation energy contained in the space volume of radiation' (Einstein, 1904, p. 361). He then proposed to test this conclusion by the following consideration. In any volume v containing blackbody radiation of temperature T, the average energy $\bar{E}$ is, according to Stefan–Boltzmann's law, Eq. (3), proportional to T^4 times the volume. The particular volume, v, in which the fluctuation term ϵ^2 is equal to the square of the average energy, $(\bar{E})^2$, may then be obtained with the help of Eq. (48). By taking the third root of this volume, Einstein arrived at a length (in millimetres) determined by the equation

$$\sqrt[3]{v} = \frac{2\sqrt[3]{\kappa/\text{const.}}}{T} = \frac{0.42}{T} , \qquad (49)$$

where 'const.' denoted the constant in Stefan–Boltzmann's law, which Einstein wrote as $\bar{E} = \text{const.} \cdot v \cdot T^4$. Now, from Wien's displacement law, Eq. (4), the wavelength of the radiation maximum at temperature T followed from the equation

$$\lambda_{\text{max}} = \frac{0.293}{T} . \qquad (50)$$

One recognizes [Einstein concluded] that not only the type of temperature dependence but also the magnitude of λ_{max} can be determined correctly with the help of the general molecular [kinetic] theory of heat; and I believe that this agreement [between the theoretical results derived from radiation theory and kinetic theory, respectively] cannot—in view of the great generality of our assumptions—be attributed to mere chance. (Einstein, 1904, p. 362)

Einstein was thrilled by the result contained in Eq. (50) and its interpretation on the basis of kinetic theory. 'I have discovered the simplest possible way of the relationship between the size of the elementary units of matter and the wavelength of radiation,' he wrote to his friend Conrad Habicht about two weeks after sending off the paper to Annalen der Physik (Einstein to Habicht, April 1904;

quoted in Seelig, 1956, p. 62). Still he did not receive much recognition for his papers on the foundation of statistical thermodynamics, because most of what he achieved had been done already by previous authors. As Max Born remarked later:

> In fact, Einstein's paper [Einstein, 1902b] is a rediscovery of all essential features of statistical mechanics and obviously written in total ignorance of the fact that the whole matter had been thoroughly treated by Gibbs a year before (1901) . . . . Einstein's method is essentially identical with Gibbs' theory of canonical assemblies. In a second paper, of the following year, entitled "*Eine Theorie der Grundlagen der Thermodynamik*" [Einstein, 1903], Einstein builds the theory of another basis not used by Gibbs, namely the consideration of a single system in course of time (later called "*Zeit-Gesamtheit*", time assembly), and proves that this is equivalent to a certain virtual assembly of many systems, Gibbs' micro-canonical assembly. Finally, he shows that the canonical and micro-canonical distributions lead to the same physical consequences. (Born, 1949, pp. 164–165)

Actually, the time ensemble had also not been invented by Einstein; it rather goes back to a paper of Ludwig Boltzmann (1871). However, Einstein presented matters in such a way that he must be considered, together with Gibbs, as the founder of the ensemble theory in statistical mechanics.[90] Though he did not obtain much recognition for his pioneering work, his efforts were not in vain. After all, in establishing the foundation of statistical thermodynamics he had had in mind only one goal—the proof of the molecular hypothesis—and he would achieve this goal very soon.[91]

Some time in the first half of the year 1905 Einstein wrote again a letter to Conrad Habicht, in which he announced that he would soon send him copies of four different scientific papers: the first dealt with radiation (Einstein, 1905b); the second with methods to determine the real dimensions of atoms (Einstein, 1905f); the third with the irregular motion of particles suspended in fluids (Einstein, 1905c); and the fourth with the electrodynamics of moving bodies

---

[90] Einstein's papers of 1902 and 1903 were hardly noticed. However, Boltzmann and J. Nabl cited them in their review of kinetic theory for the *Encyklopädie der mathematischen Wissenschaften* as demonstrating the fact that very general dynamical systems may behave in a thermodynamical manner (Boltzmann and Nabl, 1907, p. 549, footnote 145).

[91] As we have mentioned in footnote 84, in Einstein's mind Maxwell and Boltzmann had not completed this task. Also Gibbs, whose work he came to know only later, had not done so. In a review of Hendrik A. Lorentz' book, '*Les théories statistiques en thermodynamique*' (Lorentz, 1916), Einstein criticized Gibbs' approach in the following words:

> Everyone, who has studied mathematical theories, is familiar with the following painful experience: he verifies eagerly and laboriously each step of the deduction and at the end of his endeavours understands—nothing; he misses the guiding line of thought, which the author [of the text one reads] frequently suppresses, either because he cannot express it neatly or because he follows a kind of coquetry that used to be fashionable in earlier times, which appears to be quite ridiculous to the discerning [reader]. Against this evil only the unrestrained openness of the author can help; he should not be afraid of sharing with the reader even incomplete guiding ideas if they have assisted him in advancing his work. In theoretical physics there is hardly a field in which this requirement is more difficult to satisfy than statistical mechanics. Every expert will agree with me in the opinion that Gibbs has sinned heavily against this requirement in his pioneering book on the subject; many have read it, gone through its derivations and *not* understood it. (Einstein, 1916f, pp. 480–481)

(Einstein, 1905d).[92] In two of these papers Einstein succeeded in presenting his proof of the existence of molecules of a finite size. Especially, in the work entitled 'A New Determination of Molecular Dimensions' ('*Eine neue Bestimmung der Moleküldimensionen*,' Einstein, 1905f, 1906a), which he submitted as a doctoral thesis to the University of Zurich, he demonstrated that 'one can estimate the size of the molecules of the substance dissolved in a non-dissociated, dilute solution from the internal friction of the solution and the pure solvent and from the diffusion of the dissolved substance in the solvent, provided the volume of the molecule of the dissolved substance is large compared to the volume of a molecule of the solvent' (Einstein, 1906a, p. 289).[93] The other work dealing with the existence and size of molecules made use of the results of the foundations of thermodynamics (Einstein, 1905c). Einstein claimed that 'according to the kinetic theory of heat, particles which are suspended in fluids and can be seen under the microscope will have to perform—because of molecular heat motion— movements of such a magnitude, that they (the movements) may be easily observed with the help of a microscope' (Einstein, 1905c, p. 549). Einstein's procedure for arriving at this result was straightforward. From Eq. (42), which expressed the entropy of a mechanical system, he derived an equation for the 'osmotic pressure' of particles suspended in a liquid; this pressure was, of course, a consequence of the kinetic theory and it caused the particles to diffuse in the liquid.[94] Due to the irregularity of the heat motion the diffusion of the particles would be very irregular. Einstein assumed all suspended particles to move independently of each other; thus he found that their propagation, say in the x-direction, was represented by a Gaussian distribution with the characteristic exponent, $(-x^2/4Dt)$, where x was the distance the particle traversed during the time interval t and D denoted the diffusion coefficient. Hence the average displacement of the particle in the x-direction was given by the quantity, $\overline{x^2}$, or by

$$\lambda_x = \sqrt{2Dt} \, , \tag{51}$$

[92] Conrad Habicht, a former fellow student of Einstein's at the E.T.H. in Zurich, continued to pursue his mathematical studies at the University of Berne. He maintained close contact with Einstein; especially, he joined Einstein and Maurice Solovine, a young Rumanian student of science and literature, in reading and discussing important books on philosophy and science. (The three of them referred to their meetings jokingly as the sessions of the 'Olympia Academy.') In spring 1905 Habicht left Berne to accept a position as a high school teacher. Einstein's letter to Habicht was not dated, but must have been written after his departure and before the publication of Einstein's paper on the light-quanta in June 1905 (Einstein, 1905b).

[93] Einstein completed his thesis on 30 April 1905 (Einstein, 1905f). He submitted the paper based on it to *Annalen der Physik* several months later (Einstein, 1906a; it was received on 19 August 1905). Einstein was awarded his doctorate on 15 January 1906. The publication of the paper in the *Annalen* was delayed; it appeared in February 1906 with an addendum by the author signed January 1906.

[94] Originally, osmotic pressure was defined as a property of salts dissolved in a solution of water. The molecules of the dissolved substance act in the liquid as do the molecules in a dilute gas; in particular, they exert a pressure depending on their number in a given volume and the temperature of the solution. The kinetic theory of matter now demanded the existence of osmotic pressure, even if the dissolved particles had larger size and became observable under the microscope.

where

$$D = \frac{RT}{N_0} (8\pi k P)^{-1}.$$

(51a)

In Eq. (51a), R and N_0 denote the gas constant and the number of molecules per gram-atom, respectively, k the coefficient of friction in the liquid and P the radius of the dissolved or suspended particles. Clearly, Eq. (51) provided a method of determining the number N_0 from λ_x and the knowledge of the radius P of the suspended particles, or determining the radius of the dissolved molecules from λ_x and N_0, provided P is much larger than the molecules of the solvent.[95] On the basis of Einstein's theoretical work, Jean Perrin and his collaborators would carry out within a few years systematic experimental investigations of the molecular dimensions and the determination of the number N_0, finding for the latter values between 6×10^{23} and 7×10^{23}. (See Perrin in Langevin and de Broglie, 1912.)

The first paper which Einstein, in his letter to Habicht mentioned above, promised to send very soon, bore the title 'On a Heuristic Point of View about the Creation and Conversion of Light' ('*Über einen die Erzeugung und Verwandlung des Lichtes betreffenden heuristischen Gesichtspunkt*,' Einstein, 1905b). He had signed the paper in Berne on 17 March 1905 and sent it immediately to *Annalen der Physik*; it was received the following day and published in the first issue of Volume 17, which was distributed on 9 June 1905. 'It is on radiation and energy of light,' he described its content, 'and it is very revolutionary, as you will see yourself' ('*sie ist sehr revolutionär*,' Einstein to Habricht, 1905; quoted in Seelig, 1956, p. 74). What was the result that Einstein had obtained and why did he call it revolutionary? To understand this point we shall summarize his previous concern with the problem of radiation, especially heat radiation.

Einstein had become interested in the problem of heat radiation quite early; thus he had read Planck's work 'shortly after [its] appearance' (Einstein, 1949, p. 47). The 'major question' that concerned him was: 'What general conclusions can

---

[95] When Einstein postulated the irregular motion of suspended particles in liquids based on the molecular theory of heat, he was not certain whether it had already been observed or not. 'It is possible,' he wrote, 'that the motions treated here are identical with the so-called "Brownian molecular motion"; the information available to me about the latter is, however, too imprecise, hence I could not come to a proper judgment about it' (Einstein, 1905c, p. 549). Several months later he knew that he had indeed given a theory of the phenomenon which had been reported first by the Scottish botanist Robert Brown in 1827 (Brown, 1828). Brown had observed the pollens of a plant suspended in water under a microscope and found that they executed an irregular motion. Later on it was discovered that all small particles suspended in liquids or gases performed such motions and that the activity increased with decreasing size and growing temperature. The physicist Heinrich Friedrich Wilhelm Siedentopf of Jena, who had developed together with the chemist Richard Adolf Zsigmondy the ultramicroscope (an instrument to study extremely small objects), informed Einstein about the identity of his postulated motion of suspended particles and Brownian motion. In a second paper, entitled '*Zur Theorie der Brownschen Bewegung*' ('On the Theory of Brownian Motion') Einstein generalized his previous treatment to include also a rotational type of motion of the suspended particles (Einstein, 1906b).

be drawn from the radiation formula concerning the structure of radiation and even more concerning the electromagnetic foundation of physics?' (Einstein, 1949, p. 47). In connection with his work on the foundation of thermodynamics he had, of course, been most interested in Planck's application of Boltzmann's statistical definition of entropy to the radiation problem without actually being able to make real sense of it. 'It was,' he recalled later, 'as if the ground had been pulled out from under one, with no firm foundation to be seen anywhere, upon which one could have built' (Einstein, 1949, p. 45).

In his endeavour to secure a new understanding he did not receive much help from the available scientific literature; in the years following Planck's publication no articles throwing new light on the problem had appeared, for instance, in *Annalen der Physik*, until 1904. Then, in the *Boltzmann Festschrift* of February 1904, two authors dealt with heat radiation: Heinrich Kayser from Bonn wrote on the determination of temperature of radiating gases (Kayser, 1904), and Max Abraham from Göttingen discussed certain consequences of radiation pressure (Abraham, 1904). Abraham had, in particular, derived the displacement law in the special form given to it by Max Planck (i.e., $\rho_\nu = (\nu^3/c^3)f(T/\nu)$) and then, by integrating over all frequencies, the law of Stefan and Boltzmann.[96] Shortly after the publication of the *Boltzmann Festschrift*, Einstein, in his third paper on the molecular theory of heat, for the first time treated a problem connected with heat radiation, which we have mentioned above. The interesting point is that Einstein, like Abraham, dealt with Stefan–Boltzmann's law; but while Abraham had applied only electrodynamic and thermodynamic arguments, Einstein had introduced statistical arguments from kinetic theory.[97] The question then arose whether one could also obtain Planck's law for blackbody radiation from the same foundation. Of course, Einstein knew that Planck had derived his equation from a probability consideration similar to the one used by Boltzmann in kinetic theory; especially, he had given an expression for the distribution of energy among the resonators in a cavity filled with radiation. However, by applying his own statistical methods to the same problem Einstein succeeded in deriving not Planck's distribution but rather a different one. Indeed, when treating a cavity containing molecules, electrons and resonators (which he represented by electrons fixed at a given space point, and oscillating in one direction), he found the average kinetic energy of the resonator of frequency ν as

$$\overline{E}_\nu = U_\nu = \frac{R}{N_0} T, \tag{52}$$

which deviated strongly from Planck's expression, Eq. (25). That is, according to

---

[96] The distribution law, $\rho_\nu = (\nu^3/c^3)f(T/\nu)$, with $f(T/\nu)$ a function depending only on the quotient (T/ν) may be obtained immediately from Max Thiesen's form of Wien's displacement law, Eq. (31).

[97] Max Abraham had tried to extend Willy Wien's derivation of the displacement law by removing a restriction; Wien, in his adiabatic processes, had discussed only the reflection of light from slowly moving walls (see Section I.1).

kinetic theory a one-dimensional oscillator at temperature T had to possess an average kinetic energy of $\frac{1}{2}(R/N_0)T$, and the same amount of potential energy. With Eq. (52), the energy distribution in blackbody radiation would be described by the equation

$$\rho_\nu = \frac{8\pi\nu^2}{c^3} \frac{R}{N_0} T, \tag{53}$$

rather than by Planck's law. Hence there existed, as Einstein remarked explicitly, 'a difficulty pertaining to the theory of black radiation' ('*eine die Theorie der "schwarzen Strahlung" betreffende Schwierigkeit*' (Einstein, 1905b, p. 133), a difficulty which he wanted to analyze carefully and to overcome.

To begin with, Einstein recognized easily that Planck's law for the distribution of energy among frequencies in blackbody radiation, that is,

$$\rho_\nu = \frac{8\pi h\nu^3}{c^3} \left[\exp\left(\frac{h\nu}{kT} \right) - 1 \right]^{-1}, \tag{25a}$$

passed over into Eq. (53) for large values of the quotient, T/ν, i.e., for large wavelengths and large radiation densities. 'We thus arrive at the conclusion,' he noted, 'that the higher the energy density and the longer the wavelength of radiation are, the more applicable are the theoretical foundations employed by us; however, for short wavelengths and low radiation densities, these foundations fail completely' (Einstein, 1905b, p. 137). What were the theoretical foundations that had failed? Einstein was convinced that, in obtaining Eq. (53), he had applied mechanics and the thermodynamics correctly (like the expert whose duty it was to check thoroughly the physical basis of the inventions submitted to the Patent Office); he was also convinced that his statistical thermodynamics, which he had developed over the previous years, was the only one consistent with the theoretical foundations. Thus the question arose as to what had to be changed in order to arrive at Planck's law, a law described by two universal constants—the Boltzmann constant k and Planck's constant h, the latter determining the size of the finite energy elements. The constant k had already appeared in Einstein's theory of statistical thermodynamics as the constant R/N_0.[98] But there was definitely no place at all for Planck's constant h. Even more, it seemed that the very introduction of a finite value for h—and consequently of finite elements in the statistical considerations—somehow contradicted the accepted principles of kinetic theory. Therefore, the question had to be answered whether or not one

[98] Einstein proved the equality of R/N_0 with k explicitly by taking the limit of large T/ν in Planck's formula and comparing the result to Eq. (53). Then he calculated N_0^{-1}, identifying it with the mass of the hydrogen atom, whose value he found to agree completely with Planck's result of December 1900 (Planck, 1900f, p. 244).

could give h a different physical meaning without violating the foundations of the theory.[99]

Einstein knew that, to establish an interpretation of the constant h, he had to analyze the energy distribution of the heat radiation in the region where h did not drop out of Eq. (25a), but rather determined the structure of the radiation law: that is, for large values of the factor $h\nu/kT$ in the exponential. This, however, was the region where Wien's law described the observations accurately. Now Willy Wien had derived his law under the special assumptions that the radiating parts in a blackbody obeyed the Maxwellian velocity distribution and that the frequency of the emitted radiation depended only on the velocity of the emitter (Wien, 1896). Later, in 1900, when the experiments showed strong deviations from his law, Wien had argued that this had to be expected, for the statistical assumption used in 1896 was correct only for large values of the quotient ν/T and did not apply to radiation of long wavelengths (Wien, 1900b). Wien had stated, at the same time, that the heat radiation satisfying his law could not be explained on the basis of the electrodynamic theory; rather, it exhibited a structure which was different in quality from that of the long-wavelength radiation, the radiation which obeyed the laws of classical electrodynamics. Einstein, in his analysis of the foundations of Planck's radiation formula in 1905, picked up Wien's idea of 1900. That is, he assumed that the heat radiation consisted of two parts: the long-wavelength radiation, described by the known laws of electrodynamics, and short-wavelength radiation, determined by laws still unknown. He wanted to discover the nature of these unknown laws.

The type of radiation that interested Einstein was described formally by Wien's radiation law, or Eq. (11), which he wrote as

$$\rho_\nu = \alpha\nu^3 \exp\left(-\frac{\beta\nu}{T}\right), \tag{11'}$$

with two constants α and β. (In Planck's notation the constants were $\alpha = 8\pi h/c^3$ and $\beta = h/k$, respectively.) In order to analyze the physical content of Eq. (11'), Einstein (following Planck's preferred method) studied the entropy of radiation associated with it. In particular, he considered the entropy S of radiation in a volume v, having the frequency ν and energy E_ν (where $E_\nu = \rho_\nu v\,d\nu$ is the energy in the frequency interval between ν and $\nu + d\nu$), that is,

$$S = -\frac{E_\nu}{\beta\nu}\left[\ln\left(\frac{\rho_\nu}{\alpha\nu^3}\right) - 1\right], \tag{54}$$

[99] Einstein, in his review of Planck's *Lectures on Heat Radiation*, pointed out the fact that the author had not yet given a proper interpretation of the constant h by saying: 'The author points repeatedly to the necessity of introducing this universal constant h and emphasizes how important its physical significance (which is not treated in this book) is' (Einstein, 1906f, p. 766).

which he obtained from the definition of the entropy differential and the fact that the entropy of blackbody radiation takes on a maximum value.[100] Planck had derived a similar expression for the entropy of radiation (see Eqs. (7) and (8a)), but Einstein went beyond Planck in deriving physical consequences from the volume dependence of S. If v_0 and v denote the initial and final volumes, respectively, filled with heat radiation (obeying Wien's law), having the energy E_ν, then the entropy difference, $S - S_0$, was given by the equation

$$S - S_0 = \frac{E_\nu}{\beta\nu} \ln\left(\frac{v}{v_0}\right). \tag{55}$$

Einstein noted:

> This equation shows that the entropy of monochromatic radiation of sufficiently small density [such that it satisfies Wien's radiation law] varies with the volume according to the same rules as the entropy of a perfect gas or of a dilute solution. The equation just obtained [Eq. (55)] shall be interpreted in the following on the basis of the principle introduced by Mr. Boltzmann into physics, according to which the entropy of a system is a function of the probability of its state. (Einstein, 1905b, pp. 139–140)

On first inspection these conclusions appear to be rather surprising, and one has to ask the question how Einstein arrived at them. The answer must be sought in the ideas and motivations guiding him at that time. Einstein had decided, at least since 1904, that one should apply the general molecular theory of heat or statistical mechanics to the problem of heat radiation; in particular, he had shown then that fluctuation considerations yielded reasonable results (Einstein, 1904). Encouraged by his success he investigated further consequences of the equations of statistical mechanics in radiation theory. Since he was most fascinated by the problem of fluctuations—as evidenced by his interest in such phenomena as the Brownian motion—he preferred to deal with equations related to it. Now the dependence of the entropy on the volume was indeed connected with fluctuations: if one interpreted Boltzmann's entropy–probability relation ($S = (R/N_0)\log W$) as an equation for calculating the probability of a state from its entropy, then Eq. (55) allowed one to compare the probabilities of two situations, in which heat radiation (of low density and high frequency) of the

---

[100] Evidently, the total entropy of radiation of all frequencies in a volume v is given by an integral, $v \int \phi(\rho_\nu, \nu) \, d\nu$, with $\phi(\rho_\nu, \nu)$ a function of the two variables ρ_ν and ν, and this integral assumes a maximum value only if the function $\phi(\rho_\nu, \nu)$ is independent of the frequency. Then, for each frequency one has the relation: $\partial\phi/\partial\rho_\nu = T^{-1}$. And one can use Eq. (11') to obtain an expression for the inverse of the temperature, from which one calculates (by integration) the function $\phi = \phi(\rho_\nu, \nu)$ as:

$$\phi(\rho_\nu, \nu) = -\frac{\rho_\nu}{\beta\nu}\left[\ln\left(\frac{\rho_\nu}{\alpha\nu^3}\right) - 1\right].$$

Then the entropy of the radiation in the frequency interval ν, $\nu + d\nu$ will be $v\phi(\rho_\nu, \nu) \, d\nu$.

same energy per frequency interval was concentrated in the volumes v_0 and v, respectively. As a consequence Einstein's above conclusion can be perfectly understood.

Einstein thought along these lines, as is evident from the discussion in his paper (Einstein, 1905b, §§5 and 6). And he had no difficulty in recognizing the analogy of Eq. (55) to the equation expressing the volume dependence of the entropy of ideal gases or dilute solutions. In fact the latter (i.e., the volume dependence of the entropy) was well known at the time Einstein discussed heat radiation. For example, in the *Boltzmann Festschrift*, N. Schiller of Kharkov had presented a formula describing the entropy of gases in the process of diffusion (Schiller, 1904, p. 361, Eq. (20)). In it the proportionality of the entropy on the logarithm of the volume showed up explicitly. Incidentally, Einstein reviewed Schiller's paper in the *Beiblätter zu den Annalen der Physik*, and in this review he quoted Schiller's entropy formula (Einstein, 1905a, p. 238).

After this outline of Einstein's motivation, we may return to the specific presentation of the arguments in his paper. He first derived the volume dependence of the entropy of a gas consisting of molecules according to the kinetic theory. According to Boltzmann, the difference between the initial and final entropy of a system which passes from a given initial state (entropy S_0) to a final state (entropy S) is given by the equation

$$S - S_0 = \frac{R}{N_0} \ln W, \qquad (56)$$

where R and N_0 denote the universal gas constant and the number of molecules per gram-atom, respectively, and W is the relative probability of the final state.[101] Then Einstein considered n molecules (of a given type) in a volume v_0, which move about in arbitrary directions and do not exert any actions upon each other except by elastic collisions (i.e., they constituted an ideal gas); he calculated the entropy difference of this state with another, in which the n molecules were concentrated in the volume v, a fraction of the initial volume v_0. He argued that, according to Boltzmann, the 'statistical probability' for the final state would be proportional to $(v/v_0)^n$; hence from Eq. (56) the entropy difference became[102]

$$S - S_0 = \frac{R}{N_0} \ln\left(\frac{v}{v_0}\right)^n. \qquad (57)$$

Einstein commented on Eq. (57) as follows: 'It should be noted that for the derivation of this equation—from which one can easily derive thermodynamically the Boyle–Gay-Lussac law [i.e., the equation of state for ideal gases] and

[101] Note that Einstein in 1905 consistently used the ratio R/N_0 instead of his earlier universal constant 2κ.

[102] The same *Ansatz* for the statistical probability was used by Marian von Smoluchowski in his paper for the *Boltzmann Festschrift* (Smoluchowski, 1904, p. 626).

identically the same law of osmotic pressure—no other assumption need be made about the law according to which the molecules move' (Einstein, 1905b, p. 142). Now the last step was quickly performed: all Einstein had to do was to write the right-hand side of Eq. (55) in a form which could be compared to the right-hand side of Eq. (57); evidently this was

$$S - S_0 = \frac{R}{N_0} \ln\left(\frac{v}{v_0}\right)^{(N_0/R)(E_\nu/\beta\nu)}. \tag{55a}$$

Einstein interpreted this equation as follows: 'If monochromatic radiation of frequency ν and energy E is enclosed (by reflecting walls) within a volume v_0, the probability that at any arbitrary instant of time the total radiation energy is contained in a part v of volume v_0 will be: $W = (v/v_0)^{(N_0/R)(E_\nu/\beta\nu)}$,' (Einstein, 1905b, p. 143). Evidently the probability for the radiation was now described by an expression having exactly the same structure as that for the probability for n molecules, provided one took the factor $(N_0/R)(E_\nu/\beta\nu)$ in the exponent instead of n. Hence Einstein concluded finally:

Monochromatic radiation of low density (within the range of validity of Wien's radiation formula) behaves, in a thermodynamic sense, as if it consisted of mutually independent radiation quanta of magnitude $R\beta\nu/N_0$. (Einstein, 1905b, p. 143)

To some extent the procedure outlined above seems to be straightforward and immediately evident. As we have pointed out, we may assume with good reason that Einstein had Eq. (57) (for the molecules in a gas) in mind for a long time. All he had to do was to derive the expression for the corresponding entropy of radiation satisfying Wien's law, Eq. (55), say with the help of Planck's papers in *Annalen der Physik* (Planck, 1901a, c), and to realize its formal relationship to Eq. (57). Still one has to ask the question: how did Einstein come to search for a relation between a dilute gas and Wien's radiation? One reason may perhaps be seen in the formal similarity between the energy distribution in Wien's radiation law, Eq. (6), and the energy distribution in a system consisting of independent gas molecules; in fact, Wien had obtained his law by referring explicitly to Maxwell's velocity (and energy) distribution. The physical argument in favour of his approach was, as we have pointed out above, that the molecules emitting radiation in a blackbody were described by Maxwell's distribution and, therefore, this distribution had to be reflected in the emitted radiation. For this purpose, of course, an equilibrium had to be assumed to exist between radiation and molecules. If that was true, Einstein concluded, then kinetic theory implied that heat radiation, on one hand, and the molecules, on the other, must lead to the same macroscopic effects. Especially, both must create an 'osmotic pressure,' say, on a mirror in a cavity. Now the osmotic pressure, p, created by n molecules of a dilute gas (or—in the original sense of the actual definition of the osmotic pressure—by n molecules dissolved in a liquid) in a volume v at temperature T was given by the equation

$$p \, dv = T \, dS = \frac{RT}{N_0} n \frac{dv}{v} .$$ (58a)

On the other hand, the radiation satisfying Wien's law yielded for $p \, dv$ the result

$$p \, dv = \frac{E_\nu}{\beta\nu} T \frac{dv}{v} .$$ (58b)

Evidently, the right-hand sides of Eqs. (58a) and (58b) had the same structure, provided the factors $(RT/N)n$ and $E_\nu/\beta\nu$ were associated with each other. The consequence from this association agreed with the consequence derived from the expressions for probability. Therefore, the comparison of the osmotic pressure of a dilute gas with the pressure of radiation obeying Wien's law provided an immediate proof of the molecular structure of radiation.[103]

The physical interpretation of the short-wavelength radiation obeying Wien's law also implied a physical interpretation of Planck's constant h. By substituting the value of β $(= h/k)$ and observing the equality of the constant k with the ratio R/N_0, one finds for the number of energy quanta n the relation

$$n = \frac{N_0}{R} \frac{E}{\beta\nu} = \frac{E}{h\nu} .$$ (59)

Thus the quantity $(R/N)\beta\nu$ indeed represented the size of energy quanta, in agreement with Planck's fundamental assumption of 1900. Einstein did not, however, write Eq. (59) in his paper, for he wanted to keep his distance from Planck's theory. 'At that time it appeared to me,' he explained a year later, 'that Planck's theory constituted in some respects a counterpart to my own work' (Einstein, 1906c, p. 199). It appeared to Einstein in 1905 that Planck, in formulating his theory of heat radiation, had moved away from the basis of kinetic theory and had employed an extra hypothesis to arrive at Eq. (18); otherwise he would have obtained the radiation law given by Eq. (53). Indeed, it

---

[103] Although Einstein did not give the procedure outlined above in his paper, he recalled later that considerations of the osmotic pressure had played an important role in his interpretation of Wien's radiation law (see Einstein, 1949, pp. 48–52). That osmotic pressure considerations occupied Einstein in 1905 is also confirmed by their appearance in the paper on Browian motion (Einstein, 1905c), which was composed at the same time as the paper on radiation. In any case, the procedure of arriving at the interpretation of Wien's radiation law via the consideration of the osmotic pressures can be obtained from equations given in the published paper (Einstein, 1905b, p. 142, footnote 1). (Note that in the equation for $T \, dS$ a factor T is missing on the right-hand side, i.e., the expression $R(n/N_0)(dv/v)$ must be replaced by $RT(n/N_0)(dv/v)$.)

must have seemed as if the foundations of Einstein's and Planck's radiation
theories were entirely different: Planck, on several occasions, had emphasized the
fact that he had used a unified description of all wavelengths on the basis of
Maxwell's electrodynamics, while Einstein had just arrived at the conclusion that
radiation of short wavelengths possessed a structure that did not fit into the
electrodynamic theory. In any case, Einstein demonstrated his independence
from Planck by not using the latter's notation for the constants; he preferred to
denote Boltzmann's constant k (which Planck had employed) by the ratio R/N_0
of the gas constant and number of molecules per gram-atom and Planck's
constant h by the quantity $(R/N_0)\beta$. Nevertheless he knew that he had given at
least a partial answer to the question concerning the physical interpretation of
Planck's constant h. For the radiation satisfying Wien's law he had proved the
equality of h and $(R/N_0)\beta$, and the product of $(R/N_0)\beta$ and the frequency
denoted the size of the independent energy-quanta of radiation.

Einstein did not stop at this point but proceeded, as had become his custom in
scientific work, to check certain consequences of the assumption of independent
energy-quanta in radiation theory. To begin with, he expected that energy-
quanta of radiation did not play a role in the phenomena of diffraction and
propagation of light *in vacuo*; however, whenever radiation was created or
transformed—i.e., a change of frequencies occurred—then the molecular structure of light might show up.[104]

> If monochromatic radiation (of sufficiently low density) behaves, as far as the
> volume-dependence of its entropy is concerned, as a discontinuous medium consist-
> ing of energy quanta of magnitude $R\beta\nu/N_0$ [he remarked] then it is plausible to
> investigate whether the laws of the creation and transformation of light are so
> constituted as if light consisted of such quanta. (Einstein, 1905b, pp. 143–144)[105]

But what were the phenomena in which the quantum structure of radiation
intervened? Einstein selected three examples from his knowledge of the scientific
literature: first, the well-known phenomenon of fluorescence; second, the so-
called photoelectric effect; third, the ionization of gases by ultraviolet light. In all
three cases he demonstrated agreement of the observed phenomena and the
consequences from his heuristic point of view concerning the nature of radiation.

In the example of fluorescence—or photoluminescence, as Einstein called
it—George Gabriel Strokes had found that the (visible) light emitted by fluores-
cent substances was excited mainly by ultraviolet radiation; hence he had
established the rule, later called Stokes' rule, that the absorbed light had higher
frequencies than the reemitted one. Einstein now explained this rule on the

---

[104] Although Einstein's and Planck's theories were different, both authors agreed on one point: the
fact that propagation and diffraction of light were satisfactorily described by the electromagnetic
theory. In 1905, Einstein had not yet developed the full corpuscular theory of the light-quantum
including energy and directed momentum.

[105] Note that for the purpose of describing these processes Planck invoked the concept of
irreversibility, which offered him the possibility of entering into his statistical considerations.

assumption that

> both the original and the newly produced light consist of energy quanta of magnitude $(R/N_0)\beta\nu$, where ν is the corresponding frequency. [He concluded:] Unless the photoluminescent substance can be regarded as a continuous source of energy [i.e., as violating the law of conservation of energy], the energy of the light-quantum that is produced cannot—according to the law of energy conservation—be greater than that of an initial light-quantum; hence the condition, $(R/N_0)\beta\nu_2 \leqslant (R/N_0)\beta\nu_1$, or $\nu_2 \leqslant \nu_1$, must hold. (Einstein, 1905b, p. 144)

With this argumentation, in which the name 'light-quantum' occurred for the first time, he provided an elegant derivation of Stokes' rule which could not be understood on the basis of the electromagnetic theory of radiation.[106]

The next example which Einstein considered was the photoelectric effect. It had been discovered by Wilhelm Hallwachs, who in turn had been stimulated by previous experiments of Heinrich Hertz; he had shown, in particular, that some uncharged metallic surfaces acquired a positive charge if exposed to ultraviolet light (Hallwachs, 1888).[107] Joseph John Thomson and Philipp Lenard had then shown that the effect was caused by the fact that negatively charged corpuscles, the electrons, were emitted from the surfaces of metals (Thomson, 1899; Lenard, 1899). Lenard had continued to investigate the phenomenon and had presented detailed results in a long memoir published in *Annalen der Physik* (Lenard, 1902).[108] In this paper he had reported especially two important findings: first, in order to obtain electrons from a given metallic surface only the incident light of

---

[106] Einstein further concluded that his light-quantum hypothesis implied that fluorescence radiation could be excited by incident light (of higher frequencies) having arbitrarily small intensity. He also added that exceptions from Stokes' rule might occur, if either the intensity of the exciting radiation was very large (so that one newly created quantum obtained its energy from several incident quanta) or if the radiation considered did not satisfy Wien's law.

[107] Wilhelm Ludwig Franz Hallwachs was born on 9 July 1859 in Darmstadt. He studied physics at the Universities of Strasbourg and Berlin, obtaining his doctorate in 1883. He then became assistant to August Kundt in Strasbourg and Friedrich Kohlrausch in Würzburg, and later professor of physics at the *Technische Hochschule* of Dresden (1900–1922). He died in Dresden on 20 June 1922.

[108] Philipp Eduard Anton von Lenard was born in Pozsony (Pressburg), Hungary, on 7 June 1862. He studied physics successively at Budapest, Vienna, Berlin and Heidelberg and had among his teachers Robert Bunsen, Hermann von Helmholtz and Georg Hermann Quincke. After obtaining his doctorate from the University of Heidelberg in 1886 he became assistant to Quincke in Heidelberg and then to Heinrich Hertz in Bonn; in Bonn he also received his *Habilitation* in 1892. In 1894 he was appointed *Extraordinarius* of physics at the University of Breslau, and one year later he became *Ordinarius* (full professor) at the *Technische Hochschule* in Aachen. In 1896 he moved to Heidelberg as Professor of Theoretical Physics. From 1898 to 1907 he served as Professor of Physics at the University of Kiel; then he returned to Heidelberg as Professor of Physics (1907–1930). Lenard died at Messelhausen on 20 May 1947.

Philipp Lenard worked on theoretical mechanics, on the phenomena of phosphorescence and luminescence. From 1888 he investigated cathode rays; stimulated by Hertz, he invented the 'Lenard window,' which made it possible to isolate cathode rays from the source and to observe them as free objects. For his pioneering investigations in the field of cathode rays he received the Nobel Prize in Physics in 1905. He also worked on the photoelectric effect and on the nature and origin of spectral lines.

certain frequencies was effective; second, the velocity of the emitted electrons did not depend on the intensity of the incident radiation. He had argued, therefore, that

> the initial velocities of the emitted quanta [so did Lenard call the electrons] do not originate from light energy at all, but from the violent motions existing already before the illumination within the interior of the atoms; thus the resonance motions [of the electrons stimulated by the incident radiation] only play the role of a mechanism for the release [of electrons]. (Lenard, 1902, p. 170)

Einstein interpreted Lenard's results differently; especially, he took the above statement as an indication of the fact that, 'The usual idea, that the energy of light is continuously distributed over the space through which it travels, meets with especially great difficulties when one tries to explain the photoelectric phenomena' (Einstein, 1905b, p. 145). He suggested the following solution of these difficulties. If the radiation falling on a metallic body possesses the structure of energy-quanta and one assumes that each energy-quantum gives its energy to a *single* electron, then the emitted electron receives the kinetic energy,

$$E_{\text{kin}} = \Pi e = \frac{R}{N_0} \beta\nu - P, \tag{60}$$

where P is the potential barrier (characteristic for the body under investigation) which the electron (of charge $-e$) has to overcome in order to leave the metallic surface. This kinetic energy had been measured by Lenard by subjecting the metallic surface to a positive potential Π, which prevented the electrons from being emitted at all. The order of magnitude of his values agreed with Einstein's estimate from Eq. (60). Einstein further demanded: 'If the formula derived above is correct, then Π—if drawn in Cartesian coordinates as a function of the frequency of the incident light—must be a straight line, the slope of which is independent of the nature of the substance studied' (Einstein, 1905b, p.146). Evidently, the slope of the function is given by $(R/N_0)\beta$ or Planck's constant h. In 1905 Einstein could only state that: 'As far as I can see, our ideas are not in contradiction with the properties of the photoelectric action observed by Mr. Lenard' (Einstein, 1905b, p. 147).[108a] A decade later Robert Andrews Millikan would report a detailed verification of Einstein's formula, Eq. (60), and note especially: 'Planck's h has been photoelectrically determined with a precision of about .5 per cent and is found to have the value $h = 6.57 \times 10^{-27}$ [erg · sec]' (Millikan, 1916b, p. 388).

The last phenomenon which Einstein considered in order to support his conception of a quantum-like structure of radiation was the ionization of gases

---

[108a] Einstein drew attention in particular to the fact that his hypothesis of light-quanta explained Lenard's observation that the number of electrons was proportional to the intensity of the incident light. In addition, he considered the inverse of the photoelectric effect, that is, the creation of radiation when cathode rays fall on a metallic surface; in that case he derived from his light-quantum hypothesis the conclusion that the product of the frequency of the emitted radiation, ν, and the factor $R\beta/N$ ($= h$) was always smaller than the kinetic energy of the cathode rays (electrons), in agreement with the experimental results (Lenard, 1903a).

by ultraviolet light. By assuming that the ionization of each molecule was due to the energy transfer from one energy-quantum of radiation, he found values for the ionization energy that were in agreement with the empirical data available at that time (Einstein, 1905b, Section 9).

Einstein was very satisfied that all these phenomena fitted well into his new picture about the quantum structure of light. In the following years he would find further empirical facts which could be easily accounted for by the same assumption. These would confirm and develop his belief in the dual nature of radiation, an undulatory one described by the laws of classical electrodynamics and a corpuscular one described by the laws of molecular theory. He would also discover that his light-quantum theory and the blackbody radiation theory of Planck involving the quantum of action rested on the same foundation. Hence he would become convinced that both had to be combined into a new theory. This theory—the quantum theory—would have a different structure and be expressed by means of laws different from those of classical theories, i.e., of classical mechanics and electrodynamics (including special relativity theory), which were fully and completely established by the end of the first decade of the twentieth century.

I.4   Energy-Quanta and the Derivations from Classical Theories

When Planck developed the theory of his blackbody radiation law, Eq. (18), he did not think that it might contradict any of the established theories. He rather believed that all the steps he had taken remained perfectly within the scope of thermodynamics, Maxwell's electrodynamics and the statistical mechanics of Maxwell and Boltzmann. In his opinion, he had only contributed a certain amount to Maxwell and Boltzmann's statistical mechanics by giving an adequate statistical interpretation of the entropy of radiation. Thus Fritz Reiche, a student of Planck's in Berlin, recalled from Planck's lectures on heat radiation: 'I would not say that he, Planck, had the feelings: "I give you here something very new which is very funny and a very complete break." I cannot remember that he said this, that it is a complete break. One had [rather] the impression that one cannot speak of counted probability if you are not subdividing energy, at least in part' (Reiche, AHQP Interview, 30 March 1962, p. 6). Also Einstein, when he discussed the physical interpretation of Wien's law and arrived at the light-quantum point of view, did not think primarily of a break with tradition. After all, by introducing light-quanta he had restored the consistent application of the statistical methods to heat radiation. Still, Einstein called his introduction of the light-quantum hypothesis a 'revolutionary' step, for he believed that the new point of view about radiation contradicted the very foundation of Maxwell's electrodynamics, which demanded the energy of radiation to be described by a continuous function in space and time.

It was generally understood that radiation in Maxwell's theory had been considered as an electromagnetic stream of energy which extended continuously over a region of space and whose transport could be expressed by a vector, the so-called Poynting vector. While Einstein referred to this understanding in speaking about the electromagnetic theory of radiation, Joseph John Thomson, the third occupant of the Cavendish Chair at Cambridge University, held a different view.[109] Thomson, in particular, had developed the theory of moving electric tubes, by means of which Oliver Heaviside had earlier described the transport of electrical energy. In this theory a beam of light had to be regarded as a collection of tubes (of electric force) moving with the velocity of light at right angles to their own length. Thomson had pictured the tubes as threads with fibrous structure embedded in a continuous ether, and with this picture he had challenged the view that Maxwell's theory demanded a continuous, uniform distribution of energy over the surface of the wave front.[110] In 1903, in his Silliman Lectures at Yale University, he went a step further; he considered the detailed consequences that followed from the theory of electric tubes for the propagation of light. In particular, he assumed that oscillating charges set up regular trains of disturbances in the electric tubes attached to them; then light would be identified with the transverse vibrations migrating along the tubes of force and, since the tubes did not fill the ether, electromagnetic radiation would have to possess a discontinuous structure. Thus, in the process of propagation of light a swarm of 'bright specks' should appear (against a dark background) at those places where the tubes of force intersect an advancing wave front (Thomson, 1904c, p. 63). In support of this picture, Thomson cited the empirical fact that X-rays passing through gases ionize only a few molecules, which could not be understood if the energy distribution in the wave front was continuous.

[109] J. J. Thomson was born at Cheetham Hill, a suburb of Manchester, on 18 December 1856. He began to study at Owens College, Manchester, in 1870 and entered Trinity College, Cambridge, six years later. He was Second Wrangler and Second Smith's Prizeman in 1880. During the same year he became a Fellow of Trinity College, and in 1883 a lecturer there. In 1884, after Lord Rayleigh's retirement, he was appointed Cavendish Professor of Experimental Physics at Cambridge University. He stayed in this position until 1918 when he became Master of Trinity College. Thomson's early work was on theoretical mechanics and electrodynamics. His experimental researches culminated in the discovery of the electron in 1897 and, more than a decade later, he developed a special method to deflect positive ions by means of which the isotopes of a given chemical element could be separated.

Thomson received many honours, including honorary degrees from numerous universities and memberships of many learned societies, including the Royal Society of London (1884). In 1906 he was awarded the Nobel Prize in Physics for his theoretical and experimental investigations on the conduction of electricity through gases. Two years later he was knighted. He died on 30 August 1940 in Cambridge, England.

[110] Thomson presented his theory in detail in a book, entitled 'Notes on Recent Researches in Electricity and Magnetism' (Thomson, 1893), which served as a supplement to Maxwell's Treatise on Electricity and Magnetism (the third edition of which was prepared for publication by Thomson in 1891). Thomson's treatment of electromagnetic theory—like the earlier one of Poynting— was based on Michael Faraday's concepts of lines of force. Thomson discarded the magnetic lines of force—in contrast to Faraday and Poynting—as secondary phenomena, because he thought that they might arise from the motion of electric tubes. From the existence of a smallest electric charge (the charge of the electron or its positive counterpart), he further concluded that unit tubes of electric force did exist as well, and that these unit tubes led to a 'kind of molecular theory of Electricity, the Faraday [unit] tubes taking the place of molecules in the Kinetic Theory of gases' (Thomson, 1893, p. 4).

However, if the front consisted of localized spots of light intensity 'analogous to a swarm of cathode rays,' then only a small number of collisions between the molecules of the gas and X-radiation would occur and the small number of resulting ionized molecules could be explained (Thomson, 1904c, p. 63). Later Thomson applied his concept of discontinuous wave fronts of radiation also to account for the observations of Lenard on the photoelectric effect; assuming that the energy of a pulse of light remains constant, the velocity which an electron receives can depend only on this energy and not on the energy of all other pulses which add to the (total) intensity of the incident radiation (Thomson, 1907b). In the same paper he argued that the character of radiation of a given intensity appears to be more discontinuous the higher its frequency is, hence ultraviolet light will have an 'exceedingly coarse character' and might be 'pictured by supposing the particles on the old [Newtonian] emission theory replaced by isolated transverse disturbances along the lines of force' (Thomson, 1907b, p. 423). Even coarser still would be the character of the X-rays and γ-rays which are emitted from radioactive substances; the pulses of these radiations 'will have all the properties of material particles, except that they cannot move at any other speed than that of light' (Thomson, 1907b, p. 424). Altogether he arrived at practically the same conclusions as Einstein: that light of high frequencies and low density behaves like quanta of energy concentrated in space. However, his conclusions differed in two respects from those of Einstein: first, he did not invoke a definite size for his quanta of radiation, hence he could not derive any quantitative results; second, he believed strongly that his theory of tubes of force and the propagation of radiation did remain within the framework of the electrodynamic theory developed by Faraday and Maxwell.[111]

In the early years of the twentieth century the question concerning the nature of radiation was approached from still another point of view. The latter emerged from a debate about a fundamental theorem in the kinetic theory of matter; this was the equipartition theorem or the Maxwell–Boltzmann doctrine—as it was then called—which stated that the average kinetic energy distributed to every degree of freedom of a system in thermal equilibrium was the same. Lord Kelvin, one of the pioneers of kinetic theory, had argued for many years against the validity of this doctrine; for instance, in 1892 he had constructed a particular example in which it did not hold (Kelvin, 1892). Many scientists, especially in Great Britain, participated in the ensuing debate during the following decade without being able to decide the question either way. In 1900 Lord Rayleigh devoted an article to the subject; after presenting reasons why Kelvin's example did not necessarily imply a violation of the equipartition theorem, he went on to prove the theorem for systems of particles which could move in two directions,

---

[111] Thomson, in his paper of 1907, did not refer to Einstein's two-year-old paper on light-quanta (Einstein, 1905b). However, even if he knew about it, he would not have taken it seriously because he disagreed with Einstein's point of departure. Thomson recognized the connection between his corpuscular theory of light and quantum theory later on. An analysis of 'J. J. Thomson and the Structure of Light' has been given by Russell McCormmach (1967b).

thus possessing two degrees of freedom (Rayleigh, 1900a).[112] By investigating the changes in time of such systems he arrived at the result that the stationary (steady) distribution function, $f(x, y, u, v)$, which determined the number of particles having coordinates x and y and velocities u and v, implied equal average kinetic energy for the velocity components u and v. In deriving this result, Lord Rayleigh claimed to have used '*only* the application of Maxwell's assumption that all *phases*, i.e. all states, defined both in respect to configuration and *velocity*, which are consistent with the energy condition, lie on the same path, i.e. are attained by the system in its free motion sooner or later' (Rayleigh, 1900a, pp. 99–100). Still another assumption was involved in Lord Rayleigh's proof of the equipartition theorem, namely, that the kinetic energy of the system could be expressed in terms of the squares of the velocity components; but he showed that this standard form of the kinetic energy could always be achieved by choosing suitable variables to describe the system.

Although he was convinced of the soundness of his proof, Lord Rayleigh was perfectly aware that nature did not always respect the consequences that followed from the equipartition theorem. Especially, the molecules of a diatomic gas behaved as if they had only five degrees of freedom and no energy was provided, for instance, to the rotation around the axis joining the atoms. Lord Rayleigh suggested that perhaps this situation could be explained by the fact that the binding between the two atoms in a diatomic molecule was not infinitely tight (Rayleigh, 1900a, pp. 117–118). In any case, he felt that the equipartition theorem would hold for any dynamical system if applied with proper care. Thus, as a special example, he investigated heat radiation in a cavity in a paper entitled 'Remarks upon the Law of Complete Radiation' (Rayleigh, 1900b). In early 1900 the energy distribution of blackbody radiation was described, according to standard knowledge, by Wien's law, Eq. (6); that is, the distribution function as dependent on the temperature T and the wavelength of radiation λ had the form

$$\Phi(\lambda, T) = \frac{c_1}{\lambda^5} \exp\left(- \frac{c_2}{\lambda T}\right). \tag{61}$$

[112] John William Strutt, third Baron Rayleigh, was born on 12 November 1842 at Langford Grove near Maldon, Essex. He studied mathematics at Trinity College, Cambridge, from 1861 to 1865. He was Senior Wrangler in 1865 and Smith's Prizeman of the year and became a Fellow of Trinity College in 1866. After his marriage in 1871 he returned to the family estate and stayed there until 1879, when he succeeded Maxwell as Cavendish Professor. Five years later he left the Cavendish Chair, but continued his experimental and theoretical researches at his country seat at Terling, Essex. From 1887 to 1905 he was Professor of Natural Philosophy at the Royal Institution of Great Britain, having succeeded John Tyndall in that position. In 1908 he became Chancellor of Cambridge University. He was elected Fellow of the Royal Society in 1876; he served as Secretary of the Society from 1885 to 1896, and as President from 1905 to 1908. Lord Rayleigh held many other scientific and public offices. He died on 30 June 1919 at Witham, Essex.

Lord Rayleigh's researches ranged over many fields of theoretical and experimental physics, including optics, vibrating systems, theory of sound, wave theory, colour vision, electrodynamics, hydrodynamics, gas theory, elasticity and photography. He received many honours, including the Nobel Prize in Physics in 1904 for his work on the density of gases and the discovery of argon. Lord Rayleigh's papers were collected together in six volumes of *Scientific Papers*, published by Cambridge University Press, Cambridge, 1899–1920.

Rayleigh now argued that, for high temperatures, Eq. (61) attributed too little energy to long wavelengths, in disagreement with the Maxwell–Boltzmann doctrine of equipartition. 'According to this doctrine,' he remarked, 'every mode of vibration should be alike favoured, and although for some reason not yet explained the doctrine fails in general, it seems possible that it may apply to the graver modes' (Rayleigh, 1900b, pp. 539–540). That is, he took for granted that Wien's law yielded essentially the correct description for short wavelengths but not for long ones, where he expected the equipartition theorem to hold. In order to derive the distribution function from the latter, he compared the system of cavity radiation to acoustic vibrations of a cubic mass of air. Any particular vibration was characterized by a wave number k, and the number of modes with wave numbers between k and $k + dk$ was determined by the product $k^2 dk$.[113] By interpreting k as ν, the frequency of heat radiation, and assigning each frequency mode a kinetic energy proportional to T (in agreement with the equipartition theorem), Rayleigh found that the energy of the modes having frequencies between ν and $\nu + d\nu$ must be proportional to the expression $T\nu^2 d\nu$. Consequently the distribution function $\Phi(\lambda, T)$ was determined, apart from a constant factor, by the expression $T\lambda^{-4}$. Rayleigh now suggested that the 'complete expression' for the distribution function was

$$\Phi(\lambda, T) = c_1' \frac{T}{\lambda^4} \exp\left(-\frac{c_2}{\lambda T}\right). \tag{62}$$

Evidently, for large values of λT, the right-hand side of Eq. (62) satisfied the equipartition theorem; in the case of short wavelengths it passed over into Wien's distribution, Eq. (61).[114]

Lord Rayleigh's law for complete radiation, Eq. (62), was noticed immediately by the experimentalists; it was quoted, for instance, by Otto Lummer and Eugen Jahnke (1900) and by Heinrich Rubens and Ferdinand Kurlbaum (1900). However, it did not leave a lasting impression at the time, for it was soon replaced by Planck's formula, Eq. (18), which fitted the experimental data perfectly and seemed to rest on less hypothetical assumptions than were involved in the application of the equipartition theorem to the vibrational modes of the ether. Still, Rayleigh's considerations on blackbody radiation were taken up again about five years later by Rayleigh himself and James Hopwood Jeans. In the issue of *Nature*, dated 16 May 1905, a letter of Lord Rayleigh appeared, dealing with 'The Dynamical Theory of Gases and Radiation' (Rayleigh, 1905b). In this letter Rayleigh rederived his previous formula, Eq. (62), without the exponential factor, $\exp(-c_2/\lambda T)$; he also calculated the value of the constant c_1' as $64k$, with

---

[113] In order to understand Rayleigh's reasoning, one might think of a (one-dimensional) string of length L, which can vibrate only in modes of wavelength $\lambda = 2L/k$, with $k = 1, 2, 3, \ldots$. A similar condition applies to the vibrations of a cube having sides of length L, except that the possible modes are described by a set of triple integers, p, q, r, and the wavelength is given by the expression $2L(p^2 + q^2 + r^2)^{-1/2}$.

[114] Wien's distribution is established if one takes into account the displacement law, Eq. (4). The factor c_1' in Eq. (62) may then be related to the constant c_1 in Wien's law.

k the Boltzmann constant. He made two comments concerning his new radiation law. First, he asked why, in the limit of long wavelengths, it did not agree with Planck's law, although Planck had claimed his result to be consistent with kinetic theory. Second, he pointed out a difficulty which the new formula presented: it implied that no equilibrium could exist between heat radiation and molecules, because all the energy would be absorbed eventually by the radiation of higher frequencies. The problem raised by Rayleigh's two remarks were answered by James Jeans, then University Lecturer of Applied Mathematics at Cambridge University.[115]

Jeans had concerned himself deeply for several years with the problem of the equipartition theorem. Especially, he had thought about the difficulties raised by the application of this theorem—as, for example, in the theory of diatomic molecules. Already in 1901 he had argued that the equipartition theorem might indeed break down for systems in which the molecules were capable of radiating energy into the ether; for then, due to the absence of equilibrium, not every degree of freedom of the molecule need possess an average kinetic energy of magnitude $\frac{1}{2}kT$ (Jeans, 1901b). He had elaborated on the same idea in further publications and presented a detailed account of his results in the first edition of his book on the dynamical theory of gases (Jeans, 1904). In a letter to *Nature*, published in the issue of 13 April 1905, Lord Rayleigh referred to Jeans' argument and mentioned its consequences for the problem of heat radiation (Rayleigh, 1905a). He concluded, in particular, that if molecules were confined in a cavity (as Planck had assumed in his derivation of blackbody radiation), then the energy from the molecules would pass over to the infinitely many modes of radiation having high frequencies. In his second letter, referred to above, Rayleigh discussed radiation theory in detail (Rayleigh, 1905b).

---

[115] J. H. Jeans was born on 11 September 1877 at Ormskirk in Lancashire, the son of a parliamentary journalist and popular writer on science and technology. He was educated at Merchant Taylor's School, London, from where he proceeded in 1896, armed with a mathematical scholarship of Trinity College, to Cambridge University. After only two years of study he won the place of Second Wrangler, and in 1900 he shared a Smith's Prize with G. H. Hardy. The same year he was awarded the Isaac Newton Studentship, and in 1901 he became a Fellow of Trinity College. Apart from a course in practical physics at the Cavendish Laboratory (1899–1900), Jeans worked exclusively on problems of theoretical physics, especially on the application of dynamical principles to describe the behaviour of gases. He soon became an expert on the so-called equipartition theorem. In connection with his dynamical studies he also became interested in a problem considered earlier by his teacher George Howard Darwin: it concerned the behaviour of a swarm of meteorites, from which certain conclusions about the existing theories of cosmogony could be derived. Jeans discussed the 'Stability of a Spherical Nebula,' taking into account the heat radiation from the system into space (Jeans, 1902). After publishing an elaborate treatise on *The Dynamical Theory of Gases* (Jeans, 1904), he became University Lecturer at Cambridge in 1904; he was invited as Professor of Applied Mathematics to Princeton University in 1905, and later, in 1910, he was appointed Stokes Lecturer at Cambridge University. After his marriage to Charlotte Tiffany Mitchell he had become a wealthy man, and he retired from the Stokes Lectureship two years later. Still he was somewhat disappointed when Eddington, and not he, was elected to succeed G. Darwin as Plumian Professor. In 1919 Jeans became one of the Joint Secretaries of the Royal Society, whose Fellow he had been since 1906. In this position he not only helped to organize the Society's investments but also undertook to improve the standard of the *Proceedings*, which soon became the leading British scientific journal. In 1935 he

Meanwhile Jeans also dealt with the same problem, that is, the equilibrium between radiation and molecules, in two papers submitted to *Philosophical Magazine* (Jeans, 1905b) and *Proceedings of the Royal Society* (Jeans, 1905d). Prior to submitting these papers, however, he saw Lord Rayleigh's first letter in *Nature* (Rayleigh, 1905a) and soon afterwards the second letter (Rayleigh, 1905b). He then decided to answer the questions raised by Rayleigh at two places: in a letter to *Nature*, dated 20 May 1905 and published in the issue of 1 June (Jeans, 1905a), and in a postscript to his paper in the *Philosophical Magazine*, dated 7 June 1905 (Jeans, 1905b, pp. 97–98).[116] In the postscript he especially derived the following expression for the distribution function,

$$\Phi(\lambda, T) = 8\pi k \frac{T}{\lambda^4} , \tag{62a}$$

and commented: 'This is one-eighth of the amount found by Lord Rayleigh (*Nature*, May 16), but agrees exactly with that given by Planck (Drude's *Annalen*, IV. p. 533 [Planck, 1901a]) for large values of λ' (Jeans, 1905b, p. 98). Thus he claimed that Rayleigh's first difficulty arose from an error in the calculation.[117] In the letter to *Nature*, on the other hand, Jeans also discussed the second problem, namely, the existence of equilibrium between molecules and heat radiation (Jeans, 1905a). He remarked, that although Eq. (62a) represented the stationary energy distribution of radiation, this distribution would not be observed experimentally, because in a cavity the transfer of the kinetic energy to radiation modes of short wavelengths at a given temperature T proceeds very slowly.[118] As a consequence, at any finite time t, only the long-wavelength modes would possess the full amount of energy allowed by the equipartition theorem (i.e., the amount kT) but not the short-wavelength modes. The latter, instead of

---

was appointed Professor of Astronomy in the Royal Institution. He resigned from this position several months before his death, which occurred on 16 September 1946 at his home, Cleveland Lodge, Dorking.

Jeans worked on many branches of mathematical physics and theoretical astronomy. After 1905 he devoted himself for several years to the application of statistical mechanics to radiation theory. Later he concerned himself with problems of cosmogony and stellar dynamics, becoming one of the pioneers in that field. Besides his researches he served science by introducing its results in masterly essays and popular books for the public. He was knighted in 1928.

[116] In a footnote of his paper in *Philosophical Magazine* (Jeans, 1905b, p. 91), Jeans mentioned that it was composed before Lord Rayleigh's letters to *Nature* of April and May (Rayleigh, 1905a, b); only the postscript, added on 7 June (Jeans, 1905b, pp. 97–98) referred to the second letter (Rayleigh, 1905b). Lord Rayleigh had submitted three letters to *Nature* at that time, published in the issues of 13 April (1905a), 16 May (1905b) and 7 July (1905c).

[117] Rayleigh confirmed the correctness of Jeans' formula, admitting his own mistake, soon afterwards in a letter to *Nature* which appeared in the issue of 7 July 1925 (Rayleigh, 1905c).

[118] Jeans argued that, 'If the average time of collision of two molecules in a gas is a great multiple N of the period of vibration, whether of matter or of ether, then the average transfer of energy to the vibration per collision can be shown to contain a factor of the order of smallness of e^{-N}' (Jeans, 1905a, p. 101). The idea that kinetic energy is only very slowly transferred to certain degrees of freedom had first been mentioned by Ludwig Boltzmann, who had tried to explain by it the failure of the equipartition theorem in molecular theory (Boltzmann, 1895b, p. 414).

$\Phi(\lambda, T)$, would contribute to the radiation energy a time-dependent share, $f(\lambda, T, t)$, which would quickly go down to zero with decreasing wavelength λ.[119]

Though Jeans, in his papers of 1905, gave a detailed kinetic theory of blackbody radiation, especially of the distribution law expressed in Eq. (62a), which was later called the Rayleigh–Jeans law, his work did not receive immediate recognition. Several reasons were responsible for this. First, Planck's law, Eq. (18), did describe the experimental results perfectly. Second, no sign had ever been detected that the energy distribution in a cavity changed in the course of time and that, therefore, the equilibrium situation had not been established in the experiments. Third, if Jeans was right, then the accepted thermodynamic proofs for several laws of blackbody radiation, such as the Stefan–Boltzmann law, Eq. (3), and Wien's displacement law, Eq. (4), broke down, and it would be difficult to establish them in the case of a nonequilibrium situation.[120] Finally, apart from the fact that the question of the validity of the equipartition theorem, especially in radiation theory, had not yet been decided at all, Jeans, like Rayleigh, seemed to have described blackbody radiation in terms of the vibrations of an elastic ether—a concept which had become very suspect at that time. In any case, Planck, in the derivation of his radiation law, had not referred to any such questionable foundations; he had rather based his approach on the validity of thermodynamics and electrodynamics, which also implied the existence of a genuine equilibrium between matter (represented in Planck's approach by resonators) and radiation. Thus, after discussing Jeans' law and mentioning, in particular, the difficulty concerning the absence of equilibrium, he stated in his lectures on heat radiation:

> In my opinion the difficulty referred to is caused by an unjustified application of the theorem of equipartition of energy to all independent state variables. In fact, for the validity of this theorem one has to assume the essential prerequisite that the distribution of states among all possible systems, for a given total energy, will be an "ergodic" one or, in short, that the probability for the state of the system to occupy a certain small "elementary region" is simply proportional to the magnitude of this region, however small it might be. However, this condition is not satisfied in the case of stationary energy radiation; the reason is that the elementary regions cannot be taken to be arbitrarily small, for their size is finite, determined by the value of the elementary quantum of action h. (Planck, 1906, p. 178)

Only in case of a vanishing value for the quantum of action, he remarked, would

---

[119] For any given wavelength λ the function $f(\lambda, T, t)$ increases in the course of time until it reaches the value $8\pi kT\lambda^{-4}$ (Jeans, 1905b, p. 98). For small values of λT, and at a given instant of time t_0, the function $f(\lambda, T, t_0)$ may be written as

$$f(\lambda, T, t_0) = 8\pi kT\lambda^{-4}\exp(-c_2/\lambda T),$$

i.e., in the form of the law given by Eq. (62), which had been proposed by Lord Rayleigh in 1900.

[120] Jeans did provide proofs for Eqs. (3) and (4) in fall 1905 based on dimensional arguments (Jeans, 1905e).

one arrive at Jeans' law, Eq. (62a). Since, however, the observations showed that *h* assumed a finite value, the equipartition theorem must break down. The real problem of radiation theory, in Planck's opinion, was different from the one which Jeans had treated. The real problem was not the question of an incomplete equilibrium, but rather how to explain the finite value of the constant *h*. 'Of course,' Planck concluded, 'one has to assign to the element of action also a direct electromagnetic significance; but of what kind it might be remains an open question' (Planck, 1906, p. 179).

The suggestion to search for an electromagnetic interpretation of the constants in the radiation law was not new. It had been mentioned already before the discovery of Planck's law by Hendrik Antoon Lorentz when he concerned himself for the first time with the problem of heat radiation in 1900.[121] In a paper, presented to the Amsterdam Academy of Sciences, he investigated the connection between the accepted laws of radiation theory and the second law of thermodynamics (Lorentz, 1900b). He claimed that, in order to justify the derivation of the Stefan–Boltzmann law and Wien's displacement law, one had to assume the existence of a special property characterizing matter; it should explain why any blackbody of a given volume and at a given temperature *T*—independent of its specific constitution—emitted radiation of the same total energy and also why λ_{max}, the wavelength emitted with maximum intensity, was always the same. Lorentz argued that this property consisted in the fact that all ponderable bodies had among their constituents electrons bearing identical electric charges. And he proposed the following procedure to derive the displacement law: put the electrostatic energy of a spherical electron with radius *r*, whose charge is uniformly distributed, equal to the average kinetic energy of the electron at temperature *T*; then the product *rT* does not depend on the value of the temperature, and Wien's displacement law applied to the maximum wavelength λ_{max} corresponds to a temperature-independent ratio λ_{max}/r. Provided

---

[121] H. A. Lorentz was born in Arnhem, the Netherlands, on 18 July 1853. He studied at the University of Leyden, obtaining his doctorate in 1875 with a thesis on the theory of reflection and diffraction of light (Lorentz, 1875; reprinted in *Collected Papers*, Vol. I, 1934). Just three years later he was appointed to the chair of theoretical physics at Leyden, which he occupied until he retired in 1912, when he accepted the position of Curator of the Teyler Foundation in Haarlem. He died in Haarlem on 4 February 1928.

Lorentz made pioneering investigations on the application and extension of Maxwell's electrodynamics to moving bodies and on the theory of electrons. He discovered, independently of George Francis FitzGerald, the length contraction of fast-moving bodies. On the basis of his electron theory he developed a full description of electrical and optical phenomena; he also provided the explanation of the Zeeman effect, i.e., of the separation of atomic spectral lines in the presence of a magnetic field. For the latter work he received, together with Pieter Zeeman, the Nobel Prize in Physics in 1902. Besides electron theory he worked on kinetic theory, radiation theory, theory of gravitation, thermodynamics and hydrodynamics, including applications to engineering problems. (See Lorentz, Collected Papers, Vols. I to IX, Martinus Nijhoff, The Hague, Holland, 1934–1939.)

For his outstanding contributions Lorentz received numerous honours, including the Foreign Membership of the Royal Society of London, the Rumford Medal and the Copley Medal. A deep knowledge of all fields of physics, combined with great linguistic and pedagogical abilities, made him an important leader of the development of physics of his time; thus he was chosen frequently as president of important meetings of physicists, such as the Solvay Conferences.

such a universal ratio, λ_{max}/r, could be obtained from electron theory, it would also imply an electromagnetic interpretation of the elementary quantum of action h. The reason was that in this case the product $\lambda_{max}T$ could be expressed in terms of the constants of electron theory (i.e., the elementary charge e and the velocity of light *in vacuo* c) and Boltzmann's constant k. On the other hand, due to Planck's radiation law, the product $\lambda_{max}T$ was proportional to the ratio h/k, hence Planck's constant h followed in terms of the constants of electron theory.[122] Lorentz continued to work on radiation theory during the following years; thus, for instance, he demonstrated that the Stefan–Boltzmann law and Wien's displacement law might be deduced from purely thermodynamic and electrodynamic arguments without introducing—as previous authors had done—the concept of rays of radiation (Lorentz, 1901). Then, two years later, he approached the problem of the energy distribution in blackbody radiation for the first time (Lorentz, 1903).

Lorentz started out from a conclusion, which had been reached experimentally by Ernst Hagen and Heinrich Rubens in Berlin: they had investigated the reflecting power of metals with respect to radiation of wavelengths larger than 8 μ and had found that the results were perfectly described by the usual laws of electrodynamics, invoking no other property than the conductivity of metals (Hagen and Rubens, 1903). Since, as Eduard Riecke (1898) and Paul Drude (1900a, b) had shown, the conductivity of metals could be calculated from electron theory, the same was true—because of Hagen and Rubens' conclusion—for the reflective power of long electromagnetic waves. Then, however, the connection between reflective and absorptive powers allowed Lorentz to determine the absorption and, by applying Kirchhoff's laws, finally the emission of radiation (having long wavelengths) by metals. These considerations allowed him to derive an equation for the blackbody distribution function $\Phi(\lambda, T)$ as

$$\Phi(\lambda, T) = \frac{16\pi}{3\lambda^4} \alpha T. \tag{63}$$

By comparing with Planck's distribution law for long wavelengths, he determined the value of the constant α to be $\frac{3}{2}k$. Lorentz commented on his result as follows: 'Now the mean kinetic energy of a molecule of gas would be $\frac{3}{2}kT$ according to Planck and has been represented in what precedes by αT. There appears therefore to be a full agreement between the two theories in the case of long waves, certainly a remarkable conclusion, as the fundamental assumptions are widely different' (Lorentz, 1903, p. 678; *Collected Papers*, Vol. III, 1936, pp. 168–169).

This success encouraged Lorentz to continue to proceed in the same direction and to extend the theoretical investigations with the goal of deriving the distribution law for all wavelengths from electron theory. Now Planck felt that the

---

[122] In the case of Planck's radiation law the product $\lambda_{max} \cdot T$ was given by the expression $(c/4.9651) \cdot (h/k)$.

interpretation of his radiation law had to be established on the same foundation, and he wrote in 1905:

> If one introduces the assumption that resonance oscillations consist of motions of the electrons, then one invokes a new element [besides the hypothesis of natural radiation] in the theory . . . . It now appears to me that one cannot entirely exclude the possibility that from this assumption (i.e., the existence of an elementary electric quantum), there arises a connection with the existence of the elementary quantum of action h, especially since h possesses the same dimension and the same order of magnitude as e^2/c (e being the elementary electric quantum in electrical units, c the velocity of light *in vacuo*). But I am not in a position to express a definite opinion on this. (Planck to Ehrenfest, 6 July 1905)

By that time, however, Planck had many reasons to be optimistic. He had become accustomed to electron theory in the preceding years and had applied it to explain various optical properties. Even more successful in this direction had been Planck's Berlin colleague Paul Drude. But the physicist, who was most deeply concerned with electron theory, was certainly Hendrik Lorentz; he had achieved great progress in clarifying and establishing more securely the foundation of the theory in a memoir on the 'Electrodynamic Phenomena in a System Moving with any Velocity Smaller than that of Light' (Lorentz, 1904). In summer 1905, just before he made the remarks quoted above in a letter to Ehrenfest, Planck had seen another paper, entitled '*Zur Elektrodynamik bewegter Körper*' ('On the Electrodynamics of Moving Bodies'), which Albert Einstein had submitted to *Annalen der Physik* at the end of June (Einstein, 1905d). Planck recognized its importance immediately and began to work himself on the consequences of the new approach—the so-called principle of relativity—in mechanics and thermodynamics. In the course of these investigations he often used considerations involving heat radiation; he did this especially in the long paper, '*Zur Dynamik bewegter Systeme*' ('On the Dynamics of Moving Systems'), in which he developed the fundamental equations of relativity mechanics and thermodynamics (Planck, 1907, 1908). As one of the principal results, he found the relativistic invariance of the action integral,

$$W = \int_1^2 L\,dt, \tag{64}$$

where L denotes the kinetic potential or the Lagrangian function of the system under investigation, and the integration extends from the initial state 1 to the final state 2. Planck then noted:

> If one further makes the assumption that the magnitude of the action [i.e., W in Eq. (64)] is determined by a definite elementary quantum of action, $h = 6.55 \times 10^{-27}$ erg · sec, then one may also state: To any change in nature there corresponds a definite number of elements of action, independent of the choice of the coordinate system. (Planck, 1908, p. 23)

Although the consequences from the principle of relativity provided a consistent electron theory, it became clear that the new theory did not answer Planck's
question concerning the interpretation of the quantum of action h. The situation
even became worse. On 8 April 1908, at the fourth International Congress of
Mathematicians in Rome, Lorentz gave a talk, entitled '*Le partage de l'energie
entre la matière pondérable et l'éther*,' in which he discussed the result of his
endeavours to derive the law of blackbody radiation (Lorentz, 1908a). He had
arrived, especially at the law of Rayleigh and Jeans, Eq. (62a), and remarked:

> According to the theory which I have presented, Eq. [(62a)] must hold for all
> wavelengths longer than the value which I have denoted as λ_0 [i.e., a minimum
> wavelength, below which Lorentz assumed that the electromagnetic field cannot be
> expanded in a Fourier series]; in the limiting case, which one obtains by decreasing
> λ_0 further and further, the equation should even be applicable to all wavelengths,
> however small they may be. (Lorentz, 1908a; *Collected Papers*, Vol. VII, 1934, p.
> 337)

Of course, Lorentz knew perfectly that Eq. (62a) implied that no equilibrium
existed between matter and radiation and that it did not describe the observed
distribution. However, he could not offer any suggestion on how to overcome
these difficulties and concluded his talk as follows:

> In theoretical physics one cannot do anything else but examine the different
> hypotheses and evaluate the probability [of their being true] by drawing the
> consequences from each of them. Well, if one compares the theory of Planck with
> the one of Jeans, one finds that both possess their merits and faults. Planck's theory
> is the only one which has given us a formula in agreement with the experimental
> results, but we cannot accept it without profoundly changing our fundamental ideas
> about electromagnetic phenomena. One recognizes this already when one considers
> a single electron, moving in an arbitrary manner and emitting radiation of all
> wavelengths; evidently, it is impossible to apply in this case the hypothesis of energy
> elements [quanta] whose magnitude depends on the frequency. Jeans' theory, on the
> other hand, forces us to ascribe the agreement—inexplicable at the moment—
> between observations and the laws of Boltzmann and Wien to a coincidence.
> Fortunately one may hope that new experimental determinations of the radiation
> function [i.e., the energy distribution of blackbody radiation] will allow us to decide
> between the two theories. (Lorentz, 1908a; *Collected Papers*, Vol. VII, 1934, p. 341)

The last remark caused several physicists to raise serious protests. Thus Otto
Lummer and Ernst Pringsheim, who had been informed by Willy Wien—who
was a participant in the Rome conference—about the content of Lorentz' lecture,
immediately sent a note to *Physikalische Zeitschrift*, in which they wrote: 'If one
looks at the Jeans–Lorentz formula, one recognizes on first inspection that it
leads to completely impossible consequences, which not only disagree blatantly
with the results of radiation measurements but also with everyday experience'
(Lummer and Pringsheim, 1908, p. 449). If Eq. (62a) were correct, they argued, a
piece of steel should emit visible heat radiation at room temperature, in contrast

to observation. Lorentz, in a letter to *Physikalische Zeitschrift*, agreed with this conclusion, but repeated his inability to obtain a result different from Eq. (62a) (Lorentz, 1908b).[123] In any case, in 1908 one result had become definite: 'that the electron theory leads, without introducing new hypotheses, inevitably to Jeans' conclusions [i.e., to the radiation law, Eq. (62a)]' (Planck to Lorentz, 1 April 1908). The question arose, therefore, as to what these new hypotheses—which would enable one to arrive at a more consistent description of the experimental data on heat radiation—would look like; and also, whether these hypotheses would fit into the scheme of (classical) theories developed so far. Such questions had to be decided now.

Actually, a partial answer to the last question had been given already by an earlier investigation of Paul Ehrenfest, a former disciple of Boltzmann.[124] In his first paper on radiation theory he had analyzed the physical assumptions underlying Planck's theory of irreversible processes (Ehrenfest, 1905). He had arrived at the conclusion that Planck's results rested on two basic assumptions:

1. Hypothesis concerning energy distributions among resonators having equal probability; 2. hypothesis that the radiation energies for the different colours [frequencies] are constituted of smallest energy elements of magnitude: $E_\nu = \nu \times 6.55 \times 10^{-27}$ erg · sec., ν being the frequency of the colour in question. [And he added the following comments:] The hypothesis 1 possesses an analogue in Boltzmann's theory [i.e., Boltzmann's kinetic gas theory]. The hypothesis 2, which in its present version is introduced in only a formal manner, needs a better foundation. As far as I can see, there is no analogue of it in Boltzmann's theory. (Ehrenfest, 1905, p. 1313)

---

[123] Lorentz described his fruitless efforts to find a radiation law different from Eq. (63) in a letter to Wien as follows:

I had hoped . . . to be able to determine the function (E/A) [i.e., Kirchhoff's function $\Phi(\lambda, T)$] by considering the electron motions and electron collisions in a metal even for arbitrarily short wavelengths [of the incident radiation], if I would only remove the simplifications valid for long waves. Then a relation between the constant $\lambda_m \cdot T$ and certain quantities concerning the free electrons had to follow. This relation, for instance, might be of the following kind: the wavelength λ_m of the [intensity] maximum [of heat radiation] might have the value $\lambda_m = a(c^2/v^2 \cdot R)$, with R denoting the radius of an electron, v its average velocity, c the velocity of light *in vacuo*, and a a numerical constant. Now I have pondered incessantly about this problem, until I have finally recognized that I would not be able to reach my goal along this path. (Lorentz to Wien, 6 June 1908)

[124] Paul Ehrenfest was born in Vienna on 18 January 1880 into a family which hailed from Loschwitz, a small Jewish village in Moravia. He studied physics from 1899 to 1904 at the Universities of Vienna and Göttingen, obtaining his doctorate in 1904 with a thesis on '*Die Bewegung starrer Körper in Flüssigkeiten und die Mechanik von [Heinrich] Hertz*' ('The Motion of Rigid Bodies in Fluids and the Mechanics of [Heinrich] Hertz,' Ehrenfest, 1904) at the University of Vienna (under the supervision of Ludwig Boltzmann). Three years later, after spending some time in Vienna and Göttingen, Ehrenfest and his Russian-born wife Tatyana Alexeyevna Afanassjewa (whom he had met in 1902 in Göttingen and married two years later in Vienna) went to St. Petersburg. There he taught at the Polytechnic Institute. Not being able to secure an adequate position in Russia, Ehrenfest travelled in spring 1911 through Germany and Western Europe. In fall 1912, upon the retirement of Hendrik Antoon Lorentz, he was appointed to the chair of theoretical physics at the University of Leyden. He died there on 25 September 1933, having committed suicide, like his teacher Ludwig Boltzmann.

Ehrenfest made important contributions to problems of statistical mechanics and quantum theory. Together with his wife he wrote the article on 'The Conceptual Foundations of the Statistical Approach in Mechanics' (Paul and Tatyana Ehrenfest, 1912).

In a further paper, submitted to *Physikalische Zeitschrift* more than half a year later (in late June 1906), Ehrenfest continued his analysis of Planck's radiation theory (Ehrenfest, 1906b). He had just received Planck's *Lectures on Heat Radiation*, and now criticized certain points in the new book (Planck, 1906).[125] After repeating the argument (stated clearly in his previous publication, Ehrenfest, 1905) that Planck had not provided a unique definition of the entropy of the radiation in a cavity, Ehrenfest turned to compare Boltzmann's and Planck's concepts of complexions. He first pointed out the fact that Boltzmann's definition inevitably led to the radiation law of Rayleigh and Jeans. Planck, he argued, had applied a different definition in calculating the entropy of blackbody or cavity radiation; in particular, Planck had imposed an additional hypothesis on the one-dimensional resonators in a cavity: namely, the resonators' phase-space points could not assume arbitrary positions in the two-dimensional plane but were restricted to lie on definite closed curves, i.e., on discrete ellipses.[126] In Ehrenfest's opinion, by invoking this hypothesis Planck deviated clearly from the accepted principles of statistical mechanics.

At the same time as Ehrenfest, Albert Einstein also concerned himself with the theory of heat radiation. He arrived, as we have discussed in the previous section, at the conclusion that radiation obeying Wien's law, Eq. (61), consisted of energy- or light-quanta having the magnitude $(R/N_0)\beta\nu$, or $h\nu$ in Planck's notation. Since this conclusion contradicted the electrodynamic description of radiation—the very foundation of Planck's theory of heat radiation—it appeared to Einstein 'as if Planck's theory formed in some way a counterpart to my own work' (Einstein, 1906c, p. 199). In a paper, entitled '*Zur Theorie der Lichterzeugung und Lichtabsorption*' ('On the Theory of Emission and Absorption of Light'), which he submitted to *Annalen der Physik* in March 1906, he took a deeper look at the derivation of Planck's law and noticed that two steps, completely different from each other, were involved in it (Einstein, 1906c). On one hand, Planck had

[125] In his book on the development of blackbody theory, Thomas S. Kuhn claimed that Ehrenfest's analysis of the radiation theory between summer 1905 and summer 1906 had exerted considerable influence on Planck and some of his formulations in the book on *Lectures on Heat Radiation*; Kuhn also referred to letters exchanged between Planck and Ehrenfest (Kuhn, 1978, pp. 161–163). It is difficult to justify this claim, especially since the letters, on which Kuhn has based his claim, do not exist (there is only one letter of Planck to Ehrenfest, dated 6 July 1905, on file, none later, but as a rule Ehrenfest kept his letters very carefully). A closer analysis of the passages in Planck's book, which according to Kuhn exhibit Ehrenfest's influence, reveals that Planck may have arrived at the conclusions by himself. Ehrenfest, at that time, was not yet the sharp, accepted critic that he became later. On the other hand, Planck was certainly glad that his treatment of heat radiation evoked a response from Boltzmann's Institute in Vienna. Since it was the custom at that time that authors sent copies of their new books to interested colleagues, it is possible that Planck sent Boltzmann or Ehrenfest a copy of his *Lectures on Heat Radiation*. In any case, Ehrenfest's article, entitled '*Zur Planckschen Strahlungstheorie*' ('On Planck's Radiation Theory'), which was received on 2 July 1906 by *Zeitschrift für Physik* and published in the issue of 1 August 1906 (Ehrenfest, 1906b), contained numerous references to Planck's book (see, especially, Ehrenfest, 1906b, p. 528).

[126] Planck had already stated this conclusion in his book (Planck, 1906, Section 150, pp. 154–156). Though Ehrenfest was not certain that his formulation of the hypothesis agreed completely with Planck's, no difference can be recognized between the two formulations.

established a relation between U_ν, the average energy of the resonators of frequency ν, and ρ_ν, the energy density of cavity radiation, that is, Eq. (7); on the other, he had calculated the average energy U_ν of a resonator from its entropy. Einstein analyzed, in particular, Planck's definition of the entropy of the resonators and tried to relate it to his own treatment of the foundations of kinetic theory. Thus he started from the definition of entropy in Eq. (42), that is,

$$S = \frac{\overline{E}}{T} + \frac{R}{N} \ln \int \exp\left(-\frac{N}{RT} E\right) dp_1 \cdots dq_n, \qquad (42')$$

with $\overline{E}$ denoting the average value of the total energy (E) of the system at (absolute) temperature T. (E is identical with the Hamiltonian expressed as a function of the dynamical variables $p_1, \ldots, p_n, q_1, \ldots, q_n$ of the system.) The integrations extend over all values of the variables compatible with $\overline{E}$. In the case of a system composed of very many oscillators, the energy fluctuations could be considered to be negligible and the entropy S could be written as

$$S = \frac{R}{N} \ln W, \qquad (65)$$

with the probability W defined by the equation

$$W = \int_{E}^{E+\Delta E} dp_1 \cdots dq_n. \qquad (65a)$$

As Einstein realized, the right-hand side of Eq. (65a) did not depend on the quantity ΔE, provided the expression $R \ln(\Delta E)/N$ could be neglected in the calculation of W. By substituting the dynamical variables of the resonators in Eq. (65a), he obtained the result

$$W = \text{const.} \int_{E}^{E+\Delta E} dE_1 \cdots dE_n, \qquad (65a')$$

in which the integrations extended over E_α ($\alpha = 1, \ldots, n$), the values of the energies of the individual resonators, such that the total energy added to values between E and $E + \Delta E$.

> If one calculated [the entropy] S according to this formula [Einstein noted] then one would arrive again at the inadmissible radiation formula [(53), which corresponded to the Rayleigh–Jeans law, Eq. (62)]. However, one arrives at Planck's formula, if one assumes that the energy E_α [$\alpha = 1, \ldots, n$] of a resonator cannot possess every arbitrary value, but only values which are integral multiples of ϵ, with $\epsilon = (R/N_0)$ · $\beta\nu$ [with $\beta = h/k$ and $R/N_0 = k$, in terms of Planck's constants]. (Einstein, 1906c, p. 202)

He concluded his analysis as follows:

> We must therefore regard the following law as the basis of Planck's quantum theory
> of radiation: The energy of an elementary resonator can only assume values which
> are integral multiples of $(R/N_0)\beta\nu$; the energy of a resonator changes in jumps by
> absorption or emission in integral multiples of $(R/N_0)\beta\nu$. (Einstein, 1906c, p.
> 202)[127]

With those words Einstein focused sharply on what he considered to be the
main hypothesis of Planck's radiation theory, namely, the fact that the resonators
in a cavity change their energies only by finite amounts, that is, discontinuously
in discrete steps. 'If the energy of a resonator can alter only in jumps,' he argued
further, 'then for the evaluation of the average energy of a resonator in a
radiation cavity, the usual [electromagnetic] theory cannot be used, for the latter
does not admit any distinctive energy values for a resonator' (Einstein, 1906c, p.
203).

Two years after Einstein, Lorentz also came to the conclusion that Planck had
introduced an essentially new hypothesis which contradicted the usual laws of
electrodynamics. He stated this explicitly in a letter to Willy Wien after the
Rome conference:

> Even I would be prepared to accept immediately without reservation the hypothesis
> [of elementary quanta of energy], had I not met with a difficulty. This difficulty lies
> in the fact that those resonators, whose [wavelength] λ is appreciably smaller than
> λ_m [the wavelength of the maximum radiation intensity], do not receive even a
> single energy element due to Planck's formula. In other words, some of these
> resonators (under certain circumstances the majority) should possess no energy;
> and yet they are, like the others, exposed to the continuous excitation of the
> electromagnetic waves of the ether. One must realize that due to Planck's theory the
> resonators receive energy from the ether and return it to the ether in a completely
> continuous way (without a finite energy quantum being involved). Still, I do not
> wish to treat this question now in greater detail; I hope to learn soon what Professor
> Planck himself thinks about it. (Lorentz to Wien, 6 June 1908)

Lorentz' opinion on these matters had to be especially respected, for undoubtedly he was the greatest representative of the authority of the existing electrodynamic theory. Moreover, his derivation of the radiation equation (63) from
Maxwell's equations and standard kinetic theory could hardly be criticized, in

---

[127]With $\Delta H = \epsilon$, Eq. (65a′) becomes equivalent to Planck's definition of the number of complexions in the following way. If a resonator can only assume energy values of integral multiples of
$(R/N_0)\beta\nu$, then the integral on the right-hand side of Eq. (65a′) may be replaced by a sum which
expresses the number of possibilities of distributing the total energy in quanta of magnitude
$(R/N_0)\beta\nu$ among n resonators, and this sum is equal to the number of complexions as counted by
Planck in December 1900. (Einstein gave a detailed procedure of arriving at Planck's law in a later
paper: Einstein, 1906g, pp. 180–184.)

contrast to Jeans' derivation which rested heavily on the equipartition theorem and the comparison of radiation to the vibrations of the ether, a concept which had become unfashionable due to the recent development of relativity theory. Planck, therefore, accepted Lorentz' conclusions insofar as he agreed that the Rayleigh–Jeans radiation law followed from the standard theory. He did not agree, however, with Lorentz' and Einstein's opinion that the introduction of the quantum of action contradicted Maxwell's electrodynamics by causing a discontinuity in the description of electromagnetic phenomena. 'I still do not see any reason,' he wrote to Lorentz, 'to abandon the assumption of the absolute continuity of the free ether and all processes in it. As a consequence the quantum of action h is a property of the resonators' (Planck to Lorentz, 7 October 1908). In the following two years he would try to construct a modified theory of heat radiation, which would take into account the quantum of action without violating any aspect of classical electrodynamics. However, in the course of Planck's investigations, as well as the investigations of others, one fact was confirmed over and over again: it was impossible to establish the theory of blackbody radiation entirely on the classical foundation of Maxwell's electrodynamics and the statistical mechanics of Maxwell and Boltzmann. Moreover, new phenomena came to be discovered which also could not be accounted for by means of classical theories, but seemed to require a description in terms of a theory involving the quantum of action. Thus quantum theory emerged as a new, separate and important theoretical discipline.

I.5 The Search for Other Quantum Phenomena

When Planck developed his theory of blackbody radiation involving the energy-quantum, he stated that the latter arose because of the irreversible emission and absorption of electromagnetic waves. Einstein, on the other hand, defined the range of application of his conception of the light-quantum more clearly; he claimed that it played a role in all those phenomena in which light was created or transformed. As special examples he treated fluorescence (in which light of short wavelengths is absorbed by certain substances and light of longer wavelengths is reemitted) and the photoelectric effect (in which the energy of incident ultraviolet radiation is used to liberate electrons bound in a metal and to give them some kinetic energy). Einstein had thus pointed to definite physical phenomena which could be explained on the basis of the light-quantum concept, and the question arose whether there existed other phenomena of such kind. In the years following his pioneering publications (Einstein, 1905b, 1906c), new quantum phenomena were indeed discovered by Einstein himself and others.

Among the most effective early advocates of the quantum concept in Germany was Johannes Stark. Stark was extremely interested in new physical discoveries, such as radioactivity and relativity theory, and had already acquired a reputation through his discovery of the Doppler effect in canal rays (Stark, 1905).[128] He had observed, in particular, that the discrete lines emitted by the hydrogen canal rays had the following structure: a sharp 'stationary' line whose position coincided with one of the lines of the hydrogen series spectrum (H_β, H_γ, etc.), and a diffuse 'moving' line whose intensity dropped to zero sharply on the ultraviolet side while it decreased smoothly on the infrared side of the spectrum. The difference between λ_0, the wavelength of the stationary line, and λ_s, the shortest wavelength of the moving line—the latter being observed in the direction parallel to the motion of canal rays—then yielded the maximum velocity of the hydrogen ions by the equation

$$\frac{\lambda_0 - \lambda_s}{\lambda_0} = \frac{v}{c} , \tag{66}$$

where c denotes the velocity of light *in vacuo*.[129] In his investigations on the Doppler effect of canal rays, Stark also found that canal rays of small velocities did not contribute to the shifted (moving) lines, because an intensity minimum showed up between any stationary line and the corresponding moving one (Stark, 1906d, pp. 439–440). A year later he provided a theoretical interpretation for this fact on the basis of Planck's energy-quantum (Stark, 1907c). For this purpose he assumed that the lines emitted by canal rays where created by collisions among the positive ions constituting canal rays in the following way: a fraction, α, of the kinetic energy, E_{kin}, of the ions is used to excite oscillations of the electrons bound in ions, but the ions radiate only if the transferred energy

---

[128] Johannes Stark was born at Schickenhof (in the district of Amberg), Bavaria, on 15 April 1874. He studied physics, mathematics, chemistry and crystallography at the University of Munich from 1894 to 1897, obtaining his doctorate in 1897. He then continued to work in Munich as assistant of E. von Lommel until 1900, when he received his *Habilitation* at the University of Göttingen. In 1906 he was appointed *Dozent* (with the title of professor) at the *Technische Hochschule*, Hanover, and three years later he became Professor of Physics at the *Technische Hochschule*, Aachen. In 1917 he moved to the University of Greifswald and in 1920 to the University of Würzburg. In 1922 he retired from his academic career. Being a supporter of the Nazi movement, he became in 1933 President of the *Physikalisch-Technische Reichsanstalt*, Berlin, and in 1934 also President of the *Notgemeinschaft der Deutschen Wissenschaft*. Stark retired in 1939 from the *Reichsanstalt* and returned to Bavaria. A denazification court sentenced him in 1947 to four years in a labour camp. He died on 21 June 1957 at his country estate in Eppenstatt near Traunstein, Bavaria.

Stark's scientific work covered the fields of electric discharges in gases, spectroscopy and the study of chemical valency. In 1919 he was awarded the Nobel Prize in Physics for his discovery of the Doppler effect in canal rays and the splitting of spectral lines in electric fields, the so-called Stark effect. He was a member of many learned societies, including the academies in Göttingen, Rome and Vienna, and received many honours, such as the Matteucci Medal of the *Accademia dei Lincei*. He propagated information about new physical developments through the *Jahrbuch der Radioaktivität und Elektronik*, which he founded in 1904. Reviews as well as original articles on modern experimental and theoretical physics were published in the *Jahrbuch*; thus Einstein contributed in 1907 a report on relativity theory (Einstein, 1907d).

[129] Einstein (1907b) himself proposed to test the relativity principle with observations on the second-order effect in v/c, which Stark had discussed earlier (Stark, 1906b).

takes on the minimum amount,

$$E_{kin}\alpha = h\nu, \tag{67}$$

where h denotes Planck's constant and ν the frequency of the emitted line. (The value of α, entering on the left-hand side of Eq. (67), should depend on the frequency ν and on the ion under consideration.)

> From the above law [Stark noted] there follows directly the important postulate that by the impact of a moving atomic ion a characteristic electron oscillation will only then be excited and the corresponding series line emitted, if the kinetic energy of the atomic ion (canal ray) exceeds a certain threshold value, which is characteristic of the ion in question. Hence, if one increases the translational velocity of an atomic ion, starting from the value zero, then a given series line will initially possess no intensity at all; but when the threshold value of the translational velocity has been surpassed, suddenly the emission of the series line takes place. (Stark, 1907c, p. 914)[130]

Stark completed the explanation of the Doppler effect of canal rays by assuming that the unshifted 'stationary' line was caused by the collision of ions with cathode rays, i.e., free electrons. 'At this point,' he remarked, 'there occurs a transformation of the kinetic energy of the cathode rays into the radiation energy associated with the electrons bound in the atomic ion' (Stark, 1907c, p. 917). Stark now realized that, because of the small mass of cathode rays in comparison to the mass of the ions, the fraction of energy thus transferred was very small. 'If, therefore, the temperature of the gas is low,' he concluded, 'then the atomic ion, which has been hit by a canal ray, can be considered as being at rest, and therefore the series line emitted with [finite] intensity after collision with the [moving] canal ray will also be at rest' (Stark, 1907c, p. 917).

Although only little was known for certain about the structure and the radiation properties of canal rays—the Rutherford–Bohr atomic model did not exist yet—Stark's interpretation of the Doppler effect intensity of canal rays immediately aroused some criticism. Especially, Willy Wien concluded from his own experiments that the observed canal ray radiation seemed to be emitted also by neutral objects—namely, by the atoms that were created by the recombination of canal rays with electrons (Wien, 1908). On the other hand, Stark received encouraging support from Albert Einstein and Arnold Sommerfeld. The latter wrote to him on 10 October 1908: 'I would be very grateful to you for [sending me] a spare spectrogram [i.e., a photograph showing the above-mentioned structure of the line emitted by hydrogen canal rays], which I would like to use in my lectures, but more importantly for the purpose of converting myself

---

[130]Stark also concluded that the energy available for radiation—and, therefore, the intensity of the Doppler-shifted radiation—would remain constant if the quantity $E_{kin} \cdot \alpha$ increased beyond the value $h\nu$; however, it would grow discontinuously at the values $2h\nu$, $3h\nu$, etc. (Stark, 1907c, p. 915).

definitively to Planck's fundamental hypothesis.'[131] And Hendrik Lorentz, in spite of his opposition to the light-quantum hypothesis, quoted Stark's canal ray observations as providing evidence for it (Lorentz, 1910a, p. 352). Many years would pass until the question was finally settled, the result being against the quantum interpretation.[132]

In late 1907, besides the Doppler effect of canal rays, Stark mentioned two other quantum effects in a footnote to his paper on '*Elementarquantum der Energie, Modell der negativen und positiven Elektrizität*' ('Elementary Quantum of Energy, Model of the Negative and Positive Electricity'), which he submitted in October 1907—i.e., a month before the paper on the quantum interpretation of the Doppler effect (Stark, 1907b, p. 882). Stark claimed first that 'Planck's fundamental law' ('*Plancksche Elementargesetz*'), Eq. (23') [i.e., the relation $\epsilon = h\nu$, connecting the energy ϵ and the frequency ν of radiation absorbed and emitted by an oscillator], provided an explanation of the minimum wavelength of X-rays produced by the impact of electrons of kinetic energy eV, with e the charge of the electron and V the potential difference; this minimum wavelength should be given by the equation

$$\lambda_{min} = \frac{2hc}{eV} , \tag{68}$$

where h and c denote Planck's constant and the velocity of light *in vacuo*, respectively (Stark, 1907b, 1908b).[133] For instance, by applying Eq. (68) to X-rays created by electrons passing through a potential difference of 60 kV, Stark found a minimum wavelength of 6×10^{-9} cm, in agreement with available data obtained from the diffraction of X-rays (Haga and Wind, 1903).[134] As a second example of the 'fundamental law' he considered the effect inverse to the production of X-rays, namely, the creation of free electrons from surfaces hit by

---

[131] On receiving a reprint of the paper, Einstein also wrote to Stark: 'Especially your application of the light-quantum hypothesis to the curve representing the Doppler effect [of canal rays, i.e., its intensity as a function of the frequency] has pleased me extremely' (Einstein to Stark, 2 December 1908).

[132] It should be mentioned that in 1911 one of Willy Wien's associates, H. Baerwald, investigated the origin of canal rays, again confirming that most likely the objects emitting the radiation were neutral particles (Baerwald, 1911). It was further assumed that to the detailed structure of the lines emitted by hydrogen canal rays—Friedrich Paschen had found that the shifted line consisted of two separate components (Paschen, 1907)—also the hydrogen molecule-ion (H_2^+) contributed (Gehrcke and Reichenheim, 1910). With the acceptance of the Rutherford–Bohr model of atomic structure, Stark's quantum interpretation of the Doppler effect of hydrogen canal rays had to be abandoned. Empirically the gap between the stationary and the moving line was then explained by the fact that in the velocity spectrum of canal rays small velocities did not really occur.

[133] Stark obtained Eq. (68) from the following argument. The minimum wavelength of X-rays is found by assuming that the total kinetic energy of the electron, eV, is transformed into radiation energy during a period $T = \lambda_{min}/2c$. Then Planck's relation, $\epsilon = h\nu$, may be used in the form, $hT^{-1} = eV$, yielding the desired result.

[134] Independently, Willy Wien also arrived at the conclusion that Planck's relation, $\epsilon = h\nu$, determined the short-wavelength limit of X-rays (Wien, 1907). In his theory λ_{min} had only half the value given by the right-hand side of Eq. (68).

ultraviolet radiation. He argued that the maximum velocity of the electrons, arising from the photoelectric effect, must be given by the equation

$$v_{\max} = \frac{2hc}{m_e \lambda} \tag{69}$$

with m_e denoting the mass of the electron and λ the wavelength of the incident radiation. Stark did not realize—in October 1907—that this conclusion had already been obtained by Einstein more than two years earlier (Einstein, 1905b). Therefore, in his following paper, he added a remark acknowledging the priority of Einstein's application of 'a similar consideration concerning the photoelectric cathode rays' (Stark, 1907c, p. 914, footnote 1).

During the following years Stark became very enthusiastic about new applications of Planck's quantum relation, Eq. (23'). He used it, in particular, in connection with his views on atomic and molecular structure to explain the properties of series and band spectra (Stark, 1908a, d) and the fluorescence of organic substances (Stark and Steubing, 1908a, b). At the same time he continued, together with Wilhelm Steubing of the University of Greifswald, to do research on the Doppler effect of hydrogen canal rays (Stark, 1908c; Stark and Steubing, 1909).[135] 'The work,' he said in explaining his motivation, 'had been undertaken to test the light-quantum hypothesis, which was established by M. Planck and whose experimental significance was pointed out by Einstein quite early in several papers [Einstein, 1905b; 1906c, g]' (Stark, 1908c, p. 768). In this connection Stark considered the 'light-quantum hypothesis' to be synonymous with what he called 'Planck's elementary law,' and stated explicitly:

> In my opinion it is not necessary to postulate a discontinuous structure of the radiation energy, which flows with the velocity of light in ether; rather there exists in the process considered, which is governed by the elementary law, a specific type of action exerted by the electromagnetic resonators. (Stark, 1908c, p. 768)

In that sense he also applied the 'light-quantum hypothesis' to photochemical reactions, i.e., to explain chemical reactions in substances which are caused by incident radiation (Stark, 1908d). For this purpose he assumed, especially, that molecules (or atoms) consisted of ions and bound valence electrons, and that the binding between them was changed by the absorption of radiation. In the 'primary' photochemical reaction, then, either a valence electron was completely separated, leaving a free ion, or a molecule was split into two parts (basically two ions). The resulting ions, on the other hand, could react in 'secondary'

---

[135] Stark spent the year 1907–1908 at the University of Greifswald to substitute for Oskar Hermann Starke, *Extraordinarius* in physics. There he met and collaborated with Friedrich Wilhelm Steubing, who was born on 12 June 1885 at Dillenburg, Hessen-Nassau. Steubing later followed Stark to Aachen and assisted him in important experimental investigations, e.g., on the Stark effect. In 1925 he became *Ordinarius* at the University of Breslau, and in 1945 he moved to the University of Hamburg, from where he retired in 1954.

photochemical processes, in which the incident radiation played no role; for example, they might form new molecules or capture a free electron to create neutral molecules. Now the primary or 'direct' photochemical reactions were ruled by the following laws: first, since the molecules absorbed only the light of a given characteristic frequency, the reaction must be monomolecular; second, since the average kinetic energy per molecule at the usual temperatures assumes values far below the energy of a light-quantum (of visible or ultraviolet radiation), the primary photochemical reactions do not depend on temperature; third, the number of molecules involved must be proportional to the intensity of the absorbed light.[136]

While Stark initially, i.e. in 1908, *did not adhere* to Einstein's point of view concerning an atomistic constitution of radiation (see, e.g., the application of the quantum law, Eq. (23') to explain the sputtering action of radiation on surfaces in Stark, 1908e), he changed his opinion several months later. In a paper, entitled '*Über Röntgenstrahlen und die atomistische Konstitution der Strahlung*' ('On X-rays and the Atomistic Constitution of Radiation'), he developed a detailed corpuscular theory of short-wavelength radiation (Stark, 1909b). He employed the following arguments. If an accelerated electron emits radiation of frequency v in elementary quanta of magnitude hv, h being Planck's constant, then for a given amount of energy available for radiation the number of emitted energy-quanta associated with high frequencies is very small. Even in the case of very intense X-ray sources, Stark concluded:

> The average free period of the Röntgen [X-] radiation is very large compared to the period of the Röntgen-ray quantum; the emission of Röntgen-ray quanta occurs at random, fluctuating in time; the coincidence in time of the emission of several Röntgen-ray quanta occurs quite rarely as compared to the case of the emission of a single quantum. (Stark, 1909a, p. 582)[137]

---

[136] Several years later Albert Einstein gave a treatment of the photochemical laws on the basis of quantum theory (Einstein, 1912a, b). Johannes Stark, in a subsequent note (Stark, 1912a), then drew attention to his earlier treatment, to which Einstein replied by stating: 'J. Stark has written a note concerning a paper of mine, published recently [Einstein, 1912a], with the purpose of defending his intellectual property. I do not wish to comment on the question of priority implied in it, for it would hardly interest anyone, especially since the photochemical equivalence law must be regarded as a completely obvious consequence of the quantum hypothesis' (Einstein, 1912c, p. 888). He further remarked that his new derivation was different and founded on thermodynamic arguments. Stark pointed out in a second note that he agreed with the last statement; but he still claimed that he had obtained the photochemical laws on arguments different from Einstein's in his paper on the light-quantum (Stark, 1912b).

In the above analysis we have confirmed Stark's opinion that he had derived independently the laws governing photochemical processes. In 1912 the relations between Einstein and Stark had already become less cooperative than they had been in previous years, and they would deteriorate further in the future. Einstein's harsh answer in the question concerning the priority of the formulation of photochemical laws may have triggered Stark's unhappiness.

[137] Stark also emphasized that he had arrived at his corpuscular view of radiation by means of different argumentation than Einstein's in 1905. He remarked:

When I apply the light-quantum hypothesis [note that Stark's "light-quantum hypothesis" was not the same as Einstein's, but was synonymous with Planck's law, Eq. (23')] to the results on X-rays, I arrive at the same consequence as Einstein by following a path which is different from Einstein's, but perhaps less intricate and shorter. In the following, I obtain the conclusion that

It should be noted that Stark did not assume that X-rays had a structure different from that of long-wavelength radiation—as had been at the basis of Einstein's reasoning in 1905; he rather believed that radiation of all wavelengths was described by what he called the 'light-quantum hypothesis,' i.e. by the quantum law, Eq. (23′). The corpuscular structure of radiation that thus arose, did, of course, contradict the usual concept of radiation. Stark gave the latter a new formulation as what he called the 'ether-wave hypothesis' ('*Ätherwellenhypothese*'). According to this hypothesis, which he claimed to agree with the classical radiation theory, radiation—even when emitted by single electrons—should propagate symmetrically in all directions in space (Stark, 1909b, p. 584). Only for large intensities and small frequencies did both hypotheses, the 'light-quantum' and the 'ether-wave' hypotheses, yield an equivalent description of radiation; in that case, many quantum-theoretical packets of radiation would combine to produce a net description, say of the diffraction of light, similar to the one given by the classical radiation theory. Stark claimed further that one of the crucial differences between the two hypotheses was the following: due to the 'ether-wave' hypothesis the total radiation emitted by an accelerated electron possessed no momentum, but a finite momentum (of magnitude $h\nu/c$ per energy quantum) due to the light-quantum hypothesis. He concluded, therefore, that the observed asymmetry in the emission of X-rays from the anticathode decided clearly in favour of the quantum structure of radiation (Stark, 1909d).

While Stark applied his version of the light-quantum hypothesis successfully to explain various radiation phenomena (see, e.g., Stark, 1909c, 1910b), Arnold Sommerfeld raised serious objections against Stark's arguments. There developed a rather polemical debate between the experimentalist Stark and the senior, experienced theoretician Sommerfeld, which was pursued in several publications in the *Physikalische Zeitschrift* during late 1909 and early 1910 (Sommerfeld, 1909b, 1910a; Stark, 1910a, b, c).[138] This debate, in the course of which Sommerfeld pointed out that the 'ether-wave hypothesis,' as advocated by Stark, did not follow from but contradicted classical electrodynamics, led to important consequences which influenced the development of quantum physics in Germany. One was that Johannes Stark isolated himself increasingly from the other scientists who carried on the development—especially from the theoreticians such as Einstein and Sommerfeld. Sommerfeld would become the most effective advocate and promoter of the quantum concepts in years to come. We shall now briefly review his background.

Arnold Johannes Wilhelm Sommerfeld was born on 5 December 1868 at Königsberg, the son of the physician Franz Sommerfeld and his wife Cäcilie, *née*

---

the consequence of an atomistic constitution of radiation cannot be avoided *whenever the transformation of energy in the individual resonators (electrons) satisfies the law of light-quanta.* (Stark, 1908b, p. 583)

[138] The details of the debate between Stark and Sommerfeld on the 'light-quantum hypothesis,' as pursued especially in the letters exchanged by the two scientists, have been discussed by Armin Hermann (1967; 1969, Chapter IV).

Mathias. He studied mathematics at the University in his hometown from 1886 to 1890; among his teachers there were the mathematicians Ferdinand Lindemann, Adolph Hurwitz and David Hilbert, and the physicist Emil Wiechert. In 1891 he obtained his doctorate with a thesis on the arbitrary functions employed in mathematical physics. After passing the examination for the (high school) teacher's certificate and doing his military service, he moved to the University of Göttingen, becoming Felix Klein's assistant the following year and *Privatdozent* in mathematics in 1896 with a *Habilitation* thesis on the mathematical theory of diffraction (Sommerfeld, 1896). Sommerfeld's scientific career flowered quickly: in 1897 he obtained the chair of mathematics at the *Bergakademie* in Clausthal, in 1900 the chair of applied mechanics ('*Technische Mechanik*') at Aachen's *Technische Hochschule*, and in fall 1906 he was appointed Professor of Theoretical Physics at the University of Munich. During this period his scientific interests had shifted from primarily mathematical problems to those of engineering and physics, and he turned into one of the most outstanding representatives of theoretical physics in Germany, working on the fundamental problems of mechanics, fluid dynamics and electromagnetic theory. Thus Sommerfeld came into close contact with problems at the frontiers of physics, especially with the dynamical theory of the electron (Sommerfeld, 1904a, b; 1905). At first he was opposed to the principle of relativity, but he changed his opinion in 1907 and became an important supporter of Einstein's theory (Sommerfeld, 1907).

Sommerfeld first became interested in X-rays in 1899 and he wrote papers on the theory of their diffraction (Sommerfeld, 1899, 1900).[139] In these contributions he used the assumption, which George Gabriel Stokes (1896) and Emil Wiechert (1896) had proposed and Joseph John Thomson (1898a) had further explored, that X-rays represented perturbations of the ether or short electromagnetic pulses. Sommerfeld found that the ether pulse gave rise to diffraction fringes when scattered by an impenetrable screen or by a slit, similar to the fringes obtained from the scattering of light. It appeared that the Dutch physicists Hermanus Haga and C. N. Wind of Groningen University had already observed such diffraction fringes (Haga and Wind, 1899); Sommerfeld now analyzed their data and obtained wavelengths of X-rays that were of the order of magnitude of 10^{-8} cm (Sommerfeld, 1900, p. 59). The validity of the diffraction interpretation of Haga and Wind's results was questioned in the following years (see, e.g., Walter, 1902); therefore, they repeated their experiment with improved techniques and concluded: 'We have obtained diffraction phenomena that are more pronounced than previously; thus we are now more convinced than before that the Röntgen [X-] rays have to be regarded as a radiation process in the ether' (Haga and Wind, 1903, p. 305).[140] For the X-rays employed, they derived the values of wavelengths between 5×10^{-9} and 16×10^{-9} cm. Even those measurements, however, did not settle the question completely; thus B. Walter and

---

[139] Sommerfeld's work was stimulated by his former teacher and friend Emil Wiechert (1861–1928), whom he knew from Königsberg (see Sommerfeld, 1899, p. 105).

[140] It should be noted in this connection that C. N. Wind had shown earlier that previous X-ray experiments had not really demonstrated the existence of diffraction phenomena (Wind, 1898).

Robert Pohl, who carried out further experiments, remarked as late as 1909: 'The diffraction of Röntgen rays has not yet been proved even by the above experiments; hence the wave nature of the rays cannot yet be regarded as being supported by diffraction experiments' (Walter and Pohl, 1909, p. 354).

The view that X-rays were electromagnetic radiation of very short wavelengths received strong support by the work of Charles Glover Barkla in Liverpool.[141] Stimulated by J. J. Thomson's theory of the scattering of electromagnetic pulses by the electrons contained in matter (Thomson, 1903), he investigated experimentally the scattering and double scattering of X-rays and proved that X-rays could be polarized (Barkla, 1905, 1906).[142] In 1908 Barkla provided a further argument favouring the electromagnetic theory of X-rays: the analysis of secondary X-rays, which were obtained when the X-rays produced in a tube ('primary rays') passed through matter, showed that a new type of (secondary) X-rays existed that were homogeneous with respect to their penetration property (Barkla and Sadler, 1908a, b). The hardness of these 'homogeneous' X-rays, which could be related—on the basis of electromagnetic theory—to their frequency, was characteristic of the scattering substance; hence the homogeneous, characteristic X-rays could be interpreted as representing parts of a discrete spectrum of very short wavelengths emitted by atoms, similar to the line spectra observed in the visible, the infrared and the ultraviolet regions.[143]

Besides Barkla, another English scientist contributed to the discussion about the nature of X-rays: he was William Henry Bragg, then at the University of Adelaide, South Australia.[144] Bragg had become interested in 1904 in the new field of radioactivity. He was especially struck by the similarity between the

[141] C. G. Barkla was born 7 June 1877 at Widnes, Lancashire. He studied mathematics and physics at University College, Liverpool (1894–1898), and continued with an 1851 Exhibition Scholarship at Trinity College, Cambridge, working under J. J. Thomson at the Cavendish Laboratory. Then he spent two years at King's College, London (1900–1902), and returned to Liverpool in 1902 as Oliver Lodge Fellow. From 1905 to 1909 he served successively as demonstrator, assistant lecturer in physics and special lecturer in advanced electricity at the University of Liverpool. In 1909 he became Wheatstone Professor of Physics in the University of London, and four years later he accepted the Chair of Natural Philosophy in the University of Edinburgh. He held this position until his death on 23 October 1944. Barkla worked mainly on the properties of X-rays.

[142] The polarization of X-rays was confirmed by H. Haga (1907) and Eugen Bassler (1909).

[143] For the discovery of characteristic X-rays, Barkla was awarded the Nobel Prize in Physics for 1917.

[144] W. H. Bragg, born at Westward, Cumberland, on 2 July 1862, studied at King William's College, Isle of Man, and from 1881 at Trinity College, Cambridge. After passing the Mathematical Tripos (Third Wrangler in Part I of 1884 and First Class in Part II of January 1885) he worked at the Cavendish Laboratory. At the end of 1885 he was elected to a professorship of mathematics and physics at the University of Adelaide, Australia; he returned to England in 1909 as Cavendish Professor of Physics at the University of Leeds. In 1915 Bragg was appointed Quain Professor of Physics at University College, London, and ten years later Fullerian Professor of Chemistry at the Royal Institution. He died on 10 March 1942 in London.

Bragg, who did research in many fields, especially radioactivity and X-rays, received many honours. He was elected a Fellow of the Royal Society in 1907 and became its President in 1935. For his services during World War I and his scientific achievements, he was knighted in 1920. He was awarded the Rumford Medal (1916) and the Copley Medal (1930). He shared with his son William Lawrence Bragg the 1915 Nobel Prize in Physics for their joint work on the analysis of crystal structure by the X-ray diffraction method.

properties of the γ-rays emitted by radioactive substances and the X-rays, and he began to look for a unified picture to describe both of them. Thus he developed a corpuscular theory based on the following argument. Since γ-rays always seemed to be accompanied by the emission of α- and β-rays, it was suggestive to assume that γ-rays might consist simply of a pair of α- and β-particles, which were electrically bound together to form a 'neutral pair'; and the same should be true of X-rays (Bragg, 1907).[145] When Bragg's paper became known in England, it stimulated a lively discussion, which was carried on in *Nature* and *Philosophical Magazine*, Bragg's main opponent being Barkla.[146]

Barkla argued, in particular, that the angular distribution of X-rays supported the electromagnetic pulse theory, because the observed data confirmed Thomson's scattering theory; he claimed that the 'neutral-pair' theory gave wrong results (Barkla, 1908a). Bragg responded that his experiments with ionization chambers showed a considerable lack of symmetry between forward and backward scattering of X-rays, in contrast to Barkla's theory (Bragg, 1908a, b; Bragg and Madsen, 1908). On the other hand, the existence of secondary characteristic X-rays—as well as similar characteristic γ-rays—was difficult to explain on the basis of the 'neutral-pair' theory, while it seemed to fit well into the electromagnetic pulse theory; all one had to assume was that the primary pulse induces 'a disturbance in the atom, which quickly recovers its normal configuration' by emitting a characteristic X-ray (Barkla and Sadler, 1908b, p. 579). A further difficulty in Bragg's theory was to explain the fact that X-rays apparently travelled with the velocity of light (Marx, 1905). Still, experiments on the emission of β-rays from matter caused by incident γ-rays supported the 'neutral-pair' theory: not only was the emission asymmetrical and peaked forward (i.e., the emitted β-rays had mainly the direction of the incident γ-rays), but also the same γ-rays (of a given hardness) always produced β-rays of the same velocity (Bragg, 1908; Bragg and Madsen, 1908).[147] The debate between Barkla and Bragg thus ended without any agreement being reached; at the end of 1909 both opponents were still unshaken in their respective beliefs. Then Bragg discovered Johannes Stark's work on the corpuscular nature of X-rays, especially his discussion of their asymmetric production (Stark, 1909d). And he also discovered Arnold Sommerfeld's response to Stark's arguments, namely, the proof that the asymmetry could be accounted for perfectly by the electromagnetic pulse theory (Sommerfeld, 1909b).

Sommerfeld, who had followed Stark's investigations with interest and sympathy for several years and had helped to secure for him the position of a professor of physics at Aachen's *Technische Hochschule*, could not agree with Stark's

[145] The fact that the α-particles were doubly-charged objects was still unknown at that time.

[146] For a detailed account of the Bragg–Barkla debate on the nature of X-rays, see the article of Roger H. Stuewer (1971).

[147] This result agreed to some extent with the observations of the photoelectric effect, provided one used hard X-rays; then the potential difference, which the bound electrons had to pass through in order to become free electrons, could be neglected. (See Section I.3.)

theoretical arguments against the electromagnetic pulse theory.[148] Thus he wrote to Stark in December 1909:

> Your last paper on Röntgen [X-] rays has stimulated me to publish considerations, which have occupied me for some time and which I have often discussed with Röntgen. He assigned the experimental check of these ideas as a doctoral thesis [to someone] in his Institute. My considerations are rooted in the pure "ether-hypothesis" and have led inevitably to those phenomena which you have obtained experimentally. (Sommerfeld to Stark, 4 December 1909)

Sommerfeld added that already in 1906 he had requested Haga to look for the angular dependence of X-ray production, but the Dutch experimentalist had not found anything. He remarked further:

> Concerning the details I refer you to my note which I have sent today to *Zeitschrift für Physik* [Sommerfeld, 1909b] and you will receive a copy of its proof-sheets. You will, I hope, become convinced that the [*Bremsstrahlung*] theory immediately supplies the result, for which you invoke the (in any case very hypothetical and unverified) light-quantum theory. Of course, I do not doubt the importance of the quantum of action. But the interpretation, which you give to it, appears not only to me but also to Planck very audacious. (Sommerfield to Stark, 4 December 1909)

Sommerfeld, in his note on the angular distribution of X-rays produced by cathode rays hitting the anticathode, proposed the following interpretation (Sommerfeld, 1909b, pp. 970–972). An electron of velocity v is decelerated in an atom and emits X-radiation; if the velocity is small, the maximum intensity of the radiation is emitted in the direction perpendicular to the electron's motion, i.e., at an azimuthal angle $\phi = \pi/2$, but the intensity maximum shifts to smaller angles for larger velocities due to a relativistic effect. For the angular dependence of the X-ray intensity S_ϕ, he found the equation (Sommerfeld, 1909b, p. 972, Eq. (5))

$$ S_\phi = \frac{\sin^2\phi}{\cos\phi}\left[\left(1 - \frac{v}{c}\cos\phi\right)^{-4} - 1\right] \tag{70} $$

which seemed to be in agreement with the experimental findings of Stark and others.[149] Thus he arrived at the conclusion that X-ray production could be explained completely by the deceleration of electrons on the basis of the classical electromagnetic theory. However, concerning that part of the X-rays, which Stark had called 'fluorescence radiation' and which was identical with the homogeneous X-rays observed by Barkla and Sadler, he remarked: 'It is quite

---

[148] Sommerfeld, himself formerly a professor at Aachen, had proposed Stark as the successor of Adolph Wüllner (1835–1908). (See Hermann, 1967, pp. 40–43.)

[149] Besides Stark's work, Sommerfeld referred to the experimental results of W. C. Kaye (1909) and the 1908 thesis of Eugen Bassler, a student of Röntgen (Bassler, 1909).

possible that in this connection Planck's quantum of action plays a role. I would like to see the calculation of the wavelengths of Röntgen rays—which has been proposed by J. Stark [1907b] and W. Wien [1907]—restricted to this (in general important) aspect of radiation' (Sommerfeld, 1909b, p. 970).

Stark was perturbed by Sommerfeld's new findings and the formulation he gave them. On one hand, the 'deceleration theory' ('*Bremstheorie*') of X-rays destroyed his claim that the angular asymmetry in the production of X-rays could not be described in terms of classical electrodynamics; as Sommerfeld had pointed out in his letter and paper, Stark's 'ether-wave hypothesis' ('*Ätherwel-lehypothese*') did not agree with Maxwell's theory, and Stark was not happy about it. On the other, he felt that Sommerfeld had acted like a schoolmaster towards him without really good reason. He was most offended by another fact: Sommerfeld had based his conclusions on the experimental results of other physicists and had in some cases taken them more seriously (especially when they disagreed with Stark's). He wrote to Sommerfeld:

> That, on one hand, you present my observations—which I obtained by very hard work, employing extreme care and constant criticism—as being unreliable, and take, on the other hand, the work of naïve beginners—who were not able to see through the complex of phenomena under investigation—as a reliable basis for your theory, sounds strange to me in several respects. If one describes the observations of a serious author as being unreliable, then in my opinion one has to point to specific sources of error [in my experiments], and one has to stress the real advantages of the methods [of other authors] used for comparison; otherwise the criticism appears to be arbitrary, subjective and personal. (Stark to Sommerfeld, 18 December 1909)

It seemed to Stark that Sommerfeld had attacked not only his theoretical but also his experimental ability. Besides, he had taken away the priority of certain discoveries, such as the observation of the asymmetric emission of X-rays, which Stark claimed to be his own. He carried on the debate with Sommerfeld in print (Stark, 1910a, b, c), to which the latter replied (Sommerfield, 1910a). They could not come to an agreement and their previously cordial, if not hearty, relationship became worse and worse as the years went on. The episode involving Sommerfeld in late 1909 marked the beginning of Stark's isolation from the physicists who contributed primarily to the progress of physics in the following decades. Stark gradually withdrew from work in modern theory, and he made only one other important experimental discovery—the separation of spectral lines emitted by atoms under the action of an electric field (Stark, 1913b).[150]

Sommerfeld's results also bothered William Henry Bragg who, in early 1910, entered into a correspondence with Sommerfeld on the interpretation of phenomena that occurred when X-rays were created by electrons and vice versa. In his first letter he wrote:

---

[150] For the discovery of this effect, which was called Stark effect, and his previous discovery of the Doppler effect in canal rays, Stark was awarded the Nobel Prize in Physics in 1919.

I think it is not possible to work for days and months on the conversion of X-ray energy into cathode ray energy and γ-ray energy into β-ray energy without seeing how simply and completely the whole behaviour is represented by the hypothesis, . . . : "an electron may by encounter with an atom, have its charge neutralized and its mass only slightly altered (relative to its own); it then takes the form of the X-ray or the γ-ray as the case may be; it may again lose the neutralizing complement and become a secondary cathode- or β-ray, the double transformation being accompanied by no very great change of speed." (Bragg to Sommerfeld, 7 February 1910)[151]

This view, remarked Bragg, only failed to provide an obvious explanation of the observed polarization of X-rays. 'On the other hand,' he argued, 'the pulse theory fails in a far more important and fundamental particular, viz. that to which you allude in §8 (p. 100) of your last paper [Sommerfeld, 1910b]. Three years ago I felt this so strongly that I ceased to use the pulse theory as a guide to experimental research, and adopted the above theory instead, and I think I have been well rewarded, no matter whether this theory be true or not' (Bragg to Sommerfeld, 7 February 1910). By the difficulty, to which Sommerfeld had alluded, Bragg meant the production of (secondary) cathode rays of nearly identical velocities by X-rays, which could be explained easily by means of the corpuscular hypothesis, but not on the basis of the electromagnetic pulse theory.

Sommerfeld, in his reply to Bragg, admitted the weakness of the electromagnetic theory of X-rays, but he also stressed again the advantages of his 'deceleration theory' ('*Bremstheorie*') in providing an explanation of the asymmetric production and polarization phenomena of X-rays (Sommerfeld to Bragg, 8 February 1910). Nevertheless, he felt impressed by Bragg's arguments in dealing with the β- and γ-ray emission in radioactive decays, and with the transformation of X-radiation into photoelectric electrons and vice versa, as the same type of processes. And he concerned himself with this problem during the following months. In a paper, entitled '*Über die Struktur der γ-Strahlen*' ('On the Structure of γ-Rays') and presented at the Bavarian Academy on 7 January 1911, he developed a complete theory of β- and γ-emission based on the following ideas (Sommerfeld, 1911a). He assumed that β-rays, when leaving the effective range ('*Wirkungssphäre*') of an atom, were accelerated from velocity zero to their final velocity v, and that in this acceleration process the γ-radiation was emitted (in agreement with the usual electromagnetic theory). Then he calculated the ratio of the energies of β- and γ-rays, E_β / E_γ, obtaining the result (again from electromagnetic theory)

$$\frac{E_\beta}{E_\gamma} = \frac{6\pi m_e c^2 l}{e^2} \frac{\sqrt{1 - v^2/c^2}}{v/c} , \tag{71}$$

---

[151] For the letters of Bragg and Sommerfeld see: Stuewer, 1971, pp. 269–273. Note that, in 1910, Bragg had changed his 'neutral pair' theory; the 'neutral pair' now had a mass close to the mass of the electron and was not composed of α- and β-particles anymore.

where m_e and e denote the mass and charge of the electron, respectively, c the velocity of light, and l the length of the path of acceleration (which is of the order of 10^{-8} cm). Now in order to determine the value of l, the effective range of the radioactive atom for β-decay, Sommerfeld introduced the quantum concept.

> However hypothetical the considerations employed thus far may seem to be [he remarked] we shall still continue one step further by introducing a new hypothesis, in order to express the ratio E_β / E_γ only in terms of known quantities and as a pure function of the velocity. For this purpose we shall take over the fundamental hypothesis of Planck's radiation theory for the explanation of radioactive emissions [of β- and γ-rays], assuming that in any such emission process exactly one quantum of action h is transferred. (Sommerfeld, 1911a, pp. 24–25)

Since the 'action' of the emission had to be put equal to the product of the acceleration period, τ, times the total energy, he obtained the equation

$$\tau \frac{m_e c^2}{\sqrt{1 - v^2/c^2}} = h, \tag{72}$$

which determined τ and therefore, also the quantity l in Eq. (71).[152]

Sommerfeld's new theory of β- and γ-decay, though it was not inconsistent with the available data, did not satisfy Bragg. He noticed that it would not resolve the difficulties he had in mind; in particular, it provided no answer to the problem of how to produce a β-ray from a γ-ray. At that time, people saw no distinction between the phenomena occurring in atoms and in nuclei. Hence it seemed necessary to explain simultaneously the radioactive decay, on the one hand, and the photoelectric effect, on the other. Within a few years one would learn more about this difference and recognize that Sommerfeld's 1911 theory of β-decay represented a too-premature approach based upon an insufficient knowledge of the nature of the phenomena involved. While it would take several decades until a satisfactory description of the nuclear processes was achieved, one would soon be able to make progress in the problems of the production and transformation of X-rays. It would turn out that the X-rays had indeed to be considered as electromagnetic radiation of very short wavelengths, and the corpuscular viewpoint developed by Einstein, Thomson, Stark and Bragg would be abandoned for a while. This development would confirm the attitudes of Planck, Lorentz and Sommerfeld without, however, preventing the progress of the quantum theory. Sommerfeld, especially, would apply his first quantum *Ansatz*, contained in Eq. (72), to an increasing number of quantum phenomena

---

[152] The relation between τ and the acceleration length l was given by the equation, $\tau = (lv/c^2) \cdot (1 - \sqrt{1 - v^2/c^2})^{-1}$.

and become one of the most productive and powerful advocates of the quantum concept.

I.6 Specific Heats, New Quantum Hypotheses and the First Solvay Conference

In 1872 Heinrich Friedrich Weber (1843–1912) observed from careful measurements that the atomic heat of diamond, which had been found to have at room temperature a much smaller value than other monatomic substances (of about 1.8 cal per gram-atom), reached at high temperatures (i.e., at 1300°C) nearly the standard value of 6 cal per gram-atom (Weber, 1872). The latter value had been established many decades earlier by Pierre Louis Dulong and Alexis Thérèse Petit, who had observed the constancy of the product of specific heat and atomic weight of substances (Dulong and Petit, 1819).[153] Franz Ernst Neumann had later on extended the rule of Dulong and Petit, which applied only to pure substances, also to chemical compounds: the atomic (or, more accurately, molecular) heat of a compound with composition $A_a B_b C_c$, composed of a, b and c atoms of the substances A, B and C, respectively, was obtained by adding the atomic heats of the components, i.e., the terms ac_A, bc_B and cc_C, with c_A, c_B and c_C representing the atomic heats of pure substances at constant pressure (Neumann, 1832).[154] Weber's findings of 1872 were substantiated by many experiments on different substances at low temperatures; in general a decrease of the specific heats was observed as lower and lower temperatures were attained (Behn, 1898; Dewar, 1905). On the other hand, the kinetic theory of matter provided a secure foundation for the value given by the Dulong–Petit rule. Ludwig Boltzmann had shown already in 1871 that the atomic heat of monatomic solids, consisting of harmonically bound atoms, had to assume the value

$$c_v = 3R = 5.955 \text{ cal,} \tag{73}$$

where c_v denotes the atomic heat at constant volume, which is smaller than the usually observed atomic heat at constant pressure (Boltzmann, 1871). This result is very easy to understand: the kinetic energy of N_0 atoms at absolute

---

[153] P. L. Dulong (1785–1838) was professor of chemistry at the normal and veterinary school of Alfort and later (from 1820) Professor of Physics and Director (from 1830) of the *École Polytechnique*, Paris. He carried out researches on the problems of heat with A. T. Petit (1791–1820), who had been his predecessor as Professor of Physics at the *École Polytechnique*.

[154] F. E. Neumann (1798–1895) had been Professor of Mineralogy and Physics at the University of Königsberg since 1831. He worked on the analysis of crystal structures, the mathematical theory of electrodynamics (including the law of induction), the reflection and refraction of light, heat flow in metals, and on special functions in mathematics (e.g., spherical functions). His rule giving the specific heats of compounds is often referred to in literature as the 'Neumann–Kopp rule,' for it was later further substantiated by the chemist Hermann Franz Moritz Kopp (1817–1892) in a series of papers published in 1864 and 1865 (Kopp, 1864, 1865).

temperature T is given by $\frac{3}{2}RT$, with R the universal gas constant ($R = kN_0$ in terms of Boltzmann's constant k and Avogadro's number N_0); since the potential energy of N_0 atoms bound by elastic forces takes on the same value, the total energy is $3RT$, which, after differentiation with respect to the temperature T, yields the atomic heat, Eq. (73). As a consequence, any value of c_v different from $3R$ violated one of the basic assumptions of kinetic theory, notably the equipartition theorem, according to which the kinetic energy per degree of freedom must be $\frac{1}{2}kT$. Thus the large deviations of the observed atomic heats from the rule of Dulong and Petit contributed to what Lord Kelvin called in 1900 the 'nineteenth century clouds over the dynamical theory of heat and light.' However, the difficulty, which had been illuminated by Weber's experiments, was to be attacked by two of his former students, Albert Einstein and Walther Nernst. Their respective theoretical and experimental investigations provided the key to a successful explanation of atomic heats on the basis of the concept of energy-quanta.

In spring 1906 Einstein analyzed Planck's theory of radiation and concluded that it contradicted the consequences derived from conventional statistical mechanics (Einstein, 1906c). Later in the same year he returned again to this conclusion and added:

> I now believe that we should not be satisfied with this result, for the following question arises: If the elementary objects, [the existence of] which has to be assumed in the theory of the exchange of energy between radiation and matter, cannot be interpreted according to the present molecular-kinetic theory, do we then not also have to modify the theory for the other periodically oscillating objects that are used in the molecular theory of heat? In my opinion, there can be no doubt about the answer. If Planck's theory strikes the heart of the matter, then we must expect contradictions also in other areas of heat theory between the molecular theory of heat and experience, which may be removed by the method proposed. In my opinion this is actually the case, as I shall demonstrate in what follows. (Einstein, 1906g, p. 184)

What he sought to demonstrate constituted the main part of his paper on '*Die Plancksche Theorie der Strahlung und die Theorie der spezifischen Wärme*' ('Planck's Theory of Radiation and the Theory of Specific Heats'), which was received by *Annalen der Physik* on 9 November 1906 and appeared in an issue before the end of the year (Einstein, 1906g).

In dealing with the problem of specific heats, Einstein started from Boltzmann's result discussed above and he referred to the violations of the rule of Dulong and Petit, or of Neumann, which did agree with kinetic theory. Einstein claimed that the actual situation was even worse, if viewed in the light of the modern electron theory of matter. The reason was that the oscillating electrons, which caused the dispersion of light, gave rise to extra degrees of freedom, and these should also contribute to the specific heat of any substance; hence a value of c_v even larger than the one given by Eq. (73) had to be expected for monatomic solids. On the other hand, if one applied Planck's radiation theory, the result was different, because it gave to any object oscillating in space with the frequency ν at temperature T the mean energy

$$\bar{E}_\nu = 3 \frac{R}{N_0} \frac{\beta\nu}{\exp(\beta\nu/T) - 1} , \tag{74}$$

rather than $3kT$. (Remember that Einstein's β is h/k in Planck's notation, with $k = R/N_0$.) By summing over all the oscillating constituents of a gram-atom of a solid and by subsequent differentiation of the result with respect to temperature, Einstein obtained for the atomic heat (at constant volume) the expression

$$c_v = \frac{5.94}{N_0} \sum \frac{\exp(\beta\nu/T) \cdot (\beta\nu/T)^2}{\left[\exp(\beta\nu/T) - 1\right]^2} . \tag{75}$$

Equation (75) immediately provided an explanation for the observed deviations of the atomic heats from the Dulong–Petit rule as well as their temperature dependence. Einstein remarked:

> If $T/\beta\nu > 0.9$, the contribution of the [oscillating constituent] to the molecular [atomic] specific heat does not deviate appreciably from the value 5.94 [he made the assumption that the number of oscillating elements in a gram-atom of the substance under investigation is N_0], which also follows from the hitherto accepted molecular-kinetic theory of heat. The smaller ν is, the lower will be the temperatures for which this case is already achieved. If, however, $T/\beta\nu < 0.1$, then the [oscillating] object under consideration does not contribute appreciably to the specific heat. In between [these two cases] the expression [(75)] increases [with increasing temperature], first rapidly then more slowly. (Einstein, 1906g, pp. 186–187)

He determined that as a consequence, at room temperature (about 300°K) any oscillator emitting optical wavelengths above 48 μ would contribute the full amount $3R/N_0$ to the atomic heat, while oscillations with wavelengths smaller than 4.8 μ would contribute practically nothing. Hence one might understand why the electrons in solids did not give rise to any additional terms to the specific heat; it was because their optical frequencies usually lay in the visible and ultraviolet regions. On the other hand, the observation of residual rays (Rest-strahlen) from solid crystals indicated that the ions and atoms had optical frequencies with wavelengths of the order of 10^{-2} to 10^{-3} cm. Einstein now tested his theory of atomic heats on the assumption that in Eq. (75) it was mainly the frequency of the residual rays that played a role, and that in a gram-atom, N_0 elementary three-dimensional oscillators took on this frequency. Then every solid whose residual rays possess wavelengths larger than 48 μ should have an atomic heat of $3R$, while substances whose residual-ray wavelengths lie between 4.8 μ and 48 μ have smaller atomic heats. In particular, he described the specific-heat data of diamond with the help of the characteristic (residual-ray) wavelength of 11 μ.

In spite of the importance of the specific-heat problem, Einstein's paper of December 1906 did not receive attention for more than two years.[155] Several reasons could be given for this neglect. One was, of course, that still only a few people concerned themselves with quantum theory. Second, the specific heats of solids represented a rather complex problem, both experimentally and theoretically. Not only did the measurements of specific heat require subtle observational techniques, especially at low temperatures, but the quantity, which could be measured, was not specific heat at constant volume—the one calculated in theory—but the specific heat at constant pressure. So, to obtain c_v one had to determine the difference, $c_p - c_v$, with the help of additional experimental and theoretical information.[156] This task was attacked most successfully by Walther Nernst and his collaborators at the University of Berlin.

Walther Hermann Nernst was born on 25 June 1864 in Briesen, West Prussia and studied physics at the Universities of Zurich, Berlin (where Helmholtz attracted him), Graz (where Boltzmann taught) and Würzburg (where he collaborated with Friedrich Kohlrausch). In 1887 he obtained his doctorate with a thesis on a new thermoelectric effect which he had discovered in Graz together with Albert von Ettinghausen. Then he joined Wilhelm Ostwald in Leipzig as assistant and became *Privatdozent* in 1889 with a thesis on the theory of galvanic cells. He went on to Heidelberg and in 1891 to Göttingen, where he became *Extraordinarius* for physical chemistry (and *Ordinarius*, i.e., full professor, three years later). In 1905 he moved to Berlin as Director of the Physico-Chemical Institute at the University.[157]

Already early in his scientific career Nernst was attracted by important research problems lying on the borderline between physics and chemistry, such as the theory of galvanic cells and the theory of solutions. Thus he may be

---

[155] The first reference to Einstein's theory of specific heats is to be found in a paper of Max Reinganum (1876–1914), entitled '*Gesichtspunkte für eine Strahlungstheorie. I*' ('Points of View Concerning a Theory of Radiation. I', Reinganum, 1909). In this paper Reinganum discussed the problem of deriving Planck's law and he assumed, for that purpose, that for small wavelengths most resonators in matter do not radiate at all but stay at rest. Reinganum then derived an expression for the atomic heat different from Einstein's, namely,

$$c_v = 3R \exp\left(-\frac{\beta\nu}{T}\right)\left(1 + \frac{\beta\nu}{T} + \frac{\beta^2\nu^2}{3T^2}\right).$$

(See Reinganum, 1909a, p. 354, Eq. (9).)

[156] From thermodynamical arguments the difference, $c_p - c_v$, may be calculated in terms of the thermal expansion coefficient α and the cubic compressibility χ as

$$c_p - c_v = \frac{\alpha^2 v_0^2}{\chi v} T,$$

where v and v_0 denote the volumes of a gram-atom at temperatures T and T_0 (usually $T_0 = 0°C$), respectively. The factor multiplying T on the right-hand side was found for most monatomic substances—on theoretical and experimental grounds—to be approximately a constant times c_p^2. (See, e.g., Nernst and Lindemann, 1911b, p. 820.)

[157] In 1924 Nernst became Director of the Experimental Physics Institute at the University of Berlin. He retired in 1933 and died on 18 November 1941 in Muskau near Berlin.

counted, together with Wilhelm Ostwald, Jacobus Henricus van't Hoff and Svante August Arrhenius, as one of the founders of physical chemistry. Nernst also wrote one of the first textbooks on that subject, entitled *Theoretische Chemie vom Standpunkte der Avogadroschen Regel und der Thermodynamik* (Nernst, 1893). It was the problem of a physical definition of chemical properties which gave rise to a fundamental thermodynamic theorem, expounded by Nernst in late 1905. And this heat theorem led Nernst to investigate experimentally the specific heats at low temperatures.

One of the most important chemical concepts is the so-called 'affinity'; it is the ability of substances to combine with each other to yield compounds. In 1867 the French chemist Marcellin Berthelot had given the first quantitative definition of this concept by stating that a chemical reaction between substances occurs spontaneously if additional heat is produced. Van't Hoff had then improved on Berthelot's definition; it is not the heat that is produced which characterizes the affinity, he claimed, but the so-called free energy F, i.e., the energy available from the reaction to perform useful work.[158] Now the difference between the free energy F and the total energy E of a reacting system was given by the Gibbs–Helmholtz relation,

$$F - E = T \frac{dF}{dT} . \tag{76}$$

Equation (76) stated that only a part of the total energy can be converted into work. However, Nernst recognized that in galvanic reactions involving solids and concentrated solutions the difference between F and E was extremely small. From this he derived the general conclusion that for reactions taking place at temperatures close to absolute zero, both quantities coincide completely; he wrote the limiting relation

$$\lim_{T \to 0} \frac{dF}{dT} = 0, \tag{77}$$

which also implied the relation

$$\lim_{T \to 0} \frac{dE}{dT} = 0. \tag{77a}$$

Equations (77) and (77a) expressed what Nernst called a 'new heat theorem' ('*das neue Wärmetheorem*', Nernst, 1906, pp. 8 and 28). He immediately applied it to calculate the equilibrium of physical and chemical reactions and to the problems of specific heat. He concluded in particular that the atomic heat of any monatomic solid must assume at the absolute zero of temperature the same value, which need not necessarily be zero (Nernst, 1906, p. 12).

In order to test the consequences of his heat theorem, Nernst started experiments to measure the specific heats at low temperatures, assisted in particular by

---

[158] In the case of a galvanic cell the useful work becomes available as electrical energy.

his collaborators Arnold Eucken, Frederick Alexander Lindemann and Charles Lindemann.[159] They developed a new measuring device, in which the substance under investigation served as a calorimeter (Eucken, 1909), and carried out the necessary supplementary measurements—such as that of the thermal expansion coefficient (C. L. Lindemann, 1911). On 17 Februrary 1910 Nernst presented a report on the initial results on the specific heats to the Prussian Academy. In particular he noted

> If one plots the [experimentally] obtained values graphically [as a function of the temperature], then in most cases one obtains nearly straight lines, which frequently fall off very rapidly at low temperatures; thus one obtains the impression that at very low temperatures the specific heats assume a value zero or at least very low values. [And he remarked further:] This result agrees quantitatively with the theory developed by Mr. Einstein. Messrs Lindemann and Magnus are occupied with a qualitative evaluation of the data. (Nernst, 1910a, p. 276)

The subsequent analysis of the preliminary data by Alfred Magnus and F. A. Lindemann, which was submitted on 9 March 1910 to *Zeitschrift für Elektrochemie*, then showed that Einstein's Eq. (75) with one characteristic frequency v did indeed describe the situation fairly well (Magnus and Lindemann, 1910).

Nernst was thrilled by the success of both his heat theorem—which had predicted that the atomic heats of all monatomic solids approach the same value at low temperatures—and of Einstein's quantum-theoretical formula, which demanded that this limiting value be zero. He immediately travelled to Zurich to meet and discuss his results with Albert Einstein, who meanwhile (in fall 1909) had become extraordinary professor (*Extraordinarius*) at the University of Zurich. About this meeting Einstein wrote to Johann Jakob Laub: 'I am a firm believer in the quantum theory. My predictions concerning the specific heats seem to be splendidly confirmed. Nernst, who just visited me, and Rubens are eagerly occupied with the experimental test; hence we will soon know about the situation' (Einstein to Laub, 1910; quoted in Seelig, 1960, p. 197).

Although the principal advocates of the quantum theory, like Max Planck and Albert Einstein, shared the conviction that the quantum concept was needed in

---

[159] Arnold Eucken, the son of the philosopher Rudolf Eucken, was born on 3 July 1884 in Jena. He studied at the Universities of Kiel, Jena and Berlin, and obtained his doctorate under Nernst in 1906. In 1911 he became *Privatdozent* at the University of Berlin and, four years later, Professor of Physical Chemistry at the University of Breslau, from where he moved to the University of Göttingen in the same position (1930–1950). He died on 16 June 1950 in Traunstein, Bavaria.

F. A. Lindemann was born on 5 April 1886 in Baden-Baden, Germany. He studied physics at the University of Berlin under Rubens, Planck and Nernst. He received his doctorate under Nernst in 1910 with a thesis on the specific heats of metals at low temperatures. He stayed on in Berlin, returning to England only in 1914; then he served in the Royal Air Force during World War I. In 1919 he was appointed Lee Professor of Experimental Philosophy at Oxford University, where, after 1933, he attracted many Jewish physicists from Germany, and turned the Clarendon Laboratory into one of the leading centres of low temperature physics in the world. Lindemann was made Baron Cherwell in 1941 and Viscount Cherwell in 1956 for his services to the British Government (especially, as advisor to Prime Minister Winston Churchill). He died on 3 July 1957 in Oxford.

order to account for a large number of observed phenomena, they differed strongly in the details of their approach. Planck and Einstein, who had thought about the problems for many years, had even developed strongly divergent views concerning the basic concepts. Already in his first analysis of Planck's radiation law, Einstein had emphasized that its foundation was not consistent with classical electrodynamics (Einstein, 1906c). In a later paper, entitled '*Zum gegenwärtigen Stand des Strahlungsproblems*' ('On the Present Status of the Radiation Problem') and submitted in January 1909 to *Physikalische Zeitschrift*, Einstein had gone a step further with his criticism (Einstein, 1909a).[160] After emphasizing again that 'our present theoretical concepts lead inevitably to the law advocated by Mr. Jeans' and that '[his] formula does not agree with the facts' (Einstein, 1909a, pp. 186–187), Einstein discussed again the foundations of Planck's law—a law which did describe the data correctly. In particular, he drew attention to the following point. The statistical theory demanded that the entropy of any state of a given system must be defined by the equation

$$S = \frac{R}{N_0} \ln W + \text{const.}, \tag{78}$$

with W representing the probability of the state (see Eq. (20)). But Einstein also knew that, 'Neither Mr. Boltzmann nor Mr. Planck have given a definition of W. In a purely formal manner, they put W = number of complexions of the state under consideration' (Einstein, 1909a, p. 187). However, the number of complexions could only be *related* to W—i.e., put equal to W up to a constant—if one knew the probability of every complexion. The complexions had to be so chosen that 'they are found to be equal [i.e., of equal weight] on account of statistical considerations, based upon the chosen theoretical picture [of the system under consideration]' (Einstein, 1909a, p. 187). Had Planck proceeded in agreement with Boltzmann's view, Einstein went on to argue, then he would have arrived at Jeans' radiation law. 'As much as every physicist must be happy about

---

[160] During the second half of 1908 there appeared several articles and notes on the problem of radiation theory in *Physikalische Zeitschrift*. They started with a note of Otto Lummer and Ernst Pringsheim on what they called the 'Jeans–Lorentz radiation formula' (i.e., the Raylcigh–Jeans law), in which the authors pointed out the discrepancy between this formula and the empirical facts (Lummer and Pringsheim, 1908). Then Hendrik Lorentz wrote on the same question and arrived at the same conclusion (Lorentz, 1908b). In the December issues two further papers were published: one was by James Jeans, in which he responded to Lummer and Pringsheim's criticism (Jeans, 1908); in the other, Walther Ritz suggested that in the derivation of the 'Jeans–Lorentz law' an incorrect assumption had been made (Ritz, 1908c). Ritz claimed in particular: 'The *Ansatz* for the electric and magnetic forces, which has been made in the afore-mentioned proof, is too general; it contradicts the formulae for the retarded potentials, which must be satisfied by every physically admissible solution of the fundamental equations' (Ritz, 1908c, p. 903). Ritz concluded that in order to obtain the correct radiation law, the available electron theory had to be supplemented by conditions which determine unique solutions. Einstein, in the first part of his paper of January 1909 on radiation theory, discussed Ritz' arguments; he showed, especially, that supplementary conditions, such as the one proposed by Ritz, had no influence on the derivation of the classical law (Einstein, 1909a). A debate ensued between Ritz and Einstein, with neither persuading the other (Ritz, 1909; Ritz and Einstein, 1909).

the fact that Mr. Planck avoided this demand in such a successful manner,' he concluded, 'so should it also not be forgotten in the least that Planck's radiation formula is not consistent with the theoretical foundation from which Mr. Planck started' (Einstein, 1909a, pp. 187–188).

In 1909 Einstein was convinced more than ever before that the foundations of Planck's radiation theory had to be modified drastically. Still he did not know where to begin. He, therefore, proposed to take Planck's formula as being empirically given and to analyze it in terms of statistical theory. Especially, he assumed that Planck's equation (19) for the radiation entropy was also valid in the vicinity of equilibrium, and he went on to analyze it physically by employing the concept of energy fluctuations. Thus he found for $\overline{\epsilon^2}$, i.e., the average value of the square of the deviation from the mean radiation energy $\overline{E}$ in the frequency range between ν and $\nu + d\nu$, the equation

$$\overline{\epsilon^2} = \overline{(E - \overline{E})^2} = h\nu E_0 + \frac{c^3}{8\pi\nu^2\,d\nu}\,\frac{\overline{E}^2}{v}\,, \tag{79}$$

where v is the volume of the cavity containing radiation.[161] The right-hand side of Eq. (79) contained two terms. Einstein found from dimensional considerations that the electromagnetic theory yielded only the second term. He noted: 'However, we would have obtained this second term alone for $\overline{\epsilon^2}$, if we had started from Jeans' formula. The first term of the above expression for $\overline{\epsilon^2}$, which for the visible radiation—that surrounds us everywhere—provides a much larger contribution than the second one, is therefore not compatible with the present [radiation] theory' (Einstein, 1909a, p. 189). However, this term could be easily interpreted physically by assuming the radiation to consist of independent, point-like energy-quanta of magnitude $h\nu$.

Einstein immediately added a physical consideration to support the above conclusion. For this purpose he investigated the motion of a perfectly reflecting mirror suspended in a cavity filled with heat radiation of temperature T. If the mirror moves with a nonzero velocity, then it reflects more radiation of a given frequency ν from its front than from its back; hence the motion of the mirror would be damped unless it received new momentum from radiation fluctuations. Now the blackbody radiation made two contributions to $\overline{\Delta^2}$, the velocity fluctuations of the mirror accumulated in the time interval τ, that is,

$$\overline{\Delta^2} = \frac{3\tau kT}{c}\left[\rho_\nu - \frac{1}{3}\,\nu\,\frac{d\rho_\nu}{d\nu}\right]d\nu \cdot f, \tag{80}$$

---

[161] Einstein derived Eq. (79) by expanding the entropy of radiation (enclosed in a volume v, having frequencies between ν and $\nu + d\nu$) in terms of the deviations from the average energy, $\epsilon = E - E_0$; thus he obtained the fluctuation formula, $\overline{\epsilon^2} = \overline{(E - E_0)^2} = (R/N_0)(d^2\sigma/d\epsilon^2)_0$, where σ is the entropy of radiation in the volume v, and the subscript 0 refers to the average of the expression within round brackets. Then he calculated from Planck's Eq. (19) the right-hand side of the fluctuation formula and arrived finally at Eq. (79).

where ρ_ν is the energy density of radiation (in the frequency interval $\nu, \nu + d\nu$), f denotes the area of the mirror, and c the velocity of light *in vacuo*. 'Also here the formula states,' Einstein emphasized, 'that due to Planck's radiation formula the effects of both fluctuation phenomena behave like [mere] fluctuations (errors), originating from mutually independent causes (additive combination of terms constituting the square of the fluctuation)' (Einstein, 1909a, p. 190). Evidently, the second term—which was the only one arising from the electromagnetic theory—would lead to a damping of the mirror's motion, and this effect would violate the equilibrium condition unless the first term were taken into account. Einstein therefore claimed:

> From the foregoing [results] it is not enough to assume that oscillator energy with the frequency ν can only be emitted and absorbed in quanta of this size [i.e., $h\nu$], that one is only concerned with a property of emitting or absorbing matter. The [above] considerations ... show that also the fluctuations in the spatial distribution of radiation and of radiation pressure behave as if radiation consisted of quanta of the size indicated. Now it cannot be maintained that quantum theory [i.e., the theory of light-quanta] follows *immediately* from Planck's radiation law. But one can safely state that quantum theory provides the simplest interpretation of Planck's formula. (Einstein, 1909a, p. 191)

Several months later, on 21 September 1909, Einstein gave a talk at the 81st Assembly of German Scientists (*Naturforscherversammlung*) in Salzburg, entitled '*Über die Entwicklung unserer Anschauungen über das Wesen und die Konstitution der Strahlung*' ('On the Development of our Concepts concerning the Nature and Constitution of Radiation'), in which he repeated the main arguments leading up to the fluctuation formula (79) before a large audience of physicists and mathematicians.[162] He then drew attention to the fact that 'it has not yet been possible to formulate a mathematical theory of radiation, which describes both [its] undulatory structure and the structure (quantum structure) following from the first term [in Eq. (80)]' (Einstein, 1909b, p. 824). Einstein did not have such a unified theory either, but he made the following suggestion:

> Still the picture which seems to me to be the most natural of all [is] that the occurrence of the electromagnetic fields of light is connected with singular points, like the occurrence of electrostatic fields in electron theory. One cannot entirely exclude the possibility that in such a theory the total energy of the electromagnetic field can be viewed as being localized in these singularities, just as in the old action-at-a-distance theory. I suppose, say, that each such singular point is surrounded by a field of force, which essentially has the character of a plane wave, whose amplitude decreases with increasing distance from the singular point. If many such singularities exist at distances which are small compared to the extension

---

[162] Among those present in Einstein's audience were: M. Born, J. Elster, P. Epstein, J. Franck, Ph. Frank, J. R. von Geitler, A. Gockel, O. Hahn, W. Hallwachs, F. Hasenöhrl, D. Hondros, L. Hopf, H. Kayser, R. Ladenburg, M. von Laue, L. Meitner, E. Meyer, G. Mie, M. Planck, F. Reiche, H. Rubens, C. Schaefer, K. Scheel, E. von Schweidler, H. Siedentopf, A. Sommerfeld, J. Stark, W. Steubing and W. Voigt. (See Hermann, 1969, p. 79, footnote 17.)

of the field of force of one singular point, then the fields of force will overlap and together constitute an undulatory field of force, which differs only little from an undulatory field in the sense of the present electromagnetic theory. We need not, of course, especially emphasize that any value should not be attributed to such a picture so long as it does not lead to an exact theory. I just wanted to use [this example] to show that two structural properties (the undulatory structure and the quantum structure), both of which must be ascribed to radiation on account of Planck's formula, need not be considered as being mutually incompatible. (Einstein, 1909b, pp. 824–825)[163]

Einstein's talk at the Salzburg meeting was followed with great interest, though his light-quantum hypothesis did not convince people. As Paul Epstein recalled later: 'The chairman at the meeting was Planck, and he immediately said that it was very interesting but he did not quite agree with it. And the only man who seconded at that meeting was Johannes Stark' (Epstein, AHQP Interview). Planck, indeed, opened the discussion of Einstein's talk with an extended remark on the question of whether the light-quantum hypothesis was necessary or not. He claimed that the answer was no and that Einstein had based his conclusions in favour of it on an implicit assumption: he had derived the fluctuations of free radiation from the motion of matter without knowing the detailed interaction between matter and radiation.[164] According to the existing electron theory radiation was emitted by accelerated electrons; however, the same electron theory contained many difficulties—such as the ones connected with the extended structure of electrons; hence its conclusions could not be considered as entirely secure. One should, therefore, 'first try,' said Planck, 'to shift the entire

---

[163] Einstein had indicated in his previous paper how the ideas expressed above might be realized; there he had referred to the fact—observed earlier by Planck and Jeans—that the quantity e^2/c (where e is the charge of the electron and c the velocity of light *in vacuo*) and the quantum of action had the same dimension and could be related, up to a numerical factor of the order of 100, by an equation. 'It seems to me that it follows from the relation $h = e^2/c$,' he had argued, 'that the same modification [of the theory], which implies the existence of the elementary quantum [of electricity], must also lead to the quantum structure of radiation' (Einstein, 1909a, pp. 192–193). That is, he considered the existence of the electron and of the light-quantum, both of which were strangers to classical electrodynamics, as being due to the same origin. Thus the light-quantum might well follow from a nonlinear equation generalizing the fundamental equation for the propagation of light in optics (i.e., the generalization of an equation of the type $(1/c^2)(\partial^2\phi/\partial t^2) - \partial^2\phi/\partial x^2 - \partial^2\phi/\partial y^2 - \partial^2\phi/\partial z^2 = 0$, where ϕ is an amplitude in space and time) and containing the universal constant e.

Einstein continued to ponder in the following years about the possible, new fundamental equation of electromagnetic theory. In late 1910 he reported to Johann Jakob Laub about this endeavour: 'At the moment I have the great hope of solving the quantum problem, and this without using the light-quanta. I am curious how this will work. But if it does, one must abandon the energy principle [i.e., conservation of energy] in its present form' (Einstein to Laub, 4 November 1910). A few days later he admitted the failure of his efforts and wrote: 'Again I have not succeeded in solving the radiation problem. There the devil has played a dirty trick on me' (Einstein to Laub, November 1910; both letters to Laub are quoted in Seelig, 1960, p. 197).

[164] Planck said: 'Now this interaction between free electric energy *in vacuo* and the motion of matter seems still to be very little known. It is active primarily in the processes of absorption and emission of light. Also the pressure of radiation basically stems from it, at least according to the dispersion theory, which, in general, is assumed to be valid, and in which the reflection [of light] is reduced to absorption and reemission. Now, emission and absorption are exactly the dark points [of the theory], about which we know very little.' (Planck, in discussion of Einstein, 1909b, p. 825).

difficulty of the quantum theory to the question of the interaction between matter and radiant energy. The processes in pure vacuum could then, for the present, still be explained with the help of Maxwell's equations' (Planck in Einstein, 1909b, pp. 825–826).

Planck explained his point of view in some detail in an article which he submitted in January 1910 to *Annalen der Physik* (Planck, 1910a). There he reviewed the main progress in radiation theory that had taken place since 1900, and also indicated his own attitude towards it. Two extreme points of view, he said, could be assumed. The most conservative one was taken by Jeans, who had derived from the equations of classical mechanics and electrodynamics the classical radiation equation. Since this equation obviously disagreed with experiments, the necessity arose to find a way of modifying the equations of dynamics so as to include the existence of the quantum of action *h*. Planck wrote:

The most extreme attitude in this regard is taken by the English physicists J. J. Thomson and J. Larmor, and the German physicists A. Einstein and J. Stark. They tend to the view that even the electrodynamic processes in pure vacuum, even the light waves, do not propagate continuously but in discrete quanta (the light-quanta) of magnitude *hν*, where *ν* denotes the frequency. (Planck, 1910a, p. 761)[165]

[165] Joseph Larmor had discussed Planck's theory of blackbody radiation already in his article on the 'Theory of Radiation' for the *Encyclopedia Britannica* in 1902; there he drew attention to Planck's use of Boltzmann's combinatorial definition of the entropy and also mentioned the difficulty that the resonators could not change the type of radiation (Larmor, 1902). In his talk at the Belfast meeting of the British Association in September 1902, Larmor had tried to recast Planck's theory into a form which avoided the use of resonators (Larmor, 1903). Seven years later, Larmor returned to the same question in his Bakerian Lecture, entitled 'On the Statistical and Thermodynamical Relations of Radiant Energy' (Larmor, 1909). In this lecture he developed the theory of radiation on the assumption that with any blackbody or cavity radiation system there were associated 'elementary receptacles of energy . . ., which we may call *cells*: and we may establish a relation between the extents (in a generalized sense) of these cells by the condition that they shall be of equal opportunity, that the element of disturbance possessing the element of energy under consideration is as likely in its travels to occupy any one of them as any other' (Larmor, 1909, p. 86). That is, Larmor replaced Planck's statistical considerations on finite energy packets by a consideration of what he called 'cells' —which corresponded in the language of ordinary statistical mechanics to cells in the phase space of the system. He summed up by saying:

The procedure of Planck . . . depends essentially . . . on the assumption of a discrete or atomic constitution of energy; and the indivisible element of energy, as estimated from the constants of the formula for natural radiation, proves to be of considerable amount, even compared with the energy of a molecule of a gas. A somewhat similar implication survives in the present development, but it now appears in the form that the ratio of the energy element to the extent of the standard unit cell is an absolute physical quantity determined similarly by the observations on natural radiation. (Larmor, 1909, pp. 89–90)

With this generalization of Planck's method, Larmor wanted to narrow the gap between the classical and the quantum-theoretical description of radiation; this would allow one to introduce a limiting differential ratio of the energy element to the extent of the cell (i.e., both tending to zero) without any implication that the energy itself was constituted on an atomic basis.

While Larmor showed that his cell treatment of radiation avoided the problematic introduction of finite energy quanta, he also emphasized that the assumption of 'equal opportunity' applied to the 'elements of disturbance' travelling through phase space. In a system of radiation of different wavelengths, each of these elements, he figured, might consist of a short train of simple undulations. Thus the 'elements of disturbance' exhibited some similarity with Einstein's light-quanta and Joseph John Thomson's localized spots of light intensity (which we have discussed in Section I.4).

Planck then briefly outlined Einstein's fluctuation considerations; he argued, however, that these considerations might not hold because,

> even if the failure of Hamilton's differential equations of motion is acknowledged beyond doubt [for the purpose of describing the radiation phenomena], it is evident that one cannot choose statistical mechanics—which is founded on Hamilton's equations—as the starting point of the consistent radiation theory. (Planck, 1910a, p. 763)

The main difficulty of the corpuscular approach seemed to be the following:

> If one wants to stick to the electromagnetic nature of light, which the representatives of the corpuscular theory also do, then the latter theory exhibits—even when judged most sympathetically—remarkable weaknesses from the very beginning. How, for instance, should one imagine an electrostatic field? For this field, the frequency ν is equal to zero, hence the energy of the field must consist of infinitely many energy-quanta of magnitude zero. Can one then still imagine a finite field strength having a definite direction? (Planck, 1910a, pp. 763–764)

The only way out of this difficulty would be to introduce a different description of static and dynamical fields, which would imply giving up the successes of the electrodynamic theory and that seemed to be too big a sacrifice. However, Planck admitted that the introduction of energy packets of finite size into the statistical considerations did lead to a serious question.[166] But he hoped to account for it by a finite set of equations 'valid for a discrete number of time-points, which perchance are related to the (sudden) excitations of the oscillator' (Planck, 1910a, p. 766). Such a modification of the dynamical laws, he suggested, might provide a kind of excitation barrier for the oscillator in response to incident radiation. He continued to think along these lines and presented his results on 3 February 1911 to the German Physical Society in Berlin (Planck, 1911a).

The main problem, to which Planck addressed himself in early 1911, was how an oscillator (or vibrator) in a cavity filled with blackbody radiation absorbs energy-quanta from incident radiation. Did this process occur continuously or discontinuously? Now, according to (classical) electrodynamics a Hertzian resonator of frequency ν absorbs in a time interval τ the energy E_ν, where

$$E_\nu = \frac{3c^2\sigma}{16\pi^2\nu} J_\nu \tau, \tag{81}$$

---

[166] Planck showed here for the first time that he did take Einstein's criticism (Einstein, 1906c, g; 1909a, b) seriously. Einstein previously had not been certain about it; thus he had written about it to J. J. Laub:

> Planck is also very pleasant in correspondence. Only he has the fault that he finds it difficult to find his way through foreign trains of thought. Thus one can understand that he has raised totally wrong objections against my last paper on radiation. Against my criticism, however, he has said nothing; I do hope that he read it and acknowledged it. This quantum problem is so immensely important and difficult that everyone should apply himself to it. (Einstein to Laub, 1908; quoted in Seelig, 1960, p. 147)

from radiation of frequency ν having the intensity J_ν. (The factor σ in the numerator on the right-hand side denotes the damping constant of the oscillator.) For small intensities, J_ν, and large frequencies, ν, the time interval, τ, necessary to absorb an energy quantum $h\nu$ becomes very large. Therefore, Planck argued, the absorption process could not be a discontinuous one; it had to be *completely continuous*. As a result, he dropped the assumption that the oscillator was only able to take on discrete energy values; it could now have all energies between zero and infinity. On the other hand, the emission of energy from the oscillator should occur in a discontinuous manner, and Planck proposed what he called 'a new radiation hypothesis' ('*eine neue Strahlungshypothese*') as follows:

> The emission of energy occurs spontaneously in definite quanta of magnitude $\epsilon = h\nu$, and the probability for a single oscillator of frequency ν to emit an elementary quantum of energy in a sufficiently small interval of time dt (which is small compared to the average time interval between two subsequent emissions) is equal to $\eta \cdot n \cdot dt$. Here η denotes a constant depending only on the nature of the oscillator, which is determined readily [i.e., $\eta = 2\sigma\nu$]; n is the number of integral energy elements ϵ, which the oscillator possesses at a given instant of time (i.e., n represents positive integers; it may also be zero), and which renders the expression $U/\epsilon - n$ a positive real fraction (< 1). (Planck, 1911a, p. 143)

Planck's new hypothesis led to an important result. The energy, U_ν, of the oscillator of frequency ν emitting radiation was no longer an integral multiple (n) of $\epsilon = h\nu$, but rather

$$U_\nu = n\epsilon + \rho, \tag{82}$$

with ρ assuming any value between zero and ϵ. Thus, on the average—i.e., if one considers the average of a large number of oscillators of the same frequency, having an energy between $n\epsilon$ and $(n + 1)\epsilon$—the quantity ρ assumes the value $\epsilon/2$. As a result, the average energy available for spontaneous emission is not $\overline{U}_\nu$ but $\overline{U}_\nu - \epsilon/2$, and Planck's relation between ρ_ν, the energy density of blackbody radiation, and the oscillator energy, Eq. (7), must be replaced by the equation

$$\rho_\nu = \frac{8\pi\nu^2}{c^3}\left(\overline{U}_\nu - \frac{\epsilon}{2}\right). \tag{83}$$

Planck made a similar replacement in Eq. (30) for the entropy of the oscillator. Finally, he obtained for $\overline{U}_\nu$, the equilibrium energy of the oscillator at temperature T, the equation

$$\overline{U}_\nu = \frac{h\nu}{2}\frac{\exp(h\nu/kT) + 1}{\exp(h\nu/kT) - 1}. \tag{84}$$

In comparison with the earlier result, Eq. (25), the energy of the oscillator became larger by an additive constant, $h\nu/2$. While this additive constant implied no change in the radiation law, Eq. (18), it altered the behaviour of the oscillator at low temperatures considerably. Planck noted:

> For $T = 0$, $\overline{U}_\nu$ is not equal to zero but equal to $h\nu/2$. This rest-energy [zero-point energy] remains with the oscillator even at the absolute zero of temperature. It [the oscillator] cannot lose it, because it does not emit any energy if U_ν becomes smaller than $h\nu$. (Planck, 1911a, p. 146)

For high temperatures, on the other hand, $\overline{U}_\nu$ again assumed the value given by the classical theory.

Planck immediately checked all consequences of his new hypothesis of quantum emission. He found that it described very well the quantum phenomena that were known at the time, such as the photoelectric effect and the behaviour of the specific heats at low temperatures. So he finally declared it to be 'suitable not only for removing the serious contradictions of the radiation theory with the most important foundations of Maxwell's electrodynamics, but thereby also throw a brighter light on certain other phenomena which could not be brought together properly until now' (Planck, 1911a, p. 148). Planck presented the new hypothesis—which came to be called 'Planck's second quantum hypothesis'—on various occasions, such as the Easter meeting of the French Physical Society in Paris on 11 April 1911 (Planck, 1911b) and at a session of the Prussian Academy on 13 July 1911 (Planck, 1911c). He again discussed it in his report to the Solvay Conference in Brussels in fall 1911 (Planck in Langevin and de Broglie, 1912; Eucken, 1914).

Walther Nernst was not bothered by the subtleties that preoccupied Planck and Einstein, and he went ahead with the task of propagating the applications, promise and problems of the new quantum theory. In spring 1910 he encountered, at the home of the chemist Robert Goldschmidt in Brussels, the Belgian industrialist Ernest Solvay. Solvay, who had invented a special process to produce sodium carbonate and had founded the firm of Solvay & Company in 1863, was, for a long time, attracted to the study of the structure of matter in the context of a theory of gravitation.[167] In his meeting with Nernst at Goldschmidt's house, Solvay talked about his scientific ideas and wondered whether they could be brought to the attention of the great physicists like Planck, Lorentz, Poincaré and Einstein. Nernst at once saw a great opportunity, namely, to connect Solvay's desire to have his work discussed with the possibility of holding an

---

[167] Ernest Solvay was born at Rebecq-Rognon, Belgium, on 16 April 1838. (His father, Alexandre Solvay, owned a salt refinery.) Since his early youth he was interested in chemistry, but illness prevented him from pursuing university studies. Instead he took a position in the small gas factory of an uncle. There he invented a new method of producing sodium carbonate from sea salt, ammonia and carbonic acid. After being established in 1865 as a small factory at Couillet, the Solvay & Company began to flourish gradually; in 1872 the first factory was founded abroad at Dombasle, France, and soon others followed in many different countries. Ernest Solvay became a rich man and he used a large part of his wealth for educational and social purposes. He died in Brussels on 26 May 1922. (For further details of Solvay's life, see Mehra, 1975c, Chapter 1.)

international conference on the current problems of the kinetic theory of matter and the quantum theory of radiation. Thus he proposed a council, or conference, to be held for that purpose. Solvay responded favourably and charged Nernst to explore matters further with Planck, Lorentz, Einstein and other prominent scientists.

Nernst drafted a short memorandum on the plan and purpose of the conference he had in mind and sent it to Max Planck for comments. Planck gave a detailed answer; he wrote:

> To the marginal notes that I have already made, with your permission, on your manuscript allow me to add a few more generalities. Your idea corresponds, fully and completely, to the problem whose solution is envisaged, and I can only associate myself with it with full conviction. However [he continued] I am not able to hide my great concern about its execution. As I have already mentioned in my marginal notes, such a conference will be more successful if you wait until more factual material is available. (Planck to Nernst, 11 June 1910)

Planck thought that at that time most scientists, including many of those whom Nernst had suggested should participate in the conference, were not really excited about the quantum problems. 'Among all those mentioned by you,' he said, 'I believe that, other than ourselves, only Einstein, Lorentz, W. Wien and Larmor will be seriously interested in the matter' (Planck to Nernst, 11 June 1910).[168] Planck, therefore, suggested to postpone the conference. 'Let one or better two years go by,' he wrote, 'and we shall see how the gap which begins to open in the theory shall develop, and how finally those who still stand at a distance will be forced to join in' (Planck to Nernst, 11 June 1910).

In spite of Planck's cautious warning, Nernst went ahead with the plan. Already on 26 July he wrote a letter to Solvay, in which he enclosed the draft of an 'Invitation to an "International Scientific Conference to Elucidate Certain Current Questions of the Kinetic Theory."' This draft opened with the remarks:

> It appears that we find ourselves at present in the midst of an all-encompassing re-formulation of the principles on which the erstwhile kinetic theory of matter has been based. On the one hand, this theory leads to a logical formulation—which nobody contests—of a radiation formula whose validity is contradicted by all experiments; on the other, there follows from the same theory certain results on the specific heat (constancy of the specific heat of a gas with the variation of temperature, the validity of Dulong and Petit's law up to the lowest temperature), which are also completely refuted by many measurements. As Planck and Einstein in particular have shown, these contradictions disappear if one places certain limits (doctrine of energy quanta) on the motion of electrons and atoms in the case of their

---

[168] In a paper on the theory of heat radiation, which Planck had submitted to *Annalen der Physik* in January 1910, he had stated: 'Altogether one must admit that meanwhile any fundamental progress has not been achieved in the theory, and that my following considerations will also not offer any such progress' (Planck, 1910a, p. 758). Among the scientists, who supported the application of the quantum concept, Planck mentioned the names of J. J. Thomson, J. Larmor, A. Einstein and J. Stark.

oscillations around a position of rest. But this interpretation, in turn, is so far removed from the equations of motion of material points employed until now, that its acceptance would incontestably lead to a far-reaching reformulation of our erstwhile fundamental notions. (See Mehra, 1975c, p. 6.)

The draft then proposed that perhaps a solution might be found by 'a personal exchange of views on these problems between the researchers who are more or less actively concerned with them' (see Mehra, 1975c, p. 6). Nernst suggested, in the draft, certain topics to be discussed at the conference, and he further proposed a list of eighteen participants with Lord Rayleigh as chairman. Solvay approved Nernst's plan in principle; as for the date of the conference, which Nernst had put around Easter 1911, he preferred to shift it to October of the same year. After due preparations the invitations finally went out in June 1911. All the invited scientists accepted, but for Joseph Larmor—who thought he had not had the time to keep up with the recent progress—and Lord Rayleigh—who thought of himself as 'a very poor linguist' to be useful at the conference. At the end of October the following participants arrived in Brussels to take part in the first Solvay Conference (30 October to 3 November 1911): Walther Nernst, Max Planck, Heinrich Rubens, Arnold Sommerfeld, Emil Warburg and Willy Wien from Germany; James Hopwood Jeans and Ernest Rutherford from England; Marcel Brillouin, Marie Curie, Paul Langevin, Jean Perrin and Henri Poincaré from France; Albert Einstein and Friedrich Hasenöhrl from Austria; Heike Kamerlingh Onnes and Hendrik Lorentz from Holland; and Martin Knudsen from Denmark. Lorentz assumed the Chairmanship and Robert Goldschmidt, Maurice de Broglie and Frederick A. Lindemann acted as Scientific Secretaries of the Conference, which was devoted to the problems of 'Radiation Theory and the Quanta.'[169]

By fall 1911, when the Conference was held, the number of people who took an active interest in the problems of quanta had increased considerably. Thus Peter Debye (1910), Arthur Erich Haas (1910a, b, c), Friedrich Hasenöhrl (1911), Arthur Schidlof (1911), Pierre Weiss (1911) and Harold A. Wilson (1910a) had already published papers on which they contributed to this field. The rapporteurs and participants of the Solvay Conference, therefore, represented an important segment of the physics community interested in quantum problems. Moreover, the topics covered at the Conference were deeply connected with the foundations of kinetic theory and the existence of molecules, in which many physicists and chemists were interested. From the very beginning Walther Nernst, the principal instigator of the Conference, had one important goal in mind: to establish the connection between the molecular hypothesis and its experimental verifications, on the one hand, and the hypothesis of energy-quanta and its consequences, on the other. He also emphasized this goal in his address at the opening of the

---

[169]The reports and discussions of the first Solvay Conference were published in French as *La Théorie du Rayonnement et les Quanta*, edited by P. Langevin and M. de Broglie (1912), and in German as *Die Theorie der Strahlung und der Quanten*, edited by A. Eucken (1914).

Conference, for he said: 'Allow me to tell you in a few words about a congress of chemists, which was held almost exactly half a century ago at Karlsruhe. This congress was called together to discuss a question [that was] important for the foundations of the atomic theory, and I believe it [the congress] was the only one that bears some similarities with our own' (Nernst in Eucken, 1914, p. 8). In the Karlsruhe Congress of Chemists, to which Nernst referred and which took place in September 1860, about 160 chemists had participated and had discussed the problem of nomenclature in chemistry. Although this problem had not been solved completely, the general attention of chemists had been directed to the prevailing difficulties and soon they had agreed on a proposal made earlier by Stanislao Cannizzaro (1858), which was based on the systematic application of the atomic hypothesis. 'We would like to hope,' Nernst remarked in his address, 'that our conference will also exert an important influence on the development of physics, and perhaps we have an advantage over the Karlsruhe Congress insofar as the work for the "*Conseil Solvay*" has been better prepared. The numerous reports, which we have had in our hands for some time, will indicate to us the direction of our discussions and will prevent us from frittering away [our energies]' (Nernst in Eucken, 1914, p. 9).

The reports, which were prepared and distributed in advance to all participants of the Conference, were supposed not only to make everybody familiar with all the topics covered but also to provide the basis of intense and fruitful discussions. Two of the twelve reports did not deal with the theory of energy-quanta directly: Martin Knudsen spoke on the kinetic theory and the experimental properties of gases, emphasizing full agreement of the data with the theory; and Jean Perrin presented a thorough review of all available proofs of molecular reality including, of course, his own detailed tests of the Einstein–Smoluchowski theory of Brownian motion. Heike Kamerlingh Onnes, in his talk on the electric resistance of metals including the phenomena of superconductivity, referred only at one point to a formula which seemed to imply the use of the quantum concept. In Paul Langevin's report on the kinetic theory of magnetism and the magnetons, the quantum of action h entered explicitly, especially in the equation for the magnetic moment I_0 of a gram-atom of a substance, i.e.,

$$I_0 = \frac{N_0 e}{m} \frac{h}{2\pi}, \tag{85}$$

where m and e are the mass and charge of the electron, respectively, and N_0 the number of atoms in a gram-atom (see Eucken, 1914, p. 327). Hendrik Lorentz and James Jeans delivered the first two talks on derivations of the blackbody radiation law from classical considerations. While Lorentz, in his report entitled 'Application of the Energy Equipartition Theorem to Radiation,' came to the conclusion that the classical theories offered no chance to account for the empirical data, Jeans pointed out a possibility of reconciling the observations with the classical theory; however, Jeans had to assume that in all experiments on blackbody radiation no equilibrium existed between matter and radiation and, therefore, the equipartition theorem of energy also could not be applied. The main reports on the quantum theory were given by Max Planck, Walther Nernst, Arnold Sommerfeld and Albert Einstein; their reports were supplemented by two short reviews of Emil Warburg and Heinrich Rubens, respectively, dealing with the most recent empirical evidence in favour of Planck's radiation law.

Planck addressed himself, in particular, to the theoretical steps leading to his blackbody radiation law (Planck, 1912c). He listed and discussed the various methods that had been proposed for deriving this law: his original introduction of energy packets for calculating the entropy of radiation (Planck, 1900f); Einstein's method of late 1906 for determining the average energy of an oscillator that could absorb only integral energy-quanta (Einstein, 1906g); and two other methods due to Lorentz (1910b, p. 1255) and Nernst (1911b), which also made use of the hypothesis that the radiation energy will be absorbed by the resonators (existing in matter) in discrete amounts. All these methods yielded Eq. (25) for the equilibrium energy of the oscillator and, together with Planck's equation (7), the correct radiation formula. Planck agreed with the criticism, raised mainly by Einstein and Jeans, that the assumption of discrete oscillator energies contradicted the electrodynamic foundations, especially Eq. (7). He pointed out, however, that all dynamical models proposed up to that time for an oscillator, which was capable of emitting and absorbing radiant energy in quanta, such as the one of Arthur Erich Haas (1910a), suffered from some internal inconsistency. This inconsistency could only be removed by returning to the assumption that radiant energy may be emitted and absorbed in arbitrary, including arbitrarily small, amounts; but then, of course, the radiation law of Rayleigh and Jeans was obtained.[170] Thus the only way out appeared to be the application of Planck's new hypothesis: quantum emission together with continuous absorption. This new hypothesis allowed one to retain electrodynamics, and yet explain the quantum phenomena. The discontinuity brought in by the quantum of action then had to be transferred to the mechanical laws; the forces binding atoms and electrons in molecules, Planck suggested, had to be held responsible for the quantum discontinuity.[171]

Planck's report at the Solvay Conference received great attention, and many participants—especially, Einstein, Hasenöhrl, Poincaré, Jeans and Lorentz—contributed to its discussion. Henri Poincaré, for example, wanted to know whether there existed any conditions determining the shape of the finite area in

[170] Among the models which Planck refuted was also the one due to Max Reinganum (1909). Reinganum had assumed, in particular, that the bound electrons remained at rest until they could absorb an energy-quantum of radiation. The difficulty of this model was that the energy-quanta had to be absorbed instantaneously, which seemed to be impossible in the case of low intensities of the incident radiation. (See also footnote 155.)

[171] Planck assumed that these binding forces had a chemical origin. He preferred to retain the usual differential laws for what he called the 'physical forces,' such as the electromagnetic and gravitational forces.

phase space; this shape, he pointed out, might play a crucial role if one attempted to extend the quantum concept to systems having several degrees of freedom.[172] The most critical comment certainly was made by Albert Einstein, who declared openly that Planck had not introduced the statistical probability in a proper way. Einstein returned to this question in his own report, where he again emphasized the result at which he had arrived earlier: namely, that a consistent application of the statistical mechanical equations to the energy distribution of blackbody radiation led to the conclusion that radiation not only possessed the properties described by the electrodynamic theory but also the properties of light-quanta (Einstein, 1909a). The same conclusion, he now demonstrated, applied to the thermal vibrations in solids on account of the quantum formula, Eq. (75).

The status of the problem of specific heats was discussed in detail in the reports of Nernst and Einstein at the Solvay conference. Important progress had been achieved in dealing with this problem since 1910. Especially, Nernst and his collaborators had obtained new, more accurate data down to lower temperatures; and these data showed that Einstein's formula, Eq. (75), did not describe the behaviour at the lowest temperatures if one inserted *just one* characteristic frequency for any solid.[173] Nernst and Lindemann had proposed to fit the new results with the help of the equation

$$c_v = \frac{3}{2} R \left\{ \frac{(\beta\nu/T)^2 \exp(\beta\nu/T)}{\left[\exp(\beta\nu/T) - 1\right]^2} + \frac{(\beta\nu/2T)^2 \exp(\beta\nu/2T)}{\left[\exp(\beta\nu/2T) - 1\right]^2} \right\}, \qquad (86)$$

which they had found by trial and error (Nernst and Lindemann, 1911a, b). They had also given a theoretical interpretation for this equation by assuming that, 'when solid bodies are heated the potential energy increases in quanta, whose magnitude is one-half of the energy quanta hitherto assumed, while the kinetic energy grows stepwise by the amounts required by the erstwhile quantum hypothesis' (Nernst and Lindemann, 1911a, p. 501). That is, the first term on the right-hand side of Eq. (86) should represent the kinetic energy of the vibrating atoms, the second the potential energy. However, the assumption of two energy-quanta left the blackbody radiation law unchanged, because only the kinetic part of the energy of bound atoms would be in equilibrium with radiation.

In his report at the Solvay Conference, Nernst discussed both the theoretical

---

[172] Poincaré discussed the case of the three-dimensional harmonic oscillator and claimed that different phase-space separations led to different results. (See Mehra, 1975c, p. 38.) The problem of the quantum theory of systems having several degrees of freedom was attacked in detail some years afterwards, with Max Planck and Arnold Sommerfeld contributing.

[173] Nernst had presented the first results on the specific heats of solids and fluids to the Prussian Academy on 17 February 1910 (Nernst, Koref and Lindemann, 1910). These results were roughly in agreement with Einstein's formula, even if only one characteristic frequency was inserted. The experiments had then been continued in the following months, and already in the beginning of 1911 it had become clear that very close to absolute zero the specific heat did not fall off exponentially.

arguments and the empirical evidence in favour of Eq. (86) (Nernst, 1912b). He then showed that the frequencies ν, determined from the specific-heat data with the help of Eq. (86), agreed well with those of the residual rays of the same substances.[174] Nernst also presented the relation of these characteristic frequencies to certain elastic properties of solids, which had been found previously by several authors, such as Erwin Madelung (1909, 1910a, b), William Sutherland (1910) and Albert Einstein (1910)[175]; and he referred to Frederick A. Lindemann's formula expressing the characteristic frequency in terms of the melting point of a given substance (Lindemann, 1910). While these relations gave frequency values, ν, having the same order of magnitude as those resulting from the specific-heat data according to Eq. (86), Nernst finally drew attention to still another difficulty that existed in the theory: at low temperatures the vibrations of the atoms cease, which would imply that the heat conduction (at low temperatures) becomes small, in contrast to the observations of Arnold Eucken (1911a, b).[176]

In their respective reports at the Solvay Conference, Nernst and Einstein expressed general agreement on the theory of specific heats. They did not, however, share exactly the same opinions. Thus, Nernst showed no interest in the light-quantum concept; all he needed for the applications he had in mind was the concept of the energy-quantum. Einstein, on the other hand, did not believe that the Nernst–Lindemann equation, Eq. (86), could be given a deep theoretical foundation. He rather thought that 'the heat vibrations of atoms deviate strongly from monochromatic oscillations, such that these vibrations are not connected with a definite frequency, but with an [entire] frequency domain,' and the atomic heat had to be found by taking a genuine sum in Eq. (75), i.e., by summing over infinitely many ν-values in certain characteristic frequency intervals, which may be finite or infinite (Einstein, 1912d; in Eucken, 1914, pp. 336–337).

Arnold Sommerfeld gave the fourth main report on the quantum theory at the Solvay Conference. As we have mentioned earlier, he had become interested in the concept of light-quanta only the year before and had employed it first in a paper submitted to the Bavarian Academy in January 1911 (Sommerfeld, 1911a). However, in spite of his late conversion he had quickly shown great enthusiasm; thus, already at the following *Naturforscherversammlung* at Karlsruhe in September 1911 he gave a long review entitled '*Das Plancksche Wirkungsquantum und seine allgemeine Bedeutung für die Molekularphysik*' ('Planck's Quantum of Action

---

[174] The recent residual-ray data (Rubens and Hollnagel, 1910a, b) had been compared to the specific-heat data already in the papers of Nernst and Lindemann (1911a, b).

[175] For example, Einstein had derived the result that the residual-ray frequency was proportional to the product, $M^{-1/3}\rho^{-1/2}\kappa^{-1/2}$, where M is the molecular (or atomic) weight, ρ the density, and κ the compressibility (Einstein, 1910, p. 173).

[176] As Einstein pointed out in his report to the Solvay Conference, one would arrive at a related conclusion from theoretical considerations: the equilibrium of gas molecules with heat radiation, described by Planck's radiation law, gave rise to too small values of the kinetic energy of molecules (Einstein, 1912d; in Eucken, 1914, p. 339).

and its General Significance for Molecular Physics,' Sommerfeld, 1911b).[177]
While Sommerfeld was the only reviewer of quantum theory in Karlsruhe and,
therefore, tried to discuss all aspects of the theory and its applications, he
concentrated at the Solvay Conference in Brussels only on his own ideas that
were concerned with the role of the quantum of action in molecular physics
(Sommerfeld, 1912). Indeed, people had, so far, turned their attention nearly
exclusively to periodic phenomena, such as blackbody radiation and the vibrations responsible for the specific heats in solids; Sommerfeld claimed, however,
that a more general approach to the problems and difficulties of quantum theory
might be obtained by studying aperiodic processes. The fact that the fundamental quantity in the theory was a quantum of action (h) rather than an energy-
quantum provided, in his opinion, the key to a generalization of the existing
theory. Therefore he proposed to start from the following point of view:

> The universal property of all molecules (atoms), which is expressed in radiation,
> does not consist in the fact that certain characteristic energy quanta occur, but in
> the fact that the time sequence of the energy exchange is governed in a universal
> manner. To state it in complete generality, a large amount of energy is absorbed
> and emitted by matter in a short time interval, a smaller amount in longer time,
> such that the product of energy and time, or the time-integral of the energy (which
> has to be defined more precisely), is determined by the quantity h. (Sommerfeld,
> 1912; in Eucken, 1914, pp. 252–253)

Sommerfeld then formulated this fundamental assumption mathematically in the
equation

$$\int_0^\tau L \, dt = \frac{h}{2\pi} , \qquad (87)$$

where L denoted the Langrangian function of the system under investigation
(i.e., the kinetic energy minus the potential energy) and τ the duration of the
quantum process which was connected with an exchange of energy. He believed
that Eq. (87) was particularly valuable, for it could be applied both to relativistic
and nonrelativistic systems.

Sommerfeld went on to discuss several applications of his fundamental
hypothesis contained in Eq. (87). Thus he presented the mechanism of the
production of X- and γ-rays from decelerating charged particles, especially
electrons, in matter, i.e., the mechanism of 'Bremsstrahlung,' which he had
developed earlier (Sommerfeld, 1911a), in agreement with Eq. (87). He showed

---

[177] Originally, Sommerfeld had been invited to talk on relativity theory at the Karlsruhe meeting,
but then he chose to speak on quantum theory, for he argued:

> Here the fundamental concepts are still fluid and there are innumerably many problems . . .
> Nothing would add more to the progress of modern physics than a clarification of the views on
> these problems. Herein lies the key to the situation, the key not only to the theory of radiation
> but also to the molecular constitution of matter, and admittedly it still lies hidden quite deeply
> at present. (Sommerfeld, 1911b, pp. 1057–1058)

that a satisfactory description of bremsstrahlung produced by the stopped electrons could be obtained if it was assumed that the product of the energy loss of the electrons times the deceleration or stopping time had the order of magnitude of $h/2\pi$. That is, he replaced the quantity L on the left-hand side of the action integral, Eq. (87), by the kinetic energy of the electrons (or its loss). Similarly he proposed to treat the photoelectric effect with the help of his quantum hypothesis. In order to apply it properly, he started from the following picture of the process. The electrons in the atoms of the surface of a metal (exposed to the incident radiation) were subject to a quasi-elastic force, $-fx$, where x is the distance of the electron from its equilibrium position and f is a constant; in the case of an incident radiation of frequency ν the electrons were made to oscillate by this electric field, which varied in time as $E\cos(2\pi\nu t)$. Now Sommerfeld demanded that the photoelectric effect occurred if Eq. (87) was satisfied. By evaluating the action integral he found that it assumed the value $\frac{1}{2}m_e x\,dx/dt - \frac{1}{2}e\int_0^\tau E\cos(2\pi\nu t)x\,dt$, m_e and e being the mass and charge of the electron, respectively. In the case of resonance, where ν coincides with the characteristic frequency ν_0 $(= (1/2\pi)(f/m_e))$ of the electron, the integral term could be neglected compared to the first term, which corresponded just to the ratio of the kinetic energy to the frequency ν $(= \nu_0)$; hence Eq. (87) yielded basically Einstein's equation of the photoelectric effect (i.e., $E_{\text{kin}} = h\nu$).[178] Sommerfeld further believed that his new theory of the photoelectric effect—which also implied that in the case of imperfect resonance $(\nu \neq \nu_0)$ deviations would occur from the linear relation between the maximum kinetic energy of the photoelectrons and the frequency of the incident radiation—accounted even better for the observed phenomena than the light-quantum theory of Einstein. In this connection he referred especially to the so-called selective photoelectric effect, which had been investigated by Robert Pohl and Peter Pringsheim: they had shown that the number of photoelectrons was a maximum when the frequency of the incident radiation assumed a characteristic value (Pohl and Pringsheim, 1910, 1911). Equation (87) also seemed to provide a satisfactory description of the ionization data.

Sommerfeld's report stimulated—just as the reports of Planck, Nernst and Einstein had done—an extended discussion, especially about his claim that his fundamental hypothesis was consistent with classical electrodynamics in contrast to Planck's original quantum hypothesis and Einstein's light-quantum hypothesis.[179] Many participants in the Conference were attracted by Sommerfeld's

---

[178] In the case of incomplete resonance, i.e., for $\nu \neq \nu_0$, the bound electron can be liberated for $\nu \gg \nu_0$, hence Stokes' rule followed from Sommerfeld's theory. But also certain violations of Stokes' rule were obtained, in contrast to Einstein's earlier treatment (Einstein, 1905b).

[179] Sommerfeld also showed that the conclusions from Eq. (87) did not agree quantitatively with the ones derived from Planck's new quantum hypothesis. In the case of the photoelectric effect he obtained the relation, $E_{\text{kin}}\tau = \frac{3}{2}(h/2\pi)$, for the kinetic energy of the liberated electron, while from Planck's hypothesis it followed that $E_e\tau = h$, where E_e was the energy of a characteristic radiation of the atom (from which the photoelectron was removed), whose frequency lay close to $1/\tau$ (τ being Sommerfeld's accumulation time).

approach to quantum phenomena, although it implied for each application a detailed study of the molecular mechanisms involved. Naturally, critical questions were also raised. For example, Einstein immediately noticed a peculiar difficulty: since the Lagrangian L was not zero for a free particle, possible violations of the principle of relativity might be obtained. But, in general, the impression prevailed that Sommerfeld's method signified a considerable enrichment of the quantum theory. At that time this theory was not yet in a satisfactory state, as even the quantum physicists admitted, for as Einstein remarked during the discussions: 'We all certainly agree that the so-called quantum theory of today, though it is a useful tool, does not represent a theory in the usual sense of the word, in any case not a theory which can be developed in a closed form' (Einstein in Eucken, 1914, p. 353).[180]

On the other hand, it had also become clear, especially from the report of Hendrik Lorentz at the Solvay Conference, that the classical theory—i.e., classical dynamics as formulated in terms of the equations of Lagrange and Hamilton —'cannot be regarded as being more than a useful scheme for the theoretical interpretation of all physical phenomena' (Einstein in Eucken, 1914, p. 353). But in what direction should one go in order to search for a consistent quantum dynamics? To this problem Henri Poincaré addressed the following remark:

> The new investigations discussed here do not only seem to question the fundamental principles of mechanics, but also shake violently an assumption which was hitherto completely tied together with the concept of a natural law. Can we still express these laws [of quantum theory] in the form of differential equations? (Poincaré in Eucken, 1914, p. 364)

Poincaré, like his Paris colleague Marcel Brillouin, was impressed by the fact that 'we have to introduce into our physical and chemical considerations a discontinuity, a quantity which changes in jumps and of which we had no notion until recently' (Brillouin in Eucken, 1914, p. 364). Since this discontinuity seemed to imply a revolution in the available concepts of physics, it was argued on several occasions at the Conference that one must first check most carefully whether the quantum effects could not be explained by some strange, but still classical, mechanism. In any case, all the participants in the Solvay Conference, including those who had previously not been concerned with the problems of quantum theory, departed with the impression that some very fundamental principles of the classical description of nature were at stake. Thus the Conference helped considerably in establishing quantum theory, previously the occupation of a few

---

[180] In the final general discussion at the Solvay Conference, Poincaré drew attention to the necessity of investigating the question of the equations used in any application of quantum theory. 'In this context one must keep in mind,' he said, 'that one can probably prove every theorem without too much effort if one bases the proof on two mutually contradictory premises' (Poincaré in Eucken, 1914, p. 364).

scientists, as a field of serious and active research at many places and in many countries.[181]

The reports and discussions of the Solvay Conference also provided a great stimulus to those who had been involved in quantum problems for years. After all, it was the first time that a meeting of the leading experts in this field had taken place. They had been given the opportunity of meeting together personally, of presenting their respective arguments in detail, and of debating for several days all the fundamental as well as controversial aspects. Immediately on his return to Prague from the Conference, Einstein wrote to Heinrich Zangger in Zurich: 'I was able to convince Planck to a large extent about my views, now that he has resisted them for years. He is a totally honest man who does not worry about himself' (Einstein to Zangger, November 1911, see Mehra, 1975c, p. xiv).[182] To be sure, the Conference did not provide an immediate solution to all the problems that were brought up. Indeed, some participants felt that that was the case. Thus, for instance, Einstein remarked a few weeks later: 'I have not made any progress in electron theory. In Brussels also people deplored the failure of the theory without finding a remedy. The conference there indeed seemed to be like a lamentation on the ruins of Jerusalem. Nothing positive was achieved. My fluctuation considerations were received with great interest and without serious criticism. I was challenged very little, for I heard nothing that was not known to me' (Einstein to Besso, 26 December 1911). However, the full impact of the Conference, also on Einstein, would show up only gradually. Then it would become evident that the 'witches' sabbath in Brussels,' as Einstein's friend Michele Besso referred to the Solvay Conference ('*Brüsseler Hexensabbat*,' Besso to Einstein 23 October 1911), had in fact achieved its primary goal: to bring the full extent of the quantum problem to the attention of the experts and to persuade them to cooperate more closely for the future growth of quantum theory.

I.7   The Consolidation of Quantum Theory

In the years following immediately after the Solvay Conference the development of quantum physics went through a period of consolidation. People continued to work with concepts which had been established in the previous decade; they extended the application of these concepts and tried to deepen the foundation of

---

[181] The early situation in quantum theory is illustrated by a story which Philipp Frank related about his visit to Einstein in Prague. Opposite to the building of the Institute of Theoretical Physics in Prague lay the park of the mental hospital. Einstein (since April 1911, Professor of Theoretical Physics at the German University of Prague), looking out of his office window, remarked to Frank: 'There you see the madmen who do not work on the quantum theory' (Frank, 1949, p. 143).

[182] It is not quite clear about which views of his Einstein had been able to convince Planck, for he wrote several days later again to Zangger: 'Planck is untractable about certain preconceived ideas which are, without any doubt, wrong . . . , but nobody really knows' (Einstein to Zangger, 16 November 1911).

the quantum theory. One of the remarkable features of this period was the fact that certain personalities, whose contributions had determined the advance of quantum theory—such as Albert Einstein and Johannes Stark—more or less withdrew from active research in this field. Among the numerous scientific papers, which Einstein published between 1912 and 1915, only a few dealt with quantum problems: one in 1912 on the thermodynamic foundation of the photochemical equivalence law (Einstein, 1912a; see also Einstein, 1913a, and some addenda in Einstein, 1912b, c); another, written jointly with Otto Stern in 1913, on molecular motion at the absolute zero of temperature (Einstein and Stern, 1913); and a third in 1914 on certain specific questions of quantum theory (Einstein, 1914a). In these papers Einstein treated certain special points of the theory—as, for instance, which parts of it would follow from the application of thermodynamic arguments (Einstein, 1912a, 1914b).[183] He stopped contributing to the discussion of the really fundamental aspects of the theory, and when Sommerfeld tried to obtain his opinion about some recent development in the theory of the specific heat of solids, which had occurred in 1912, he replied: 'I assure you that I have nothing new to say in the matter of the quanta that may be of interest' (Einstein to Sommerfeld, 29 October 1912). Instead, he increasingly occupied himself with the great task of generalizing the principle of relativity in such a way as to obtain a consistent field-theoretical description of gravitation. In contrast to Einstein, Stark continued to publish various papers invoking the quantum concept, without, however, arriving at particularly new conclusions. Basically, he shifted his interests from theoretical to experimental questions. Finally, in early 1914 he discovered, what seemed to him, a serious difficulty with the energy-quantum hypothesis: by considering simultaneously the description of the effects caused by external electric fields on the spectral lines emitted by atoms, on the one hand, and external magnetic fields, on the other, Stark concluded that the energy available for the spectral line in the visible region was much smaller than the energy-quantum associated with the same frequency (Stark, 1914b, c).

> In order to retain Planck's light-quantum hypothesis [Stark meant, of course, the hypothesis of energy-quanta] in the face of the above mentioned consequence from experience, there remains nothing else but to invoke a further hypothesis, namely the assumption that several electrons participate in the emission of the series line, and that a system of numerous series electrons must be regarded as a resonator in the sense of Planck's hypothesis. (Stark, 1914c, p. 304)

In addition, he claimed, that the sharpness of the lines emitted by canal rays

---

[183] It is of interest to note that Einstein, in the few papers on quantum theory which he wrote between 1911 and 1916, did not emphasize the light-quantum point of view but rather tried to avoid it; instead he used thermodynamical arguments similar to Planck's. On the basis of such arguments he treated the photochemical equivalence law (Einstein, 1912a, 1913a) and the blackbody radiation law (Einstein, 1914b). Einstein was also sympathetic to Planck's idea of the zero-point energy. (See his paper with Stern—Einstein and Stern, 1913—on the specific heat of molecular hydrogen.)

implied that any 'light-cell' ('*Lichtzelle*'; he meant 'light-quantum') must have an extension of more than 1000 wavelengths, and remarked:

> These consequences concerning the extension of a light-cell as compared to the wavelength [of light] remove from Einstein's light-cell hypothesis the advantage of visualizability and the heuristic value [which is important] for the experimentalist. The same applies to Planck's light-quantum hypothesis in connection with the consequence that numerous series electrons combine to [emit or absorb] one light-quantum. (Stark, 1914c, p. 306)

It was the loss of visualizability (*Anschaulichkeit*) in the more recent quantum-theoretical description of atomic structure and their spectra, which made Stark revise his former positive attitude towards quantum theory.

While Stark retired completely, and Einstein partially, from the forefront of research in quantum theory, many new people joined the ranks of quantum physicists. One of them was Peter Debye, a former student of Sommerfeld's.[184] His first contribution to quantum theory was a new derivation of Planck's radiation law in a paper submitted in October 1910 to *Annalen der Physik* (Debye, 1910).[185] In this paper Debye sought to avoid the main inconsistencies of Planck's earlier derivations, which Einstein had criticized. For instance, he did not make use of electromagnetic theory to arrive at the relation (7) between the radiation density, ρ_ν, and the average energy of the oscillator, $\overline{U}_\nu$. He rather pursued a path 'which with the help of the elementary quantum hypothesis made it possible to calculate the probability for a given state of radiation and therefore, as is well known, the entropy, using the properties of the state alone without employing resonators' (Debye, 1910, p. 1428). To achieve his goal, Debye just calculated, in agreement with the method of Rayleigh and Jeans, the number

[184] Peter Joseph William Debye (Petrus Josephus Wilhelmus Debije) was born on 24 March 1884 in Maastricht, Netherlands. He studied at the *Technische Hochschule*, Aachen, from 1901 to 1905, and graduated with a *Diplom* in electrical engineering. Then he became Sommerfeld's assistant. When Sommerfeld moved to Munich (in 1906) as Professor of Theoretical Physics, Debye went with him; there he obtained his doctorate in 1908 and his *Habilitation* in 1910. In 1911 he succeeded Einstein as the *Extraordinarius* for theoretical physics at the University of Zurich. In 1912 he moved to Utrecht as Professor of Theoretical Physics, and two years later to Göttingen (where Woldemar Voigt took his retirement to make way for Debye). In 1920 he returned to Zurich, this time at the E.T.H. (Swiss Federal Institute of Technology). In 1927 he accepted a call from the University of Leipzig as Professor of Experimental Physics and Director of the Physical Institute. In 1935 he moved to Berlin as Director of the *Kaiser-Wilhelm-Institut für Physik*. He left Germany in 1940 and became a professor of chemistry at Cornell University, Ithaca, New York. Debye was made Professor Emeritus in 1950. He died in Ithaca on 2 November 1966.

Debye worked in many areas of mathematics and theoretical, experimental and applied physics. He made important contributions to the theory of dipole moments of molecules, X-ray studies of crystals, the theory of electrolytes, and the studies of polymers. He received many honours, including the Nobel Prize in Chemistry for 1936.

[185] Debye's interest in the quantum problem developed when he prepared to give his first lecture course at the University of Munich (Debye, AHQP Interview, p. 2). He wrote about it to Sommerfeld in Göttingen: 'With my considerations on radiation I have now more or less arrived at a definite point of view, and since I think that some of it may possibly interest you—also for Göttingen [where, in March 1910, Sommerfeld was visiting]—, I want to write to you how I have dealt with the thing . . . ' (Debye to Sommerfeld, 2 March 1910).

$N_\nu \, d\nu$ of elementary states—or ether modes—in the interval, ν and $\nu + d\nu$, contained in a volume v. He found

$$N_\nu \, d\nu = \frac{8\pi v \nu^2}{c^3} \, d\nu, \qquad (88)$$

with c the velocity of light *in vacuo*. This number had to be multiplied with the average energy, $h\nu f(\nu)$, where the distribution function, $f(\nu)$, was obtained by considering the state of maximum probability; the calculation indeed yielded Planck's law.[186] After his first success in radiation theory, Debye became a partisan of the quantum concept and he continued to publish further papers dealing with it.[187] His next important contribution, however, dealt with the specific heats of solids.

On 9 March 1912, at a meeting of the Swiss Physical Society in Berne, Debye gave a talk on certain peculiarities of the specific heats at low temperatures (Debye, 1912b). He informed Sommerfeld about its contents in a letter, in which he wrote:

> My Berne lecture should appear soon in the *Archives de Genève* as a short note. I do not have the galleys yet; however, in order to inform you about my goal, I shall emphasize here the main points of this matter. Einstein had proceeded as follows: He takes a solid body, considers an atom in it and claims that he can treat it as a resonator of frequency ν. Then he inserts for the energy of the single atom the value $3h\nu/(\exp(h\nu/kT) - 1)$. His assertion is incorrect, as he himself recognized later on, for one cannot speak of an oscillating motion [of the atom in the solid] with constant frequency. Therefore I proceed like this: The entire body is in my opinion like a composite molecule. It [the solid body] is able to perform oscillations, infinitely many according to the theory of elasticity. In reality the latter is not true; the body possesses only $6N$ degrees of freedom because it consists of N atoms. Having recognized this fact, one has to do the following: 1. To calculate the frequencies of the body, taking into account its atomic structure, just as Jeans does it for the empty cavity; 2. to assume that each of the two degrees of freedom [of the oscillation] possesses the energy $h\nu/(\exp(h\nu/kT) - 1)$. The combination [of both steps] yields the energy content of the body.

With respect to the details of the calculations, Debye remarked:

> The problem 1 can be approximated by elasticity theory. This approximation is sufficient for very low temperatures, because the highest frequencies do not play a

[186]The probability of any state was given by the expression,

$$\prod_\nu \frac{[N_\nu \, d\nu + N_\nu f(\nu) \, d\nu]!}{(N_\nu \, d\nu)! \, [N_\nu f(\nu) \, d\nu]!},$$

in agreement with Planck's original *Ansatz*. The maximum of this expression had to be calculated under the requirement of a fixed total energy. Debye obtained $f(\nu) = [\exp(h\nu/kT) - 1]^{-1}$.

[187]For example, in 1911 Debye treated two quantum problems: the quantum structure of energy (Debye, 1911a) and an application of the quantum hypothesis to explain deviations of the magnetic behaviour of substances from the Curie–Langevin law (Debye, 1911b).

role. One finds that the number of oscillations [having frequencies] between ν and $\nu + d\nu$ is, as for the radiation according to Jeans, proportional to $\nu^2 d\nu$; hence the energy content [of the solid] becomes proportional to T^4. (Debye to Sommerfeld, 29 March 1912)

These remarks contained all the crucial ideas which went into Debye's paper, entitled '*Zur Theorie der spezifischen Wärmen*' ('On the Theory of Specific Heats') and submitted to *Annalen der Physik* in July. 1912 (Debye, 1912c). According to the continuum theory of elasticity the number of vibrational modes, z, in a volume v, having frequencies smaller than ν, is given by the equation

$$z = \nu^3 \cdot v \cdot F, \tag{89}$$

where

$$F = \frac{4\pi}{3} \rho^{3/2} \kappa^{3/2} \left[2\left(\frac{2}{3}\right)^{3/2} \left(\frac{1+\sigma}{1-2\sigma}\right)^{3/2} + \frac{1}{3^{3/2}} \left(\frac{1+\sigma}{1-\sigma}\right)^{3/2} \right], \tag{89a}$$

where ρ and κ denote the density and compressibility, respectively, of the substance, and σ is the ratio of transverse to longitudinal (elastic) expansion coefficients (Debye, 1912c, p. 795, Eq. (2)). The terms on the right-hand side of Eq. (89a) may be identified, apart from a factor 4π, with $2/c_t^2$ and $1/c_l^2$, respectively, where c_t and c_l denote the velocities of propagation of transverse and longitudinal vibrations. Evidently, the continuum theory of elasticity, which admitted arbitrarily high frequencies in the spectrum, had to be modified in order to describe the behaviour of N atoms in a solid. Debye proposed, therefore, to introduce a maximum frequency, ν_{max}, given by the equation

$$3N = \nu_{max}^2 \cdot v \cdot F, \tag{90}$$

because the number of modes in a monatomic substance could not exceed the degrees of freedom. The same maximum frequency then played a crucial role in the expression for the total energy of the solid, from which Debye derived the equation for the atomic heat, given by[188]

$$c_v = \frac{9R}{x^3} \int_0^x \frac{\xi^4 \exp\xi \, d\xi}{(\exp\xi - 1)^2}, \tag{91}$$

where

$$x = \frac{h\nu_{max}}{kT} = \frac{\Theta}{T}. \tag{91a}$$

[188] By assigning each mode of frequency ν an average energy, $h\nu[\exp(h\nu/kT) - 1]^{-1}$, Debye obtained for the total energy of the field containing N atoms the expression

$$U = \frac{9N}{\nu_{max}^3} \int_0^{\nu_{max}} \frac{h\nu}{\exp(h\nu/kT) - 1} \cdot \nu^2 \cdot d\nu.$$

Equation (91) then followed by taking the derivative $dU/dT = c_v$.

Thus the specific heat turned out to be a universal function of the ratio, Θ/T, with Θ a characteristic temperature for each substance. For large values of T, which imply small x, the right-hand side of Eq. (91) approaches the classical value very fast. For very low temperatures, on the other hand, Debye noticed that the integral could be replaced by a constant, hence c_v was given by the equation

$$c_v = \text{const.} \frac{T^3}{\Theta^3}, \qquad (92)$$

or 'for sufficiently low temperatures the specific heat becomes proportional to the third power of the absolute temperature' (Debye, 1912c, p. 800).

Debye immediately noticed that Eq. (91) described the specific-heat data of diamond, copper, silver and lead (Nernst and Lindemann, 1911b) perfectly. The T^3-law was tested especially carefully at low temperatures and was found to be correct (Eucken and Schwers, 1913). This result may be easily understood because of the basic assumption underlying Debye's theory, namely, the approximation of a solid by a continuum. Since any piece of matter was known to consist of atoms of small but finite size, it could be regarded as being continuous only for vibrations having long wavelengths. However, for low temperatures, the latter contribute mainly to the specific heats, hence the T^3-law should be valid in any case, even if Eq. (91) did not represent the behaviour at higher temperatures. Still, the overall description of the specific heats of many substances by Debye's functions was amazing.

At the same time as Debye worked out his theory of specific heats, Max Born and Theodore von Kármán worked out another approach in which the atomic structures of solids entered explicitly.[189] In their first paper on the problem,

---

[189] Theodore (Theodor) von Kármán was born in Budapest on 11 May 1881. He studied at the Technical University of Budapest, where, after completing his military service, he continued as an assistant professor (1903–1904). Then he went to Germany and took the position of a research engineer with Ganz & Co. In 1906 he resumed his studies at the University of Göttingen; he obtained his doctorate under Ludwig Prandtl in 1908 and his *Habilitation* a year later. In 1912 he was appointed Director of the Aeronautics Institute at the *Technische Hochschule*, Aachen. He stayed in this position until 1930, when he went to the United States; there, at the California Institute of Technology, he continued his career as one of the outstanding pioneers in aeronautics. Von Kármán died in Aachen on 6 May 1963.

Theodore von Kármán knew his colleague Max Born—they were both *Privatdozenten* at the University of Göttingen—rather well, because both took their meals in the same boarding house and later, in fall 1911, even moved together into the same house. They had ample opportunities of discussing scientific problems. Born later recalled the beginning of their collaboration on crystal-lattice theory as follows:

> That our attention was drawn to crystals and their lattice structure was connected with some work by [Erwin] Madelung, who was an assistant in the physics department. Lattice structure of crystals was at that time (1911, 1912) much under discussion, but still hypothetical; Madelung's work [Madelung, 1909; 1910a, b] came just a little before the discovery of X-ray diffraction by von Laue, Friedrich and Knipping (1912). He obtained a numerical relation between the elastic constants and the wavelength of infrared absorption (Rubens' *Reststrahlen*) by studying the vibrations of a model lattice for rock salt . . . He [von Kármán] suggested to me that we should reconsider the problem from the standpoint of mechanics of a system of coupled particles. Which of us connected this line of thought with Einstein's papers on specific heat of solids, I cannot remember. (Born, 1978, p. 141)

entitled '*Über Schwingungen in Raumgittern*' ('On Vibrations in Space Lattices') and submitted to *Physikalische Zeitschrift* in March 1912, Born and von Kármán discussed in detail the frequency spectrum of oscillations of atoms bound in regular crystal lattices (Born and von Kármán, 1912). Thus they obtained the following expression for the atomic heat:

$$c_v = 3R \frac{3}{(2\pi)^3} \int_0^{2\pi} \frac{(\beta\nu_0/T)^2 \sin^2(\omega/2)\exp((\beta\nu_0/T)\sin(\omega/2))}{\left[\exp((\beta\nu_0/T)\sin(\omega/2)) - 1\right]^2} \cdot \omega \cdot d\omega, \quad (93)$$

with ν_0 denoting the average value of the limiting (highest) frequencies in the lattice.[190] The authors concluded:

> This expression shares with the one of Einstein [i.e., Eq. (75)] the property that it converges towards zero for decreasing temperatures and, at large temperatures, approximates the value $3R = 5.94$ given by the law of Dulong and Petit. The difference between our formula and Einstein's arises from the fact that we have not distinguished from the beginning any eigenfrequency, but have taken into account all frequencies of the spectrum as prescribed by the number of degrees of freedom. The qualitatively similar form of both laws [of Einstein and Born–von Kármán] follows from the fact that the number of degrees of freedom accumulate in the region of the limiting [i.e., the highest] frequencies [of the spectrum]. In addition, one easily recognizes that the corrections [to Einstein's formula, as derived from Eq. (93)] lie in the same direction as demanded by the systematic deviations in the data, and as the formula of Nernst and Lindemann seeks to represent. (Born and von Kármán, 1912, p. 308)

Of course, the accurate evaluation of the right-hand side of Eq. (93) turned out to be a cumbersome task, and only few immediate consequences could be drawn from it. Thus, in a second paper, which they submitted in November 1912, Born and von Kármán showed that for low temperatures there followed an expression consisting of three terms *à la* Debye—i.e., given by Eq. (91)—and further terms of the type considered by Einstein in Eq. (75). (See Born and von Kármán, 1913a.)[191]

Although the lattice theory of Born and von Kármán did not provide an immediate possibility of comparison with experimental data, it soon became clear that it described the situation in solids better than Debye's continuum

---

[190] In each of the main directions in the crystal (or the three-dimensional lattice) the frequencies of oscillations are given by the formula $\nu_j = \nu_{0j} \sin(\omega/2)$ ($j = 1, 2, 3$), where ν_{0j} are the limiting frequencies. (The limiting frequency is 1, the corresponding wavelength to which equals twice the distance to the next lattice point.)

[191] Born and von Kármán considered the crystals as being built up of basic cells, each cell consisting of p (identical or different) atoms (or ions). The resulting low-temperature expression for the specific heats could then be split up into 3 Debye terms and $3(p-1)$ Einstein terms. The frequencies associated with the latter lay in the infrared region. In a third paper, Born and von Kármán finally completed their method of calculating eigenfrequencies (Born and von Kármán, 1913b).

approach.[192] The reason was that almost exactly at the time of the publication of Born and von Kármán's work the lattice structure of crystals was proven: on 21 April 1912 Walter Friedrich and Paul Knipping at Sommerfeld's Institute in Munich, acting upon the proposal of Max von Laue, observed the diffraction patterns of X-rays from crystals (Friedrich, Knipping and Laue).[193] These diffraction patterns could only be interpreted in the following way: the X-rays, which were electromagnetic radiation, were scattered by the three-dimensional grating of atoms constituting a crystal.[194] Thus two important issues were settled at once: the nature of X-rays and the discrete structure of solids. This discovery immediately aroused the Braggs—William Henry Bragg and his son William Lawrence—in England to pursue investigations of their own in the field of crystal structure analysis by X-rays. William Lawrence Bragg, in particular, studied von Laue's interpretation of the diffraction patterns and got a new idea. 'I tried to attack the problem from a slightly different point of view,' he explained in his Nobel lecture, 'and to see what would happen if a series of irregular [X-ray] pulses fell on diffracting points arranged on a regular space lattice. This led naturally to the consideration of the diffraction effects as a reflexion of the pulses by the planes of the crystal structure' (W. L. Bragg in *Nobel Lectures in Physics*,

---

[192] Debye's theory was expounded a little before Born and von Kármán's. Thus Born and von Kármán stated in their second paper:

> Shortly before the appearance of our communication of April 1912 Mr. Debye has, as he told us afterwards, reported his results in the March meeting of the Swiss Physical Society and published a short note in the *Archives de Genève*, March 1912, p. 256. Further, Mr. Natanson claims—based on a communication to the February session of the Cracow Academy—that he first stated the idea which lies at the basis of the treatments of Debye and ourselves. It seems to us that the priority for giving an exact formulation and an approximate solution of the problem belongs to Mr. Debye by several days. (Born and von Kármán, 1913a, p. 15, footnote 1 in the right column)

Naturally, Born and von Kármán were disappointed that Debye had the priority for the work, but they consoled themselves with the fact that their results 'are more satisfactory than Debye's, which applies only to quasi-isotropic substances [apart from the continuity approximation]' (Born, 1978, p. 142).

[193] Walter Friedrich, born on 25 December 1883 in Salbke, close to Magdeburg, was Sommerfeld's experimental assistant. (Later, in 1923, he would become a professor of physics at the University of Berlin.) Paul C. M. Knipping, born on 21 May 1883 in Neuwied, was at the same time a doctoral student of Röntgen's. (He obtained his doctorate in 1913; in 1928 he became *Extraordinarius* at the *Technische Hochschule*, Darmstadt; he died in a motorcycle accident on 26 October 1935.)

Max Theodor Felix von Laue was born on 9 October 1879 in Pfaffendorf, near Koblenz. He studied at the Universities of Strasbourg, Göttingen, Munich and Berlin, and obtained his doctorate under Planck in Berlin in 1903. He spent the following two years at the University of Göttingen, then went back to Berlin as Planck's assistant. He obtained his *Habilitation* in 1906 and became *Privatdozent* in Berlin (1906–1909). Then he went to Munich, and in 1912 succeeded Debye as *Extraordinarius* for theoretical physics at the University of Zurich. From 1914 to 1919 he was full professor at the University of Frankfurt-am-Main, then he joined Planck again at the University of Berlin as a colleague. From 1921 onwards he also acted as one of the directors of the *Kaiser-Wilhelm-Institut für Physik* in Berlin. In 1951 he became Director of the *Fritz-Haber-Institut* of the *Max-Planck-Gesellschaft*. He retired in 1959, and died in Berlin on 23 April 1960. Among the many honours von Laue received was the Nobel Prize in Physics for 1914 for his discovery of the interference of X-rays.

[194] An account of the discovery of Friedrich, Knipping and von Laue was first presented by Sommerfeld to the Bavarian Academy of Sciences on 8 June 1912 (Friedrich, Knipping and Laue, 1912). For details of the story, see the article by Paul Forman (1970a).

1967, p. 371).[195] William Henry Bragg constructed an instrument, the so-called X-ray spectrometer, suitable for observing the reflections, and both Braggs carried out investigations leading to a detailed knowledge of many crystal structures (Bragg and Bragg, 1913). The importance of the results obtained from X-ray diffraction and reflection was immediately acknowledged by the community of physicists. As a result, Max von Laue and William Henry Bragg were invited to talk about their results at the second Solvay Conference, which took place in Brussels from 27 to 31 October 1913 and dealt with 'The Structure of Matter.'[196] Besides the reports of von Laue and Bragg, two other reports—one by W. Barlow and William Jackson Pope and another by Marcel Brillouin—also dealt with crystal structure; in other words, this subject dominated the second Solvay Conference.[197]

While Debye, Sommerfeld's former student, made a successful application of quantum theory to the specific heat of solids, his extended collaboration with Sommerfeld was not equally successful (Debye and Sommerfeld, 1913). Based on the fundamental hypothesis, Eq. (87), they proceeded along the path indicated by Sommerfeld at the first Solvay Conference in fall 1911; in particular, they derived an expression for the accumulation time τ of the energy from radiation of frequency ν and electric amplitude E, i.e.,

$$\tau = \frac{\sqrt{16m_e\nu h}}{eE} , \qquad (94)$$

[195] W. Lawrence Bragg was born on 31 March 1890 in Adelaide, Australia. He was educated at St. Peter's College and the University of Adelaide. He went to England with his father in 1909 and entered Trinity College, Cambridge. After graduating with first class honours in 1912, he collaborated with his father on X-ray interference phenomena during the following couple of years. Lawrence Bragg's scientific career was interrupted by World War I; in 1919 he was appointed Langworthy Professor of Physics at the University of Manchester. In 1935 he became Director of the National Physical Laboratory; in 1938 he was appointed Rutherford's successor as Cavendish Professor in Cambridge. He retired from Cambridge in 1953 and then served as Fullerian Professor at the Royal Institution (1953–1966). Among his many honours, he shared the Nobel Prize in Physics for 1915 with his father for their work on the analysis of crystal structures by X-rays.

[196] Because of the great success of the Conference of 1911 Ernest Solvay, following the initiative of Hendrik Lorentz, established on 1 May 1912 an International Institute of Physics at Brussels. This institute (a foundation), run by an Administrative Council (including E. Solvay or his representative) and an International Scientific Committee (consisting initially of H. A. Lorentz as President, Marie Curie, M. Brillouin, R. B. Goldschmidt, H. Kamerlingh Onnes, M. Knudsen, W. Nernst, E. Rutherford and E. Warburg), was endowed by funds which were to be used for two purposes: first, to encourage research on a deeper understanding of natural phenomena, with grants and subsidies going to research workers; second, to organize periodic conferences at Brussels with selected participants.

[197] At the second Solvay Conference, the participants (besides the members of the Scientific Committee mentioned in footnote 196) were: G. Gouy and P. Langevin from France; W. Barlow, W. H. Bragg, J. H. Jeans, W. J. Pope and J. J. Thomson from England; E. Grüneisen, H. Rubens, A. Sommerfeld, W. Voigt and W. Wien from Germany; A. Einstein, M. von Laue and P. Weiss from Switzerland; F. Hasenöhrl from Austria; and R. W. Wood from the United States. The Secretaries of the Conference were: R. B. Goldschmidt (Brussels), M. de Broglie (Paris) and F. Lindemann (Berlin). Eight reports were presented; besides the ones already mentioned, there were talks by J. J. Thomson on the structure of atoms, by Woldemar Voigt on the temperature dependence of pyroelectricity, and by Robert W. Wood on resonance radiation and its spectra. (For more details of the second Solvay Conference, see Mehra, 1975c, pp. 75–92.)

where m_e, e and h denote the mass and charge of the electron and Planck's constant, respectively. This formula could not be justified experimentally; no relation between τ and E of the type of Eq. (94) was found to hold empirically (Gerlach and Meyer, 1913). Thus Sommerfeld's (action-integral) quantum hypothesis did not turn out to be as helpful as had been foreseen in 1911. On the other hand, Walther Nernst's approach led to several successful applications of the quantum concept: in particular, to a theory of the so-called chemical constant (Sackur, 1911), to an interpretation of the specific heats of molecules (Nernst, 1911b), to an understanding of molecular spectra (Bjerrum, 1912), and to the prediction of degeneracy of ideal gases at low temperatures (Nernst, 1912a).

In his paper on the new heat theorem, read before the Göttingen Academy of Sciences on 23 December 1905, Nernst had given an expression for the kinetic constant K_c (occurring in the law of mass action of Guldberg and Waage); he had written it in the form

$$\ln K_c = \frac{Q_0}{RT} + \frac{\sum c_v}{R} \ln T + J, \tag{95}$$

with Q_0 denoting the energy produced as heat at the absolute zero of temperature, R the gas constant, and c_v the atomic heats of the substances involved (see Nernst, 1906, p. 4, Eq. (7)). The additional term J—which he later called the 'chemical constant'—determined, together with Q_0, the chemical equilibrium of the process under investigation. Equation (95) could be applied also to physical equilibria, such as the one between the solid and the gaseous state of a substance. Otto Sackur, a former pupil of Nernst, did so and obtained an equation for the vapour pressure of the monatomic solid, that is,

$$\log p = -\frac{\lambda_0}{RT} + \frac{c_p}{R} \ln T + \frac{1}{RT} \int c_f dT + \frac{1}{R} \int \frac{c_f}{T} dT + J + \ln R, \tag{96}$$

where p is the vapour pressure of a monatomic solid, λ_0 the heat of vapourization (at zero temperature), and c_p and c_f denote the atomic heat at constant pressure of the vapour and the solid, respectively (Sackur, 1911, p. 965).[198] Since the chemical constant was related to S', the entropy constant of the vapour (defined as the entropy of a gram-atom, which assumes at the temperature $T = 1°K$ the volume $v = 1$ cc), through the equation

$$S' = RJ + c_v, \tag{97}$$

[198] Otto Sackur was born on 28 September 1880 in Breslau. He studied at the University of Breslau and obtained his doctorate in 1901. After spending two years working with Rudolf Ladenburg in Breslau, he went to London (to William Ramsay) and to Berlin (to Nernst). In 1905 he obtained his *Habilitation* at the University of Breslau and in 1911 he became Professor of Chemistry there. Three years later he joined the *Kaiser-Wilhelm-Institut für physikalische und Elektro-Chemie*, Berlin, as the director of a department. As a result of an accident in the laboratory, he died in Berlin on 17 December 1914.

c_v being the atomic heat per gram-atom of the vapour, one could apply quantum theory to calculate the value of J. For this purpose Sackur separated the phase space of the vapour atoms in cells of size h^3, and obtained the result

$$S' = \frac{3}{2} R + \ln \frac{(2\pi mk)^{3/2}}{N_0 h^3} ,$$ (97a)

with m the mass of the atoms, k and h Boltzmann's and Planck's constants, respectively, and R and N_0 denoting the gas constant and Avogadro's number (Sackur, 1911, p. 968, Eq. (12)). A year later Otto Stern rederived the same result by employing a special molecular-kinetic model of the process of vapourization (Stern, 1913).[199] In order to do so, however, he had to assume that each atom of the solid possessed a zero-point energy of magnitude $\frac{3}{2} h\nu$.[200] Several years later he repeated the derivation with an improved model of the solid, describing the latter by the lattice theory of Born and von Kármán (Stern, 1919).[201]

The question of the validity of the second quantum hypothesis of Planck, especially the existence of the zero-point energy, also entered into the discussion of another type of quantum phenomena, the ones connected with the rotational motion of atoms and molecules. Early in 1911 Nernst had remarked in an article on specific heats:

> We have seen that there arises a deviation from the laws of statistical mechanics if the atoms rotate around a position of equilibrium. If we generalize the quantum hypothesis—which is not far-fetched—in such a way that the energy is always absorbed in fixed quanta not only in the case of oscillations around an equilibrium position but also in the case of an arbitrary rotation of the mass, then we arrive at further conclusions, which may perhaps be able to explain certain contradictions of the old theory [i.e., statistical mechanics]. (Nernst, 1911b, p. 270)

From it he had concluded that the molecule of a monatomic gas at usual temperatures could not assume a quantum of rotational energy, as the frequency of rotation might be very high, while a diatomic molecule would be endowed with a finite rotational energy. Nernst's proposal to quantize rotational energy had been received favourably at the first Solvay Conference; thus Einstein had

---

[199] Stern assumed, in particular, that the atoms in a solid were bound by harmonic forces; under normal conditions they performed vibrations with characteristic frequencies, which entered into Einstein's formula for the specific heat of the solid. Outside a sphere of finite radius, however, the atoms were not bound anymore, but behaved like free atoms (i.e., they were atoms of the vapour). With this assumption for his model, Stern calculated both the entropy of the solid and of the vapour. His expression for the latter was identical with Eqs. (97) and (97a).

[200] Thus, instead of λ_0, the heat of vapourization at zero temperature, Stern had to insert the expression $\lambda_0 - \frac{3}{2} Nh\nu$.

[201] It should be mentioned at this point that Frederick A. Lindemann used the same *Ansatz* (as Stern did in 1913 for the heat of vapourization) in order to explain the negative result of his attempts to separate, together with Francis William Aston, the isotopes Ne^{20} and Ne^{22} (Lindemann and Aston, 1919).

suggested using two equations,

$$\overline{E}_{rot} = \frac{h\nu}{\exp(h\nu/kT) - 1} \tag{98a}$$

and

$$\overline{E}_{rot} = \tfrac{1}{2}A(2\pi\nu)^2, \tag{98b}$$

to describe the temperature dependence of $\overline{E}_{rot}$, the average energy of a molecule (Einstein in Eucken, 1914, p. 351). These two equations allowed one to eliminate ν, the unknown frequency of the molecules, with the help of the quantity A, denoting the moment of inertia of the molecule.

Einstein's proposal, contained in Eqs. (98a) and (98b), implied, however, that the rotational frequency ν of a molecule did not change with temperature, which seemed difficult to understand, especially at low temperatures. To ensure such a behaviour, Einstein and Stern assumed that the rotating molecules possessed a zero-point energy; hence Eq. (98a) could be replaced by

$$\overline{E}_{rot} = \frac{h\nu}{\exp(h\nu/kT) - 1} + \frac{h\nu}{2}. \tag{98a$'$}$$

With this *Ansatz* they also succeeded in fitting together the new data on the specific heat of molecular hydrogen.[202] However, Paul Ehrenfest pointed out immediately that Eq. (98a$'$) did not guarantee a fully temperature-dependent frequency; he proposed rather to use a quantization prescription similar to Planck's original one for the oscillator, namely,

$$E_{rot} = \frac{1}{2}A(2\pi\nu)^2 = n\frac{h}{2}, \tag{98b$'$}$$

in agreement with an earlier suggestion of Lorentz (Ehrenfest, 1913b).[203] The rotating system then assumed the discrete frequencies

$$\nu_n = \frac{nh}{4\pi^2 A} \tag{99a}$$

and the rotational energy of a molecule the discrete values

$$E_{rot} = \frac{n^2 h^2}{8\pi^2 A}. \tag{99b}$$

---

[202] According to the classical theory, the rotational motion of a diatomic molecule should contribute a term of R per gram-molecule to the specific heat of hydrogen. Arnold Eucken found, however, that the rotational contribution to the specific heats of molecular hydrogen would start from very low values at low temperatures and increase steadily, taking on the value R at room temperature (Eucken, 1912).

[203] Lorentz had made the suggestion at the Solvay Conference in 1911 during the discussion of Einstein's report (see Eucken, 1914, p. 362). The factor $\frac{1}{2}$ stemmed from the fact that the rotator—in contrast to Planck's resonator—possesses only kinetic energy.

Ehrenfest then calculated the expression for the average energy at a given temperature and from it the specific heat of molecules.[204] Also his curve for the temperature dependence of the specific heat, i.e., the rotational part of it, fitted the hydrogen data reasonably well. Hence a clear decision between Planck's first and second quantum hypothesis could not be reached.[205]

The Lorentz–Ehrenfest *Ansatz* of discrete frequencies for the rotational motion was also used to explain another property of molecules, namely, their discrete spectra. Especially, Niels Bjerrum, who worked in 1912 as a guest in Nernst's Institute in Berlin, developed a theory of the infrared absorption spectra of diatomic molecules, i.e., of spectra having relatively short wavelengths (Bjerrum, 1912).[206] For that purpose, he assumed that atoms (or ions) in the molecule perform an oscillation with frequency ν_0 (in the near-infrared region), and that the discrete rotational motions combine with the oscillation to yield equidistant absorption frequencies ν, where

$$\nu = \nu_0 \pm \nu_n = \nu_0 \pm n\,\frac{h}{4\pi^2 A}\,. \tag{100}$$

Exactly such equidistant absorption frequencies were found subsequently by Eva von Bahr (1913). These molecular spectra in the near infrared thus seemed to confirm Planck's original quantum hypothesis.[207]

So far the applications of the quantum concept concerned only periodic phenomena. However, certain phenomena in kinetic theory suggested a possible role of the quantum of action also in the translational motions of atoms or molecules. The earliest hints of such a generalization had been given by Planck. In connection with his second hypothesis—the hypothesis of quantum emission

---

[204] The contribution of the rotational degrees of freedom of N molecules to the specific heat of hydrogen, according to Ehrenfest (1913b, p. 455, Eqs. (12)–(14)), was

$$c_v^{\text{rot}} = 2R\sigma^2 \cdot \frac{d \ln Q(\sigma)}{d\sigma}\,,$$

where $\sigma = h^2/8\pi^2 A k T$ and $Q = \sum_{n=0}^{\infty} \exp(-n^2\sigma)$.

[205] The problem of the correct quantum-theoretical description of the rotational motion in the case of a hydrogen molecule was not solved until much later. After the advent of quantum mechanics it was shown that—in contrast to the assumption of Einstein and Stern—half-integral quanta did not play a role. Rather, one had to take into account the fact that molecular hydrogen was a mixture of two different kinds of molecules, ortho-hydrogen and para-hydrogen, having different nuclear spins.

[206] N. Bjerrum was born on 11 March 1879 in Copenhagen. He studied chemistry under Julius Thomsen at the University of Copenhagen from 1897 to 1908 (assistant, 1902–1908), obtaining his doctorate in 1908. In 1912 he became a lecturer at the University of Copenhagen and, two years later, Professor of Chemistry at the Agricultural High School, Copenhagen. He retired in 1949 and died on 30 September 1958 in Copenhagen. Bjerrum published many papers on electrochemistry, theory of acids and bases, electrolytes, and infrared spectra of polyatomic molecules.

[207] Several years later, Elmer S. Imes studied these spectra again in the case of hydrogen chloride, hydrogen bromide and hydrogen iodide; he found that the middle line (of frequency ν_0) was missing, and this fact seemed to speak indeed for the possible existence of a zero-point rotation, and thus in favour of Planck's second quantum hypothesis (Imes, 1919).

and continuous absorption—he had stated that because of the interaction between radiation and electrons, the latter may assume discrete velocities. Then he had remarked:

> Perhaps in this way can be explained the important problem of kinetic theory, as to why the "free" electrons of a metal do not yield an appreciable contribution to the specific heat. Because, according to the view presented here, the electrons possess no independent degrees of freedom at all; the reason being that they assume only certain velocities, and their motions do not at all participate when the energy of the metal as a whole is distributed among the various degrees of freedom. (Planck, 1911a, p. 147)

A year later Nernst took up this idea; while discussing the equation of state of monatomic gases, he concluded that there had to occur deviations from the classical theory which could be explained by quantum theory (Nernst, 1912a).[208] Otto Sackur took up Nernst's suggestion and tried to construct a theory of gases by employing the concept of finite cells in phase space, whose size was determined by Planck's constant (Sackur, 1912a, b). Such consequences of the quantum theory for the behaviour of gas molecules and electrons were discussed in detail at a meeting in Göttingen, held from 21 to 26 April 1913, which was later on referred to as the 'Kinetische Gaskongress' or the 'Gaswoche.' At this meeting the participants—including Hendrik Lorentz, Max Planck, Arnold Sommerfeld and Peter Debye—reviewed critically many aspects of the theory of gases, solids and metal electrons and mentioned the new perspectives opened up by the application of quantum theory.[209]

The period following the first Solvay Conference was thus characterized by a

[208] In particular, Nernst pointed to the fact that at low temperatures monatomic gases cannot emit radiation of short wavelengths, although their atoms collide with each other and the constituent electrons should—in agreement with classical electrodynamics—emit such radiation. Hence, he argued, that the motions of atoms at low temperatures could not be described by classical laws (Nernst, 1912a, p. 1066).

[209] The meeting in Göttingen was announced as a 'series of lectures in the field of the kinetic theory of matter' and sponsored by the *Kommission der Wolfskehlstiftung der Königlichen Gesellschaft der Wissenschaften [zu Göttingen]* (the Wolfskehl Foundation); the announcement was made in issue No. 6 (of 15 March 1913) of *Physikalische Zeitschrift* (Vol. 14, pp. 258–264; see also the previous short announcement on p. 88). As the principal organizer of the meeting David Hilbert signed the announcement. From Monday, 21 April, to Saturday, 26 April 1913, the lectures were given by: Max Planck on '*Gegenwärtige Bedeutung der Quantenhypothese für die Gastheorie*' ('Present Significance of the Quantum Hypothesis for Gas Theory'); Peter Debye on '*Die Zustandsgleichung auf Grund der Quantenhypothese*' ('The Equation of State on the Basis of the Quantum Hypothesis'); Walther Nernst on '*Theorie des festen Aggregatzustandes*' ('Theory of the Solid Aggregate State'); Marian von Smoluchowski on '*Gültigkeitsgrenzen des zweiten Hauptsatzes der Wärmetheorie*' ('Limits of Validity of the Second Law of Heat Theory'); Arnold Sommerfeld on '*Probleme der freien Weglänge*' ('Problems of the Free Path'); and Hendrik Lorentz on '*Anwendung der kinetischen Theorien auf Elektronenbewegung*' ('Application of the Kinetic Theories to Electron Motions'). Brief outlines of the lectures were given in the above-mentioned announcement in order to give the participants the opportunity of preparing for fruitful discussions. The lectures and discussions were published as a book entitled *Vorträge über die kinetische Theorie der Materie und Elektrizität* (*Lectures on the Kinetic Theory of Matter and Electricity*) in the following year (Planck, Debye, Nernst, Smoluchowski, Sommerfeld, Lorentz, etc., 1914; with a preface by D. Hilbert).

continuous increase in the number of topics, for the treatment of which the quantum of action seemed to be necessary. Together with the extension of the applications of quantum theory—of which we have discussed only the most important in the foregoing—efforts were made to deepen its theoretical foundations. However, in spite of the sustained work of certain people these problems proved to be very difficult and only slow progress could be achieved. For instance, Max Planck never ceased to think about the fundamental basis of his radiation theory, especially about the problem of combining the statistical and the electrodynamic aspects leading to the law of blackbody radiation in a way which would be free of inconsistencies. In a paper, presented to the German Physical Society in Berlin at its meeting on 12 January 1912, he proposed a detailed mechanism of the discontinuous emission of radiation (Planck, 1912a). He introduced the hypothesis that the emission might be described in terms of probability arguments; especially, he claimed: 'The ratio of the probability that no emission takes place, to the probability that emission occurs, is proportional to the intensity of the vibration exciting the oscillator' (Planck, 1912a, p. 116).[210] More than two years later, in a paper read before the Prussian Academy on 23 July 1914, Planck went a step further (Planck, 1914a). In what he called a 'modified formulation of the quantum hypothesis' ('*eine veränderte Formulierung der Quantenhypothese*') he proposed that both emission and absorption occurred in a *continuous* manner. This new hypothesis implied that 'the quantum action does not take place at all between the oscillators and the wave radiation [in a cavity], but only between the oscillators and the free particles (molecules, ions, electrons), which exchange energy in collisions with the oscillators' (Planck, 1914a, pp. 918–919). Hence the entire quantum behaviour should originate from the special mechanism by which energy was exchanged between the real constituents of chemical substances and the hypothetical oscillators assumed to be present in a cavity; and this mechanism had to function in such a way that any particle would transfer energy only in multiples of $h\nu$ to an oscillator having the frequency ν. These assumptions sufficed, as Planck showed, to derive his radiation law.

Planck's new ideas concerning the *continuous* exchange of energy in absorption and emission processes remained quite isolated, although at that time numerous physicists tried to improve on the derivation of the radiation formula, Eq. (18). Perhaps the most remarkable attempt in this direction had been Debye's derivation in 1910, which we have already discussed. Debye had based it on the assumption that the heat radiation in a given volume could be regarded as consisting of modes of many frequencies, and that every mode of frequency ν had to be endowed with the energy quantum $h\nu$. However, this assumption seriously contradicted Planck's views, while it supported Einstein's light-quantum

---

[210] It should be noted that Planck did not intend to give up the causal description of the emission process. He emphasized: 'It is not that we assume the lack of causality for the emission; but the processes which determine the emission are of such a hidden nature that the laws describing them can thus far be determined only by statistical methods' (Planck, 1912b, p. 15).

hypothesis. Debye then applied the same arguments to the vibrations of atoms in solids, i.e., he treated them like frequency modes endowed with appropriate energy packets; and by calculating the energy distribution as a function of time he derived his successful formula, Eq. (91), for the atomic heats. The fact that the radiation energy existed only in packets of finite size, i.e., $h\nu$, $2h\nu$, etc., for the frequency ν, also entered into the derivation of Planck's law proposed by Ladislas (Władysław) Natanson: he obtained the equilibrium energy distribution of blackbody radiation by considering the maximum entropy of a distribution of indistinguishable energy packets (of size $h\nu$) among a given number of 'receptacles of energy' which could be identified (Natanson, 1911).[211] While Natanson's fellow countryman, Mieczysław Wolfke, even stressed the light-quantum aspects further—he spoke about localized light-atoms (Wolfke, 1913a, b, 1914a)—the Swede Carl Benedicks tried to interpret the quantum features appearing in Planck's radiation law and the specific heats as being due to an agglomeration of atoms in matter. He argued that at the absolute zero of temperature any atoms, which could oscillate, were not available, but that their number increased with the temperature; thus he believed that the violation of the equipartition theorem could be avoided (Benedicks, 1913). However, Benedicks' agglomeration hypothesis met with an insurmountable obstacle: if at absolute zero the atoms were completely agglomerated, the compressibility of the solid should also disappear in contrast to observation.

Any treatment of the foundations of quantum theory had necessarily to answer the following questions, which had already been raised at the first Solvay Conference: Did quantum theory lie within the range of accepted (classical) dynamics? Or, was classical dynamics a special case of a wider theory (embracing quantum theory as well) which could still be formulated in terms of differential equations? Or, did the formulation of quantum theory deviate totally from the hitherto accepted dynamical laws by involving an explicit discontinuity? The reports presented at Brussels, especially those of Hendrik Lorentz and James Jeans, had persuaded most scientists that the answer to the first question was definitely no. Thus, for instance, Henri Poincaré—who had previously shown no interest in quantum theory—upon Jeans' desperate attempts to rescue the classical equipartition theory in the problem of specific heats by referring to a hydrodynamical analogue consisting of a complicated system of tanks, pipes and leakages, had remarked:

It is quite obvious that Mr. Jeans can explain—by a suitable choice of those connecting pipes between the vessels, on one hand, and of the amount of losses, on the other—any experimental fact. But this is not the task of physical theories. The latter must not introduce as many arbitrary constants as are necessary to represent

[211] Natanson's derivation did not deviate formally from the one given by Einstein (1906g) or Debye (1910) as far as the application of the hypothesis of energy packets (or quanta) was concerned. However, he claimed that his assumption about considering the distribution of indistinguishable energy packets (quanta) among distinguishable receptacles of energy (which replaced Planck's resonators) provided the proper definition of equally probable states in radiation theory.

the phenomena; their goal is rather to establish a connection between different experimental facts and, above all, to predict still unknown phenomena. (Poincaré in Eucken, 1914, p. 64)

Immediately after the Solvay Conference Poincaré sat down to analyze the question of whether quantum theory could be formulated in terms of differential equations or not. On 4 December 1911 he presented his answer to the Academy of Sciences in Paris (Poincaré, 1911); his extended paper on the problem appeared in the January issue of *Journal de physique théorique et appliquée* (Poincaré, 1912a). Poincaré came to the conclusion that any theory from which Planck's law followed would necessarily have to contain an essential discontinuity. In order to prove his conclusion, Poincaré treated the interaction of Planck's resonators from the point of view of statistical mechanics. Thus he described each resonator by a probability density, $w(\eta)$, where η denoted its energy, and calculated the distribution of a given energy among a large number of resonators (having different frequencies), which could interact with each other by collisions. He found that the probability of the total system was just the product of the probabilities of the individual resonators and that the distribution of energy among the resonators of different frequencies did not depend on their respective numbers. Then, by examining the expression obtained for the average energy of resonators having a large frequency ν, Poincaré showed that Planck's result leading to the experimentally confirmed blackbody radiation law, Eq. (18), could be reproduced only if he assumed that $w(\eta)$ was a *discontinuous* function of its argument, η; that is, it had to be zero for all values of η, except when $\eta = 0$, $h\nu$, $2h\nu$, etc.[212]

Poincaré's proof that the quantum phenomena could not be accounted for by a classical description and that the discontinuous description was necessary appeared to be so conclusive that it even persuaded James Jeans, the most obstinate opponent of quantum theory. On 12 September 1913, at the Birmingham meeting of the British Association, Jeans—in the address with which he opened the special discussion on radiation theory—remarked that now, because of Poincaré's work, he felt 'logically compelled to accept the quantum hypothesis in its entirety' (Jeans, 1914a, p. 318). Soon afterwards, in his '*Report on Radiation and the Quantum Theory*,' prepared for the Physical Society of London, he again referred to Poincaré's proof of the discontinuity in the description of radiation phenomena (Jeans, 1914b). He even gave there, in a few pages, a simplified formulation of it, which he based on certain ideas which he had developed in an earlier paper (Jeans, 1910). The basic assumption, Jeans claimed, was 'to suppose that the density of the swarm of points [in the phase-space of the system] must be

---

[212] However, the discrete function $w(\eta)$ had to yield a finite integral transform, $\Phi(\alpha) = \int w(\eta) \cdot \exp(-\alpha\eta)\,d\eta$; that is, $\Phi(\alpha)$ had to be given by the discrete sum, $\sum_n \exp(-\alpha n h\nu)$.

Poincaré went beyond this result to prove that blackbody radiation, whatever its energy distribution may be, could only have a finite energy if $w(\eta)$ exhibited a discontinuity at least at the value $\eta = 0$. (A historical review of Poincaré's work on the quantum theory has been given by Russell McCormmach, 1967a. For a discussion of Poincaré's paper see also Lorentz, 1921, and Planck, 1921.)

zero throughout the whole, except for infinitesimally small regions of the generalized space [i.e., the phase-space].' He explained further:

> There must be isolated small regions $R_1, R_2, \ldots$ in the general space occupied by dense swarms of points; in all other regions the density of points must be zero or infinitesimal. And, in order to satisfy the hydrodynamic equation of continuity in the generalized [phase-] space, the motion of the points must consist of sudden jumps from one of the regions $R_1, R_2, \ldots$ to another. It appears in this way that, as soon as we seek to avoid the equipartition formula, we are compelled to assume motion involving discontinuities of some kind. (Jeans, 1914b, p. 34)

And this assumption agreed, of course, with Poincaré's conclusion.

Still there appeared to be an escape from the proofs of Poincaré and Jeans. Especially, Poincaré had employed in his derivation certain principles of dynamics, such as the principles of energy and momentum conservation, which one might be tempted to sacrifice in order to retain a continuous description. Thus Eucken stated the situation more cautiously than either Poincaré or Jeans as follows: 'In order to avoid the assumption of a discontinuity in establishing the foundation of Planck's formula, at least one of the fundamental laws of the present mechanics (most probably the conservation of momentum) must be formulated in another, extended manner' (Eucken, 1914, p. 373). However, at about the same time as this possibility was mentioned, there came the announcement of observations which provided strong evidence in favour of the discontinuities. In May 1914 Albert Einstein wrote to Paul Ehrenfest from Berlin: '[James] Franck and [Gustav] Hertz have discovered that electrons will be reflected elastically from mercury atoms, as long as they have velocities [he meant kinetic energy] of up to 4.8 volts. At the latter velocity they lose their entire kinetic energy [when colliding with atoms] and emit monochromatic light, such that the relation, kinetic energy $= h\nu$, is valid to within a few percent.' He added: 'Wonderful reversal of the photoelectric phenomenon' and 'brilliant confirmation of the quantum hypothesis' (Einstein to Ehrenfest, 25 May 1914). The experiments of Franck and Hertz, referred to by Einstein, confirmed a theory which Niels Bohr had published a year earlier: in it the quantum theory had been used to explain the structure of atoms (Bohr, 1913b).

Chapter II
The Bohr–Sommerfeld Theory of Atomic Structure

In 1913 the first phase of the development of quantum theory came essentially to an end. This phase was characterized by the establishment of the atomistic structure of matter and the recognition that for the description of many properties of matter quantum theory plays a fundamental role. Although a vast amount of research would still be performed on problems concerning the structure of matter—thus, for instance, Max Born would develop a detailed theory of the properties of solid matter—the main interest of the quantum physicists shifted to the problems of the structure of atoms and molecules. Such a shift did not come about unexpectedly, because the discrete frequencies of the emission and absorption lines of atoms seemed to be related to the fact that, according to quantum theory, the energy of an oscillator assumed values that were integral multiples of the energy packets. Already in 1901, immediately after the derivation of the law of blackbody radiation dealing with the emission and absorption of radiation having continuous frequencies, Planck had directed attention to the problem of explaining the occurrence of characteristic discrete lines in the spectra of elements and had remarked:

> If the question concerning the nature of white light may thus be regarded as being solved, the answer to a closely related but no less important question—the question concerning the nature of light of the spectral lines—seems to belong among the most difficult and complicated problems, which have ever been posed in optics or electrodynamics. (Planck, 1902a, p. 400)

From the point of view of electrodynamics, the emission of any radiation, having continuous or discrete frequency, originated from the (accelerated) motion of charged objects in the atom; that is, it depended on the atom's constitution. Now, in the discussions concerning the foundations of quantum theory during the first decade of the twentieth century it had been concluded at a rather early stage that the structure of atoms and the very nature of the quantum of action h were intimately connected. Especially Hendrik Lorentz had advocated this view at the first Solvay Conference and defended Arthur Haas' theoretical relation between the dimensions of the hydrogen atom and the constant h. Missing were the ideas about how to explain the discrete emission spectra of atoms from their constitution with the help of quantum theory. Such ideas were provided by Niels Bohr in a fundamental paper in the middle of 1913.

The quantum theory of atomic structure, as developed by Niels Bohr and his followers, made detailed use of the properties of atoms, which had been investigated in the course of the past century. Much information about the spectra of atoms and molecules had accumulated since the discovery of dark lines in the solar spectrum in 1802. Important steps forward included the recognition— around 1860—that each chemical substance emits a characteristic discrete spectrum, and the description of some types of such spectra by simple mathematical formulae, the first one being Balmer's formula for the hydrogen spectrum published in 1885. Other researches, spectroscopic as well as nonspectroscopic, revealed important features of the structure of atoms: thus the Zeeman effect showed that electrons seemed to be responsible for the emission of spectral lines, and the scattering of electrons and α-particles by atoms provided a picture of their constitution. Bohr used quantum-theoretical ideas to arrive at a description of atomic structure and spectra. In particular, he constructed a theory of the simplest atom—hydrogen—and derived Balmer's formula. Bohr's results were immediately tested by experiments and confirmed. Two years later Arnold Sommerfeld generalized Bohr's theory and described atoms as multiply periodic systems. The resulting Bohr–Sommerfeld theory of atomic structure, together with three basic quantum principles (the adiabatic principle, the statistical nature of emission and absorption processes, and the correspondence principle), was applied with great success to analyze the principal features of atomic spectra.

II.1   The Spectra of Atoms and Molecules: The Empirical Foundations[213]

The physics of discrete atomic and molecular spectra began with the observation of the English physician and scientist William Hyde Wollaston, who discovered dark lines in the spectrum of sunlight (Wollaston, 1802).[214] Twelve years later Joseph Fraunhofer, employed at the optico-mechanical institute in Benediktbeuern, Bavaria, rediscovered the dark solar lines while measuring the dispersive powers of various kinds of glass for light of different colours (Fraunhofer, 1814).[215] At the same time he noticed that the strong 'orange' streak, which

[213] A detailed account of the development of spectroscopy in the nineteenth century can be found in the book of William McGucken (1969).

[214] W. H. Wollaston was born on 6 August 1766 at East Dereham, Norfolk. He was educated at Charterhouse and Caius College, Cambridge, and received his medical doctor's degree in 1793. In the same year he was elected a Fellow of the Royal Society and set up a private research laboratory at the Royal Society. From 1797 to 1800, when he became partially blind, he practiced medicine. Then he lived on the income from his discovery of making platinum malleable, devoting his time to the investigation of physical and chemical problems, such as the voltaic cell, the solar spectrum (besides the dark lines in it, Wollaston also discovered the ultraviolet region), the structure of crystals, problems of electrochemistry, and the theory of multiple proportions in chemistry. He died in London on 22 December 1828.

[215] Joseph von Fraunhofer was born on 6 March 1787 in Straubing, Bavaria. In 1806 he became employed as an optician in Utzschneider's Optical Institute, located in the buildings of the former monastery of Benediktbeuern. Later he was promoted to manager and finally coproprietor of the Institute, which moved to Munich in 1819. In 1823 Fraunhofer was appointed Conservator of the Physical Cabinet at the Bavarian Academy of Sciences in Munich, and in the following year he received the Civil Order of Merit from the King of Bavaria. He died in Munich on 7 June 1826.

he found in the light of all flames, lay at the position of the dark D-line of the solar spectrum.[216] Fraunhofer's discovery represented the beginning of what later came to be called chemical spectral analysis, the development of which was associated with the names of David Brewster, John Herschel, William Henry Fox Talbot, Charles Wheatstone, Antoine-Philibert Masson, Anders Jonas Ångström and William Swan. These investigators examined the origin of the dark lines in the solar spectrum—the so-called Fraunhofer lines—and suggested that they might be created by the selective absorption of the light emitted by the sun in its atmosphere (Brewster, 1832, p. 320). The question then arose as to which chemical substances emitted which particular discrete lines.[217] The final and conclusive steps towards chemical spectral analysis, however, were taken by the chemist Robert Bunsen and the physicist Gustav Kirchhoff, then colleagues at the University of Heidelberg (Kirchhoff and Bunsen, 1860, 1861).[218] The two of them constructed a standard apparatus for analyzing the spectra of elements contained in salts, which were made incandescent by being sprinkled on a colourless gaseous flame (produced by the so-called Bunsen burner)[219]: the light passed through a narrow slit, was dispersed by a prism, and the spectrum was then observed through a telescope. Bunsen and Kirchhoff examined systematically the chlorides, bromides, iodides, hydrated oxides, sulfates and carbonates of potassium, sodium, lithium, strontium, calcium and barium, and concluded: 'The different bodies with which the metals were combined, the variety in the nature of chemical processes occurring in the several flames, and the wide differences of temperature which these flames exhibit, *produce no effect upon the position of the*

[216] Fraunhofer isolated altogether 574 dark lines in the solar spectrum, the strongest of which he named with capital letters from A to G. The D-line was found to consist of two separate components, like the corresponding emission line of flames; its position was determined from refraction experiments as lying in the yellow part of the spectrum.

[217] The problem of relating spectral lines to chemical elements was first considered by D. Brewster (1781–1868) in the 1820s; the idea was picked up by the astronomer John Frederick William Herschel (1792–1871), the son of William Herschel, and finally by William Henry Fox Talbot (1800–1877), the pioneer in photography. Charles Wheatstone (1802–1875), Antoine-Philibert Masson (1806–1880) and Anders Jonas Ångström (1814–1874) discovered and studied the discrete spectra emitted by the sparks between metal electrodes. William Swan, professor at the Scottish Naval and Military Academy, showed that Fraunhofer's emission line (at the position of the solar D-line) was produced by the presence of sodium in the flames (Swan, 1857).

[218] Robert Wilhelm Bunsen was born in Göttingen on 31 March 1811. He received his doctorate from the University of Göttingen in 1830. Then he spent some time travelling abroad (e.g., Paris). He received his *Habilitation* at the University of Göttingen in 1834. Two years later he moved to Kassel, where he taught at the vocational school; in 1838 he was appointed Professor of Chemistry at the University of Marburg, in 1851 at the University of Breslau and the following year at Heidelberg. Bunsen became a Foreign Member of the Royal Society (1858) and a Corresponding Member of the Academy of Sciences, Paris (1853). He died in Heidelberg on 16 August 1899.

Bunsen worked in many fields of chemistry and physics. His early researches were devoted to organic chemistry (cacodyl compounds) and to the study of the composition of gases in furnaces. Then he investigated a new kind of galvanic cell and certain properties of the electric arc. He succeeded in obtaining metallic magnesium for the first time by electrolytic decomposition. By using chemical spectral analysis he discovered the alkali elements cesium and rubidium.

[219] Bunsen developed the flame together with his student Henry Enfield Roscoe (Bunsen and Roscoe, 1857). Similar flames, however, had been used earlier.

bright lines in the spectrum which are characteristic of each metal' (Kirchhoff and Bunsen, 1860, p. 92). That is, every metal gave the same spectrum, no matter whether or how it was bound in a chemical compound.

In Bunsen and Kirchhoff's method of spectral analysis were combined the special experimental skill of the chemist with the insight of the physicist. During his collaboration with Bunsen, Kirchhoff arrived at an explanation of the Fraunhofer lines (Kirchhoff, 1859a). By letting the rays of sunlight of moderate intensity pass through a flame containing lithium chloride, which fall on the slit (of a Bunsen–Kirchhoff apparatus for spectral analysis), he observed the following: 'One sees at the specific position [where the sharply defined lithium line had to appear] a bright line on a dark background; for a greater intensity of the incident sunlight, however, there appears at the same place a dark line, having exactly the same character as Fraunhofer's lines' (Kirchhoff, 1859a; *Gesammelte Abhandlungen*, 1882, p. 565). He concluded that flames, in whose spectra bright lines occur, will absorb the same frequencies from the continuous radiation (e.g., sunlight) that passes through them, and that 'the dark lines of the solar spectrum, which are not caused by the earth's atmosphere, originate from the presence of those substances in the glowing solar atmosphere, which cause bright lines at the same place in the spectrum of a flame' (Kirchhoff, 1859a; *Gesammelte Abhandlungen*, p. 565). Thus, from the fact that the dark *D*-lines of sodium were observed in the solar spectrum, he concluded the existence of sodium in the atmosphere of the sun, and from the nonexistence of a dark lithium line he concluded that only little or no lithium should be found there. These conclusions agreed perfectly with his theory of the absorption and emission of radiation: each substance will absorb radiation of the same frequencies as it emits (Kirchhoff, 1860).

The results of Bunsen and Kirchhoff encouraged many scientists—physicists and chemists—to concern themselves with the study of spectra emitted by various substances. Thus Bunsen's former collaborator, Henry Enfield Roscoe (1833–1915), now Professor of Chemistry at Owens College, Manchester, found a strange behaviour of the spectra emitted by some strontium and barium salts: at low temperatures of the flame (or for weak discharges, if the spectra were created by discharges between two electrodes) broad bands occurred, which disappeared at higher temperatures, with the usual kind of discrete strontium and barium spectra emerging (Roscoe and Clifton, 1862). Alexander Mitscherlich (1836–1918), the son of the renowned Berlin chemist Eilhard Mitscherlich, independently arrived at a similar result while investigating certain barium compounds; he concluded that the band spectrum arising at low temperatures was characteristic of the oxide (Mitscherlich, 1862). Other observations pointed in the same direction. For example, Julius Plücker (1801–1868) and Johann Wilhelm Hittorf (1824–1914) discovered that certain gases, like nitrogen or the vapours of sulfur and selenium, which were excited electrically in so-called Plücker tubes (specially shaped evacuated tubes having two electrodes, containing the gas or vapour under consideration) emitted two types of spectra: for small currents passing through the tubes a band spectrum appeared, which changed into a discrete spectrum when the current was increased and it heated the gas under investigation (Plücker, 1862). Plücker's successor as professor of physics at the University of Bonn, Adolph Wüllner (1835–1908), and the French chemist Georges Salet provided further evidence for the existence of multiple spectra (i.e., line spectra and band spectra); however, Ångström and Arthur Schuster, then a student under Roscoe and Balfour Stewart at Owens College, opposed their findings. They argued that any spectra different from the usual discrete ones arose because of impurities. Finally, however, the facts decided in favour of multiple spectra.

The complication, which was brought into spectral analysis by the existence of multiple spectra, called for an explanation. Wüllner proposed the first one: he claimed that the spectra associated with low temperatures arose from a thick layer of gas, while the discrete spectra (occurring especially for discontinuous or spark discharges) arose from a thin layer (Wüllner, 1872). In light of later experiments this explanation could not be upheld. However, the English astronomer Joseph Norman Lockyer proposed a different one.[220] He drew attention to the fact that the dissociation of chemical substances might lie behind the multiple spectra. This spectroscopic characterization would, in turn, give a clear separation between what had so far been called atoms and molecules. 'If you allow,' Lockyer said, 'that in the line spectrum an atom is at work, in channelled spectra and continuous spectra molecular aggregates, you will see at once that Professor Maxwell and others will be able to get a sharper definition of atom and molecule than they have now' (Lockyer, 1874, p. 70). Georges Salet, in his report at the 1875 meeting of the *Association Française pour l'Avancement des Sciences*, went a step further. He addressed himself to the existence of two different types of spectra connected with one chemical substance by the following considerations.

As heat has ordinarily the effect of simplifying chemical compounds and of reducing them to their elements; as, on the other hand, elementary channelled spectra are observed at the lowest temperatures and are very similar to those of compound bodies, one can suppose that they are produced by aggregations of homogeneous atoms. This assumption will appear natural above all to chemists . . . molecules are themselves capable of being resolved into atoms: it is to this *allotropy of elevated temperatures* that the variation of spectra seems due. (Salet, 1875, quoted in McGucken, 1969, p. 79)

[220] J. N. Lockyer was born on 17 May 1836 in Rugby, England. He was educated in private schools and on the Continent. He became a clerk in the War Office in 1857. He embarked upon a scientific career as late as 1870, when he was appointed Secretary to the Duke of Devonshire's Commission on Science; in 1871 he became Rede Lecturer at Cambridge University; in 1881 professor of astronomical physics at the Royal College of Science; in 1885 Director of the Solar Physics Observatory, South Kensington; and in 1913 Director of Hill Observatory, Salcombe Regis, Devon, where he died on 16 August 1920.

Lockyer headed eight governmental eclipse expeditions between 1870 and 1905; he became a Fellow of the Royal Society (1869) and a Corresponding Member of the Academy of Sciences, Paris (1873). He pioneered spectroscopic studies of the sun and discovered helium in its atmosphere. In 1869 he founded the weekly scientific journal *Nature*.

That is, he associated line spectra with chemical atoms and band spectra with chemical molecules of the same substance.[221]

Simultaneously with the investigations of the characteristic spectra of an increasing number of substances, one also began to think about the theoretical description of these spectra. Thus Alexander Mitscherlich examined the spectra of the halogen compounds of the alkaline earths—calcium, strontium and barium—which consisted of single lines; he recognized that the spacings of the two principal lines in the spectra of chlorides, iodides and bromides, were related to each other as the atomic weights of the compounds (Mitscherlich, 1864).[221a] His idea was not pursued further immediately, but five years later the physicist Eleuthère-Élie-Nicolas Mascart, who extended spectral analysis to the ultraviolet region, remarked: 'An important problem, which emerges necessarily from spectral analysis, is to know whether there exists a relation between the different lines of the same substance or also between the spectra of analogous substances' (Mascart, 1869, p. 338).[222] After pointing out that he had observed such relations himself in the doublet spectra of sodium and the triplet spectra of magnesium, he continued: 'It appears to me difficult [to conclude] that the reproduction of such a phenomenon should be an accidental effect; is it not more natural to admit that these similar groups of lines are harmonics, which are related to the molecular constitution of the radiating gas?' (Mascart, 1869, p. 338).[223]

In the following decade chemists and physicists concerned themselves with the search for numerical relations within the spectrum of a given substance and between the spectral lines of different substances. The French chemist Paul Émile Lecoq de Boisbaudran was the first.[224] In three papers, published in 1869

---

[221] The idea of dissociation played an important role in Lockyer's theoretical treatment of stellar spectra. He imagined, for instance, that the chemical elements occurring in stellar atmospheres are created by successive aggregations of the most elementary atom; the fact that stellar atmospheres having a high temperature emit simple line spectra, and those of lower temperatures emit complex (as well as band) spectra, supported such a conclusion, which also agreed with Dalton's atomic hypothesis (Lockyer, 1881). Lockyer's keen speculation could not, however, be upheld in the light of more accurate data, hence it disappeared from discussions of such questions by 1890. Still his basic idea of dissociation of molecules and atoms at higher temperatures was revived several decades later and then contributed to a deepened understanding of certain problems of atomic structure and the theory of stellar spectra.

[221a] Alexander Mitscherlich was born on 28 May 1836 in Berlin. His father was the chemist Eilhard Mitscherlich (1794–1863). He obtained his doctorate in chemistry in 1861 at the University of Berlin (where his father was professor of chemistry). He became a professor at a private *Chemieschule* in Hannoversch-Münden near Kassel (1868–1883). Around 1870 he developed a process to extract cellulose from wood. Mitscherlich died in Obersdorf on 31 May 1918.

[222] E. Mascart was born on 20 February 1837 at Quarouble, France. He entered the *École Normale*, Paris, in 1858 and obtained his doctorate in 1864. After serving as a physics teacher at the high school in Versailles, he became Director of the Central Meteorological Bureau in 1871, and in 1872 Professor at the *Collège de France*. He died in Paris on 26 August 1908.

[223] The idea that the spectra might give the key to the inner structure of atoms and molecules had also been proposed by Mitscherlich (1864).

[224] Paul Émile (François) Lecoq de Boibaudran was born on 18 April 1832 in Cognac, Charante. He studied physics and chemistry with Kirchhoff, Bunsen and Crookes, and continued to do research later on in his own laboratory. He made many discoveries, including those of the rare earths: samarium, dysprosium, gadolinium and gallium. In spectroscopy he worked in particular with the spark spectra of substances. He died on 28 May 1912 in Paris.

—soon after Mascart's—in the *Comptes Rendus*, he studied the relation between the spectra of homologous substances (e.g., alkali metals) and their atomic weights; further, he proposed a numerical relation for the four principal lines of potassium (Lecoq de Boisbaudran, 1869a, b, c). Then, in a fourth paper, he turned to the spectrum of nitrogen, which—as was known already—consisted of two series of bands; he claimed that it might be considered as being composed of two systems of harmonics, one represented by doublet lines and the other by shaded bands (Lecoq de Boisbaudran, 1869d). The suggestion that harmonics might be involved in the structure of line (and band) spectra was also made by George Johnstone Stoney (Stoney, 1871).[225] He had already concerned himself earlier with the problem of the origin of discrete spectra and claimed that they arose from the internal motions of the molecules (Stoney, 1868). He now presented a detailed mathematical theory of spectra, based on the assumption that the periodic motions of the gas molecules couple with the ether, causing it to vibrate and 'one periodic motion in the molecules of the incandescent gas may be the source of a whole series of lines in the spectrum of the gas' (Stoney, 1871, p. 293). The numerical relations obtained by Stoney on the basis of this theory were tested in detail in the following years and finally rejected. Especially Arthur Schuster made a careful numerical analysis of the available data on many spectra and concluded: 'Most probably some law hitherto undiscovered exists, which in special cases resolves itself into the law of harmonic ratios' (Schuster, 1881, p. 343).[226] A year later, in his progress report on spectra presented at the meeting of the British Association for the Advancement of Science, he stated:

It is the ambitious object of spectroscopy to study the vibrations of atoms and molecules in order to obtain what information we can about the nature of forces which bind them together . . . But we must not too soon expect the discovery of any grand and very general law, for the constitution of what we call a molecule is no doubt a very complicated one, and the difficulty of the problem is so great that

[225] G. J. Stoney was born at Oakley Park, Ireland, on 15 February 1826. He was educated at Trinity College, Dublin, receiving his B.S. in 1848 and M.A. in 1852. He became an assistant at the Parsonstown Observatory in 1848, and Professor of Natural History at Queen's College, Galway, Ireland, from 1852 to 1857. After that he joined the civil service. He died in London on 5 July 1911. Stoney worked and published on many topics: wave motion, acoustics, optics, atomic structure and theory of spectra, kinetic theory of gases, constitution of stars and planetary atmospheres. He had been elected a Fellow of the Royal Society in 1861.

[226] Arthur Schuster was born in Frankfurt-am-Main on 12 September 1851. As a youth he went to England and became a British subject in 1875. He received his education at Owens College, Manchester, and his doctorate in 1873 from the University of Heidelberg, where he was a student of Kirchhoff's. Then he went to Göttingen and Berlin to work with Wilhelm Weber, Eduard Riecke and Hermann von Helmholtz. From 1876 to 1881 he was a member of the staff of Cavendish Laboratory. He returned to Owens College as Professor of Applied Mathematics (1881–1907), and then was appointed Langworthy Professor of Physics. Schuster worked in many fields of physics, both experimentally and theoretically, e.g., on an absolute determination of the ohm, on spectroscopy, discharge of electricity through gases, terrestrial magnetism, and on optical and astrophysical problems. In 1875 he headed the Royal Society Expedition to observe the solar eclipse in Siam. He received many honours, including the Fellowship of the Royal Society (1879), the Rumford Medal (1926) and the Copley Medal (1931). Schuster died in Yeldall, near Twyford, Berkshire, on 14 October 1934.

were it not for the primary importance of the result which we may finally hope to obtain, all but the most sanguine might well be discouraged to engage in an inquiry which, even after many years of work, may turn out to have been fruitless. We know a great deal more about the forces which produce the vibrations of sound than about those which produce the vibrations of light. To find out the different tunes sent out by a vibrating system is a problem which may or may not be solvable in certain special cases, but it would baffle the most skillful mathematician to solve the inverse problem and to find out the shape of a bell by means of the sounds which it is capable of sending out. And this is the problem which ultimately spectroscopy hopes to solve in the case of light. In the meantime we must welcome with delight even the smallest step in the desired direction. (Schuster, 1882, pp. 120–121)

Schuster's rejection of Stoney's approach to describe the spectra in terms of harmonic ratios discouraged many scientists to look further in that direction. 'Still the hope of finding a simple law, as that of musical harmonics, is the sign of a preconceived idea which is important to discard immediately,' remarked the French physicist Alfred Cornu, who was himself concerned with investigating the law of spectra (Cornu, 1885, p. 1182).[227] While Cornu did not succeed in obtaining from the data a general law governing the spectra, one solution—the law for the spectrum of hydrogen—was already at hand. On 25 June 1884 Johann Jakob Balmer reported on the law of the spectral lines of hydrogen to the *Naturforschende Gesellschaft in Basel* (Balmer, '*Die Spectrallinien des Wasser-stoffs*,' *Verh. d. Natf. Ges. Basel* 7, p. 894, 1885). On this occasion he presented a formula to describe the available data, expressing the wavelength λ of the hydrogen lines in terms of two integers m and n. It stated

$$\lambda = \text{const.}\left(\frac{m^2}{m^2 - n^2} \right), \tag{101}$$

with m and n assuming the values $2, 3, 4, \ldots$ and $1, 2, 3, \ldots$, respectively. Balmer was then a nearly sixty-year-old schoolteacher in a high school for girls (*Höhere Töchterschule*) in Basle.[228] Balmer's interest in mathematical and physi-

---

[227] Marie Alfred Cornu was born at Orléans on 6 March 1841. He studied at the *École Polytechnique* and at the *École de Mines*, Paris. After receiving his doctorate, he worked as head engineer of mines, and then (from 1867) as Professor of Physics at the *École Polytechnique*, Paris. He did research, both experimental and theoretical, mainly in optics and spectroscopy. Cornu died at Romorantin on 12 April 1902.

Cornu worked mainly on optical problems. For example, he made a new measurement of the velocity of light (1877), which won him the Lacaze prize of the *Académie des Sciences* (1878) and the Rumford Medal of the Royal Society (1884). He studied the line spectra emitted by celestial and terrestrial sources. His name is also connected with the 'Cornu spiral,' a curve which he used for describing the intensity of the Fresnel diffraction of light. Cornu was elected a member of the *Académie des Sciences* of Paris (1878), and a Foreign Member of the Royal Society of London (1884).

[228] J. J. Balmer was born on 1 May 1825 at Lausen, Canton Basle. He was educated at the *Pädagogium* in Basle, at the *Technische Hochschule* in Karlsruhe, and at the University of Berlin. In 1849 he received his doctorate at the University of Basle with a mathematical dissertation on cycloids. He taught at the *Pädagogium* until he received a position at the *Höhere Töchterschule* (high

cal proportions brought him to consider the relations in the spectra of atoms.[229] He was thrilled by the fact that the wavelengths of hydrogen could be represented by integral numbers. He published his results in two notes submitted to the *Verhandlungen der Naturforschende Gesellschaft in Basel* (Balmer, 1885a, b). In the first, '*Notiz über die Spektrallinien des Wasserstoffs*' ('Note on the Spectral Lines of Hydrogen'), which was based on his lecture, he not only gave the formula, Eq. (101), but also discussed the significance of the constant, whose value turned out to be 3645.6×10^{-7} mm.

> One might call this number the *fundamental number* of hydrogen; and if one should succeed in finding the corresponding fundamental numbers for other chemical elements as well, then one could speculate that there exist between these fundamental numbers and the atomic weights [of the substances] in question certain relations, which could be expressed as some function. (Balmer, 1885a, pp. 552–553)

Balmer further noticed that the 'constant' ('*Grundzahl*') determined the limiting wavelength of the lines described by Eq. (101) with $n = 2$. He argued:

> If the formula for $n = 2$ is correct for all the main lines of the hydrogen spectrum, then it implies that towards the ultraviolet end these spectral lines approach the wavelength 3645.6 in closer and closer sequence, but cannot cross this limit; while at the red end [of the spectrum] the C-line [later called H_α] represents the line of longest possible [wavelength]. Only if in addition lines of higher order [i.e., those described by Eq. (101) with $n = 3, 4, \ldots$] existed, would further lines arise in the infrared region. (Balmer, 1885a, p. 559)

---

school for girls) where he continued to give instruction in arithmetic and calligraphy until advanced age. He was also *Privatdozent* for descriptive geometry at the University of Basle. Balmer's main interest was in geometry, including its application to technical problems. He died in Basle on 12 March 1898.

[229] Markus Fierz (Wolfgang Pauli's successor as Professor of Theoretical Physics at the E.T.H. in Zurich, 1960–1978) told me (in Brussels, May 1973, when we were both attending the bicentennial celebrations of the Royal Belgian Academy) that when he was in Basle (Professor of Theoretical Physics, University of Basle, 1944–1960) he had tried to ascertain a number of things about the Balmer legend. He learned, for instance, that Balmer had reconstructed the design of the Temple from the measurements given in Ezekiel's vision in the *Old Testament*. (Ezekiel was an Hebrew prophet of the exile, who lived in the sixth century B.C. In the 25th year of Ezekiel's exile and 14 years after Jerusalem was destroyed, Ezekiel saw in a vision the city and the kingdom restored. He is guided round the new Temple by an angel architect equipped with a cord and measuring rod. The detailed plans of the Temple and the adjacent structure are given, with exact measurements of the wooden altar. See *The Holy Bible* (King James Version), *The Book of the Prophet Ezekiel*, Chapters 40–43.) (J. Mehra.)

G. P. Thomson has reported that when he visited Switzerland shortly after World War I, he was told by a young relative of J. J. Balmer that he (Balmer) was devoted to numerology and interested in such things as the number of sheep in a flock or the number of steps of the Pyramid. One day, chatting with a physicist or chemist friend (probably Eduard Hagenbach), Balmer complained that he had 'run out of things to do.' The friend replied: 'Well, you are interested in numbers, why don't you see what you can make of this set of numbers that come from the spectrum of hydrogen?' and he gave him the wavelengths of the first few lines of the hydrogen spectrum (G. P. Thomson, AHQP Interview, 20 June 1963).

At about the same time as Balmer, other attempts at deriving formulae for spectra were also successful. For example, Alexander S. Herschel, son of John Herschel, and Henri Deslandres found a mathematical description for various band spectra.[230] For obtaining the necessary data, Deslandres made use of the high resolution of lines achieved by a Rowland grating, which had been presented by Henry Rowland himself to the École Polytechnique. By analyzing the bands of nitrogen, he observed that the intervals between the inverse wavelengths, or vibration numbers, within a particular band, were nearly an arithmetical progression (Deslandres, 1886). In the following year he obtained a formula describing the wavelengths of the lines which formed the heads of different bands—the band spectrum of nitrogen, just as of any other substance, could be regarded as consisting of bands (each composed of lines whose distances became shorter towards the end), marked by a particular line, the band head—that is,

$$\lambda^{-1} = Bn^2 + \beta, \tag{102}$$

where n assumed integral values, $n = 1, 2, 3, \ldots$, and B and β denoted constants (Deslandres, 1887). Parallel to these efforts to describe the band spectra, one looked for an extension of the Balmer formula, Eq. (101), for the spectra of other elements. In particular, Heinrich Kayser, a former assistant of Helmholtz' in Berlin and since fall 1885 Professor of Physics at the *Technische Hochschule* in Hanover, devoted himself fully to spectroscopic work.[231] He received help from Carl Runge, who arrived as Professor of Mathematics in Hanover the following year.[232] Beginning in 1887 Kayser and Runge systematically investigated the line spectra of elements and developed formulae generalizing Balmer's Eq. (101) (Kayser and Runge, 1890). At the same time the Swedish mathematical physicist

---

[230] Henri Alexandre Deslandres was born in Paris on 24 July 1853. He studied at the *École Polytechnique*. Later he became Director of the Meudon Observatory (1907) and of the Paris Observatory (1927). He died in Paris on 15 January 1948. Deslandres made important contributions to molecular spectroscopy and the physics of the solar atmosphere. He was a member of the Academy of Sciences in Paris (since 1902, elected President in 1920); he was elected a Foreign Member of the Royal Society in 1921.

[231] Heinrich Gustav Johannes Kayser was born at Bingen on the Rhine on 16 March 1853. He studied at the Universities of Strasbourg and Berlin, where he obtained his doctorate in 1879. From 1878 to 1885 he was assistant to Helmholtz at the University of Berlin. Then he became Professor of Physics at the *Technische Hochschule*, Hanover (1885–1894), and at the University of Bonn from 1894 to 1920. Kayser published many articles and several monographs on spectroscopy and became an authority on it. He died in Bonn on 14 October 1940.

[232] Carl David Tolmé Runge was born in Bremen on 30 August 1856. He studied physics, mathematics and chemistry at the University of Munich (1876–1877) and mathematics (under Karl Weierstrass) at the University of Berlin (1877–1880), obtaining his doctorate in 1880. He continued his studies in Berlin under Weierstrass and Leopold Kronecker and became *Privatdozent* at the University of Berlin in 1883. In spring 1886 he was appointed professor of mathematics at the *Technische Hochschule* in Hanover. In Hanover, Runge investigated, in collaboration with the physics professor Heinrich Kayser, the series spectra of chemical elements. Thus Runge, who had previously worked on problems of analytic geometry, number theory and algebra, became deeply involved in spectroscopy. In 1906 he accepted the professorship of applied mathematics at the University of Göttingen. He died in Göttingen on 3 January 1927.

Johannes (Janne) Robert Rydberg published the results of his analysis of spectra (Rydberg, 1890).[233]

Rydberg already had started his studies of the relations between spectral lines before Balmer published his formula. Like Balmer he tried to introduce integral numbers in the description of the spectra of sodium, potassium, magnesium, calcium and zinc. 'In the spectra of all the elements analyzed,' he remarked 'there exist series of lines, whose wavelengths are determined by consecutive integers' (Rydberg, 1890, p. 33). In describing the data, he referred to the classification of alkali spectra, which had been obtained especially by George D. Liveing and James Dewar (1883) for the principal, sharp and diffuse series, and to the observations of Walter Noel Hartley, who had observed that the separation of the two components of each doublet (as well as the separations of the components of triplets), when measured in wave numbers (i.e., inverse wavelengths) was often a constant throughout a given series (Hartley, 1883).[233a] Rydberg then showed that the wave numbers n of the lines in these series could be expressed in terms of integers m in the following way: the components of the principal series satisfied the equations

$$n_i = \lambda_i^{-1} = n_0 - \frac{N_0}{(m + p_i)^2} \quad (i = 1, 2), \tag{103}$$

where N_0, n_0 and p_i denoted constants for either the components with the longer wavelengths in each doublet (denoted by the suffix 1) or those with the shorter wavelengths (denoted by the suffix 2). A similar relation applied to the components of the diffuse and sharp series. (In these cases the constant n_0 in Eq. (103) was replaced by two constants n_0' and n_0'', referring to the longer and shorter wavelength components, respectively; and the constant p_i in the denominator was replaced by other constants, namely, d for the diffuse and s for the sharp

---

[233] Johannes (Janne) Robert Rydberg was born at Halmstad, Sweden, on 8 November 1854. He studied at the University of Lund (1873–1879), where he received his doctorate in mathematics in 1879. In 1880 Rydberg was appointed a docent in mathematics, and in 1882 a docent for physics, at the University of Lund. From 1892 to 1901 he served as assistant at the Physics Institute, and from 1901 to 1919 as Extraordinary Professor and Director of the Physics Institute, all at the University of Lund. He was elected a Foreign Member of the Royal Society in 1919. Rydberg died in Lund on 28 December 1919.

[233a] Liveing and Dewar had separated the spectra of a given element in diffuse and sharp series (Liveing and Dewar, 1883), and Rydberg followed their classification (Rydberg, 1890). By 1890 the alkali spectra were known to exhibit three series: one series composed of distinct lines, which also appeared in absorption spectra, was called the 'principal series' by Kayser and Runge (1890) and Rydberg (1890); of the two other series, one was composed of broader and brighter lines than the other, hence Rydberg named it 'diffuse series' and the other 'sharp series,' while Kayser and Runge spoke of the 'first' and the 'second subordinate series.' Later a fourth series (having wavelengths in the infrared) was discovered by Arno Bergmann (1907), and it was called either 'Bergmann series' or 'fundamental series.' The names of the series appeared also in the formulae for wavelengths: thus p denoted the variable term in the equation for the principal series, and d, s, b or f the corresponding terms in the diffuse (first subordinate), sharp (second subordinate), and Bergmann or fundamental series, respectively.

series.) Several year later, Rydberg (1896, 1900) and, independently of him, Schuster (1896) discovered further regularities. Now the lines of the principal series were described by the equations

$$n_{p_i} = \lambda_{p_i}^{-1} = N_0\left[(1+s)^{-2} - (m+p_i)^{-2}\right] \quad (i=1,2), \tag{104a}$$

with m assuming the values 2, 3, 4, etc.; the components of the diffuse series by the equations

$$n_{d_1} = \lambda_{d_1}^{-1} = N_0\left[(2+p_2)^{-2} - (m+d)^{-2}\right]$$

and

$$n_{d_2} = \lambda_{d_2}^{-1} = N_0\left[(2+p_1)^{-2} - (m+d)^{-2}\right], \tag{104b}$$

with $m = 3, 4, 5$, etc.; and those of the sharp series by

$$n_{s_1} = \lambda_{s_1}^{-1} = N_0\left[(2+p_2)^{-2} - (m+s)^{-2}\right]$$

and

$$n_{s_2} = \lambda_{s_2}^{-1} = N_0\left[(2+p_1)^{-2} - (m+s)^{-2}\right], \tag{104c}$$

with $m = 3, 4, 5$, etc. Ten years later, in his doctoral thesis at Jena, Arno Bergmann established the existence of a fourth series in the infrared spectra of potassium, rubidium, and cesium—which was later called the 'fundamental' or 'Bergmann' series—whose components had wave numbers satisfying the equation

$$n_f = \lambda_f^{-1}\left[(3+d)^{-2} - (m+f)^{-2}\right], \tag{104d}$$

with $m = 4, 5, 6$, etc. (Bergmann, 1907).[234]

While Rydberg had obtained his results from the data on the structure of line spectra of elements, the Swiss Walther Ritz tried to derive series formulae from theoretical arguments, both in his doctoral thesis (Ritz, 1903) and in a later paper (Ritz, 1907).[235] He arrived at a representation for the series more general than Eqs. (104a–d). His most general contribution to spectroscopy, however, was contained in what he called 'a new law of spectral series' in an article submitted to *Physikalische Zeitschrift* in June 1908 (Ritz, 1908b). Ritz wrote:

[234] In writing Eq. (104d), we have neglected the difference between the doublet components in the Bergmann series; it is much smaller than in the principal series.

[235] Walther Ritz was born in Sion, Switzerland, on 22 February 1878. He studied physics and mathematics at the E.T.H. in Zurich (1897–1901) and at the University of Göttingen (1901–1903), where he received his doctorate under Woldemar Voigt. Then he went to Leyden, Bonn, Paris, and Tübingen (to collaborate with Friedrich Paschen). After his return to Göttingen he obtained his *Habilitation* in 1909. His health had been greatly damaged by tuberculosis, of which he died on 7 July 1909 in Göttingen.

In the following we wish to demonstrate that from the known spectral series of an element new series can be derived, by means of which nearly all series and lines of the alkali elements that were discovered recently by [Philipp] Lenard, [Heinrich Matthias] Konen and [Eduard] Hagenbach, [Frederick Albert] Saunders, [W. J. H.] Moll, [Arno] Bergmann and others, can be described—without employing any new constant. To other spectra, especially of helium and the alkaline-earth elements, the new *combination principle* may be applied as well. One then obtains more definite relations with the atomic weight than hitherto known, and the totality of these new relations makes it easier to discover spectral series for those elements, for which they are still unknown. (Ritz, 1908b, p. 521)

Ritz' new combination principle, which had been anticipated earlier by J. R. Rydberg (1900), stated that every spectral line of a given element could be expressed as the difference of two terms, the so-called 'spectral terms,' each of which depended on an integer m besides the constants (such as the ones occurring in Eqs. (104a–b)). Thus, for instance, Ritz wrote the frequencies of the sharp series as

$$\pm \nu = (2, p_2, \pi_2) - (m, s, \sigma) \quad \text{and} \quad (2, p_1, \pi_1) - (m, s, \sigma), \quad (105)$$

where the labels in round brackets denoted the terms, which generalized Rydberg's expressions, $N_0(2 + p_2)^{-2}$, $N_0(m + s)^{-2}$, and $N_0(2 + p_1)^{-2}$ in Eqs. (104c). As Ritz had expected, the combination principle indeed helped the experimentalists in organizing unknown spectra. Thus Friedrich Paschen, who in 1908 started a systematic search for the infrared spectra of elements (Paschen, 1908, 1909a, b), remarked: 'By applying this, probably exactly valid, principle, one succeeds finally in organizing all the more intensive lines, of which there is a really large number in thallium, into the series scheme' (Paschen, 1909a, p. 627).

During the first fifty years of spectral analysis the observational techniques were gradually greatly improved. Not only were the hitherto unknown regions of wavelengths explored, such as the infrared (up to micrometres in orders of magnitude, see Paschen, 1908) or the ultraviolet (down to about 1000 Å), but in these extensions new spectral series were discovered, e.g., the infrared series of hydrogen described by Balmer's Eq. (101) with $n = 3$ and $m = 4, 5, 6$, etc., by Paschen (1908), and the ultraviolet series with $n = 1$ and $m = 2, 3, 4$, etc., by Theodore Lyman (1914). An essential role in these observations of the invisible lines was played by the concave diffraction gratings of Henry Augustus Rowland, which replaced the prisms used earlier in the spectral apparatus; they not only enlarged the resolution of wavelengths but also avoided the absorption that occurred in the material of the dispersion prisms.[236] After Charles Glover

---

[236] H. A. Rowland was born at Honesdale, Pennsylvania, on 27 November 1848. He first worked in railroad surveys (1871), before teaching at the College of Wooster (1871–1872), at Rensselaer Polytechnic Institute (1872–1875) and as the first Professor of Physics at Johns Hopkins University (from 1876). Besides spectroscopy, he performed research on the absolute units of resistance and the mechanical equivalent of heat; he also discovered the magnetic action arising from electric convection currents. Rowland received many honours, including the Foreign Memberships of the Royal Society (1889) and the Paris Academy of Sciences (1893), and the Rumford, Draper and Matteucci Medals. He died in Baltimore on 16 April 1901.

Barkla's discovery of characteristic X-rays in 1908, the X-ray spectra also became an object of study. To organize the enormous amount of spectroscopic data, it became inevitable to employ increasingly theoretical considerations. Thus Friedrich Paschen, one of the foremost experimental spectroscopists, collaborated closely with Walther Ritz and later with Arnold Sommerfeld and Alfred Landé. Through such highly successful collaborations between experimentalists and theoreticians the rapid progress in the understanding of atomic structure—which occurred during the second and third decades of the twentieth century—was made possible.

II.2   Ideas Towards a Model of Atomic Structure[237]

The idea that atoms might have a constitution contradicted the very meaning of the concept of an atom, denoting an object which cannot be separated into parts. However, soon after the old atomic hypothesis was revived at the beginning of the nineteenth century, scientists thought about endowing the atom with a structure. Thus, for instance, André Marie Ampère developed in 1814 a geometrical model, in which he pictured the atoms of chemical elements as being composed of subatomic particles (Ampère, 1814).[238] In the following decades several authors, especially at the *École Polytechnique*, Paris—such as Jean Baptiste Biot, Augustin Louis Cauchy and Siméon Denis Poisson—employed models in which the atom consisted of a massive nucleus surrounded by an atmosphere of imponderable ether particles. Gustav Theodor Fechner of the University of Leipzig, who translated Biot's textbook on experimental and theoretical physics into German (Biot, 1816)—in which the author had also mentioned views on atomic structure similar to Ampère's—then constructed a dynamic atomic model

---

[237] We have obtained useful information on the development of the ideas on atomic structure in the nineteenth century from Siegfried Wagner (private communication). The evolution of the nuclear atom from the middle of the nineteenth century up to the Rutherford–Bohr atom has been reviewed by G. K. T. Conn and H. D. Turner (1965).

[238] A. M. Ampère was born in Lyon on 20 January 1775. Ampère was educated at home; he became a private tutor in mathematics, chemistry and languages in Lyon from 1796 to 1801. He became Professor of Chemistry, Physics and Astronomy at Bourg in 1801, Professor of Mathematics at the *Lycée* of Lyon the following year, Assistant Lecturer of Mathematical Analysis at the *École Polytechnique*, Paris, from 1805 to 1809, then Professor at the same institution in 1809, and finally Professor of Experimental Physics at the *Collège de France* in 1824. He worked on problems of mathematics (theory of games), electromagnetism (for instance, he demonstrated the attraction and repulsion between wires carrying electric currents in opposite and the same directions, respectively, and the fact that a solenoid when carrying current behaves like a magnet; he formulated Ampère's law describing mathematically the force between two electric currents, and developed techniques to measure electric properties like the magnitude of an electric current), and on heat and light. Ampère, who gave the name 'electrodynamics' to the theory of currents, was honoured by memberships of the most prestigious scientific societies, such as the Paris Academy of Sciences (1814) and the Royal Society of London (1827). He died on 10 June 1836 at Marseilles.

It should be noted that the idea of circular currents, which Ampère introduced later to explain the magnetic properties of matter, had nothing to do with subatomic particles; he connected it with an electric fluid added to atoms (Ampère, 1822).

(Fechner, 1826, 1828).[239] Thus Fechner claimed that 'the atoms simulate in small dimensions the situations of the astronomical objects in large dimensions, being animated in any case by the same forces; and each body may be regarded as a system of innumerably many small suns, floating at comparatively large distances from one another, such that each or several of them together are surrounded by orbiting planetary atoms' (Fechner, 1828, pp. 275–276). He assumed, therefore, that the fundamental force in atoms, as in astronomy, was like gravitation; by calculating the dynamics of several such atomic systems according to Newton's laws he arrived at a qualitative picture of atoms and more or less tightly bound molecules.[240] Fechner's model of the atom was taken up by Wilhelm Eduard Weber with an important change: he assumed the (heavy) solar atoms and the (almost massless) planetary atoms to be electrically charged; hence he replaced Fechner's gravitational attraction by electric forces (Weber, 1871; 1875).[241] Weber, however, did not work with the exact Coulomb's law, but modified it in order to account for a finite velocity of propagation of electrical action. Only in the case of circular motion of the charged particles could the corrections to Coulomb's law be neglected. Thus he arrived at the following picture of atoms:

> Let e be the positively charged electrical particle, and let the negative particle, carrying an opposite charge of equal amount, be denoted by $-e$. Let only the latter be associated with the massive atom, whose mass is so large that the mass of the positive particle may be considered as negligible. The particle $-e$ may then be considered as being at rest, while just the particle e moves around the particle $-e$. (Weber, 1871, p. 44)

Weber had developed his atomic model in connection with an electrodynamic

---

[239] G. T. Fechner was born on 19 April 1801 at Groß-Särchen, Lower Lusatia. He was educated at the University of Leipzig and received his doctorate in biological science in 1822. He then became interested in mathematics and physics and was appointed Professor of Physics at Leipzig in 1834. Because of illness and partial blindness he retired after five years, and then turned to the study of philosophy, aesthetics and psychophysics. He recovered from his illness and became Professor of Natural Philosophy and Anthropology at the University of Leipzig. He attempted to make psychology an exact science. His name is especially connected with an equation stating the proportionality of the intensity of sensation and the logarithm of the intensity of the stimulus—the so-called Weber–Fechner law—which Ernst Heinrich Weber (1795–1878) had postulated and Fechner established on empirical grounds. Fechner died on 18 November 1887 at Leipzig.

[240] Interestingly enough the French scientists had not thought about a dynamical model of the atom. However, this idea had been mentioned earlier by the English philosopher Thomas Hobbes (1588–1679).

[241] W. E. Weber, the younger brother of Ernst Heinrich Weber (of the Weber–Fechner law), was born at Wittenberg in Thüringen on 24 October 1804, and was educated at the Universities of Halle and Göttingen. In 1828 he became *Extraordinarius* in Halle, and three years later he was called to the Chair of Physics at the University of Göttingen. He was among the seven Göttingen professors ('*Die Göttinger Sieben*') who protested in 1837 against the suspension of the constitution in the Kingdom of Hanover and was, therefore, dismissed from his professorship. From 1843 to 1849 he served as a professor at Leipzig; then he returned to Göttingen, again as Professor of Physics. He died there on 23 June 1891.

theory which, for some time, competed with Maxwell's until it became untenable because of the existence of transverse electromagnetic waves.[242] A major difficulty of Weber's model consisted in the fact that the constituents of the atom were hitherto purely hypothetical objects. And yet, one of the charged subatomic particles had already been seen in experiments by Julius Plücker and Johann Wilhelm Hittorf on the nature of the glow surrounding the cathode in a tube, in which an electric discharge took place. Especially, Plücker found that during the discharge, glass walls of the tube near the cathode emitted phosphorescent light and that the point of emission changed when a magnetic field was switched on (Plücker, 1858).[243] His pupil Hittorf then showed that if the tube was evacuated, the cathode glow extended into space towards the anode along a straight line (Hittorf, 1869).[244] In the following decades the cathode rays, as they came to be called, became the subject of many scientific investigations, carried out especially by Eugen Goldstein (1876) in Berlin and by Cromwell Varley and William Crookes in England.[244a] Varley put forward the hypothesis that the cathode rays may be composed of particles of matter ejected by the cathode (Varley, 1871), a

[242] It should be noted that Weber's moving charges did not lose energy; hence his atomic model was stable in contrast to the later ones developed on the basis of Maxwell's electrodynamics.

[243] Julius Plücker was born at Elberfeld, Germany, on 16 June 1801. He studied at the Universities of Bonn, Heidelberg, Berlin and Paris. He obtained his *Habilitation* at the University of Bonn in 1825 and was appointed *Extraordinarius* for mathematics there three years later. After serving as Professor of Mathematics at the *Friederich-Wilhelm-Gymnasium*, Berlin (1833–1834), then at the University of Halle (1834–1836), he returned to Bonn as Professor of Mathematics (1836–1847). In 1847 he assumed the Chair of Physics at the University of Bonn, which he occupied until his death on 22 May 1868. Plücker contributed both to mathematics, especially to analytic geometry and algebra, and to physics (magnetic properties of matter, electric discharges in gases and spectroscopy). One of his students was the mathematician Felix Klein.

[244] J. W. Hittorf was born at Bonn on 27 March 1824. He studied at the Universities of Bonn and Berlin. He became *Privatdozent* at the Academy of Münster, until the latter was turned into a university and he was appointed Professor of Physics and Chemistry. When in 1879 the Institute for Physics and Chemistry at Münster was split into two separate institutes, Hittorf became Director of the Physics Institute. He retired in 1889 because of ill health and died on 28 November 1914 at Münster. Hittorf worked on the migration of electrolytes, the spectra of gases, and the properties of cathode rays.

[244a] Cromwell Fleetwood Varley was born in London on 6 April 1828. He studied telegraphy at St. Saviours, Southwark, England. He worked with the International Electric and Telegraph Company from 1846 to 1868 and later as an independent inventor. Varley did research on many problems of electricity and telegraphic communication; he invented the forerunner of the telephone and many other kinds of electrical apparatus. He died at Bexleyheath in September 1883.

William Crookes was born on 17 June 1832 in London. He studied chemistry at the Royal College of Chemistry, London (under August Wilhelm von Hofmann). In 1854 he became assistant in the meteorological department of the Radcliffe Observatory, Oxford, and in 1855 he obtained a chemical teaching post at Chester. After his marriage, and upon inheriting a large fortune from his father, he moved to London in 1856 and lived only for his scientific work. Crookes worked on many problems of chemistry and physics; he constructed the radiometer and studied electric discharges through rarified gases (Crookes' space), the properties of cathode rays and later of radioactive substances. Crookes founded the journal *Chemical News* in 1859, which he edited until 1906. He was elected a Fellow of the Royal Society in 1863 (President, 1913–1915) and received its Royal, Copley and Davy Medals. William Crookes died in London on 4 April 1919.

view which was supported by Crookes, who investigated the phenomenon in detail and tried to determine the velocity of the 'molecules rebounded from the excited negative pole' (Crookes, 1879, p. 60). Others, including Heinrich Hertz, claimed that they might be electromagnetic pulses, especially since they did not seem to be affected by static electric fields. The decision in favour of one or the other hypothesis came only after more experiments were performed. In October 1892 Philipp Lenard, on a suggestion by Hertz, succeeded in obtaining cathode rays separated from the discharge; for that purpose he got the rays (produced in a low pressure discharge tube) to penetrate through a thin foil of aluminum—later called the 'Lenard window'—and thus leave the tube (Lenard, 1893). The behaviour and properties of free cathode rays were then studied by many scientists including—apart from Lenard—Joseph John Thomson in Cambridge and Jean Perrin in Paris. Thomson, for instance, in his lecture at the Royal Institution on 30 April 1897, stated the conclusion: 'Thus, from Lenard's experiments on the absorption of the rays outside the tube, it follows on the hypothesis that the cathode rays are charged particles moving with high velocities, that the size of the carriers must be small compared with the dimensions of ordinary atoms or molecules' (Thomson, 1897a, pp. 430–431). The assumption of small, charged particles—the sign of the charge was found to be negative because of the observed deflection of cathode rays in magnetic fields—explained the earlier observation that cathode rays experienced the same amount of deflection in a magnetic field, independent of the gas in the discharge tube or the material of the cathode. Soon afterwards Thomson removed the main obstacle raised against the corpuscular hypothesis by showing that electrostatic forces also influenced cathode rays: by using crossed electric and magnetic fields (at right angles to each other), such that the effect of the magnetic field was balanced by that of the electric field and no deflection resulted, he obtained a value for the ratio m_e/e (where m_e and e denoted the mass and charge, respectively, of the cathode ray particles) a value of the order of 10^{-7} (Thomson, 1897b). Approximately the same value was obtained independently by Emil Wiechert in Königsberg (Wiechert, 1897) and by Walther Kaufmann in Göttingen (Kaufmann, 1897); and this value, apart from the sign, was smaller by at least a factor of 1000 than the corresponding ratio for 'canal rays'—i.e., of the rays which Eugen Goldstein had discovered behind the cathode (Goldstein, 1886), and which now were found to be positively charged particles having about the same mass as the ions in electrolysis (Wien, 1897). By the turn of the century it was clear that cathode rays consisted of what George Johnstone Stoney had previously called 'electrons' (Stoney, 1894); these electrons bore a negative charge of the order of 10^{-10} electrostatic units—the same as a negative single-valued ion in electrolysis—and possessed a mass which was roughly the thousandth part of the value for the hydrogen atom (Thomson, 1899).

At the same time as the nature of cathode rays was settled, another kind of radiation was also recognized to consist of electrons. This radiation emerged from the so-called radioactive substances, such as uranium and thorium. Ernest Rutherford then identified two kinds of radioactive radiations, α- and β-rays, distinguished by their different absorptions in matter (Rutherford, 1899). In 1899 several scientists, including Friedrich Otto Giesel, Stefan Meyer and Egon von Schweidler, and Henri Becquerel showed that β-rays could be deflected by a static magnetic field in the same way as cathode rays (Giesel, 1899; Meyer and Schweidler, 1899; Becquerel, 1899). Further researches, carried out especially by Becquerel, Pierre and Marie Curie, and Wilhelm Wien, confirmed soon afterwards the hypothesis that β-rays, like cathode rays, consisted of electrons; they had, however, greater velocities than cathode rays.[245]

Although the emission of cathode rays and β-rays from matter strongly suggested that electrons must occur in atoms, the most convincing evidence arose from a different source: namely, the effect of a static magnetic field on the spectral lines emitted by atoms. In March 1862, Michael Faraday had attempted to demonstrate that magnetism influenced the frequencies of light emitted by a sodium flame without, however, achieving a positive result (see Jones, 1870, Volume II, p. 449). Nearly a quarter of a century later Pieter Zeeman, working at the Physics Institute of the University of Leyden, succeeded in obtaining a definite effect: by studying, like Faraday, the sodium lines emitted by a flame, which was placed between the poles of a strong electromagnet—Zeeman used a Ruhmkorff electromagnet run with a current of 27 amp, and the lines were obtained with the help of a Rowland grating of high resolving power (having a 10-ft radius and 14,983 lines/in.—he discovered that each of the two D-lines was broadened (Zeeman, 1896).[246] On investigating this phenomenon further, Zeeman found that a cadmium line, under the influence of a magnetic field of about 32,000 G, resolved into doublets for longitudinal and triplets for transversal observation (Zeeman, 1897a, b). At about the same time Albert Abraham Michelson at Chicago also managed to separate the components of the sodium and cadmium lines (under the influence of a magnetic field) by means of his

---

[245] It was found that β-rays were also deflected by electrostatic fields and that the ratio of their mass over electric charge was of the same order of magnitude as for electrons. Thus in 1901, when Walther Kaufmann started to make his measurements exhibiting the velocity dependence of the electron mass, he used as objects the β-rays emitted from radium (Kaufmann, 1901).

[246] P. Zeeman was born on 25 May 1865 at Zonnemaire, Isle of Shouwen, Zeeland. At the age of twenty he entered the University of Leyden and became a student of Heike Kamerlingh Onnes. In 1890 he was appointed assistant to Lorentz and obtained his doctorate in 1893. He then spent a year at Friedrich Kohlrausch's Institute at the University of Strasbourg. In 1895 he returned to Leyden and became *Privaat-Docent*; two years later he became a lecturer at the University of Amsterdam, where he was promoted in 1900 to an extraordinary (i.e., associate) professorship and in 1908 to a full professorship, succeeding Johannes Diderik van der Waals. He died on 9 October 1943 in Amsterdam.

Zeeman started his scientific career with investigations of the Kerr effect (which was also the subject of his thesis). After the discovery of the magnetic splitting of spectral lines he devoted his life to studying it in detail. Zeeman received many honours, including the Nobel Prize in Physics for 1902 for his discovery of the Zeeman effect (which he shared with H. A. Lorentz); he was elected a Foreign Member of the Royal Society of London and an *Associé Etranger* of the Academy of Sciences, Paris.

interferometer technique (Michelson, 1897). Zeeman's discovery started an intensive research in which many scientists joined, including Alfred Cornu in Paris, Thomas Preston in Dublin and Woldemar Voigt in Göttingen. Thus Preston, in December 1897, showed that the sodium D-lines exhibited quartet and sextet structure in the presence of a magnetic field (Preston, 1898), a result which was soon afterwards confirmed by Cornu (1898).[247]

The change caused by a magnetic field on the structure of spectral lines provided the key to the understanding of the structure of emitting atoms. Hendrik Antoon Lorentz, to whom Zeeman reported his discovery, immediately saw a possibility of interpreting it on the basis of his electron theory of matter. He assumed, in particular, that the electrons in atoms were bound quasi-elastically. He decomposed their motion—which caused them to emit radiation—into a component in the direction of the magnetic field and two components of opposite circular polarizations in a plane perpendicular to the field. While ν_0, the frequency of the component parallel to the field, would not be affected by the field, the frequency of the perpendicular components would be changed to ν due to the action of the magnetic field on the electrons moving perpendicular to the field direction, where

$$\nu = \nu_0 \pm \frac{eH}{4\pi m_e c} , \tag{106}$$

where H denotes the strength of the magnetic field (i.e., the absolute value of the field vector), m_e and e are mass and charge of the electron, and c is the velocity of light *in vacuo* (Lorentz, 1897). Equation (106) described the observed data for varying field strengths H; it also yielded a value for m_e/e of about the same magnitude as found for cathode rays. Zeeman also confirmed (Zeeman, 1897b)—as did C. G. W. König and A. Cornu—that the shifted line components were circularly polarized, and finally König and Cornu concluded (from the fact the shorter wavelength component was circularly polarized in the same sense as the current in the electromagnet) that the charge of the bound objects emitting spectral lines was negative (König, 1897; Cornu, 1897). Hence the existence of bound electrons in atoms seemed to be justified on empirical grounds, and the only difficulty that remained was to explain the complicated structure of the magnetic components observed by Preston, Cornu and others (Lorentz, 1899).

At the same time as Lorentz developed the theory of the Zeeman effect—as the magnetic splitting of the spectral lines was called—on the basis of elastically bound electrons, Joseph Larmor considered a slightly different problem: he showed that the effect of a magnetic field of strength H on rotating particles with charge e and mass m_c was to superimpose on the rotational motion a precessional motion about the direction of the magnetic field, having an angular velocity ω_L

---

[247] See Section IV.4, footnote 712, for a biographical note on T. Preston, and footnote 227 for a note on A. M. Cornu.

given by the equation (Larmor, 1897b)

$$\omega_L = 2\pi\nu_L = \frac{eH}{2m_e c} \, . \tag{107}$$

The frequency ν_L obtained from Eq. (107) thus turned out to be identical to the frequency shift of the lines observed by Zeeman. As a result, the magnetic splitting of spectral lines could be explained on the basis of two assumptions concerning the origin of spectral lines: either they were emitted by oscillating electrons or by rotating electrons (Larmor, 1900a).

The Zeeman effect, for the discovery and explanation of which Zeeman and Lorentz received the Nobel Prize in Physics for the year 1902, thus allowed one to have the first view into the structure of atoms. The next step towards a model of atomic constitution was taken by Joseph John Thomson: in his Silliman Lectures of 1903 at Yale University, New Haven, Connecticut, and in a paper published in *Philosophical Magazine* in March 1904, he proposed 'the view that the atoms of elements consist of a number of negatively charged electrified corpuscles enclosed in a sphere of uniform positive electrification' (Thomson, 1904a, p. 237).[248] In particular, Thomson calculated the configurations of electrons in spheres of uniformly distributed positive charge and obtained the following results: when the number of electrons was below four, they disposed themselves in a regular arrangement, all having the same distance from the centre; however, when their number was increased further, the electrons tended to form rings or spherical shells, and the number of electrons in these shells was somehow reminiscent of the periods in the periodic system of chemical elements. Furthermore he showed that if one electron was displaced from the equilibrium configuration, it would be driven back to it—as had been previously assumed by Lorentz, Drude and others in order to account for the dispersion of light in matter. Thomson did not attempt to determine from his theory the number of electrons in chemical atoms, but referred instead to empirical methods: to the dispersion of light in gases, to the scattering of X-rays in gases, and to the absorption of β-rays in gases (Thomson, 1906). From the formula which he derived for the refractive index of light and further empirical information on hydrogen, he concluded that the number of electrons in a hydrogen atom could not greatly exceed unity; he also found for the number of electrons in a molecule of air roughly the value 25 which corresponded to its atomic weight. Charles Glover Barkla, who continued the investigation in detail of X-ray scattering by atoms, obtained a slightly different result: he noted that the number of scattering

---

[248] Thomson did not refer to a similar suggestion of Lord Kelvin's (Kelvin, 1902), for he had developed his own preliminary ideas earlier (Thomson, 1897b, 1899). At that time he had been attracted by the experiments of Alfred Marshall Mayer (1836–1897), professor at the Stevens Institute of Technology in Hoboken, New Jersey: Mayer had magnetized sewing needles and put each of them on a small cork, letting the corks float in a bowl of water and observing the arrangements formed by the magnets, upon which a central attraction was superimposed (caused by the action of an external magnetic field of a pole placed above the bowl) (Mayer, 1878, 1879). These arrangements seemed to Thomson to be suggestive in connection with the periodic system of chemical elements, provided one replaced the arrangements of magnetized needles by arrangements of electrons in the atoms.

electrons in a given atom was about half of its atomic weight, with the possible exception of hydrogen (Barkla, 1911).[249]

Thomson's model of atomic constitution was essentially consistent with the results from the scattering of electrons by atoms, obtained earlier by Philipp Lenard. The latter had studied in some detail the absorption of cathode rays in gases (air, carbon dioxide, argon) as a function of their velocity, and from the fact that very fast electrons were much less absorbed than slower ones he had concluded that the 'atoms must exhibit within their volumes a further structure, consisting of smaller parts, having much space in between' (Lenard, 1903b, p. 736).[250] However, other experimental results presented serious difficulties. In 1909 Hans Geiger and Ernest Marsden working at Manchester observed that α-rays, when scattered by thin layers of gold foil, showed an unexpected behaviour: some α-particles suffered a deflection by an angle of 90° and more (Geiger and Marsden, 1909).[251] A couple of years later their chief, Ernest Rutherford, who, together with Geiger, had earlier established the result that the α-particles emitted by radioactive substances consisted of doubly-charged helium atoms (Rutherford and Geiger, 1908), drew the following conclusion: 'In order to explain these and other results, it is necessary to assume that the electrified [α-]particle passes through an intense electric field within the atom. The scattering of the electrified particle is considered for a type of atom which consists of a central [positive] electric charge concentrated at a point and surrounded by a uniform spherical distribution of opposite electricity equal in amount' (Rutherford, 1911a, p. 19).[252] A month after delivering this preliminary note, Rutherford

---

[249] The combined efforts of Thomson and Barkla settled the question concerning the number of electrons in atoms to a large extent. Earlier James Jeans had assumed that atoms should consist of very many, if not an infinite number, of positive and negative ions, in order to account for the emitted spectra (Jeans, 1901a).

[250] Lenard called these constituents '*Dynamiden*' and estimated their size to be so small that they filled only one-billionth of the atomic volume; he further assumed them to be neutral objects, composed of pairs of electric charges (at distances smaller than 10^{-11} cm); the pairs should perform very fast rotations in the atom, hence matter appears to be impenetrable for slow electrons.

The concept of '*Dynamiden*' had already been used by Ferdinand Jacob Redtenbacher in the middle of the nineteenth century; he pictured the atoms as consisting of a massive atomic nucleus surrounded by 'ether' particles, both of which were point-like objects (Redtenbacher, 1857).

[251] The layer of gold foil was so thin that one had to conclude that the scattering occurred between single α-particles and single atoms of gold.

[252] E. Rutherford was born at Spring Grove (later Brightwater) near Nelson, New Zealand, on 30 August 1871. He was educated at Nelson College (1887–1889) and Canterbury College, Christchurch, where he received his B.A. in 1892 and M.A. the following year. He went to England with a scholarship to continue his studies, entering Trinity College, Cambridge, in 1895 and working under J. J. Thomson at the Cavendish Laboratory. After Wilhelm Conrad Röntgen's discovery of X-rays, Thomson wished to exploit their ionizing effect for the investigation of the phenomena connected with the discharge of electricity through gases and invited Rutherford to join him. However, Rutherford soon became involved in radioactivity, which Becquerel had discovered in 1896, and became a pioneer in this field. In 1898 he was appointed Professor of Physics at McGill University, Montreal, Canada, and gathered a large number of very active collaborators from many countries around him. During his stay in Montreal came the discovery of thorium emanation, the investigation of thorium decay and the radioactive families—which led to the establishment of the laws of radioactive transformation. Finally, in 1907, Arthur Schuster invited Rutherford to become his successor as Langworthy Professor of Physics at the University of Manchester.

—in April 1911—submitted a detailed paper on 'The Scattering of α and β Particles by Matter and the Structure of the Atom' to *Philosophical Magazine* (Rutherford, 1911b). In this paper he derived—by assuming the action of Coulomb forces between the nucleus of charge $Z|e|$ and α- or β-particles of charge E (E being $2|e|$ in the case of α-particles and e in the case of β-particles) —the formula expressing y, the number of scattered particles falling on a unit area of a screen, which was placed at a distance r from the scattering substance and at an angle ϕ measured frcm the direction of the incident particles, that is,

$$y = \frac{1}{2} nt \frac{Z^2 e^2 \cdot E \cdot Q}{m^2 u^4 r^2} \operatorname{cosec}^4 \left(\frac{\phi}{2} \right). \tag{108}$$

In Eq. (108), n and t denote, respectively, the number of atoms in the unit volume of the scattering foil and its thickness, and m, u and Q the mass, velocity and total number of the incident α- or β-particles. This formula described the data from the scattering of α-particles by thin foils, especially the angular dependence, extremely well. For the charge of the atomic nucleus of gold, Rutherford obtained about 100 times the absolute value of the charge of the electron. Rutherford further convinced himself that his theory did not contradict the results obtained from the scattering of β-particles.

Thus Rutherford suggested a model, in which the atom had to be pictured as consisting of a point-like nucleus having a positive charge $Z|e|$, surrounded at some distance by Z electrons, similar to the model of Wilhelm Weber. Rutherford was not aware of Weber's model of the atom; he rather referred to the 'Saturnian' atom, which Hantaro Nagaoka had proposed in December 1903 at the Physico-Mathematical Society of Tokyo and published soon afterwards in *Philosophical Magazine* (Nagaoka, 1904).[253] In this paper Nagaoka had assumed that the atom resembled the system of the planet Saturn with its rings—he took this idea from an essay of James Clerk Maxwell on the stability of Saturn's rings (Maxwell, 1859); in particular, the negative electrons should move in rings around the attractive centre of a positively charged mass. The electron rings (or tori) in the Saturnian atom would vibrate and thus give rise to radiation; the oscillations perpendicular to the plane of a ring were then shown to lead to a spectrum having a band-like structure and the oscillations in the plane to a kind of line spectrum.[254] Nagaoka also explained the emission of α- and β-rays; they

---

[253] H. Nagaoka was born at Nagasaki, Japan, on 15 August 1865. He graduated from the Physics Section of Tokyo University and became Assistant Professor there the same year (1887). After further studies in Germany (1893–1896), he was promoted to a professorship at Tokyo University. Nagaoka became one of the founders of Japanese physics. His research ranged over many fields, especially spectroscopy, electromagnetism, atomic mechanics, properties of ferromagnetic substances, seismic disturbances, and mathematics. He died in Tokyo on 11 December 1950.

[254] Nagaoka thus explained the fact that no Zeeman effect had been observed so far in band spectra: the reason being that he assumed that any external field caused an orientation of the axis of rotation of the electron ring in the direction of the field. Later on, however, the magnetic splitting was shown to exist also in band spectra.

came about when the electron rings and the atomic nucleus broke up due to large disturbances.[255]

Many objections could be raised to Nagaoka's derivation of discrete spectra from the Saturnian atom as well as to Rayleigh's later attempt (Rayleigh, 1906) to obtain spectral lines from Thomson's model with many electrons (Thomson, 1906), for they showed several deficiencies: first, it was difficult to obtain more than one series of lines; second, since one dealt with a second-order differential wave equation to describe the vibrations, the square of the frequency rather than the frequency itself—as in the Balmer and Rydberg formulae—satisfied a simple relation involving integers. Finally, there arose a further difficulty which James Jeans, when considering the origin of spectra from atoms consisting of infinitely many negative and positive ions, had already noticed (Jeans, 1901a): how could an atom have a definite, invariable structure—as appeared necessary in order to explain the fact that it always emits the same lines—in spite of the fact that atoms collide so frequently and also interact with radiation?

Various attempts were made to remove or resolve the above-mentioned difficulties. Walther Ritz, for example, assumed that the atoms were composed of several permanent elementary magnets. He then calculated the interaction between the poles of the magnets and the oscillating electrons and derived from it formulae for the spectra that very closely resembled the empirical ones obtained by Rydberg; thus the Balmer formula resulted if he assumed a particularly simple structure for the hydrogen atom (Ritz, 1908a).[256] However, the problem still remained of how to relate the known constituents of atoms, namely, electrons and positive ions, to these *ad hoc* assumed magnets. Moreover, at that time it became increasingly obvious that the emission of line and band spectra from atoms and molecules had to do with the quantum hypothesis. Especially, Johannes Stark tried to visualize the emission of lines of definite frequency as transitions of electrons between different configurations of the atoms, in such a way that the differences of energies associated with these configurations corresponded (up to a factor h, i.e., Planck's constant) to the frequencies of the emitted spectral lines (Stark, 1908a, c).[257] Stark's ideas, though very imaginative, were not

---

[255] Already earlier Jean Perrin had used an atomic model in which the electrons moved around a positively charged nucleus in order to account for the emission of cathode rays and β-rays from matter (Perrin, 1901).

[256] Ritz' magnetic interaction gave rise to a linear relation for the frequency of the oscillating electrons, in contrast to the second-order wave equations in the models of Thomson and Nagaoka.

[257] In Stark's model, the assumptions concerning the origin of lines were the following: He assumed that the series spectra and the band spectra could be emitted by the same atomic structures. While a bound electron in an atom had to be excited with a minimum energy—obtained, say, from electron impact—in order to emit one line of a series spectrum, the band spectrum occurred when a valence electron returned, after having been completely removed from the atom, to its original state. That is, the complete band spectrum was emitted by a single electron. Stark imagined, in particular, that the band lines were emitted exactly when the electrons in eccentric orbits passed the points of maximal accelerations; thus two bands arose, one emitted from the points closest to the atomic center—in which the wavelengths increased with every successive orbit—and another emitted from the most distant points, in which the wavelengths decreased. He thus explained the observed structure of band series.

accepted generally.[258] More widely discussed, however, was a theory in which the quantum of action was introduced into Thomson's atomic model. Its author, Arthur Erich Haas, had been looking for an electromagnetic interpretation of Planck's constant h; that is, he saw the opportunity of relating h to the fundamental parameters of the electron theory by studying the situation related to the most well-founded atomic model at that time (Haas, 1910a, b, c).[259] Haas' ideas were the following: the electron in the Thomson atom moves on a circular orbit of radius r within a uniformly positively charged sphere of radius a; this sphere exerts a force on the electron, which grows with increasing radius of the electron's orbit (as long as it remains within the positively charged sphere), hence the binding of the electron and the phenomena of dispersion of light and of emission and absorption of spectral lines could be understood qualitatively; the absolute value of the total energy of the electron (the energy is negative) assumes a maximum value, e^2/a, when the electron moves on a sphere with radius a, and by putting this value equal to $h\nu^*$, i.e.,

$$h\nu^* = \frac{e^2}{a} , \tag{109}$$

where ν^* is the limiting frequency of the Balmer spectrum, a relation for h was obtained. Finally, Haas assumed that in this relation ν^* was also the frequency of the electron in the orbit with radius a; hence he was able to eliminate ν^* from the relation expressing the equilibrium between the Coulomb attraction, e^2/a^2, and the centrifugal force, $4\pi^2\nu^{*2}m_e a$, and arrive at an equation,

$$h = 2\pi|e|\sqrt{m_e a} , \tag{110}$$

between h and the constants e and m_e, denoting the charge and mass of the electron, and a the atomic radius. By substituting into it the known values of e, m_e and a ($\approx 2.8 \times 10^{-8}$ cm), he showed that Eq. (110) was indeed properly satisfied. At the first Solvay Conference, Lorentz remarked in connection with Haas' result: 'The constant h is determined by these [atomic] dimensions (Haas), or the dimensions ascribed to the atoms depend on the quantity h' (Lorentz in Eucken, 1914, p. 103). While Haas preferred the first interpretation, most people —including, for instance, Sommerfeld—turned to the second one: that is, eventually Planck's constant determines the size of the atom, or the existence of the constant h guarantees the stability of atoms.[260]

---

[258] In a slightly modified version Stark repeated his theory of the emission of line and band spectra in the second volume of his book *Atomdynamik*, which appeared in 1911 (Stark, 1911).

[259] A. E. Haas was born in Brünn, Moravia, on 30 April 1884. He studied at the Universities of Göttingen and Vienna, where he obtained his doctorate in 1906. He became *Privatdozent* at the University of Vienna in 1912, extraordinary (i.e., associate) professor at the University of Leipzig the following year, and finally Professor of Physics in Vienna (1923–1936); he then moved to Notre Dame University in Indiana (1936–1941). He died in Chicago on 20 April 1941.

[260] In the beginning Haas' ideas on atomic structure were not well received even by his colleagues in Vienna. But a year later, Friedrich Hasenöhrl, one of his critics, and Arthur Schidlof published

The first effort by Arthur Haas in 1910 to establish a theory of atomic structure and the emission of spectral lines on the basis of quantum theory was followed by an even more ambitious one by the mathematical physicist John William Nicholson of Cambridge.[261] Beginning in late 1910 he had turned his attention to problems connected with the line spectra of celestial bodies. Since the time of Norman Lockyer there existed the belief that matter occurring in stars had perhaps a more elementary structure than the known chemical atoms. Nicholson, therefore, considered the latter as compounds of primary atoms, which could not be found on earth (Nicholson, 1911b). He imagined the primary atoms to consist of small spheres of negative electricity, rotating in a ring about an even smaller sphere of positive electricity, the atomic nucleus.[262] In order to avoid the dissipation of energy from an accelerated electron, demanded by classical electrodynamics, Nicholson had to assume that the vector sum of the accelerations of all the electrons rotating in the ring of a primary atom was zero: hence the simplest atom, which he called 'coronium,' contained two electrons; the next primary atom was 'hydrogen,' which possessed three electrons; then followed 'nebulium' with four electrons and 'protofluorine' with five electrons. From these four primary elements he constructed the chemical atoms: for example, hydrogen in Nicholson's model consisted of two atoms of primary hydrogen, helium of one nebulium and one protofluorine atom, etc.[263] Primary

papers in which they basically accepted Haas' theory (Hasenöhrl, 1911; Schidlof, 1911). Especially Schidlof concerned himself with the problem of absorption and emission of light, arguing that these processes were connected with the loss of an electron by an atom and its recapture, respectively.

[261] John William Nicholson was born at Darlington, England, on 1 November 1881. He studied at the University of Manchester (1898–1901), where he graduated B.Sc. and later M.Sc., and at Trinity College, Cambridge (1901–1904). He passed the Mathematical Tripos (in 1904) as Twelfth Wrangler, was Isaac Newton Student in 1906 and Smith's Prizeman in 1907. He lectured at the Cavendish Laboratory until 1912, when he was appointed Professor of Mathematics in King's College of the University of London, where C. G. Barkla was still Professor of Physics. He left London in 1921 to become Fellow and Tutor of Balliol College, Oxford. There he remained as the director of studies in mathematics until 1930, when grave ill health forced him to retire. He died in Oxford on 10 October 1955.

Nicholson worked on many topics of mathematical physics, especially wave theory, electromagnetic theory, electrical conduction, spectroscopy and structure of atoms. He became a Fellow of the Royal Astronomical Society in 1911 and a Fellow of the Royal Society in 1917. He served on many scientific councils, and was President of the Röntgen Society and Vice President of the Physical Society of London.

[262] Though by 1911 Nicholson knew about Rutherford's views on atomic structure, he had himself started out from Thomson's model. He just assumed that the sphere of positive electricity had a very small radius, for he believed—in agreement with Thomson—that practically all the mass of the atom was attached to the positive sphere and that this mass was electromagnetic in origin. (Thus the mass of the atom corresponded, apart from a factor c^{-2}, to the energy of a charge e concentrated within a radius a, i.e., $\frac{2}{3}e^2/a^2$.)

[263] Nicholson was able to construct all chemical atoms out of his primary ones, to which he assigned the atomic weights 0.51282 (coronium), 1.008 (hydrogen), 1.6281 (nebulium) and 2.3615 (protofluorine). Thus uranium possessed the composition $8\{Nu_2(Pf\,H)_3\}4\{He_2Nu_2(Pf\,H)_3\}\cdot 2\{He(Pf\,H)_3\}$, where the symbols H, Nu, Pf, and He denoted the atoms of primary hydrogen, nebulium, protofluorine and helium ($=$ H Pf). At the time, the agreement between the atomic weights of chemical elements calculated from Nicholson's formulae and the observed ones was quite remarkable.

elements should exist only in astronomical objects, as in the nebulae or the solar corona; these objects emitted only a few spectral lines, the frequencies of which could be calculated by studying the small vibrations of the electron rings in primary atoms; thus Nicholson obtained the spectra of 'nebulium' (Nicholson, 1911a) and of protofluorine (Nicholson, 1911c, 1912b, c). These calculations of spectra proved to be very successful; especially, Nicholson had predicted a nebulium line having a wavelength of 4685.7 Å (Nicholson, 1911a), which was discovered at about the same time in the spectrum of the ring nebula in Lyra (Nicholson, 1912a). To obtain not only the ratios of wavelengths, Nicholson had to introduce the dimensions of his primary atoms. For that purpose, he considered the ratio of the potential energy of the electron rings (which equalled $2\pi v^2 nm_e a^2$, n and m_e being the number and mass of the electrons, and a the radius of the ring) to the rotational frequency v, obtaining in the case of protofluorine the result

$$\frac{5m_e a^2 \cdot 2\pi v^2}{v} = 154.94 \times 10^{-27} \, \text{erg} \cdot \text{s} \approx 25h, \tag{111}$$

with h denoting Planck's constant (Nicholson, 1912b, p. 679). Since the ratio in Eq. (111) corresponded to 2π times the value of the angular momentum of the electron ring in protofluorine, Nicholson concluded that in all primary atoms the angular momentum assumed values that were integral multiples of $h/2\pi$. He noted:

> If, therefore, the constant h of Planck has, as Sommerfeld has suggested, an atomic significance, it may mean that the angular momentum of an atom can only rise or fall by discrete amounts [i.e., integral multiples of $h/2\pi$] when electrons leave or return. It is readily seen that this view presents less difficulty to the mind than the more usual interpretation, which is believed to involve an atomic constitution of the energy itself. (Nicholson, 1912b, p. 679)[264]

All these results led Nicholson to believe that in this atomic model he had followed the right path to a complete explanation of spectra. True, he could not account easily for the Balmer spectrum of hydrogen, because he ran into the usual difficulty: namely, that the squares of the frequencies rather than the frequencies themselves satisfied a simple relation involving integers. In his 1914 report on quantum theory, James Jeans made the following comments on Nicholson's model:

> The sharpness of the observed spectral lines demands that the radius of the ring of electrons should remain invariable, or should have only a definite number of

---

[264] In the same paper, Nicholson described 21 lines occurring in the solar corona as being due to the protofluorine atoms (the neutral and the ionized ones, i.e., atoms having two, three and four electrons in the ring), leaving only six lines unexplained. (Nicholson suggested that they arose from the negatively charged protofluorine atoms.)

possible values, jumping from one to the other instantaneously, perhaps, as the energy is lost by radiation. Whatever view is taken it is obvious that something quite different from the Newtonian mechanics will be required to explain the motion. Nicholson believes he can find quite distinct evidence of the existence of Planck's quantum from a study of the radii of the different rings, but his work is hardly complete enough yet to admit of abstraction as a finished theory. We may take leave of it with the remark that, in addition to whatever it may achieve when more fully developed, it has probably already succeeded in paving the way for the ultimate explanation of the phenomenon of the line spectrum. (Jeans, 1914b, p. 50)

The explanation, to which Jeans hinted here, had already made its appearance: it was due to Niels Bohr from Copenhagen.

II.3   Niels Bohr and the Origin of the Quantum Theory of Line Spectra[265]

In issue No. 151 (July 1913) of the renowned British journal *Philosophical Magazine* there appeared an article entitled 'On the Constitution of Atoms and Molecules'; it was communicated by Ernest Rutherford of Manchester and written by a so-far-unknown physicist, a Dr. phil. Niels Bohr of Copenhagen (Bohr, 1913b). The author announced that the article was the first of a series dealing with a theory of the constitution of atoms and molecules on the basis of Rutherford's atomic model and Planck's elementary quantum of action. At the close of the introduction to his paper Bohr remarked: 'In the present first part of the paper the mechanism of the binding of electrons by a positive nucleus is discussed in relation to Planck's theory. It will be shown that it is possible from the point of view taken to account in a simple way for the law of the line spectrum of hydrogen' (Bohr, 1913b). Who was this man who had solved this great problem of atomic theory?

Niels Henrik David Bohr was born on 7 October 1885 in Copenhagen as the first son of Christian Bohr, Professor of Physiology at the University of Copenhagen, and his wife Ellen Adler, daughter of a Jewish banker.[266] Niels had an

---

[265] The origin of Niels Bohr's theory of line spectra is one of the most frequently treated topics in the literature on the development of quantum theory. The literature includes the articles of Léon Rosenfeld (in Bohr, 1963) and John A. Heilbron and Thomas S. Kuhn (1969), and the monograph of Ulrich Hoyer (1974). An authoritative account of certain events was also given by Niels Bohr himself in his Rutherford Memorial Lecture of 1958 (Bohr, 1961a). Finally, there exists the record of interviews with Bohr himself and a Manchester colleague, the chemist George de Hevesy, in which the events of the second decade of the twentieth century leading up to Bohr's model of atomic structure have been discussed.

[266] For the life of Niels Bohr we have consulted the biography by Ruth Moore (1966) and the articles of David Jens Adler, and Léon Rosenfeld and Erik Rüdinger in the book edited by Stefan Rozental (Rozental, 1967, pp. 11–37 and 38–73), as well as the introductions to the early papers (preceding the ones on atomic constitution) by J. Rud Nielsen, contained in *Niels Bohr, Collected Works*, Vol. 1 (Bohr, 1972).

elder sister, Jenny, and a brother, Harald, who was two years younger. He received an excellent education, both at home—where learned professors, the colleagues of his father, frequently assembled—and at school.[267] After attending Gammelholm School, where, despite some difficulties in composition, he passed with distinction, especially in mathematics and physics, Niels entered the University of Copenhagen to study mathematics, philosophy and physics.[268] He stood out as an unusually perceptive investigator: with his first research work, a precision measurement of the surface tension of water by the observation of a regularly vibrating jet, Niels Bohr won a gold medal of the Academy of Sciences and Letters in Copenhagen (Bohr, 1906).[269] He then decided to take his master's and doctor's degrees in physics, while his brother Harald concentrated on mathematics.[270] Niels completed his thesis for the master's degree in July 1909 and continued to work on the same subject—the interpretation of the physical properties of metals on the basis of electron theory—for his doctoral dissertation, which he defended on 13 May 1911 (Bohr, 1911).

In September 1911, supported by a fellowship from the Carlsberg Foundation, Bohr went to Cambridge to study under J. J. Thomson.[271] He entered Trinity College and attended the lectures of Thomson, Joseph Larmor and James Jeans; at the suggestion of Thomson, he also did experiments on positive rays in the Cavendish Laboratory. He anxiously waited for Thomson's comments on his doctoral thesis, a rough English translation of which he had brought along with

[267] Besides other English customs, Christian Bohr also admired the game of soccer; thus he founded *Akademisk Bold Klub*, i.e., the University Soccer Club. Harald Bohr became one of the best soccer players in Denmark; he was even on the Danish team which won the silver medal at the Olympic Games in London in 1908.

[268] Niels Bohr took courses in philosophy and logic from Harald Høffding, a close friend of his father's, who taught the main systems of philosophy from the sixteenth to the eighteenth century and aroused the interest of his students in the philosophy of Baruch Spinoza and Søren Kierkegaard. It was in the circle of philosophy students that Bohr learned his method of developing new ideas in discussion with others. Most of the physics courses were taught by Christian Christiansen, a physicist who belonged to the group of friends of Bohr's father.

[269] In this work Bohr also improved upon a theory of Lord Rayleigh. Later he published an English translation of his first work in *Philosophical Transactions of the Royal Society of London* (Bohr, 1909), with a shorter version of this paper appearing in *Proceedings of the Royal Society* (Bohr, 1910).

[270] Harald Bohr was born on 22 April 1887 in Copenhagen. He obtained his master's degree in 1909, before Niels, and then left Copenhagen to continue his studies with Edmund Landau in Göttingen. He returned to Copenhagen in fall 1909 and defended his doctoral thesis on the theory of Dirichlet series in January 1910, after which he joined the faculty of the University of Copenhagen. Five years later he was appointed Professor of Mathematics at the College of Technology, and in 1930 he became head of the newly founded Institute of Mathematics at the University of Copenhagen. Harald often visited Göttingen to collaborate with Edmund Landau, hence the name 'Bohr' had a good reputation there even before Niels ever went there. Besides Göttingen, Harald Bohr had fruitful contacts with the English mathematicians Godfrey Harold Hardy and John Edensor Littlewood at Cambridge and Oxford, respectively. He died in Copenhagen on 22 January 1951.

[271] The reason for this choice was the fact that Thomson was considered as one of the outstanding experts in electron theory and the conduction of electricity, which was the subject of Bohr's thesis.

him.[272] But Thomson was very difficult to approach for the somewhat shy Danish research student; he only suggested that the paper should be submitted to the *Transactions of the Cambridge Philosophical Society*.[273] On the occasion of a visit to Lorrain Smith, a friend and colleague of his father's in Manchester, Bohr was introduced to Ernest Rutherford. Rutherford had just returned from the Solvay Conference in Brussels, and he impressed Bohr deeply. Bohr was happy to be invited to spend some time in Rutherford's laboratory in Manchester; he went there in the middle of March 1912 and enrolled immediately in an experimental course on radioactivity. 'When I came up there,' he recalled later, 'I can say that Rutherford was very nice and arranged that I could take part in courses by [Hans] Geiger. But it was so that just a few weeks later on I said to Rutherford that it would not work to go on making experiments and that I would better like to concentrate on the theoretical things' (Bohr, AHQP Interview, 11 January 1962, p. 11).

Rutherford's laboratory in Manchester was a most active place at that time. The experiments of Hans Geiger and Ernest Marsden on testing Rutherford's formula, Eq. (108), and the related picture of atomic structure were still going on. Besides, a number of young and very good collaborators—including Charles Galton Darwin, Henry Gwyn Jeffreys Moseley and George de Hevesy—worked with Rutherford. Bohr felt very happy in the midst of these people. 'And Hevesy was most extraordinary,' he remarked many years later, 'he had such fine manners. He knew how to be helpful to a foreigner' (Bohr, AHQP Interview, 11 January 1962, pp. 11–12). Hevesy, who had just come to Manchester from

---

[272] In his doctoral thesis Bohr had criticized some of Thomson's work. In particular, he had argued that in Thomson's calculation of the absorption of heat radiation by metals (Thomson, 1907a) the motion of the individual electrons had not been taken into account properly (Bohr, 1911; see English translation p. 358).

[273] About his encounters with J. J. Thomson, Bohr reported to his brother Harald as follows:
... Thomson has so far not been as easy to deal with as I thought the first day. He is an excellent man, incredibly clever and full of imagination (you should hear one of his elementary lectures) and extremely friendly; but he is so immensely busy with so many things, and he is so absorbed in his work that it is very difficult to get to talk to him. He has not yet had time to read my paper, and I do not know yet if he will accept my criticism. He has only talked to me about it a few times for a couple of minutes, and only about a single point, namely about my criticism of his calculation of the absorption of heat rays. You may remember that I pointed out that, in his calculation of the absorption (contrary to the emission), he does not take into account the time taken by the collisions and, therefore, finds a value for the ratio between emission and absorption that is of the wrong order of magnitude for small periods of vibration. Thomson first said that he couldn't see that the duration of the collisions could have such a large influence upon the absorption. I tried to explain it to him and gave him the next day a calculation of a very simple example ... which showed it very clearly. Since then I have only talked with him about it for a moment, about a week ago, and I believe that he feels that my calculation is correct; however, I am not sure but that he thinks that a mechanical model can be found which will explain the law of heat radiation on the basis of the ordinary laws of electromagnetism, something that is obviously impossible, as I have shown indirectly, and as it has, moreover, later been proved directly by McLaren .... (Bohr to H. Bohr, 23 October 1911)

Bohr's hope of having his doctoral thesis published in the *Transactions of the Cambridge Philosophical Society* did not materialize. After waiting for several months he received the request to cut it down to half of its size. At that time, in May 1912, he was occupied with other work, hence he did not do so.

Karlsruhe in order to learn experimental techniques in physics from Rutherford, was soon impressed by Bohr as well.[274] Especially, Bohr concerned himself deeply with the structure of the atom. As Hevesy recalled many years later: 'When I first talked to him in 1912, he had quite clear[ly] the idea that the chemistry [i.e., the chemical properties of the atoms] resides in the outer electron arrangement, and mass and atomic charge in the nucleus. . . . It was a Sunday afternoon. I was at the house of Rutherford. Bohr was also present. I asked Rutherford: "Alpha-particles clearly come from the nucleus, no doubt. But where do the beta-particles come from?" Rutherford answered, "Ask Bohr." Bohr was present and with no difficulty answered that electrons involved in transmutation processes come from the nucleus, and all other electrons come from the exterior of the atom' (Hevesy, AHQP Interview, 25 May 1962, p. 3). It was in the summer of 1912 that Niels Bohr finished the first draft of a paper on the constitution of atoms and molecules, in which he worked out his ideas.[275]

In his memoir, Bohr started out from the 'Atom-model proposed by Professor Rutherford in order to explain the "big scattering" of α-particles,' which he described as follows: 'The atoms consist of a positive charge concentrated in a point (in an extension very small compared with the dimensions of the atoms) surrounded by a system of electrons, which total charge is equal to that of the positive "kern" [nucleus]; the kern is also assumed to be the seat of the mass of the atom' (Bohr, 1963, p. XXI). He immediately noticed that 'in such an atom there can be no equilibrium configuration, without motion of the electrons' and he set out to discuss the rotations of n electrons moving in a ring around the nucleus, finding that it had no 'stability in the ordinary mechanical sense,' because 'a ring, if only the strength of the central charge and the number of electrons in the ring are given, can rotate with an infinitely great number of different times of rotation, according to the assumed different radius of the ring; and there seems to be nothing (on account of the instability) to allow from mechanical considerations to discriminate between the different radii and times of vibration' (Bohr, 1963, pp. XXII–XXIII). At that point he had to make an

---

[274] George Charles de Hevesy (or Georg von Hevesy) was born on 1 August 1885 at Budapest. He studied chemistry at the *Technische Hochschule*, Berlin, and at the University of Freiburg, where he obtained his doctorate in 1908. Then he went to E.T.H., Zurich, as assistant to Richard Lorenz (1908–1910), and to Karlsruhe as assistant to Fritz Haber. On Haber's suggestion he studied electron emission from liquid sodium alloys, and he was sent to Manchester in 1912. From Manchester, Hevesy returned to the University of Budapest as lecturer (1913–1914) and, after his military service during World War I, as professor (1918–1920). In 1920 he joined Bohr's Institute of Theoretical Physics in Copenhagen as Associate; in 1926 he became Professor of Physical Chemistry at the University of Freiburg and later, after 1934, Professor at the University of Copenhagen. From 1943 to 1956 he worked at the Research Institute for Organic Chemistry in the University of Stockholm. Hevesy received the Nobel Prize in Chemistry for 1944 for his discovery of the use of radioactive tracers in the study of chemical processes. He died in Freiburg on 5 July 1966.

[275] Bohr's manuscript, six carefully written and numbered sheets in Danish—which Bohr later entitled 'First draft of considerations contained in the paper "On the Constitution of Atoms and Molecules" (written up to show these considerations to Prof. Rutherford) (June or July 1912)'—is reproduced in Bohr, 1963, pp. XXI–XXVIII. Bohr sent this memoir to Rutherford, accompanied by a letter, dated 6 July 1912.

assumption, which allowed him to fix the dimensions of the electron ring in the atom. In particular, he stated:

> This hypothesis is: that there for any stable ring (any ring occurring in the natural atom) will be a definite ratio between the kinetic energy of an electron in the ring and the time of rotation. This hypothesis, for which there will be no attempt of a mechanical foundation (as it seems hopeless), is chosen as the only one which seems to offer a possibility of an explanation of the whole group of experimental results, which gather about and seems to confirm conceptions of the mechanisms of radiation as the ones proposed by Planck and Einstein. (Bohr, 1963, p. XXIII)[276]

With the above hypothesis Bohr evidently referred to the quantum hypothesis of Arnold Sommerfeld, which the latter had presented at the Solvay Conference of 1911.[277] Bohr now believed that he could add further strong support to this hypothesis from four empirical 'facts': the periodic law of atomic volumes; Richard Whiddington's relation between the minimum velocity of cathode rays, which were able to excite characteristic X-rays, and the atomic weights of the substances (Whiddington, 1911); William Henry Bragg's conclusion that the stopping power of different metals for α-rays was proportional to the square root of their atomic weights (W. H. Bragg, 1905); and the heats of chemical combination of compounds. As for the second of these 'facts,' a theoretical investigation seemed to be relevant, which Bohr was just about to complete. In this work he showed that the energy loss of charged particles, especially α-particles, in matter might be ascribed analogously to the dispersion of light and, by comparison with empirical data, he derived two important consequences for atomic theory: first, the hydrogen atom contains one electron, the helium atom two, and the lighter elements about $\frac{1}{2}A$ electrons, A being their atomic weight; second, the absorption of α-particles per unit area of equal weight of matter decreases for elements of increasing weight (Bohr, 1913a).[278] Bohr dealt further with such considerations

---

[276] Bohr had become interested in quantum theory in late 1911. Thus he wrote to his friend Carl Wilhelm Oseen on 1 December 1911: 'I am at the moment very enthusiastic about the quantum theory (I mean its experimental side), but I am still not sure this is not due to my ignorance. I can say the same, in a far higher degree, about my relation to the theory of magnetons. I very much look forward to trying to get all these things straight next term.' (This part of the letter is quoted in Heilbron and Kuhn, 1969, p. 230.)

[277] Since Rutherford attended the Solvay Conference, he brought back copies of the reports to England. In his first paper on atomic constitution Bohr quoted the French version of the proceedings of the Solvay Conference (Langevin and de Broglie, 1912; see Bohr, 1913b, p. 2, footnote †).

[278] Bohr's interest in this problem had been aroused by a paper of Charles Galton Darwin, who had calculated a curve fitting the data on the absorption of α-rays rather well (Darwin, 1912). Darwin's paper appeared in June 1912, and Niels Bohr wrote about it to his brother: 'Things are not too bad at the moment; a few days ago I had a little idea about understanding the absorption of α-rays (it so happened, that a young mathematician here, C. G. Darwin (the grandson of the right Darwin) has just published a theory on this question and it seemed to me that it was not only not quite right mathematically (this was however rather trifling) but quite unsatisfactory in its basic conception, and I have worked out a little theory about it, which, however modest, may perhaps throw some light over a few things concerning the structure of atoms' (Bohr to Harald Bohr, 12 June 1912; see Bohr, 1963, p. XVII). The two mathematical approximations of Darwin, which Bohr criticized, were: the neglect of the effect on the α-particles when they did not penetrate the atoms and the forces exerted on the atoms when a rapidly moving α-particle collided with an atom.

in his memoir on atomic structure. Especially, he calculated the stability of electrons in rings; he found that a ring containing up to seven electrons will be stable, but a ring with eight electrons possesses positive total energy, hence the eighth electron can leave the atom. He concluded:

> This, together with the fact that inner rings of electrons in Professor Rutherford's atom-model will have only very little influence (and always to the worse [i.e., towards an instability]) on the stability of outer rings, seems to offer a very strong indication of a possible explanation of the periodic law of the chemical properties of elements (the chemical properties is [sic] assumed to depend on the stability of the outermost ring, the "valency electrons") by help of the atom-model in question. (Bohr, 1963, p. XXII)[279]

In the latter part of his memoir Bohr also calculated electron configurations in molecules; thus he proposed a model of the hydrogen molecule as consisting of two separated nuclei plus a ring, in which two electrons rotate around the axis connecting the nuclei (Bohr, 1963, pp. XXV–XXVII). It was in this part that Bohr introduced for the first time a 'special hypothesis,' i.e.,

$$E = K \cdot \nu, \tag{112}$$

where E denotes the kinetic energy of an electron, ν its frequency of rotation, and K a constant of the order of magnitude of Planck's constant.[280]

During the second half of the year 1912 Niels Bohr made only little progress with his work on the constitution of atoms and molecules. In July he left Manchester and went back to Copenhagen, where he married Margarethe Nørlund on 1 August.[281] The young couple spent their honeymoon in England and Scotland and then returned home, where Niels had to give lectures at the University of Copenhagen.[282] But he also continued to work on atomic constitution. Thus he reported in a letter to Rutherford: 'I have made some progress with regard to the question of dispersion. The number of electrons in a Hydrogen- and a Helium-Atom calculated from the dispersion seems thus to work out nearer to respectively 1 and 2 if the forces acting on the electrons are assumed to vary inversely as the square of the distance, than if they, as in Drude's theory, are

---

[279] Evidently, the above conclusions explained the periodic law of atomic volumes, i.e., the fact that the ratio of atomic weight over density is a periodic function of the position of the chemical elements in the periodic system (L. Meyer, 1870).

[280] For the exact value of K, Heilbron and Kuhn—by using the numerical values given by Bohr in the memoir (Bohr, 1963, p. XXVII) and in his paper on the absorption of α-particles (Bohr, 1913a)—found the result $0.6h$ (Heilbron and Kuhn, 1969, pp. 250–251).

[281] Margarethe Nørlund was the sister of the mathematician Niels Erik Nørlund, a good friend of Harald Bohr's.

[282] Bohr was first appointed assistant to Martin Knudsen, the successor of his former teacher Christiansen, and he gave a series of lectures on the mechanical foundations of thermodynamics from 16 October to 18 December 1912. (See Bohr, 1963, p. XXXV.)

assumed to be of the elastic type' (Bohr to Rutherford, 4 November 1912, quoted in Bohr, 1963, p. XXXV).[283] On the other hand, he became concerned again with the problem of stability. In order to devote full attention to the problem, he asked for a leave from his university duties and retired, in December 1912, with his wife to the countryside with the intention of writing a long and detailed paper on the constitution of atoms and molecules. At this time he also studied thoroughly the literature on that subject, especially the papers of John William Nicholson.[284] A Christmas card to his brother Harald noted that 'one of us [i.e., Niels] would like to say that he thinks that Nicholson's theory is not incompatible with his own' (M. and N. Bohr, 23 December 1912, quoted in Bohr, 1972, p. 563), and about a month later he wrote a long letter concerning this matter to Rutherford, in which he also enclosed the final version of his α-particle paper. In this letter he remarked: 'The theory of Nicholson gives apparently results which are in striking disagreement with those I have obtained; and I therefore thought at first that the one or the other necessarily was altogether wrong. I have however now taken the following point of view' (Bohr to Rutherford, 31 January 1913, quoted in Bohr, 1963, p. XXXVII). Then he outlined the points of atomic theory in which he agreed with Nicholson and also those about which they disagreed. Both Nicholson and he, Bohr claimed, used Rutherford's model of atomic constitution, and in determining the dimensions and energy of atoms they both sought 'a basis in the relation between the energy and the frequency suggested by Planck's theory of radiation' (Bohr to Rutherford, *ibid.*). In contrast to Nicholson, however, Bohr assumed that these considerations just determined the permanent or natural state of atoms, i.e., the state of lowest energy, but Nicholson also treated atomic states of less stable character (i.e., the states of higher energy) in the same way. Thus Bohr, for the moment, just concentrated upon treating the atom in the ground state, even when he referred to the problem of spectral lines emitted by the atom. He took the point of view that the problem could not be fully discussed on the basis of the ordinary mechanics, but that one had to take into account Planck's theory of radiation as follows:

If we thus assume that the systems considered [i.e., the atoms] are formed by a successive binding of the electrons by the nucleus until the whole system is neutral (compare the formation of a helium-atom from an α-particle), and further assume that the energy emitted as radiation by this binding is equal to Planck's constant multiplied by the frequency of rotation of the electron considered in its final orbit,

---

[283] Already in the letter accompanying the memoir to Rutherford, Bohr remarked that he had come across certain difficulties in explaining the results from dispersion: especially, the number of electrons, computed by Clive and Maude Cuthbertson (1910) were not integers but, e.g., 0.85 for hydrogen. Bohr drew attention to this difficulty also in his paper on the passage of α-particles through matter (Bohr, 1913a, p. 23).

[284] Bohr had met Nicholson earlier in Cambridge and discussed with him the problems of electron theory. About his meeting with Nicholson, Bohr had written to Oseen: 'I also had a discussion with Nicholson; he was extremely kind but I scarcely agree with him about much' (Bohr to Oseen, 1 December 1911; quoted in Heilbron and Kuhn, 1969, p. 258).

we get results which seem to be in conformity with experiments. (Bohr to Hevesy, 7 February 1913, quoted in Bohr, 1963, pp. XXXII–XXXIII)[285]

These remarks show clearly that Bohr was not dealing yet with the calculation of the frequencies corresponding to the lines in the visible spectrum of an atom, again in striking contrast to Nicholson. However, only a month later he sent his new paper to Rutherford, whose 'first chapter is mainly dealing with the problem of emission of line-spectra, considered from the point of view sketched in my former letter [of 31 January 1913] to you' (Bohr to Rutherford, 6 March 1913). The new paper contained the full account of a complete theory of atomic constitution, including the concept of stationary states and the relation between the frequencies of the emitted spectral lines and the energy difference between stationary states. How had Bohr come to such far-reaching conclusions in such a short time?

When Léon Rosenfeld, one of Bohr's closest collaborators in later years, inquired about the motivation for the dramatic development, Bohr told him: 'As soon as I saw Balmer's formula the whole thing was immediately clear to me' (Bohr, 1963, p. XXXIX). However, the actual process of arriving at the new results—going far beyond what Bohr had done previously—must have been much more involved. It is known that in those days of February 1913 Bohr had the opportunity of talking about spectroscopic questions with Hans Marius Hansen, who had worked earlier at Woldemar Voigt's Institute in Göttingen on the inverse Zeeman effect and knew spectroscopic data as well.[286] From information obtained personally from Bohr, Rosenfeld later reconstructed the events as follows:

> Hansen asked Bohr how on his theory he would account for the spectral regularities. Bohr had not been interested so far in this aspect of the question, because he thought those spectra too complicated to give any clue to the structure of atomic systems. At this stage Hansen was able to contradict him by calling his attention to the great simplicity with which the spectral lines had been represented by Rydberg, who was more preoccupied to find formulae likely to express general laws than to achieve the greatest possible accuracy for single series. (Rosenfeld in Bohr, 1963, p. XL)

By comparing the equation for the frequency of the Balmer lines,[287] i.e.,

$$\nu = \text{const.}\left(\frac{1}{4} - \frac{1}{m^2}\right), \tag{113}$$

[285] In a footnote to this sentence, Bohr remarked: 'The constant entering in the calculation [i.e., K in Eq. (112)] is not exactly equal to Planck's constant, but differs from it by a numerical factor, in conformity with what was to be expected from theoretical considerations' (Bohr to Hevesy, 7 February 1913).

[286] Hansen had spent the period from April 1911 to August 1912 in Göttingen. After his return to Copenhagen he wrote his doctoral thesis, which was accepted on 30 June 1913. (See Bohr, 1963, pp. XXXIX.)

[287] According to Bohr's testimony, he found the Balmer formula for hydrogen in the book of Johannes Stark (1911, pp. 44–45). (See Bohr, AHQP Interview, also quoted in Hermann, 1969, p. 172, and Heilbron and Kuhn, 1969, p. 265.)

and the formula for W, the total energy of the electron in a hydrogen atom,

$$W = -\frac{\pi^2 m_e e^4}{2K^2} \tag{114}$$

(m_e and e are, respectively, the mass and charge of the electron, and K a constant). Eq. (114) followed from mechanics and an application of Eq. (112). Bohr could have been brought to the idea that K might assume values that were integral multiples of Planck's constant. Then an interpretation of the right-hand side of Eq. (114) in terms of energy differences might indeed have been evident.[288] Still, on the way to arriving at this result, Bohr had to establish an intermediate step: that is, he had to assume that there was a discrete set of stable or stationary states in an atom, and that these states could be calculated in accordance with ordinary mechanics supplemented by certain quantum rules. Of course, indications of the existence of stationary states could be found in the work of Johannes Stark and also in his book on atomic dynamics (Stark, 1911) which Bohr consulted. However, it is more likely that he found stimulation in the papers of Nicholson; especially in the second paper on the constitution of the solar corona, Nicholson had drawn attention to the fact that the angular momentum of an electron ring might rise and fall only by integral multiples of $h/2\pi$, and that this change also occurred in connection with the process of emission of spectral lines (Nicholson, 1912b, p. 679). All Bohr had to do was to take over Nicholson's assumption concerning the discreteness of angular momentum and apply it systematically to determine all possible states of the hydrogen atom; this procedure would then yield Eq. (114), with K assuming the values $\frac{1}{2}nh$, $n = 1, 2, 3 \ldots$.[289] The relation between the energy differences of the stationary states, i.e., of states with different values of K, and the frequencies ν of the emitted spectral lines then followed by assuming that the energy loss of the atom was radiated away in an energy-quantum $h\nu$.[290]

On 6 March 1913 Bohr sent the 'first chapter' of his work on the constitution of atoms and molecules to Rutherford and announced that further chapters would follow within a few weeks. In the following weeks he made a few additions and alterations in the first chapter, which became part one of a trilogy: part one, signed 5 April and containing mainly the theory of the one-electron atom and its line spectra, appeared in the July issue of *Philosophical Magazine* (Bohr, 1913b); part two, dealing with systems having only one nucleus, was sent off to Rutherford on 10 June and appeared in September (Bohr, 1913c); and part three, on

[288] Rosenfeld argued that Bohr had probably proceeded in this way in order to arrive at his theory of the Balmer formula. (See Bohr, 1963, pp. XL–XLI.)

[289] It should be noted that Bohr, in his paper, explicitly quoted Nicholson's above-mentioned assumption (Bohr, 1913b, p. 15, footnote). He also knew that in an atom, besides the normal state, or the state of lowest energy, other states exist. (See Bohr to Rutherford, 31 January 1913.)

[290] This idea may be found already in Einstein's light-quantum paper of 1905 and in several papers of Stark; it appeared again in several papers of Thomson at that time (e.g., in Thomson, 1912, pp. 455–456). It is most likely that Bohr took it over from Thomson, whose works he studied carefully.

systems of several nuclei (i.e., on molecules), was ready by August and appeared in the November issue of the same journal (Bohr, 1913e).[291] The first paper opened with general considerations concerning the binding of electrons in an atom and also stated the principal assumptions used in the calculations, namely, the existence of stationary states and the frequency condition: '(1) That the dynamical equilibrium of the [atomic and molecular] systems in the stationary states can be discussed by help of the ordinary mechanics, while the passing of the systems between stationary states cannot be treated on that basis. (2) That the latter process followed by the emission of a *homogeneous* radiation, for which the relation between the frequency and the amount of energy emitted is the one given by Planck's theory' (Bohr, 1913b, p. 7).

Bohr calculated the energy of the electron in the electric field of a nucleus with charge $Z|e|$ by assuming the equality of the attractive force and the centrifugal force on the elliptical orbit with a major-axis $2a$. Thus he obtained equations expressing the frequency of revolution ω and the quantity a (which in the case of a circular orbit is the radius) in terms of the (negative) energy W and the parameters of the electron, that is,

$$\omega = \frac{\sqrt{2}}{\pi} \frac{|W|^{3/2}}{Ze^2\sqrt{m_e}} \quad \text{and} \quad 2a = \frac{Ze^2}{|W|} . \tag{115}$$

In order to find the expression for the energy of the stationary states he used the quantum condition

$$|W| = nh\frac{\omega}{2} \quad (n = 1, 2, 3, \ldots), \tag{116}$$

which he motivated by saying: 'If we assume that the radiation emitted is homogeneous, the second assumption concerning the frequency of the radiation suggests itself, since the frequency of revolution of the electron at the beginning of the emission is 0' (Bohr, 1913b, p. 5).[292] With the help of the above equations, he finally obtained the energy states as

$$W_n = - \frac{2\pi^2 m_e e^4 Z^2}{n^2 h^2} \quad (n = 1, 2, 3, \ldots). \tag{117}$$

[291] Concerning the dates and procedure of submission of Bohr's three papers on the constitution of atoms and molecules, we may refer to Rosenfeld's introduction in Bohr, 1963, pp. XLIII–XLVI. Rutherford criticized originally the length of the first paper. 'I really think it desirable that you should abbreviate some of the discussions to bring it within more reasonable compass,' he wrote to Bohr and added, 'As you know, it is the custom in England to put things very shortly and tersely in contrast to the German method, where it appears to be a virtue to be as long-winded as possible' (Rutherford to Bohr, 25 March 1913). Bohr, however, went to Manchester to fight for every word of his manuscript and he succeeded in doing so.

[292] Evidently, Bohr hinted here at Planck's second quantum hypothesis of 1911, which he knew from Planck's report at the Solvay Conference.

From this, by substituting the empirical values for the mass and charge of the electron, m_e and e, and for Planck's constant h, respectively, he obtained an ionization energy $|W_1|$ of 13 eV for the hydrogen atom ($Z = 1$) and a diameter $2a_1$ (in the lowest state) of 1.1×10^{-8} cm. Then he applied Planck's theory to determine the frequency of the homogeneous radiation emitted by the hydrogen atom in the course of a transition from a state n_1 to a state n_2, arriving at the result

$$\nu = \frac{1}{h}\left(W_{n_1} - W_{n_2}\right) = R\left(\frac{1}{n_2^2} - \frac{1}{n_1^2}\right). \tag{118}$$

In Eq. (118) the constant R ($= 2\pi^2 m_e e^4/h^3$), which was called Rydberg's constant, assumed the value 3.1×10^{15} (if the mass was given in grams, the charge in electrostatic units, and Planck's constant in erg $\times$ s). 'The agreement between the theoretical and observed values,' Bohr noted, 'is inside the uncertainty due to experimental errors in the constants entering in the expression for the theoretical value' (Bohr, 1913b, p. 9).

Equation (118) described, of course, the Balmer spectrum for hydrogen (for $n_2 = 2$ and $n_1 = 3, 4, 5, \ldots$) and the Paschen series of the same atom (for $n_2 = 3$ and $n_1 = 4, 5, 6, \ldots$), and predicted an ultraviolet series (for $n_2 = 1$ and $n_1 = 2, 3, 4, \ldots$) which was found a year later by Theodore Lyman (1914).[293] In addition, Bohr noticed that on the basis of his theory, a series discovered some time earlier in the spectrum of the star ζ-Puppis (Pickering, 1896, 1897) and another found more recently by Alfred Fowler in a mixture of hydrogen and helium, both of which had been ascribed to hydrogen, were actually emitted by the ionized helium atom.[294] Bohr knew the importance of these results and wrote to Rutherford, already in his letter of 6 March 1913, that he should request Alfred Fowler to check his helium hypothesis. Rutherford, who was very interested in this experimental consequence, had Evan Jenkin Evans in his Manchester laboratory perform the experiments, which confirmed Bohr's conclusion (Evans, 1913). To

---

[293] Bohr also discussed the conditions under which the spectral lines show up (Bohr, 1913b, pp. 9–10). Thus he remarked, in particular, that in celestial bodies more lines of the Balmer series should show up from the following argument: for the higher states the diameter of the hydrogen atom increases rapidly, thus for $n_1 = 12$ it would be 1.6×10^{-6} cm, and for $n_1 = 33$ it would be 1.2×10^{-5} cm. The latter dimension can only be achieved for low pressures. Under normal conditions, however, the yet higher Balmer lines might appear in the absorption spectrum.

[294] If he put R in the case of helium equal to $8\pi^2 m_e e^4/h^3$ and took $n_2 = 3$, he obtained Fowler's series, and with $n_2 = 4$ he got Pickering's series. Formerly these series had been described by the formula,

$$\nu = R\left[\frac{1}{(n_2'/2)^2} - \frac{1}{(n_1'/2)^2}\right],$$

R being the Rydberg constant for hydrogen. Bohr wrote instead the equation

$$\nu = \frac{8\pi^2 m_e e^4}{h^3}\left[\frac{1}{n_2^2} - \frac{1}{n_1^2}\right].$$

Evans' letter to *Nature*, dated 11 August 1913, Fowler replied in another letter, dated 13 September, that 'Dr. Bohr's theory does not at present seem to me to give much evidence to helium, as the origin of lines in question' (Fowler, 1913a, p. 96). Fowler's main argument was that Bohr's frequency formula for ionized helium did not describe the data at all better than the old Balmer–Rydberg frequency formula with half-integral numbers. Upon this criticism Bohr himself answered immediately in a letter to *Nature*, dated 8 October (Bohr, 1913d). In it he refined his theory by introducing explicitly the motion of the atomic nucleus; thus the Rydberg constant in the case of hydrogen was actually $R_H = 2\pi m_e e^4/h^3 \cdot (M_H/(M_H + m_e))$ rather than $2\pi m_e e^4/h^3$ (which corresponded to the approximation of an infinitely heavy hydrogen nucleus, i.e., $M_H = \infty$), and for helium he did not have to take four times the constant for hydrogen, but

$$R_{He} = \frac{8\pi^2 m_e e^4}{h^3} \cdot \frac{M_{He}}{M_{He} + m_e} , \tag{119}$$

where M_{He} is the mass of the helium nucleus. With it he described ten lines, which Fowler had attributed to three different 'helium' series, by one helium formula, predicting at the same time several hitherto unobserved lines.[295] Upon this development Fowler withdrew his objection and, in a further letter to *Nature*, wrote: 'I am glad to have elicited this interesting communication from Dr. Bohr, and I readily admit that the more exact form of his equation given above is in close accordance with the observations of the lines in question' (Fowler, 1913b, p. 232).

While the spectra of one-electron atoms followed very easily from the new theory, much less could be said about other atoms.[296] Bohr investigated the theory of many-electron atoms, especially their stability, in part two (Bohr, 1913c). There he achieved a few qualitative results: he concluded, also motivated by the chemical properties of atoms and their organization in the periodic system of elements, that there could be only two, four or eight electrons in the inner rings surrounding the atomic nucleus. Moreover, he claimed that characteristic X-radiation arises when an electron, which is removed from an inner ring, returns to the ground state; the velocity of a cathode ray particle, should it be able to excite characteristic X-radiation of atoms with nuclear charge $Z|e|$, must exceed the value $(2\pi e^2/h)Z$ (or $2.1 \times 10^8 Z$ cm/s). This result agreed with Whiddington's earlier observations if one substituted for Z about half of the value of the atomic weights of substances. Finally, in part three, Bohr treated the structure of molecules and their stability, taking as examples the hydrogen molecule and a molecule consisting of two helium atoms (Bohr, 1913e).

---

[295] These lines (having wavelengths of 6560.3 Å, 4859.5 Å, and 4338.9 Å, respectively) were subsequently found by Evans (1915), A. Fowler (1914) and Paschen (1916).

[296] Basically Bohr could only calculate the lines due to transitions of an electron from a large orbit. Then he found the formula $\nu = F_1(n_1) - F_2(n_2)$ with $\lim_{n\to\infty} F(n) = 2\pi^2 m_e e^4/h^3$; that is, the same constant appeared in all spectra in agreement with the empirical data (Bohr, 1913b, p. 12).

Bohr's theory of atomic structure was soon to be discussed publicly. Rutherford proposed that his former Danish disciple be invited to participate in the discussion on radiation, which was planned to be held at the 83rd Meeting of the British Association for the Advancement of Science in Birmingham. Bohr indeed went to Birmingham and attended the discussion on radiation theory on 12 September 1913.[297] It opened with a report by James Hopwood Jeans, who gave 'a masterly and concise' review of the problems starting from blackbody radiation and its theoretical analysis by Poincaré, going over the most important quantum phenomena on to 'the recent work by Dr. Bohr, who has arrived at a convincing and brilliant explanation of the laws of spectral series.' (See the report on 'Physics at the British Association' in *Nature* 92, issue of 6 November 1913, especially p. 305.) Niels Bohr spoke after Jeans, Lorentz and Pringsheim. The report in *Nature* described his talk as follows:

> His [Bohr's] work had been referred to by previous speakers, and he gave a short explanation of his atom. His scheme of the hydrogen atom assumes several stationary states for the atom, and the passage from one state to another involves the yielding of one quantum ... Professor Lorentz intervened to ask how the Bohr atom was mechanically accounted for. Dr. Bohr acknowledged that this part of his theory was not complete, but the quantum theory being accepted, some sort of scheme of the kind suggested was necessary. (*Nature*, Volume 92, p. 306)

After Bohr, other speakers, including Joseph Larmor, J. J. Thomson and Oliver Lodge, addressed the Association on other problems of radiation and quantum theory. Altogether Bohr could be quite content with the reception of his theory, especially by Jeans, who referred to it as 'a most ingenious and suggestive ... convincing explanation of the laws of spectral lines' (Jeans, 1914a, p. 379). The response from Thomson, on the other hand, was far less encouraging; he did not believe at all in the new quantum theory of the atom, especially since he had just presented a day earlier a new model of the atom, which—as he claimed—described many observed features (even the photoelectric effect) without the help of a quantum hypothesis (Thomson, 1913b); also, on another occasion at the Birmingham meeting he and Bohr had a little argument about the correct interpretation of a positive ion observed in experiments.[298] But his negative attitude, though it did not make Bohr happy, did not really hurt him. 'Thomson went out of physics,' he remarked later, 'and he never came into it really again, because he objected to the only positive way' (Bohr, AHQP Interview, 11 January 1962, p. 10). What counted most was that the scientific public got to know about his (Bohr's) model of atomic structure, which was the quantum-theoretical completion of Rutherford's atom. And soon, strong support would be

---

[297] Among the distinguished foreign scientists, who attended the Birmingham Meeting, were Madame Curie, Hendrik Lorentz, Ernst Pringsheim, Svante Arrhenius and Robert W. Wood. Lorentz and Pringsheim participated in the discussion on radiation.

[298] Thomson presented his new model of the atom in some detail also at the second Solvay Conference, which was held in Brussels from 27 to 31 October 1913. On this occasion Bohr's theory was not discussed at all.

registered in favour of his theory. This support came from Henry Gwyn Jeffreys Moseley, a member of Rutherford's Manchester group.[299]

The analysis of the radioactive emanations in the first decade of the twentieth century had revealed the nature of α- and β-rays. This recognition, together with the investigation of the chemical properties of the members of the radioactive families—of the uranium as well as the thorium family—led to the so-called 'radioactive displacement law': it stated that a daughter element arising from β-decay was located one place to the right of the parent in the periodic table (Fajans, 1913a, b; Soddy, 1913). Even before the displacement law was formulated, Antonius Johannes van den Broek had, in an article entitled '*Die Radioelemente, das periodische System und die Konstitution der Atome*' ('The Radio-Elements, the Periodic System and the Constitution of Atoms'), organized the decay substances according to their weights and chemical properties in a scheme, which he related to the known periodic system of elements (van den Broek, 1913). Referring to the available models of atomic constitution, especially the models of Thomson and Rutherford, and to the empirically established hypothesis that the atomic weight of a substance was proportional to the number of what he called the 'intra-atomic charges,' i.e., the electrons, he had arranged the elements in the periodic system according to increasing atomic weight, with an average difference of two between the weights of two successive elements.[300] The main point of his organization was, however, that he attributed to each chemical element a number Z, starting from 1 for hydrogen to 118 for uranium, which he believed to be identical with the number of intra-atomic charges. In Manchester, where the experiments on the scattering of α-rays from matter had yielded reliable information about the nuclear charges of atoms, which, within experimental errors, should be identical to the ones assumed by van den Broek, his ideas were immediately accepted.[301] The number of intra-atomic charges,

---

[299] H. G. J. Moseley was born in Weymouth, England, on 23 November 1887. He was educated at Trinity College, Oxford, where he received his M.A. in 1910. Then he went to the University of Manchester as Demonstrator (1910–1912) and Lecturer (1911–1912). In late 1913 he joined the Clarendon Laboratory, Oxford, but after the outbreak of World War I he did military service with the Royal Engineers. Moseley was killed in action during the British Gallipoli Campaign at the Dardanelles on 10 August 1915.

[300] Van den Broek had discussed the periodic system, especially the position of the radioactive elements in it, earlier (van den Broek, 1911). He had noticed already then that the average difference in atomic weights of two successive elements in the periodic system was two.

[301] Thus C. G. Darwin recalled later:

At this stage [i.e., after the establishment of Rutherford's atomic model in 1911] the whole Manchester laboratory believed in the undoubted existence of the atomic number, defined as the nuclear charge. The idea of what the number meant was perfectly precise, but the value of it could not yet be known precisely . . . To those working at the time in Manchester the work on nuclear scattering made the idea quite convincing and it was fully accepted there, though in fact the proposal of atomic number was first published rather later elsewhere [i.e., by van den Broek]. (Darwin, 1955, pp. 6–7)

Also Bohr in Copenhagen agreed more or less fully with van den Broek, as he wrote in a letter: 'When I returned home I found two papers in the *Physikalische Zeitschrift*. One was by [George de] Hevesy dealing with the radioactive properties of atoms; its results agreed totally with my ideas. The other was by van den Broek, who arrived in an empirical way at the same conceptions concerning the periodic system of elements as I did deductively with my theoretical conceptions' (Bohr to Oseen, 5 February 1913; German translation in Hoyer, 1974, p. 139).

especially their binding, seemed to be connected with the emission of the characteristic X-rays, which Barkla and Sadler had discovered earlier in 1908 (see Thomson, 1912a). Hence Moseley, in summer 1913, began to investigate systematically this characteristic radiation of different emitting substances.[302] After a few months of trying different experimental methods, in which he employed both the ionization method and the photographic method to determine the wavelengths of the emitted X-rays, Moseley was able to photograph the principal lines of the X-ray spectra of most elements within a short time. In a letter, dated 16 November 1913, he reported to Niels Bohr about his results concerning the K-lines of elements from calcium to zinc. He described these results as 'extremely simple and largely what you would expect' (Moseley to Bohr, 16 November 1913). That is: each element provided two main lines, denoted by K_α and K_β, of which the α-lines were about five times more intense; further, the frequencies of the β-lines were about 10% higher than those of the α-lines, their ratio having nearly, though not exactly, a constant value.

Moseley put together his results in a frequency formula for the α lines,

$$\nu_\alpha = \nu_0(Z-1)^2\left(\frac{1}{1^2} - \frac{1}{2^2}\right), \tag{120}$$

which he also reported to Bohr. In Eq. (120), ν_0 denoted a constant (for all elements) which appeared in Rydberg's formulae. In his letter he gave a table containing the values derived for the atomic number Z. Especially, he drew attention to the finding that the correct sequence in the periodic table should be Fe-Co-Ni rather than Fe-Ni-Co (which appeared to follow from the higher atomic weight of cobalt), in agreement with the chemical properties. Thus he remarked in his published paper:

We can confidently predict that in the few cases in which the order of the atomic weights A clashes with the chemical order of the periodic system, the chemical properties are governed by Z [the atomic number, which on account of the Rutherford–Bohr atomic model could be identified with the number of electrons in the atom]; while A is probably a complicated function of Z. (Moseley, 1913, p. 1031)

Moseley continued to work on X-ray spectra in Oxford, where he had moved in late 1913, and completed his programme of measuring the frequencies of most

---

[302] C. G. Darwin, who assisted Moseley in the early stages of the work, reported later:

He [Moseley] departed from the laboratory's exclusive devotion to radioactivity, and took up and developed some important points in the quite new science of X-ray spectroscopy. He then applied his technique to the discovery of the wavelengths of the characteristic X-rays, which had been studied by Barkla several years before. Working day and night by himself with a very characteristic excess of energy, and in spite of constantly pulling his apparatus to pieces in order to improve it, he quickly got his main results. (Darwin, 1955, p. 8)

Moseley used a method, in which the secondary characteristic radiation was reflected (according to the procedure of the Braggs) by a selected crystal; he observed the third-order pattern from a potassium ferro-cyanide crystal. This method had the advantage that the exposure time of the photoplates was comparatively short, only about 5 min.

elements. He published the results on the characteristic X-ray lines of another thirty pure substances, running from aluminum to gold (Moseley, 1914b). Because of experimental difficulties (especially, the disturbing influence of the continuous X-radiation) he was not able to observe the hardest, i.e., K-lines, for elements beyond silver; instead he had to register the softer L-lines. For the frequency of the L_α-lines, he obtained the equation[303]

$$\nu_\alpha = \nu_0(Z - b)^2\left(\frac{1}{2^2} - \frac{1}{3^2}\right) \quad \text{with } b = 7.4. \tag{121}$$

Moseley's results—especially Eq. (120)—evidently spoke clearly in favour of the atomic model of Rutherford and Bohr. Hence Rutherford, in a letter of 6 December to *Nature* drew attention to them. In particular, he wrote:

> The original suggestion of van den Broek that the charge of the nucleus is equal to the atomic number and not to half the atomic weight seems to me very promising. This idea has already been used by Bohr in his theory of the constitution of atoms. The strongest and most convincing evidence in support of this hypothesis will be found in a paper by Moseley in the *Philosophical Magazine* of this month. He there shows that the frequency of the X-radiations from a number of elements can be simply explained if the number of unit charges on the nucleus is equal to the atomic number. It would appear that the charge of the nucleus is the fundamental constant which determines the physical and chemical properties of the atom, while the atomic weight, although it approximately follows the order of the nuclear charge, is probably a complicated function of the latter depending on the detailed structure of the nucleus. (Rutherford, 1913, p. 423)

Thus Moseley had given the first detailed proof of the nuclear atom, confirming at the same time Bohr's use of the quantum theory.[304]

While Moseley's work on the characteristic X-ray spectra was motivated by the existence of the Rutherford–Bohr atomic model, another experimental inves-

---

[303] Moseley, like Bohr, denoted the atomic number by N. We have replaced it here by the later notation Z.

[304] Moseley's work on the characteristic X-rays of elements, the importance of which was quickly recognized, stimulated a lively discussion on atomic models in England. Thus, in a letter dated 28 December 1913 and published in the issue of *Nature* of 1 January 1914, Frederick Alexander Lindemann stated that atomic theories including the quantum concept, but differing from Bohr's, might explain the data as well (Lindemann, 1914a). Bohr replied immediately that the dimensional arguments, used by Lindemann, did not suffice to calculate the spectrum (Bohr, 1914a). Moseley supported Bohr with a similar argument (Moseley, 1914a); but Lindemann reemphasized his point in a further letter, dated 25 January, and published in the issue of *Nature* of 5 February (Lindemann, 1914b). Again he claimed that Bohr's connection between atomic dimensions and the quantum of action was not as unique as he (Bohr) had assumed. Although Bohr did not reply publicly to this argument, he stated in a note—which has been preserved in the *Bohr Archives* in Copenhagen—that Lindemann's example did not represent the physical situation properly (see Hoyer, 1974, pp. 202–203). Also John William Nicholson participated in the discussion; while he found Bohr's models of hydrogen and ionized helium attractive, he argued against the models of lithium, beryllium and boron—i.e., against coplanar rings assumed by Bohr in Part II (Bohr, 1913c)—on dynamical grounds (Nicholson, 1914a, b). Van den Broek, on the other hand, quoted facts in favour of Bohr's and his own ideas (van den Broek, 1914).

tigation, which provided a strong evidence for one of Bohr's basic assumptions— the existence of stationary states in atoms and molecules and the frequency condition—was not connected with it. In spring 1912 Ernst Johann Gehrcke and Rudolf Seeliger had found that the light accompanying cathode rays moving in a tube filled with gas—for instance, nitrogen or mercury vapour—changed its colour towards longer wavelengths when the velocity of the electrons was reduced by a decelerating potential (Gehrcke and Seeliger, 1912). Their observation was clarified by the results, which James Franck and Gustav Hertz obtained two years later after a long series of experiments, stimulated by the earlier work of Emil Warburg—Franck's former teacher—on gaseous discharges.[305] Franck and Hertz had studied the connection between the ionization potential—i.e., the potential difference which had to be applied to an electron in order to enable it to ionize atoms in a collision—of a given gas and quantum theory (Franck and Hertz, 1911). Thus they obtained the values of the ionization potentials of several gases, such as helium, neon, hydrogen and oxygen, according to a method of Philipp Lenard (Franck and Hertz, 1913a). Later they investigated specifically the collisions of electrons—which, after passing through a potential difference, possessed a definite, controlled kinetic energy—with atoms of inert gases (Franck and Hertz, 1913b, c). Continuing along these lines Franck and Hertz arrived at unexpected results, which Hertz presented at a meeting of the German Physical Society in Berlin on 24 April 1914. The report, which was published on 30 May in the *Verhandlungen* of the Society, closed with the following summary:

1. It will be shown that the electrons in mercury vapour undergo elastic collisions with the molecules until they obtain a critical velocity. 2. A method is described to measure this critical velocity up to an accuracy of 0.1 volts. This velocity is equivalent to the one obtained by electrons that have gone through a potential difference of 4.9 volts. 3. It will be shown that the energy of a 4.9 volt-beam [of electrons] equals the energy-quantum of the mercury lines [having wavelength] 253.6 $\mu\mu$. 4. Reasons are given for the conclusion that a fraction of the collisions, in which the 4.9 volt-rays lose their energy, leads to ionization; hence 4.9 volts might

[305] James Franck was born on 26 August 1882 at Hamburg. After attending *Wilhelms-Gymnasium* he left his hometown to study mainly chemistry at the University of Heidelberg (1901–1902) and physics at the University of Berlin. He received his doctorate under Emil Warburg in 1906, after which he became an assistant at the University of Frankfurt, and shortly afterwards an assistant to Heinrich Rubens at the University of Berlin (1906–1911). In 1911 he received his *Habilitation* in physics and in 1916 he was appointed extraordinary (i.e., associate) professor in Berlin.

Gustav Ludwig Hertz, a nephew of Heinrich Hertz, also came from Hamburg, where he was born on 22 July 1878. He studied physics at the Universities of Göttingen (1906–1907), Munich (1907–1908) and Berlin (1908–1911), where he received his doctorate under Rubens in 1911. He then served as assistant to Rubens (1911–1914). After his military duty in World War I (1914–1917), and being seriously wounded, he returned to Berlin and became *Privatdozent* there in 1917. In 1920 he joined the physics laboratory of the Philips Incandescent Lamp Factory in Eindhoven as a research physicist. In 1925 he was appointed Professor of Physics and Director of the Physics Institute at the University of Halle; three years later he moved to a similar position at the *Technische Hochschule*, Berlin. He resigned this position in 1935 and became director of the research laboratory of the Siemens Company in Berlin. From 1945 to 1954 he worked in the Soviet Union; then he was appointed Professor of Physics at the University of Leipzig (1954–1961). He died in Berlin on 30 October 1975.

represent the ionization potential of mercury vapour. Another part of the collision seems to stimulate light emission, and we suppose that it is connected with the emission of the line 253.6 μμ. (Franck and Hertz, 1914a, p. 467)

In the footnote appended to the last statement Franck and Hertz added: 'Meanwhile we have performed experiments testing the excitation of the 253.6 μμ-line by 4.9 volt-rays, and have found a positive result. We shall report about it in the following issue of these *Verhandlungen*.'

The experimental apparatus used by Franck and Hertz was very simple. Electrons, emitted from a heated platinum wire, were accelerated by a potential difference applied between the wire and a symmetrically placed cylindrical net (also of platinum) surrounding the wire (so that the wire formed the axis of the cylinder); outside the cylindrical net was fixed a cylinder of platinum foil, connected to a galvanometer, so that the current of the electrons falling on it could be measured; a tube of glass, which could be filled with any gas or vapour, covered the entire equipment. In the case of mercury vapour Franck and Hertz observed that the registered electron current increased with rising potential difference until a value of 4.9 V was reached, then it dropped suddenly; by further increasing the potential difference beyond 4.9 V the current again increased, but dropped at about 9.8 V; these sudden fall-offs occurred at all integral multiples of the critical 4.9 V. Franck and Hertz found a similar, though less pronounced, behaviour if their apparatus was filled with helium; there the critical potential difference was about 21 V. These phenomena, together with the occurrence of the characteristic radiation of 2536 Å units when the critical potential difference was applied to the electrons in a mercury tube, was interpreted by Franck and Hertz as follows:

1. The electrons are reflected by the mercury atoms without losing energy as long as their kinetic energy is below the amount $h\nu$, where ν is the frequency of the resonance line. 2. As soon as the kinetic energy of an electron reaches the value $h\nu$, this energy quantum will be transferred in one of the following collisions to the spectrum of frequency ν [that arises from] within the atom. 3. The energy transferred is partly used for ionization, and partly it is emitted as light of frequency ν. 4. The constant h calculated from these experiments is 6.59×10^{-27} erg · sec, with a possible error of 2 percent. (Franck and Hertz, 1914b, p. 517)

In arriving at the interpretation of their results Franck and Hertz referred explicitly to the ideas of Johannes Stark on the origin of discrete spectral lines. Stark had suggested several years earlier that line spectra were created in connection with the process of ionization of atoms and molecules, and that their frequency ν was related to the ionization potential V through the quantum-theoretical equation

$$h\nu = V, \qquad\qquad (122)$$

where h is Planck's constant (Stark, 1908a, p. 85; 1909c, p. 616).[306] Franck and Hertz stuck to Stark's interpretation also in the following years, for they believed to have shown conclusively that the potential difference of 4.9 V led to the ionization of mercury atoms. Thus, in their review article in *Physikalische Zeitschrift*, after mentioning the empirical evidence for the ionization, they stated:

> The applicability of quantum theory to ionization processes is essentially substantiated by these facts and [they added, that] it is not possible to make the assumption, as Bohr has done recently, that in helium the 20.5 volt-beams and in mercury the 4.9 volt-beams lead only to secondary ionization, such that the short-wavelength radiation [resulting from inelastic collisions] causes a photoelectric effect at the electrodes or at the impurities present in the gas. (Franck and Hertz, 1916, p. 438)

Bohr's different interpretation of the Franck–Hertz experiment had originated from a remark in his first paper on the constitution of atoms. After mentioning some 'instructive' calculations on the energy of β-particles emitted from radioactive substances (Rutherford, 1912a, b), he had argued:

> These calculations strongly suggest that an electron of great velocity in passing through an atom and colliding with the electrons bound will lose energy in distinct finite quanta. As is immediately seen, this is very different from what we might expect if the result of the collisions was governed by the usual mechanical laws. The failure of the classical mechanics in such a problem might also be expected beforehand from the absence of anything like equipartition of kinetic energy between free electrons and electrons bound in atoms. From the point of view of the "mechanical" states we see, however, that the following assumption—which is in accord with the above analogy—might be able to account for the result of Rutherford's calculation and for the absence of equipartition of kinetic energy: two colliding electrons, bound or free, will, after the collision as well as before, be in mechanical states. Obviously, the introduction of such an assumption would not make any alteration necessary in the classical treatment of a collision between two free particles. But, considering a collision between a free and a bound electron, it would follow that the bound electron by the collision could not acquire a less amount of energy than the difference in energy corresponding to successive stationary states, and consequently that the free electron which collides with it could not lose a less amount. (Bohr, 1913b, p. 19)

According to this view the 4.9 V did not necessarily correspond to the ionization potential of mercury. Thus, in 1915, Bohr argued:

> Franck and Hertz assume that 4.9 volts corresponds to the energy necessary to remove an electron from the mercury atom, but it seems that their experiments may

---

[306] The ionization potential V, if multiplied by e, the charge of the electron, corresponds to the kinetic energy of the ionizing electrons; it is 4.9 V for the mercury vapour.

possibly be consistent with the assumption that this voltage corresponds only to the transition from the normal state to some other stationary state of the neutral atom. On the present theory we should expect that the value for the energy necessary to remove an electron from the mercury atom could be calculated from the limit of the single line series of Paschen, 1850, 1403, 1269 [Paschen, 1911]. For since mercury vapour absorbs light of wavelength 1850 [Stark, 1913a], the lines of this series as well as the line 2536 must correspond to a transition from the normal state of the atom to other stationary states of the neutral atom (see I, [i.e., Bohr, 1913b] p. 16). Such a calculation gives 10.5 volts for the ionization potential instead of 4.9 volts [McLennan and Henderson, 1915]. If the above considerations are correct it will be seen that Franck and Hertz's measurements give very strong support to the theory considered in this paper. If, on the other hand, the ionization potential of mercury should prove to be as low as assumed by Franck and Hertz, it would constitute a serious difficulty for the above interpretation of the Rydberg constant, at any rate for the mercury spectrum, since this spectrum contains lines of greater frequency than the line 2536. (Bohr, 1915b, pp. 410–411)

A few years later Franck and Hertz finally agreed with Bohr's views and spoke, in a new review for *Physikalische Zeitschrift*, about the 'confirmation of Bohr's atomic theory by investigations of inelastic collisions between slow electrons and gas molecules' (Franck and Hertz, 1919, p. 132).[307] In 1925 they received the Nobel Prize in Physics for their discovery of the law governing the impact of an electron upon an atom, especially the verification of Bohr's hypotheses of the stationary states and the frequency condition.

Though it took some time until Bohr's theory of atomic constitution was recognized by other scientists, Rutherford knew that his former disciple had accomplished an important piece of work. In 1914, after Charles Galton Darwin was called to war service, he invited Niels Bohr to succeed Darwin as Reader at the University of Manchester. Bohr accepted and joined Rutherford. At Manchester he continued to extend his atomic theory. Just two years later the University of Copenhagen invited him back to take up a new professorship of theoretical physics.

II.4   Atoms as Conditionally Periodic Quantum Systems

Bohr's theory of the constitution of atoms and molecules received a mixed response on the Continent. His friends Carl Wilhelm Oseen from Uppsala and George de Hevesy from Vienna wrote enthusiastically about the progress brought about by it. Hevesy had even more encouraging news for Bohr, because he could report about a favourable opinion of Albert Einstein, whom he met at the 85th

---

[307] In their new paper Franck and Hertz especially admitted that the ionization effect, which they had found to accompany the inelastic collisions of electrons with the mercury atoms at 4.9 V, had been proven to be a secondary photoelectric effect, as Bohr had assumed in 1915. (See van der Bijl, 1917.)

Assembly of German Natural Scientists (*Naturforscherversammlung*).[308] He wrote:

> This afternoon I spoke with Einstein . . . then I asked him about his view on your theorie. He told me, it is a very interesting one, important one if it is right and so on and he had very similar ideas many years ago but had no pluck to develop it; I told him than that is established now with certainty that the Pickering–Fowler spectrum belongs to He. When he heard this he was extremely astonished and told me: "Then the frequency of the light does not depend at all on the frequency of the electron"—(I understood him so??) And this is an *enormous achiewement*. The theory of Bohr must be then wright. I can hardly tell you how pleased I have been and indeed hardly anything else could make me such a pleasure than this spontaneous judgement of Einstein.' (Hevesy to Bohr, 23 September 1913)

However, other physicists reacted much less favourably to the theory. For example, when Bohr's theory (the first part of Bohr's 1913 paper) was discussed shortly after its publication at a joint physics colloquium of the University and the Institute of Technology (E.T.H.) at Zurich, Max von Laue protested: 'This is all nonsense! Maxwell's equations are valid under all circumstances. An electron in a circular orbit must emit radiation' (F. Tank to Jammer, 11 May 1964; quoted in Jammer, 1966, p. 86).[309] From Göttingen, Niels Bohr's brother Harald reported: 'People are very interested in your papers; but I have the impression that most of them—though not Hilbert—especially among the youngest, Born, Madelung, etc., do not believe in the objective correctness [of your ideas]; they find the assumptions "bold" and "full of fancy"' (Harald Bohr to Niels Bohr, fall 1913).[310] Bohr had also sent the reprints of his three papers on the constitution of atoms and molecules to a number of selected scientists in order to learn their opinion about his work. Of these, especially Arthur Erich Haas, Heinrich Matthias Konen, Hantaro Nagaoka, and Arnold Sommerfeld congratulated

---

[308] The 85th *Naturforscherversammlung* took place in Vienna from 21 to 28 September 1913. Among the speakers were Einstein (who delivered a talk on the status of the gravitational problem on the morning of 23 September 1913), James Franck (on ionization by collisions), Max von Laue and Walter Friedrich (on X-ray interferences), Johannes Stark (on electrical and optical changes of chemical atoms), Charles Glover Barkla (on characteristic X-rays), and Max Born (on application of the quantum theory to the law of Eötvös). (See *Verh. d. Deutsch. Phys. Ges. (2) 15*, No. 20, 30 October 1913, pp. 919–923.)

[309] On the same occasion at the Zurich Colloquium Einstein was reported to have remarked: 'Very remarkable! There must be something behind it. I do not believe that the derivation of the absolute value of the Rydberg constant is purely fortuitous.' ('*Sehr merkwürdig, da muß etwas dahinter sein; ich glaube nicht daß die Rydbergkonstante durch Zufall in absoluten Werten ausgedrückt richtig herauskommt.*' Tank to Jammer, 11 May 1964; quoted in Jammer, 1966, p. 86). The colloquium on Bohr's theory was attended by Franz Tank, then a student at the Swiss Federal Institute of Technology (E.T.H.) in Zurich.

[310] According to Harald Bohr's letter (which was undated, but probably written in late October or early November 1913), the spectroscopist Carl Runge also doubted the correctness of Bohr's theory. In particular, he claimed that Bohr's correction to the Balmer formula due to the finite mass of the nucleus (Bohr, 1913e) did not describe the observed data correctly.

him.[311] In any case, Bohr's work on atomic structure had immediately received some attention from the scientific community; moreover, by no means did the younger scientists respond more positively than older ones. Thus, for example, the sixty-seven-year-old President of the *Physikalisch-Technische-Reichsanstalt*, Emil Warburg, gave a favourable report on Bohr's theory to the German Physical Society in Berlin on 5 December 1913; he not only outlined the content of the theory but also applied it to explain the effects of magnetic and electric fields on the frequency of spectral lines (Warburg, 1913b).[312]

It might seem somewhat surprising that the one scientist in Germany—Johannes Stark—who had been most concerned with atomic and molecular structure in previous years and who had come close in many respects to the views entering Bohr's theory, stayed away from the discussion.[313] But at that time he was preoccupied with his most important experimental discovery, that of the influence of a static electric field on the frequency of the emitted spectral lines. Since Zeeman had observed the influence of a static magnetic field on the frequency of spectral lines in 1896, the question had been asked whether or not a similar effect existed due to the electric field.[314] In 1901 Woldemar Voigt in Göttingen had tried to calculate the order of magnitude of the effect of an electric field on the spectral lines by assuming that the lines were emitted by bound electrons, with binding forces including nonharmonic terms—these ensured also that the molecular forces extended over a finite distance (Voigt, 1901a).[315] In this way he described both the Kerr effect and the still undiscovered effect on the frequency of spectral lines; by suitably fitting the constants of the theory (especially the anharmonicity parameter of the molecular force) in case of the Kerr effect, he found that even for electrostatic fields of about 300 V/cm the effect on the frequency remained below 5×10^{-5} times the separation of the two sodium *D*-lines. 'By this,' Voigt concluded, 'the failure of the previous attempts

---

[311] Some of these responses have been quoted in Bohr, 1963, pp. LI–LIII, and in Hoyer, 1974, pp. 182–186.

[312] Warburg's paper contained the first application of Bohr's theory to a new problem. Like Warburg, the senior physicist Wilhelm Wien at Würzburg also considered Bohr's theory with proper respect, especially after his student Heinrich Rau had interpreted his results dealing with the excitation of helium lines by collisions with electrons in agreement with Bohr's theory (Rau, 1914).

At about the same time as Warburg's paper was published (on 15 December), the essential results of Bohr's theory also became available through the detailed report which Paul Ewald gave in issue No. 25 of *Physikalische Zeitschrift* on the Birmingham Meeting of the British Association; he gave, especially, an account of James Jeans' review on the radiation theory including the report on Bohr's atomic model and the following discussion (Ewald, 1913, pp. 1298–1300).

[313] To some extent this fact may have prevented Stark from entering into a discussion concerning the priority of the ideas involved.

[314] A strong hint concerning the existence of such an effect arose from the arguments of W. Voigt; he had established a theory connecting the Zeeman effect with the Faraday effect—i.e., the rotation of the plane of polarization of light under the influence of a magnetic field (Voigt, 1898). Now it was known that a static electric field also influenced the propagation of light by causing double refraction in crystals (the so-called Kerr effect), hence Voigt expected a definite influence of the electric field also on the frequency of emitted and absorbed discrete spectral lines (Voigt, 1899a, pp. 308–310, p. 318).

[315] For an account of W. Voigt's scientific background, see Section III.1.

to discover an electrical analogue to the Zeeman effect is fully accounted for' (Voigt, 1901a, p. 208). This result discouraged people for many years from searching for the electric effect. However, in 1913 Stark began to look for it, for he believed that Voigt's argument could not be upheld; since he pictured the atom as consisting of a collection of charged particles (i.e., positive and negative ions and electrons) and assumed that the spectral lines were caused by an ionization of the atoms, he concluded that electric fields must alter their charge distribution and, therefore, also influence considerably the emitted frequencies.[316] In the fall of that year, finally, Stark succeeded in obtaining a positive result: on 20 November 1913 Heinrich Rubens presented Stark's paper, entitled 'Beobachtungen über den Effekt des elektrischen Feldes auf Spektrallinien' ('Observations on the Effect of the Electric Field on Spectral Lines'), at a meeting of the Prussian Academy (Stark, 1913b). At about the same time Stark sent a letter to Nature announcing his discovery (Stark, 1913c). The main results of his observations were: when a strong electrostatic field of up to 31,000 V/cm acted on radiating hydrogen canal rays (behind the cathode of a discharge tube), one found that the lines H_α and H_β, if viewed in a direction perpendicular to the electric field, were split into five polarized components; the electric vector of the middle components was perpendicular to the direction of the static electric field, while the outer ones had an electric field vector in its direction; in the case of helium lines at wavelengths of 4472 Å and 4026 Å there appeared six components, respectively, three of which were polarized perpendicular and three parallel to the direction of the applied field. Stark further reported: 'It seems . . . that the electric separation of a spectral line (the distance between components), when measured in wavelengths, is in first approximation proportional to the first power of the field strength' (Stark, 1913b, p. 941).[317] A few weeks later, on 20 December 1913, he reported to the Göttingen Gesellschaft der Wissenschaften on an effect that was observed if one looked in the direction of the electric field, the so-called longitudinal effect ('Längseffekt', Stark and Wendt, 1914). In the case of H_α- and H_β-lines of hydrogen, Stark and Wendt found three unpolarized components, while in the case of helium the same number of components showed up at the same place as in the transverse effect, except that the components were unpolarized.

[316] Stark expressed this motivation clearly in his first paper, in which he presented the discovery of the effect of electric fields on spectral lines (Stark, 1913b, pp. 933–934). In his Nobel lecture, delivered on 3 June 1920, Stark gave, however, a different reasoning. Thus, after referring to Voigt's theory, he stated:

> However, I was unable to accept the presupposition of the theory—namely, the assumption that the emission of a spectral line on the part of an atom was the work of only one single independently moving electron in the atom. In my view the structure of the whole atom was that of an individual, with all its parts interconnected, and the emission of a spectral line appeared to me the result of the coherence and cooperation of several electric quanta. (Stark, in Nobel Lectures in Physics, 1901–1921, p. 433)

It would seem that this point of view came in only later, i.e., after the discovery of the effect. (See Stark, 1914c.)

[317] The proportionality of the separation of the components to the strength of the electric field was confirmed in further experiments by Stark and his assistant Heinrich Kirschbaum (1914a).

While Stark was extending his researches and investigated, in collaboration with his students, the effect of the electric field on other spectral lines and on band spectra (Stark and Kirschbaum, 1914b), a note appeared in *Physikalische Zeitschrift*. Its author, Antonino LoSurdo of Florence, referred to his observation in summer 1913 that the lines emitted from hydrogen canal rays in thin tubes 'appear to be resolved in a remarkable manner,' and concluded by saying that he had seen Stark's note in *Nature* and that he had also 'observed the same effect [i.e., the longitudinal effect]' (LoSurdo, 1914, p. 122).[318] Stark continued to work on his effect and, by late 1914, arrived at complete separations of the first four lines of the Balmer series of hydrogen; while he found no further splitting of the H_α-line, the H_β-line was decomposed into 13, H_γ into 15, and H_δ into 17 components, both in transverse and longitudinal observations; in all cases the individual separation increased linearly with the increasing strength of the applied electric field (Stark, 1914d).

The first to attempt a theoretical explanation of the Stark effect was Emil Warburg in Berlin in the paper quoted earlier (Warburg, 1913b).[319] He based his approach on Bohr's atomic model, and considered the deformation of circular electron orbits by the electric field; especially, he claimed that the frequency condition, Eq. (118), had to be changed to

$$h\nu - A = W_{n_1} - W_{n_2},\qquad (118a)$$

where A was an additional energy term denoting the work which the electrons had to do in order to restore their original orbit in the presence of the electric field.[320] Thus he found that every line of the hydrogen atom would become smeared out symmetrically with respect to the original frequency ν, the width

[318] When, in the same issue of *Physikalische Zeitschrift*, LoSurdo's teacher Antonio Garbasso proposed to associate the new effect with the names of both Stark and LoSurdo (Garbasso, 1914, p. 123), Stark defended his priority vigorously in a letter, which appeared in the following issue (Stark, 1914a). LoSurdo published the details of his results in *Atti della Reale Accademia dei Lincei*, the first publication having been presented at the session of 21 December 1913 (LoSurdo, 1913). It is possible that LoSurdo indeed saw the effect first without knowing its origin.

[319] Emil Warburg was born on 9 March 1846 at Altona. He studied chemistry and physics at the Universities of Heidelberg (with Bunsen, Helmholtz and Kirchhoff) and physics at the University of Berlin (with Magnus and Kundt). In 1867 he received his doctorate under Gustav Magnus, and in 1870 he became *Privatdozent* at the University of Berlin; two years later he was appointed extraordinary (i.e., associate) professor at the University of Strasbourg, where his friend August Kundt was the *Ordinarius*. They collaborated for many years and published important papers, e.g., on the friction and heat conduction in dilute gases and on the specific heat of mercury. In 1876 Warburg became Professor of Physics at the University of Freiburg; there he discovered and explained the phenomena of hysteresis in ferromagnetic materials. Moving to Berlin in 1895 as the successor to August Kundt at the University, he became President of the *Physikalisch-Technische Reichsanstalt* in 1905. He retired in 1922 and died on 28 June 1931 at Gut Grunau near Bayreuth. During his Berlin years Warburg worked mainly on the phenomena of electric conduction, on the radiation problem and on photochemical reactions (proving the Einstein–Stark equivalence law).

[320] Warburg calculated this work (A) in the approximation that the deformation was small ($A \ll W_{n_1}, W_{n_2}$) as $A = \pm a_{n_1} e |\mathbf{E}|$, where a_{n_1} is the radius of the initial orbit, e the charge of the electron and $|\mathbf{E}|$ the strength of the electric field.

being $2|\Delta\nu|$, with

$$\Delta\nu = \pm \frac{hn_1^2}{4\pi^2 m_e e} |\mathbf{E}|, \tag{123}$$

where n_1, m_e and e denote the integral number of the initial orbit, the mass and the charge of the electron, respectively, h Planck's constant and $|\mathbf{E}|$ the strength of the static electric field. Equation (123) yielded the right order of magnitude in the case of the separations of the H_β- and H_γ-lines. Because of the appearance of Planck's constant in the expression for the width, Warburg argued that it was 'probable that the electric effect [Stark effect] belonged to those phenomena which could not be explained on the basis of classical electrodynamics' (Warburg, 1913b, p. 1266). He also calculated, with the same model, the magnetic splitting of spectral lines, obtaining a result depending upon the quantum numbers of the stationary states—in contrast to observation. Bohr, to whom Warburg wrote about his results, replied that he had also meanwhile tried to obtain the Zeeman and Stark effects of hydrogen lines. 'In agreement with your considerations,' he wrote, 'I find that my theory allows me to compute the effects in question correctly. In contrast to your results, however, I arrived at the conclusion that the theory explains the line separations observed by Stark, and also the observations on the Zeeman effect, quantitatively' (Bohr to Warburg, 8 January 1914).[321] In a paper, which appeared in the March issue of *Philosophical Magazine*, Bohr presented these results (Bohr, 1914b). According to his treatment an external electric field also deformed the orbits of the electrons in free atoms. However, he claimed that only two stable motions remained, i.e., the electrons passed either parallel or antiparallel to the direction of the external electric field through the nucleus; thus the original frequencies were shifted by amounts of $\pm(3h/4\pi^2 m_e e)|\mathbf{E}|(n_1^2 - n_2^2)$, respectively.[322]

With respect to the influence of a magnetic field on spectral lines, Bohr simply postulated that the (classical) Larmor precession was superimposed on the motion of the electron around the nucleus and this fact, although it did not alter the energy of the stationary states, changed the frequency condition to

$$h(\nu \pm \nu_L) = W_{n_1} - W_{n_2}, \tag{124}$$

where ν_L denotes the Larmor frequency. Bohr's treatments of the Zeeman and Stark effects did not describe any finer details; in the case of the Stark effect only a part of the transverse effect could be explained (the lines polarized perpendicular to the external fields should not exist), and the complex Zeeman effect discovered by Thomas Preston did not follow either. (The classical Larmor

---

[321] Bohr's attention had been drawn to the Stark effect by a letter from Ernest Rutherford (Rutherford to Bohr, 11 December 1913).

[322] Similar results were obtained by Garbasso and Ernst Gehrcke; while Garbasso's treatment was close to Bohr's (Garbasso, 1914), Gehrcke considered a kind of one-dimensional version of Bohr's atomic model (in which the electrons moved in a line: Gehrcke, 1914).

precession had to be added to all electron orbits.) Still, Bohr was very optimistic in spring 1914 about the possibility of finally accounting quantitatively for all observed separations of spectral lines. Thus, for example, he claimed that the extremely small doublet splittings of the hydrogen lines had their origin in terms of the perturbation of the Coulomb force between the nucleus and the electrons (Bohr, 1914b). In searching for such terms he arrived later at the conclusion that relativistic corrections to the motion of the electrons might be the reason (Bohr, 1915a). However, he did not really get hold of this problem properly. There occurred a standstill in Bohr's work on atomic structure at Manchester. When, in spring 1916, he returned to Copenhagen, he found several publications of Arnold Sommerfeld's on his desk. Sommerfeld had achieved important progress by extending the theory to electron motions having several degrees of freedom.

In the first decade of quantum theory physicists had dealt with systems of just one degree of freedom, especially the harmonic oscillator. The quantum of action entered into the description of these systems in the form of a condition, which Planck formulated as

$$\int \int dp\, dq = nh;\qquad (125)$$

that is, the integral over the phase space (with p and q denoting the momentum and position variables, respectively) took on finite values.[323] While Planck had used Eq. (125) only for the harmonic oscillator, Paul Ehrenfest applied it also to the rigid rotator in a plane (Ehrenfest, 1913b). By studying the boundaries of the phase-space regions, he discovered that the angular momentum p assumed there the values

$$p = n\frac{h}{2\pi},\qquad (126)$$

with $n = 0, 1, 2$, etc. Ehrenfest, who favoured Planck's first quantum hypothesis, claimed that in the motion of the quantum-theoretical rotator p_ϕ was always given by Eq. (126). The same condition also entered into Nicholson's and Bohr's theories of atomic constitution. Bohr, in particular, noticed that his original condition quantizing the energy of an electron around a nucleus was equivalent to John William Nicholson's assumption that the angular momentum of the electron ring satisfied Eq. (126) (Bohr, 1913b, p. 15).[324]

Although there had existed some concern previously with the treatment of

---

[323] Planck used Eq. (125) in the derivation of his second quantum hypothesis (Planck, 1911a). Later, it frequently became the starting point of the discussion of systems other than the one-dimensional harmonic oscillator. Thus Peter Debye applied it in his Wolfskehl lecture of 1913 at the University of Göttingen; he was interested, in particular, in the anharmonic oscillator, hence he concluded that the areas in phase space were not bounded by ellipses (as in the case of Planck's oscillators) but by curves of different shapes (Debye, 1914, p. 29).

[324] Bohr restated the quantum condition at the end of his first paper on the constitution of atoms and molecules as follows: '*In any molecular system consisting of positive nuclei and electrons in which the nuclei are at rest relative to each other and the electrons move in circular orbits, the angular*

systems of one degree of freedom in quantum theory, the serious discussion of this problem began only in 1911.[325] At the first Solvay Conference Henri Poincaré raised, during the discussion of Planck's report, the question of whether the particular shape of the finite areas in the phase plane of a one-dimensional quantum system played any role. After Planck denied it, giving as an example the one-dimensional rotator (in which case the energy values uniquely determined the shape of the phase cells), Poincaré asked further:

What if there are more degrees of freedom? One may think of a resonator vibrating in all directions, which therefore possesses three degrees of freedom but is isotropic, i.e., it has the same period (of vibration) along all three axes. If one decomposes its motions in the direction of the axes, one will have to take for the motion parallel to the x-axis an energy $\alpha h\nu$, with α an integer; for the [motions parallel to the] y-axis and z-axis, the energy will be $\beta h\nu$ and $\gamma h\nu$ (β and γ being integers), respectively. Now, if one introduces new axes, then one will have different energies, say $\alpha' h\nu$, $\beta' h\nu$, $\gamma' h\nu$, in the direction of the new axes. However arbitrarily one chooses the new axes, α', β', γ' must be integers; and that is impossible. (Poincaré in Eucken, 1914, p. 99)

Planck replied that so far no quantum hypothesis had been considered in the case of systems of several degrees of freedom, adding: 'Of course, I consider it by no means impossible to establish such a one' (Planck in Eucken, 1914, p. 99). On the other hand, Einstein (in the discussion of Nernst's report at the Solvay Conference) emphasized that certain difficulties might arise in dealing with that problem.[326]

---

momentum of every electron round the centre of its orbit will in the permanent state of the system be equal to $h/2\pi$, where h is Planck's constant' (Bohr, 1913b, pp. 24–25). He added a footnote at the end of this sentence, which stated: 'In the considerations leading to the hypothesis we have assumed that the velocity of the electrons is small compared with the velocity of light' (loc. cit., p. 25). In part II Bohr discussed the extension of the hypothesis to relativistic motions; he claimed that also in these cases the angular momentum should still be constant, but different from the ratio of energy W over the frequency of revolution of the electrons (Bohr, 1913c, p. 479).

[325] For example, in his theory of specific heats of solids Einstein treated the elastic vibrations of a gram-atom as being caused by N three-dimensional oscillators, and endowed each of these with energy-quanta $3nh\nu$, $n = 0, 1, 2, \ldots$ (Einstein, 1906g).

[326] Einstein gave the following argument: 'If one modifies the statistical mechanical equation for the average energy $\bar{E}$ of a three-dimensional oscillator,

$$\bar{E} = \frac{\int E^3 \exp(-E/kT)\,dE}{\int E^2 \exp(-E/kT)\,dE},$$

in such a way that one replaces the integrals by sums, i.e., one assigns to $\bar{E}$ a sequence of values 0, $h\nu$, $2h\nu$, etc., then one does not obtain three times the value of the energy of the linear Planck oscillator. In its present formulation, quantum theory therefore leads to contradictions as soon as one tries to apply it to objects having several degrees of freedom' (Einstein in Eucken, 1914, pp. 235–236). Fritz Hasenöhrl had mentioned a similar difficulty in his talk at the Karlsruhe Naturforscherversammlung in September 1911. He had then discussed in particular the problem of obtaining a quantum-theoretical expression for the entropy of systems having several degrees of freedom, which satisfied the theorem of the addition of entropy (Hasenöhrl, 1911, p. 933).

In the years following the first Solvay Conference Planck concerned himself with answering Poincaré's question concerning the structure of the phase space in quantum theory and the extension of the quantum hypothesis to systems of several degrees of freedom. He took this question especially seriously, for in his thinking a particular aspect played a role: the fact that the finite areas of phase space were connected with the probability for physical processes. If he, therefore, knew their structure, he could interpret the discontinuity occurring in the quantum-theoretical description as a *discontinuity in the probability* rather than as a discontinuity in energy or in the light-quantum. (See Planck, 1912a.) In his Wolfskehl lecture in April 1913 at the University of Göttingen he discussed a particular system of several degrees of freedom, the molecules of a monatomic gas (Planck, 1914b). These molecules had three translational degrees of freedom, and—according to the previous treatment of Otto Sackur and others (see Sackur, 1911)— a six-dimensional phase space consisting of finite cells of volume h^3 (h being Planck's constant) was associated with each molecule. Planck claimed, however, that the situation was much more complex:

> In order that this quantity [i.e., $\int dx\, dy\, dz\, dp_x\, dp_y\, dp_z$, the finite volume in phase space] may possess a definite value, as required by quantum theory, the limits of the integral must be quite definite; and it is a task of fundamental importance for the theory to reveal precisely the significance of these limits. (Planck, 1914b, p. 7)

For the moment he could give only preliminary indications as to how these boundaries might be found; he felt strongly that they depended on the interaction between the gas atoms.[327] More than two and a half years later, Planck finally found himself in a position to provide a definite answer, and he presented it in two contributions, entitled '*Die Quantentheorie für Molekeln mit mehreren Freiheitsgraden*' ('The Quantum Theory of Molecules with Several Degrees of Freedom'), at the meetings of the German Physical Society on 5 November and 3 December 1915 in Berlin (Planck, 1915b, c).

In these papers Planck considered atomic systems of f degrees of freedom, whose phase space had to be divided, according to the known procedure, into finite cells of volume h^f, h being Planck's constant. Planck first proposed the following quantum condition: he required that the phase-space integral over all variables, $p_1, \ldots, p_f$ and $q_1, \ldots, q_f$, compatible with energy values smaller or equal to a limiting value u_n, should satisfy the condition

$$\int_0^{u_n} dp_1 \cdots dp_f\, dq_1 \cdots dq_f = (nh)^f, \qquad (127)$$

n being a positive integer ($= 0, 1, 2$, etc.) (Planck, 1915b). Thus, for example, in the case of the spatial rotator (which possesses two degrees of freedom), he obtained the discrete energy states, $n^2 h^2 / 8\pi^2 A$, where A denotes the moment of

---

[327] Planck emphasized that the case of a monatomic gas differed in that sense from that of heat radiation, in which the resonators could be assumed as being strictly independent of one another.

inertia and $n = 0, 1, 2 \ldots$.[328] In his second contribution he tried to generalize Eq. (127); he especially assumed the existence of systems, in which the phase space did not just depend on the energy of the system under investigation (or on the number n denoting it), but also on other variables (Planck, 1915c). If these characteristic variables (including the energy) were denoted by $g, g', g'', \ldots$, and their discrete values determining the boundaries of phase cells by $g_n, g'_q, g''_r, \ldots$, the general quantum condition became

$$\int_0^{g_n} \int_0^{g'_q} \int_0^{g''_r} dp_1 \cdots dp_f dq_1 \cdots dq_f = (nh)^i (qh)^{i'} (rh)^{i''} \ldots, \qquad (127')$$

with $i + i' + i'' + \cdots = f$. Evidently, in Eq. (127') several 'ordering numbers' ('*Ordnungszahlen*') $n, q, r, \ldots$ entered. Further, the f total degrees of freedom of the system could be adjusted in such a way to the characteristic variables that i were associated with g, i' with g', i'' with g'', etc.; and Planck called the degrees of freedom belonging to one particular variable 'coherent,' and those belonging to different variables 'incoherent' (Planck, 1915c, p. 441). As in the simpler quantum systems, in which only one ordering number n occurred, also in general cases the discrete values of the characteristic variables, $g_n, g'_q, g''_r, \ldots$, determining the boundaries of the phase-space regions, could be calculated.[329] Planck applied the new quantum conditions to the case of a mass point, which moves in space under the attraction of a central force whose strength increases linearly with the distance of the mass from the centre. He was not interested, however, in treating the Bohr atom; he did not believe in its basic assumption of discrete stationary states, but rather connected the origin of radiation with probability jumps at the boundaries of the quantum cells in phase space (Planck, 1915a, d). Only several months later, when he prepared a paper for *Annalen der Physik* summarizing the content of the two earlier notes, did he also give the results in the case of an electron moving in the Coulomb field of a positive nucleus, achieving basically the same results as did Sommerfeld in his theory of the relativistic hydrogen atom, with which he had become acquainted in the meantime (Planck, 1916a).

---

[328] From the discrete energy values u_n, Planck also calculated the value which the oscillator assumes on the average in the phase-space region n, which is bounded by hyper-areas characterized by $u = u_n$ and $u = u_{n+1}$, i.e.,

$$\bar{u}_n = \frac{\int_{u_n}^{u_{n+1}} u \, dp_1 \ldots dp_f dq_1 \ldots dq_f}{\int_0^{u_n} dp_1 \ldots dp_f dq_1 \ldots dq_f}.$$

(Planck evaluated the phase-space integral in the denominator to be h^f times $\{(n+1)^f - n^f\}$.) Thus, in the case of the spatial rotator, $\bar{u}_n$ became equal to $(h^2/8\pi^2 A)(n^2 + n + \frac{1}{2})$.

[329] Planck proposed the following procedure: One takes the classical dependence, say, of g on the dynamical variables and considers a motion of the molecules only in the coherent degrees of freedom associated with g. It should be noted that the case of completely incoherent degrees of freedom denotes a nondegenerate system.

The papers, which Planck presented to the German Physical Society in November and December 1915, did not contain the first extension of the quantum hypothesis to systems of several degrees of freedom. Two other publications on this subject had appeared earlier that year, both in journals which, because of war conditions, were not directly accessible in Germany. On 4 April 1915, Jun Ishiwara of the Physical Institute at Sendai University presented a paper to the Mathematical-Physical Society of Tokyo, entitled 'Die universelle Bedeutung des Wirkungsquantums' ('The Universal Significance of the Quantum of Action,' Ishiwara, 1915a). In this paper Ishiwara, who had spent some years in Europe previously and had acquainted himself with the progress in relativity and quantum theories, attempted to generalize the quantum hypothesis in such a way that it embraced Planck's hypothesis of finite elementary regions (cells) in phase space, the Nicholson–Bohr–Ehrenfest quantization of angular momentum, Eq. (126), as well as Sommerfeld's action-integral hypothesis of 1911.[330] In particular, he proposed the following new, fundamental hypothesis:

Suppose an elementary object of matter, or a system of an enormously large number of elementary objects is in a state of stationary, periodic motion or in statistical equilibrium. Let its state be completely determined by the coordinates $q_1, q_2, \ldots, q_f$, and the corresponding momenta $p_1, p_2, \ldots, p_f$. *Then the motions in nature always occur in such a way that the decomposition of each state plane, $q_i p_i$, [$i = 1, \ldots, f$], is permitted into those regions of equal probability, the average value of which at a given point of the phase space,*

$$\frac{1}{f} \sum_{i=1}^{f} \int p_i \, dq_i = h, \qquad\qquad [(128)]$$

is equal to a universal constant. (Ishiwara, 1915a, p. 108)[331]

Ishiwara immediately convinced himself that the new hypothesis yielded the previously well-established results of quantum theory, such as the expression for

---

[330] Jun Ishiwara (or Ishihara) was born in Tokyo on 15 January 1881. He was educated at the University of Tokyo from where he graduated in theoretical physics in 1906. In 1911 he was appointed assistant professor in the faculty of science at Tohoku University, Sendai. Shortly afterwards he went abroad for further studies at the University of Munich (with Sommerfeld), at the E.T.H. in Zurich (with Einstein), and at the University of Leyden. He worked on relativity and electron theory, contributing articles to *Annalen der Physik* (Ishiwara, 1913) and to *Jahrbuch der Radioaktivität und Elektronik* (Ishiwara, 1912, 1914). After his return to Japan he concentrated on relativity and quantum theory. He was promoted to full professor at Tohoku University, but retired already in 1921 to devote himself to poetry. Ishiwara acted as Einstein's host during his visit to Japan in late 1922 and also translated several papers of Einstein's into Japanese. He died on 19 January 1947. (See the information provided by Pierre Speziali in *Albert Einstein–Michele Besso: Correspondance 1903–1955*, pp. 542–543.)

[331] It should be noted that Eq. (128) represented the first appearance in literature of what Sommerfeld later called the 'phase integral.' (Different statements, e.g., in Hund, 1967, pp. 33–34, cannot be substantiated.) Ishiwara motivated his quantum hypothesis on the basis of Sommerfeld's action integral hypothesis of 1911. (We have changed Ishiwara's notation slightly: he called q_i the momentum and p_i the position coordinates, and he referred to systems of j degrees of freedom.)

the chemical constants (Sackur, 1911), the Sommerfeld–Debye theory of the photoelectric effect (Debye and Sommerfeld, 1913) and Bohr's theory of atomic structure (Bohr, 1913b). However, in an important detail, he obtained a different result from Bohr's theory: if he assumed that the motion of an electron around the nucleus possessed two degrees of freedom, then the Balmer formula for the hydrogen spectrum followed quantitatively from Eq. (128) only if the nucleus of the hydrogen atom bore two positive charges and, therefore, the neutral atom contained two electrons.[332]

Even earlier than Ishiwara, namely, in March 1915, John William Nicholson communicated a paper on 'The Quantum Theory of Radiation and Line Spectra' by his colleague William Wilson to *Philosophical Magazine*, which was published in the June issue (Wilson, 1915). Wilson, then working at King's College, London, wanted to establish the 'possibility of deducing the results of Planck and Bohr from a single form of quantum theory' (Wilson, 1915, p. 795).[333] He based his approach on two assumptions: first, that interchanges between dynamical systems (atoms) and the ether occur in a discontinuous manner; second, that between these discontinuous changes the systems are described by Hamiltonian mechanics. The steady motions of periodic systems, however, satisfy the equa-

---

[332] Evidently Ishiwara's result arose from associating the electron with two degrees of freedom; hence the double of Bohr's value for the angular momentum states entered into the energy expressions. Ishiwara was satisfied with it, as he subscribed to some extent to Nicholson's opinion. (He had become familiar with Nicholson's theory of atomic constitution in London.) Ishiwara continued to discuss the theory of quantum systems in later papers; in one of them he changed his quantum hypothesis (128) to the following: the stationary states of atomic systems are given by the fact that their energy multiplied with the period of motion assumes integral multiples of Planck's constant (Ishiwara, 1915b, p. 318).

[333] W. Wilson was born on 1 March 1875 in Goody Hills, near Mawbray, Cumberland. From 1893 to 1896 he studied geology, mathematics, physics and astronomy at the Royal College of Science, London; among his teachers were Arthur Rücker, William Watson, Norman Lockyer and Alfred Fowler. Then he taught in English schools and at the Berlitz School of Languages at various places in Rhineland. His savings allowed him to enroll in summer 1902 at the University of Leipzig to continue his studies in mathematics (under Carl Neumann, who gave him a thorough course in Hamiltonian mechanics) and physics. In 1906 he earned his doctorate *summa cum laude*, the physicist Otto Wiener being his supervisor. After returning to England, Wilson got a position as assistant lecturer in the Wheatstone Laboratory at King's College, London. He worked especially on the photoelectric effect and related phenomena and also contributed papers on theoretical problems, the first one on an application of quantum theory to explain the discharge of hot bodies (Wilson, 1913). Later he also wrote on relativity theory and again on quantum theory. In 1919 he was promoted to a readership at King's College, and two years later he was elected to the Hildred Carlile Chair of Physics at Bedford College, London. He retired in 1944 and died on 14 October 1965. Wilson was elected a Fellow of the Royal Society in 1923.

In a letter, dated 7 June 1920, Wilson drew Sommerfeld's attention to the fact that he had published the quantum conditions for systems having several degrees of freedom already in his first paper of June 1915. Sommerfeld admitted in his reply: 'You are right in that I should have quoted your 1915 papers in my book [*Atombau und Spektrallinien*], as I did in my paper in the *Annalen der Physik*, Volume 51, 1916, p. 9' (Sommerfeld to Wilson, 14 July 1920). And he promised to cite Wilson in the second edition of *Atombau und Spektrallinien*. Sommerfeld further stated in his letter to Wilson that he had developed the extension of Bohr's theory—independently of Wilson—in winter 1914–1915, but that he had delayed the publication until Paschen had tested the fine-structure formula experimentally. He concluded by stating: 'The priority for the publication of the rule $\int p \, dq = nh$ belongs to you without doubt' (Sommerfeld to Wilson, 14 July 1920).

tions

$$\int p_i\, dq_i = n_i h, \quad i = 1, 2, 3, \ldots, \tag{129}$$

where the n_i denote positive integers and the paths of integration extend over all values of the dynamical variables p_i and q_i that are assumed during the period v_i^{-1} (v_i being the frequency and h Planck's constant). Using this hypothesis Wilson derived, in the case of N resonators, Planck's formula for the average energy of the resonator (entering into the fundamental equation of radiation theory). In the second part of his paper Wilson applied Eq. (129) in order to obtain (in the case of the circular motion of an electron around the nucleus) Bohr's formula for the kinetic energy of the electron. In addition, he found that the radiation emitted from any quantum system satisfied a generalization of Bohr's frequency condition, Eq. (118), to systems having several degrees of freedom.

   In a second paper, communicated again by Nicholson in November 1915, Wilson further developed his views on the emission of spectral lines (Wilson, 1916). He also referred to Ishiwara's paper of April 1915 and stated the main points in which the two approaches differed: especially, that he, Wilson, pre-ferred to assume Planck's original hypothesis of 1900, while Ishiwara tended towards Planck's second hypothesis of 1911; also Ishiwara extended the path of the integral (on the right-hand side of Eq. (128)) over the full period of motion of the total system, while in Wilson's Eqs. (129) the integral for each degree of freedom extended over the particular period associated with that degree.[334] Then he applied his scheme to the case of an elliptical electron orbit in an atom by writing down two phase-space integrals of the type of Eqs. (129): one for the pair of dynamical variables, p_ϕ (the angular momentum) and ϕ (the angle), the other for the pair of radial variables, p_r and r. He established a relation between their respective integral numbers, n_1 and n_2, and the eccentricity (ϵ) of the elliptical orbit,

$$(n_1 + n_2)\sqrt{1 - \epsilon^2} = n_2. \tag{130}$$

At this point Wilson stopped and did not draw any further conclusion, in contrast to Sommerfeld, who independently at about the same time had Eqs. (129) available.

   Arnold Sommerfeld, well-established Professor of Theoretical Physics at the University of Munich, had contributed to the quantum theory since 1911 by proposing a new quantum hypothesis involving the action integral. The application of this quantum hypothesis to aperiodic processes, such as the creation of photoelectrons or the motion of molecules in gases, seemed to be promising though it did not yield really conclusive results. Stimulated by Friedrich Paschen and Ernst Back's investigations of the Zeeman effects in strong external magnetic

---

[334] Wilson explicitly criticized Planck's second hypothesis. (See Wilson, 1916, pp. 159–160.)

fields (Paschen and Back, 1912, 1913), Sommerfeld turned in 1913 to establish a theoretical description of the anomalous Zeeman effects by generalizing Lorentz' original theory of the elastically bound electrons; he assumed in particular that electrons were isotropically bound in atoms (Sommerfeld, 1913). He continued to work on this subject, in competition with Woldemar Voigt of Göttingen (Voigt, 1913a, c, d), and submitted a second paper in March 1914 to the Göttingen Academy (Sommerfeld, 1914a). In the latter paper he simplified Voigt's classical theory considerably. Sommerfeld referred to a quantum-theoretical argument only at a minor point concerning the intensities of the Zeeman components. This did not mean that he was not interested in Bohr's quantum theory of atomic structure. Thus, when Bohr sent him a reprint of his paper in the July issue of *Philosophical Magazine* (Bohr, 1913b), he immediately wrote to him:

> Many thanks for sending me your highly interesting paper, which I read already in *Philosophical Magazine*. Since long I have thought of the problem of expressing the Rydberg–Ritz constant in terms of Planck's constant h. I discussed it several years ago with Debye. Although I am still somewhat skeptical about atomic models in general, [I must say that] the calculation of that constant represents without question a great achievement. By the way, the numerical agreement becomes even better with the new value of Planck's constant, $h = 6.4 \times 10^{-27}$. (Sommerfeld to Bohr, 4 September 1913)

At the end of his letter Sommerfeld asked Bohr whether he planned to calculate the Zeeman effect from his model. Bohr replied: 'I hope soon to publish a short note on the phenomena of magnetism and the Zeeman effect. I have been working for a long time with this problem, which seems promising on account of the close analogy between the hypothesis of the universal constancy of the angular momentum of the electron and the theory of magnetons' (Bohr to Sommerfeld, 23 October 1913).

It might have been the skepticism against atomic models per se or, what was more likely, Sommerfeld's general attitude not to interfere with young people when they were on a promising track; in any case, Sommerfeld did not concern himself with the application of Bohr's atomic model to the problem of the Zeeman effect. However, he did use Bohr's model of molecules for the purpose of obtaining a generalization of Drude's dispersion formula (Sommerfeld, 1915d). And soon he became deeply involved in the problem of atomic constitution and in Bohr's theory; especially, he announced a course of lectures on '*Zeemaneffekt und Spektrallinien*' ('Zeeman Effect and Spectral Lines') for the winter semester 1914–1915 (see *Physikalische Zeitschrift* **15**, p. 860, 1914). It turned into a course on Bohr's theory, and Sommerfeld also talked about an extension of the latter, including a theory of elliptical orbits and the relativistic fine structure.[335] Still he

---

[335] Sommerfeld mentioned this fact in his *Autobiographische Skizze* by recalling: 'My extension of Bohr's theory (elliptical orbits, fine structure) was already presented in winter 1914–1915 in my lecture course, but was first published in the beginning of 1916' (Sommerfeld, 1968, Volume IV, p. 678).

waited with the publication of his results because of an important reason. At that time, i.e., between fall 1914 and fall 1915, Einstein worked on the final steps towards the theory of general relativity, and Sommerfeld was very interested in it.[336] Since he was not certain whether the new theory of gravitation would also change his relativistic extension of Bohr's atomic model, he sent Einstein the manuscripts of two papers containing his results and asked for Einstein's opinion of them. Einstein replied to Sommerfeld in the beginning of December 1915:

> Herewith you get back both manuscripts, which I have looked through with interest. Planck is also working on a problem similar to yours (the quantization of phase space of molecular systems). He is concerned with spectroscopic questions as well. [Concerning the relativistic calculation of electron orbits he added:] General relativity will hardly be of any help to you, because its results coincide for these problems with those of special relativity theory. (Einstein to Sommerfeld, 9 December 1915)

Sommerfeld was satisfied with this answer and he quickly went on to publish his papers on the extension of Bohr's atomic theory. The first paper, entitled '*Zur Theorie der Balmerschen Serie*' ('On the Theory of the Balmer Series') and containing the nonrelativistic theory of the hydrogen atom, he presented already on 6 December 1915 to the Bavarian Academy (Sommerfeld, 1915b); he also made sure that the second paper was still included in the proceedings of the Academy for the year 1915, although he was not able to present it until the session of 8 January 1916 (Sommerfeld, 1915c).

Sommerfeld opened his first paper by referring to the successes of Bohr's theory in explaining the hydrogen spectrum and the X-ray spectra of heavier elements, and wrote:

> Nevertheless I want to demonstrate that even the theory of the Balmer series exhibits in certain respects a gap, namely as soon as one admits non-circular (i.e., even in the case of hydrogen, elliptical) orbits. [He added:] I shall fill this gap by deepening the quantum hypothesis, and I shall at the same time illuminate the special situation of the hydrogen spectrum. While the other elements exhibit a sequence of different series types (singlet, doublet and triplet series), hydrogen possesses (if one neglects the still little known multiple-line spectrum) only a single Balmer series. According to the concepts presented here, this fact can be explained as follows: in the Balmer series there coincides a sequence of series, i.e., each of its

---

[336] On one hand, Sommerfeld took a great personal interest in Einstein's work; on the other, he was also charged with editing the third edition of a collection of papers on relativity theory, which had first appeared in 1913 under the title *Das Relativitätsprinzip* (edited by Otto Blumenthal). For that purpose he wanted to include some papers on general relativity and he wrote Einstein a letter concerning the matter. In his reply, dated 28 November 1915, Einstein wrote about his recent progress in deriving the equations of the theory of gravitation, which yielded in particular a satisfactory explanation of the observed shift of the perihelion of the planet Mercury.

We may mention in passing that Sommerfeld had himself worked on electron theory early in the 1900s (see especially, Sommerfeld, 1904a, b). He had even discussed the case of electrons moving with speeds faster than light *in vacuo* (Sommerfeld, 1905).

lines arises in several different ways, not only from circular orbits but also from elliptical orbits of given eccentricities. (Sommerfeld, 1915b, p. 425)

This simplicity originated, according to Sommerfeld, in the simple constitution of hydrogen; for all other chemical atoms the different elliptical orbits led to different types of line spectra, such that in general each spectral line depended at least on two sets of integral numbers, n, n' and m, m', and its frequency was described by the formula

$$\nu = \text{const.} \left[\phi(n, n') - \phi(m, m') \right], \tag{131}$$

where $\phi(n, n')$ and $\phi(m, m')$ denoted the energy terms of the atom.[337] The task which Sommerfeld set out to accomplish in the first paper consisted, therefore, in deriving Eq. (131) from the assumption that electrons in atoms move on certain elliptical orbits.

Sommerfeld did not refer in December 1915 to the earlier publications of Ishiwara and Wilson on the generalization of the quantum hypothesis to systems of several degrees of freedom. On one hand, these publications were not available to him; on the other, he had arrived independently—and earlier—at the same results. He rather referred to the much earlier work of Planck, especially to phase-space considerations (Planck, 1906), and to the extension of Planck's considerations to the anharmonic oscillator (Debye, 1914) and the rotator (Ehrenfest, 1913b). From such examples he concluded that in the quantum-theoretical phase space of momentum and position variables (p and q) certain cells of finite size existed, enclosed by boundaries—which in the case of the harmonic oscillator were concentric ellipses of increasing semiaxes—whose areas were given by integral multiples of Planck's constant. This condition, the 'quantum condition' ('*Quantenforderung*') as Sommerfeld called it, could be reexpressed as the 'phase integral' ('*Phasenintegral*,' Sommerfeld, 1915b, p. 429)

$$\int \int dp \, dq = \int p \, dq - nh, \tag{132}$$

i.e., as a line integral. Sommerfeld then showed the close relation of the phase integral to his earlier action-integral quantum condition and derived from it the positivity of the right-hand side of Eq. (132). Hence the 'quantum number' n must be a positive integer ('*Quantenzahl*', Sommerfeld, 1915b, p. 430).[338] Sommerfeld extended the phase-integral approach to describe also the quantum-theoretical motion of systems having two degrees of freedom, especially to the

---

[337] In the special case of hydrogen, the function $\phi(n, n')$ should depend only on the sum of the two integers, $n + n'$; hence the energy term becomes proportional to $(n + n')^{-2}$.

[338] Sommerfeld found that the integral, $\int p \, dq$, could be written as $\int p \, dq = \int p\dot{q} \, dt = \int \text{const.} \, \dot{q}^2 \, dt$ in the case of a conservative system whose kinetic energy is proportional to $\dot{q}$, the time derivative of the position variable q. Since the integrand of $\int p \, dq$ is proportional to the kinetic energy, the phase integral is indeed closely related to Sommerfeld's earlier action integral (provided the Lagrangian is replaced by the kinetic energy) and it assumes a positive value.

elliptical orbits of an electron in the Coulomb field of a positively charged nucleus. An essential feature of the (classical) Kepler motion was the existence of a particular constant, the so-called 'area constant' ('*Flächenkonstante*'), which was identical with the angular momentum variable p_ϕ. Now the quantum condition (132), when applied to the azimuthal motion of the electron, i.e., to the canonically conjugate pair of dynamical variables p_ϕ and ϕ (where ϕ is the azimuthal angle), yielded the equation

$$\int_0^{2\pi} p_\phi \, d\phi = 2\pi p_\phi = nh. \qquad (133)$$

Hence the angular momentum p_ϕ had to assume values that were positive integral multiples of the quantity $h/2\pi$.[339] However, a new difficulty arose with the definition of the phase integral, a difficulty which did not exist in the one-dimensional motions considered thus far and which Sommerfeld realized immediately: 'For the Keplerian motion, however, we have two degrees of freedom, hence the concept of the phase integral ceases to be uniquely defined' (Sommerfeld, 1915b, p. 432).

In order to obtain a hint concerning the removal of the above-mentioned ambiguity in the Kepler motion, Sommerfeld studied the expression for the energy of the electron in the elliptical orbit in the case of the hydrogen atom, i.e.,

$$W_n = -\frac{2\pi^2 m_e e^4}{h^2} \cdot \frac{1 - \epsilon_n^2}{n^2}, \qquad (134)$$

with n denoting the quantum number of the angular momentum (see Eq. (133)) and ϵ_n the eccentricity of the ellipse under consideration. From Eq. (134) there followed the frequencies ν,

$$\nu = \frac{2\pi^2 m_e e^4}{h^3} \left(\frac{1 - \epsilon_n^2}{n^2} - \frac{1 - \epsilon_m^2}{m^2} \right), \qquad (135)$$

which were emitted during the transitions of the electron (mass m_e and charge e) from an initial orbit with azimuthal quantum number m and eccentricity ϵ_m to a final orbit with quantum number n and eccentricity ϵ_n. Evidently, the frequencies depended explicitly on the values of the eccentricities ϵ_n and ϵ_m. From the fact that the hydrogen lines were sharp, Sommerfeld concluded: 'If we do not want to disallow from the very beginning that the electrons move on elliptical orbits besides circular ones, then inevitably the requirement arises that the eccentricities have also to be treated arithmetically by connecting them with certain integral numbers according to the quantum condition' (Sommerfeld, 1915b, p. 435). Such

---

[339] Sommerfeld drew attention to the difference between his treatment and Bohr's. Bohr had established a relation between nh and the quantity $2\pi m_e a^2 \omega^2$, where a is the semi-major axis of the orbit and ω the frequency of the revolution (Bohr, 1913b, pp. 3–4).

a quantum condition, however, seemed to be available immediately. One had just to introduce (in accordance with the motto: 'What is fine with ϕ should also be all right with r,' as Sommerfeld used to say, see Jammer, 1966, p. 93, footnote 17) the phase-integral for the radial degree of freedom, i.e.,

$$\int p_r \, dr = n'h, \tag{136}$$

where p_r denotes the radial momentum variable ($p_r = m_e(dr/dt)$) and r the distance of the electron from the nucleus (which changes in case the orbit has nonzero eccentricity). By inserting for p_r the expression obtained from the equation of motion (i.e., $\dot{r} = dr/dt = (e^2/p_\phi)\epsilon \sin \phi$) into the left-hand side of Eq. (136), Sommerfeld derived an equation expressing the eccentricity in terms of the two quantum numbers n and n', i.e.,[340]

$$1 - \epsilon^2 = \frac{n^2}{(n + n')^2}. \tag{137}$$

Evidently, ϵ now depended on both quantum numbers separately. By combining Eqs. (134) and (137) Sommerfeld found a simple formula for the energy terms of the hydrogen atom, namely,

$$W_n = -\frac{2\pi^2 m_e e^4}{h^2} \cdot \frac{1}{(n + n')^2}, \tag{134'}$$

on which he commented as follows:

> This result is surprising in the highest degree and [at the same time] of penetrating certainty ('*schlagende Bestimmtheit*'). Not only the additional allowed energy values have become discrete and are determined by integral numbers, but the previous denominator n^2 has just cancelled such that the result depends only on $n + n'$. Hence the energy is uniquely determined by the sum of the quanta of action, which we may distribute arbitrarily among the azimuthal and radial coordinates. It seems to me to be quite out of the question that such a far-reaching result should be ascribed to an algebraic accident. I rather see in it a convincing confirmation of the extension of the quantum hypothesis to the radial component, or of the separate application of this hypothesis to the two degrees of freedom of our problem. (Sommerfeld, 1915b, p. 439)

In any case, the above treatment also yielded Bohr's formula for the hydrogen spectrum.[341]

---

[340] Equation (137) agreed, of course, with Wilson's Eq. (130) upon identifying Wilson's quantum numbers n_1 and n_2 with Sommerfeld's n' and n, respectively.

[341] Also the mass correction could be incorporated in the same way as in Bohr's theory; i.e., in considering the finite ratio of the mass of the electron over the mass of the nucleus, m_e in Eq. (134') had to be replaced by the reduced mass $m_e M/(M + m_e)$, M being the mass of the nucleus.

Altogether Sommerfeld believed confidently that he was on the right track and that his generalized quantum conditions, Eqs. (133) and (136), indeed allowed him to describe the hydrogen atom satisfactorily: that is, the electrons in stationary states moved in ellipses, whose parameters—the semimajor axis a $(= h^2(n + n')^2/4\pi^2 m_e e^2)$ and eccentricity ϵ—were completely determined by the quantum conditions. This situation justified the opinion, which he had already expressed earlier in 1915, ' . . . that the laws governing the interior of the atom do not deviate so much from those of classical mechanics and electrodynamics, as one might think according to Bohr's assumptions' (Sommerfeld, 1915d, p. 549). The quantum conditions did not restrict the validity of the classical laws but determined the initial values of the dynamical problem: while the quantum number $n' = 0$ signified, in agreement with Eq. (137), circular orbits, the choice $n = 0$ denoted linearly degenerate orbits (in which the electron describes a linear pendulum line). Sommerfeld regarded the latter orbits as 'most problematical' ('*höchst problematisch*'), because ' . . . the electron will then approach the nucleus infinitely closely and probably be repelled by it' (Sommerfeld, 1915b, p. 446). Excluding such degenerate orbits, he found that to any line in the hydrogen spectra several transitions may contribute: for example, in the case of the Balmer H_α-line, with initial state $(m + m') = 3$ and final state $(n + n') = 2$, six transitions from one stable orbit to the other were possible; in the case of the H_β-line, with initial and final states denoted by the quantum number sums $(m + m') = 4$ and $(n + n') = 2$, respectively, eight transitions were possible, etc. 'One may, however, doubt,' Sommerfeld added, 'whether each of these transitions is possible, or whether perhaps only those transitions are admitted that are associated with a loss of quanta' (Sommerfeld, 1915b, p. 447). He expected, in particular, that the Stark effect of the lines would throw some light on the question of allowed transitions, and argued: 'According to our interpretation, in each Balmer line an entire sequence of frequencies having different origins coincides. The [external] electric field will influence the different elliptical orbits in a different manner; hence it will separate the originally coinciding frequencies. In the case of hydrogen the influence will be stronger than for other elements, where such a coincidence [of the field-free lines] cannot be expected' (Sommerfeld, 1915b, p. 449). His prediction agreed with the known data on the Stark effect (Stark, 1914d).

For atoms having a more complex structure than the hydrogen atom—and the one-electron ions, He^+, Li^{++}, etc.—more spectral series should show up, because their frequencies depend on four quantum numbers, n, n' and m, m' due to Eq. (131). These frequencies could be written as

$$\nu = \text{const.} \left[f_{n'}(n) - f_{m'}(m) \right], \tag{131'}$$

where $f_{n'}(n) = \phi(n, n')$ and $f_{m'}(m) = \phi(m, m')$. Thus different series should arise because of different values of the parameters n' and m'. Sommerfeld suggested especially the following relations: the spectroscopic 'principal' series arose from the combinations $f_s(n) - f_p(m)$, with $n = 1$ and $m = 2$, 3, 4, etc.; the 'first subordinate' series from combinations $f_p(n) - f_d(n)$, with $n = 2$ and $m = 3$, 4, 5, etc.; and the 'second subordinate' series from combinations $f_p(m) - f_s(n)$, with $n = 2$ and $m = 3, 4, 5$, etc. (The f_p, f_d and f_s denote functions with special values for the quantum numbers n' and m'.) At that point Sommerfeld was aware of the possibility that the structure of orbits might be even more complicated than considered so far. He noted: 'At any rate, when one considers the structure of atoms and the shape of electron orbits, the restriction [of the motion] to the plane —which so far has been extended from the hydrogen atom to other elements— cannot be upheld ultimately' (Sommerfeld, 1915b, p. 453). In order to describe spatial orbits in quantum theory he proposed to introduce a third quantum number, n'', which determines the phase integral involving the coordinate z perpendicular to the original plane.

Sommerfeld concluded his first paper on the structure of the hydrogen atom by answering the question which Poincaré had posed in 1911: how far did the quantum conditions, Eqs. (133) and (136), depend on the coordinate system? For this purpose he started from the relation of the phase integral to his older action integral, that is,

$$2 \int T \, dt = \sum_{i=1}^{f} \int p_i \, dq_i = h(n + n' + \cdots), \tag{138}$$

where T denotes the kinetic energy of the system under investigation. Since in a periodic system the left-hand side ($= 2\tau \overline{T}$, where $\overline{T}$ is the average kinetic energy) is a constant of motion, the sum of quantum numbers, $n + n' + \cdots$, cannot depend on the coordinate system. In the case of the hydrogen atom, which he described by two quantum numbers, n and n', Sommerfeld argued that both quantum conditions, Eqs. (133) and (136), did not depend on the coordinate system: on one hand, the angular momentum p_ϕ was a constant of motion; on the other, the sum, $(n + n')$, should also be an invariant according to Eq. (138); hence also the quantum condition for the radial variables had to be an invariant. It became clear to Sommerfeld, however, that in the general case the choice of proper coordinates played an important role in proving the invariance of the associated quantum conditions; thus, if rectangular coordinates were chosen for the hydrogen atom, the corresponding quantum conditions would not be invariant.[342]

The important progress in the treatment of atomic systems by the phase-integral method showed up especially in the second paper, in which Sommerfeld presented a theory of the relativistic one-electron atom (Sommerfeld, 1915c). As we have mentioned above, Bohr had already attempted to deal with the relativistic effects in the hydrogen atom. He had proposed to interpret the very small deviations in the position of the hydrogen lines from his original formula by including the relativistic mass correction, and the fact that hydrogen lines

---

[342]The important point in the treatment of the hydrogen atom, as Sommerfeld noticed, was that the angular coordinate ϕ represented a cyclic coordinate; i.e., ϕ did not occur in the Hamiltonian function describing the hydrogen atom (Sommerfeld, 1915b, pp. 454–455).

actually occurred in very narrow doublets due to the relativistic motion of the electron (Bohr, 1915a).[343] In February 1915 Friedrich Paschen in Tübingen became very interested in the problem of separating helium from hydrogen lines and in their fine structure, and he began to work on it himself. He took up contact with Arnold Sommerfeld in Munich to discuss theoretical questions.[344] By the end of the year Paschen reported his first results on the separation of hydrogen doublets: 'My measurements yield: $\Delta\lambda_{H_\beta} = 0.09$ Å' (Paschen to Sommerfeld, 12 December 1915). At the same time he discussed possible experimental errors, such as the ones caused by Doppler effect and Stark effect. He also did not exclude the possibility, mentioned earlier by Bohr (1914b, p. 521), that the duplicity of the hydrogen lines might come from a Stark effect within the atom. Only detailed calculation, thought Paschen, would decide this question. Several weeks later, after having received copies of Sommerfeld's reports to the Bavarian Academy, he wrote again: 'The results of your theory, which you have reported, are very attractive. I believe your notion [concerning the origin of doublet structure] is more probable than mine' (Paschen to Sommerfeld, 27 December 1915). Paschen carried out further experiments, and he finally concluded: 'My measurements are now finished, and they agree everywhere most beautifully with your fine structures' (Paschen to Sommerfeld, 25 May 1916). The experimentalist Paschen was completely satisfied, for now the experimental and theoretical results agreed within an error of 10^{-3} Å units.[345]

Sommerfeld had based his theory of fine structure, to which Paschen referred above, on the assumption that the doublets observed in the spectrum of hydrogen arose from 'the finitely different, discrete eccentricities of our quantized ellipses' (Sommerfeld, 1915c, p. 461). The calculation of this effect was straightforward. Sommerfeld started simply from the equation of motion of a relativistic

---

[343] The doublet structure of the hydrogen lines had been discovered by Albert A. Michelson with the help of interferometric methods (Michelson, 1891). In the following decades several experimentalists had measured the frequency differences.

[344] Thus Paschen wrote to Sommerfeld on 7 February 1915:

I am enclosing a calculation of the Pickering series of hydrogen and note that Fowler—using a Geissler tube filled with hydrogen and helium—obtained at high voltages the series which have nearly the same values as Pickering's. On the basis of Bohr's theory Fowler thinks that they are possibly helium lines. Because of the deviations of Pickering's wavelengths [from Fowler's], however, one cannot exclude the possibility that Pickering's stellar lines are not identical with Fowler's and represent hydrogen lines. The question is still open, and it is treated here by one of my students who, at present, is [on military duty] at the front. I wish to inform you meanwhile that also according to our own [experimental] researches, Fowler's lines are probably helium lines, but we leave open the possibility that Pickering's lines are different and belong to hydrogen. We shall publish the proofs only after all the experiments have been completed. Unfortunately this cannot be done now because of the war.

The main difficulty in the experimental investigation consisted in establishing reliable absolute values for the observed frequencies.

[345] Paschen published a full account of his results on the spectra of hydrogen and ionized helium in a paper, entitled '*Bohrs Heliumlinien*' ('Bohr's Helium Lines') and submitted it in late June 1916 to *Annalen der Physik*. The final remarks in this paper were: 'That this work has advanced beyond the stage of a not-understood-contradiction with Bohr's theory, is due to the helpful encouragement of my colleague Sommerfeld, whose untiring efforts enabled us to recognize in the incomplete experimental data the wonderful laws of his beautiful theory. For this I wish to express my grateful thanks to Mr. Sommerfeld' (Paschen, 1916, p. 940).

electron (with mass $m_e / \sqrt{1 - v^2/c^2}$, where m_e is the rest mass of the electron, v its velocity, and c the velocity of light *in vacuo*) in a Coulomb field of a stationary nucleus of charge $Z|e|$, and obtained the total energy W, expressed as an expansion in rising powers of a quantity b^2, as

$$W = \frac{W_0}{\gamma} \left[1 - \frac{3}{4} \frac{b^2}{\gamma^2} (1 - \epsilon^2) + \frac{5}{8} \frac{b^4}{\gamma^4} (1 - \epsilon^2) + \cdots \right], \qquad (139)$$

where W_0 $(= -(m_e e^4 Z^2 / 2p_\phi^2)(1 - \epsilon^2))$ is the nonrelativistic energy, and the quantities b^2 $(= Z^2 e^4 / p_\phi^2 c^2)$ and γ^2 $(= 1 - b^2)$ become, in the nonrelativistic limit (for small velocities of the electron), equal to zero and unity, respectively. In applying the quantum conditions to the relativistic Kepler motion, Sommerfeld had to take into account the fact that the orbit in this case was not stationary (or closed), but that its perihelion advanced after each revolution by an angle $(2\pi/\gamma) - 2\pi$.[346] Hence he replaced the quantum conditions (133) and (136) by

$$\int_0^{2\pi/\gamma} p_\phi \, d\phi = p_\phi \frac{2\pi}{\gamma} = nh \qquad (133')$$

and

$$\int_0^{2\pi/\gamma} p_r \, dr = nh\gamma^2 \left(\frac{1}{\sqrt{1 - \epsilon^2}} - 1 \right) = n'h. \qquad (136')$$

As a consequence, the eccentricity ϵ was given not by Eq. (137) but by

$$1 - \epsilon^2 = \frac{\gamma^2 n^2}{\left(n' + \gamma^2 n \right)^2}. \qquad (137')$$

Thus the energy W, Eq. (139), finally depended on both quantum numbers n and n' separately (due to the factor γ^2 in front of n in the denominator of the expression for the eccentricity). Sommerfeld rewrote it as an expansion,

$$W = - \frac{2\pi^2 m_e e^4 Z^2}{h^2 (n + n')} \left[1 + \frac{\alpha Z^2}{(n + n')^2} \left(\frac{1}{4} + \frac{n'}{n} \right) + \cdots \right], \qquad (140)$$

where α represented a small dimensionless parameter,

$$\alpha = \frac{2\pi e^2}{hc}, \qquad (140a)$$

which assumed the value 0.7×10^{-3} (see Sommerfeld, 1916b, p. 51, Eq. (12a)).

---

[346] Sommerfeld knew about the perihelion shift from the corresponding astronomical problem. According to special relativity the perihelion shift, e.g., of the planet Mercury, would be 7″ of arc per century, while in general relativity Einstein obtained 43″ per century. In the case of the hydrogen atom the force of attraction is the Coulomb force; the gravitational forces may be neglected in comparison with it, hence general relativity does not give any sensible correction to the perihelion shift in the electron orbit in the hydrogen atom. (See also Sommerfeld, 1915c, pp. 467–468.)

It should be noted that for obtaining Eq. (140) Sommerfeld did not use the quantum condition (133′) but rather the condition (133).[347] The reason for his procedure was that in the other case the factor multiplying the term linear in α became such that it disagreed with the spectroscopic data. In his final paper for *Annalen der Physik* (Sommerfeld, 1916b), Sommerfeld excluded still another possibility. Since the quantity γ^2 had to be positive, it seemed to follow that the angular momentum p_ϕ assumed a minimum value, i.e., Ze^2/c. Hence, instead of Eq. (133), one might expect the quantum condition

$$2\pi\left(p_\phi - \frac{Ze^2}{c}\right) = nh \qquad (133'')$$

for the azimuthal degree of freedom. Such a condition also followed from Max Planck's quantum theory of systems having several degrees of freedom (Planck, 1916a, §11). Karl Schwarzschild commented on it in a letter to Sommerfeld: 'Since the lower limit of p_ϕ is not zero, but e^2Z/c, it seems to me that the discrete values of p_ϕ must be put equal to $e^2Z/c + (h/2\pi)n$ ($n = 1, 2, 3, \ldots$). This does not change the doublets, but spoils the Balmer formula' (Schwarzchild to Sommerfeld, 21 March 1916). Sommerfeld definitely wanted to keep the Balmer formula. The relativistic correction only had to determine the fine structure of each Balmer line. In that connection he added an historical remark:

> The quantity α, which will play an important role in all the following formulae, is also of some historical interest. Einstein remarked once [Einstein, 1909a] that the quantity e^2/c has the same dimension and order of magnitude as h, and thought of deriving h from this quantity. Later Jeans [in his report to the Birmingham Meeting of the British Association] thought that the accurate form of the numerical factor was given by the equation $h/2\pi = (4\pi e)^2/c$. Now we see that this numerical factor, namely our α above, becomes the fundamental quantity for describing the fine structure of spectral lines. (Sommerfeld, 1916b, pp. 51–52)

The consequences of the relativistic energy formula (140) were quite evident. An energy term having $(n + n')$ equal to 2 occurred as a doublet, due to the two possibilities of distributing 2 among n and n' (excluding $n = 0$, of course); a series term with quantum number 3 occurred as a triplet, etc. This multiplicity of terms

---

[347] Sommerfeld motivated this step in an addendum to his fine-structure paper, dated 10 February 1916, with the following words:

> The new *Ansatz* may also be understood perfectly well as a quantum *Ansatz* from the point of view of a coordinate system corotating with the rotation of the perihelion, by the introduction of which [i.e., of the new quantum *Ansatz*] the problem of the distribution of quanta in the relativistic case is reduced to the same problem in the non-relativistic case, as described in our paper I [Sommerfeld, 1915b]; thus any arbitrariness or uncertainty occurring in the *Ansatz* has been removed. (Sommerfeld, 1915c, p. 500)

By combining Eq. (133) with Eq. (136′) one found the result, $1 - \epsilon^2 = \gamma^2 n^2/(n' + \gamma n)^2$, rather than Eq. (137′), and this—when substituted in Eq. (139)—yielded Eq. (140). (Note that Sommerfeld, in his paper for the Bavarian Academy, defined a different parameter α, namely, $(\pi^2 e^2/hc)^2$, hence his expression for W, the relativistic hydrogen energy (Sommerfeld, 1916b, p. 56, Eq. 22) looks very different from our Eq. (140).)

would result in a rather large number of fine-structure components of the Balmer lines. Sommerfeld, however, reduced the number of possible transitions by additional conditions, stating that the quantum numbers of the initial state, m and m', always had to be larger than or equal to the corresponding ones, n and n', of the final state, that is,[348]

$$m \geqslant n \qquad \text{and} \qquad m' \geqslant n'. \tag{141}$$

Thus he obtained the following fine structure in the case of Balmer lines. The constant term (with $n + n' = 2$) possesses a doublet structure and causes a frequency separation $\Delta\nu_H = 2\pi^2 m_e e^4 \alpha c / 16 h^2$ (or 0.31 cm$^{-1} \cdot c$). Each doublet of the Balmer series is accompanied by satellites; i.e., each H_α-component by one, each H_β-component by two, each H_γ-component by three satellites, etc. 'Because of the large uncertainty of the H-lines and the small distances of these additional components, there is little prospect of observing them,' wrote Sommerfeld (1915c, p. 480).[349] Similarly the so-called Paschen series of hydrogen, whose constant term possesses the quantum number $n + n' = 3$, consists essentially of triplet lines with satellites, the triplets being separated by frequency differences $(8/81)\Delta\nu_H$ and $(8/27)\Delta\nu_H$, respectively. In the case of the corresponding series of ionized helium, the latter differences had to be multiplied by a factor of 16; Paschen's data on these helium 'triplets' indeed agreed perfectly with Sommerfeld's theoretical prediction. Also in the X-ray spectra of many-electron atoms the fine structure had to show up; the separation of the components there became very large due to the fact that the effective charge entering, say, the K-spectra was $(Z - 1)$ from Moseley's observations.[350]

Sommerfeld's quantization method of periodic systems with several degrees of freedom immediately obtained an important application. His former student, Paul Sophus Epstein, then interned as an enemy alien in Munich but in a position to work on theoretical problems, attacked with its help the calculation of the Stark effect of spectral lines.[351] His interest in this problem had been aroused

---

[348] Sommerfeld derived the relations (141) from the positivity of the phase integrals (Sommerfeld, 1915b, p. 448).

[349] In the case of H_α the satellites are separated from the components by the differences $(8/81)\Delta\nu_H$ and $(8/27)\Delta\nu_H$, respectively.

[350] Sommerfeld similarly discussed the L-spectra, where the effective charge is $Z - l$, where l is of the order of 8 (Sommerfeld, 1915c, pp. 496–497).

[351] P. S. Epstein was born in Warsaw on 20 March 1883. He studied at the University of Moscow from 1901 to 1906, where Petr Nikolaevich Lebedev was his teacher. Then he became assistant (1906–1909), docent (1909–1910) and assistant professor (1910–1911) at the Agricultural Institute in Moscow. In 1911 he went to the University of Munich to continue his studies in theoretical physics under Sommerfeld. He was particularly interested in relativity theory and published his first paper on this subject in *Annalen der Physik* (Epstein, 1911). In 1914 he obtained his doctorate with a thesis on an optical problem: the diffraction of light from a parabolic cylinder. After the outbreak of World War I he stayed on in Munich and worked for his *Habilitation*, which he received in 1919 at the University of Zurich. Two years later, Epstein went to the United States and became Professor of Physics at the California Institute of Technology. He retained this position until his retirement in 1953. He died on 9 February 1966 in Pasadena.

a couple of years earlier, when Stark had come to Munich to talk about his experimental discovery. Now, in early 1916, Epstein succeeded in integrating the equations of motion of an electron, moving in the Coulomb field of a nucleus of charge $Z|e|$ and an external homogeneous field of strength $|E|$, by using parabolic coordinates, ξ and η, in the plane defined by the direction of the electric field (Epstein chose the x-direction) and the radius vector joining the nucleus (at the origin) and the electron.[352] As a third variable of the problem, Epstein chose ψ, the angle between the above-mentioned (moving) plane and a fixed plane also going through the origin. In these variables, the equations of motion of the electron were separable; in particular, there followed two equations, one involving only the coordinate ξ and the canonically conjugate momentum p_ξ, and another involving η and p_η, which could be integrated to yield the equations

$$p_\xi = \sqrt{f_1(\xi)} \quad \text{and} \quad p_\eta = \sqrt{f_2(\eta)} \,, \tag{142}$$

$f_1(\xi)$ and $f_2(\eta)$ being functions of the third order in ξ^2 and η^2, respectively. Also the Hamiltonian function could be written as a sum of three separate terms, i.e.,

$$H = \sqrt{m_e}\, \alpha\psi + \int_0^\xi \sqrt{f_1(\xi)}\, d\xi + \int_0^\eta \sqrt{f_2(\eta)}\, d\eta, \tag{143}$$

involving the same functions $f_1(\xi)$ and $f_2(\eta)$ and a constant α connected with the angular rotation described by the angle ψ.[353] By quantizing the three degrees of freedom according to the phase-integral method with quantum numbers n_1, n_2 and n_3 for the ξ-, η-, and ψ-phase integrals, Epstein obtained for the energy W the expression

$$W = -\frac{2\pi^2 m_e e^4 Z^2}{h^2 (n_1 + n_2 + n_3)^2} - \frac{3h^2 |E|}{8\pi^2 m_e Z |e|} (n_1 - n_2)(n_1 + n_2 + n_3). \tag{144}$$

On the right-hand side he neglected terms of higher order in the electric field strength $|E|$. He identified the first term on the right-hand side of Eq. (144) with the energy of the atom in the absence of an external electric field.[354] The effect of the electric field on the spectral lines, therefore, yielded a separation of the frequencies, according to the formula

$$\Delta\nu = \frac{3h|E|}{8\pi^2 m_e Z |e|} \left[(m_1 - m_2)(m_1 + m_2 + m_3) - (n_1 - n_2)(n_1 + n_2 + n_3) \right], \tag{145}$$

---

[352] The parabolic coordinates are related to Cartesian coordinates x and y in this plane by the equations $x = \frac{1}{2}(\xi^2 - \eta^2)$ and $y = \xi\eta$.

[353] p_ψ, the momentum variable corresponding to ψ, is given by $\sqrt{m}\,\alpha$.

[354] The fact that three quantum numbers replace the two quantum numbers of the one-electron atom in the field-free case does not change its energy. Evidently, Epstein's sum, $n_1 + n_2 + n_3$, must be equal to Sommerfeld's, $n + n'$, in that case.

where the sets of triple numbers, (m_1, m_2, m_3) and (n_1, n_2, n_3), refer to the initial and final states of the one-electron atom, respectively. By substituting positive integral values for these triple numbers into Eq. (145) Epstein found that the Stark components of a hydrogen line lay symmetrically on both sides of the original frequency. He then used Sommerfeld's 'selection rule' ('*Auswahlprinzip*,' i.e., $n_1 \leqslant m_1$, $n_2 \leqslant m_2$, and $n_3 \leqslant m_3$), and arrived at a complete description of Stark's data. Finally, from the data he concluded the empirical rule that '*even-numbered differences* $m_3 - n_3$ *give rise to p-polarization* [i.e., polarization in the direction of the external field], *while odd-numbered to the s-polarization* [i.e., polarization perpendicular to the electric field]' (Epstein, 1916a, p. 148, footnote 3).

Epstein submitted a short note on his results on 26 March 1916 to *Physikalische Zeitschrift*, where it appeared in the issue of 15 April (Epstein, 1916a). He knew that he had to hurry because at the same time Karl Schwarzschild was working on the same problem. Hence he wanted to establish his priority, even before the detailed publication of the theoretical derivation was ready; it appeared finally in the issue of 25 July of *Annalen der Physik* (Epstein, 1916c).[355] The astrophysicist Schwarzschild had become interested in the problem two years earlier and had discussed the Stark effect from the point of view of classical theory (Schwarzschild, 1914a).[356] In it he obtained the correct order of magnitude

[355] Epstein later recalled his competition with Schwarzschild on the quantum theory of the Stark effect as follows:

> I started to work [on the theory of the Stark effect] . . . Now after some time again I saw Sommerfeld. It was already in the middle of the war. I had no key to the Institute anymore, being an alien, but Sommerfeld was interested in my access to literature, because I wrote an encyclopedia article [for the *Encyklopädie der mathematischen Wissenschaften*—Epstein had already written an article in 1915 on special diffraction problems in optics]. He was the editor of this volume [V]. So at one of the reunions, he told me, "I wrote Schwarzschild, that he should work on this article." Now I was a little crestfallen, because I regarded this as a stab in the back, since he knew that I was writing it. And Schwarzschild was a mathematician of unbelievable energy; he could do everything in a twinkling. I of course couldn't reproach him, but I decided, "Now I have no prospects unless Schwarzschild should go to heaven." And the next day when I was going to bed, I had the idea of the limit. You see, I knew already how the electron moves, and I knew how to do it. I got up at 5 o'clock the next morning and by 10 I had the formula. And then the same morning I brought it to Sommerfeld. And what do you know, the same afternoon he got a letter from Schwarzschild, and Schwarzschild had the wrong formula. It was the same order of magnitude, but didn't agree, on the positions of the lines. Sommerfeld wrote Schwarzschild, "This morning Epstein brought me the formula of the Stark effect, and this afternoon we got your letter. But Epstein's formula agrees with the observation." When Schwarzschild got [his result], he immediately announced in the Berlin Academy that he would speak about it. And he did, before he wrote the letter to us, so he reported it wrong in the Academy. By that time however I had already sent my announcement that came out just one day before he delivered that lecture in the Academy. But in the lecture he corrected it, and in the proof he removed all the discrepancies, and it came out correct also. Of course the two final papers came out much later. So I had the priority by one day. (Epstein, AHQP Interview, 25 May 1962, p. 11)

[356] Karl Schwarzschild was born on 11 October 1873 at Frankfurt-am-Main. He studied astronomy at the Universities of Strasbourg (1891–1893) and Munich (after 1893), obtaining his doctorate under Hugo von Seeliger in 1896. Three years later he became *Privatdozent* in Munich, and in 1902 Professor of Astronomy at the University of Göttingen. There he became interested in problems of astrophysics, especially the radiation emitted from the sun, and in electron theory. He continued both of these interests after his appointment as Director of the Potsdam Observatory in 1909. During the last year of his life he published two important papers on Einstein's theory of gravitation. Schwarzschild died on 11 May 1916 in Berlin. (See Schwarzschild's obituary notice by Sommerfeld, 1916a.)

of the separation of the components.[357] On 30 March 1916 he presented his paper, entitled '*Zur Quantenhypothese*,' at a session of the Prussian Academy; it was published on 11 May, the day he died. In it, Schwarzschild again studied the motion of an electron in an atom under the influence of an electric field, but in contrast to his previous paper he now used quantum theory explicitly, especially Planck's and Sommerfeld's quantum treatment of systems having several degrees of freedom.[358] For the 'conditionally periodic' systems ('*bedingt periodisch*,' Schwarzschild, 1916, p. 548) of f degrees of freedom the integration of the equations of motion led in general to solutions for the position coordinates, which depended only on f angle variables w_i (which increased linearly in the course of time, i.e., $w_i = 2\pi\omega_i t + $ const.) and f constants α_i.[359] The latter also represented the momenta, canonically conjugate to w_i, and were called action variables. The f pairs of action and angle variables thus described the system fully, and Sommerfeld's phase-integral conditions could be applied, yielding discrete values for the action variables,

$$\alpha_i = n_i \frac{h}{2\pi} + c_i \quad (i = 1, \ldots, f), \tag{146}$$

with the constant c_i possibly depending on the degree of freedom. In order to obtain the solution of the equation of motion of a quantum-theoretical multiply periodic system one had, therefore, first to introduce action-angle variables by a (classical) canonical transformation of the original variables, say with the help of the Hamilton–Jacobi differential equation. The energy function of the system could then be expressed by action variables alone, and through the quantum conditions the discrete energy values followed immediately. Schwarzschild used

---

[357] Schwarzschild found, in particular, the solutions of the classical equations of motion and expanded them into a Fourier sum; he identified the frequencies which occurred there with the frequencies of emission. Thus he arrived at the result that each spectral line of the free atom would split into a symmetrical triplet under the influence of a homogeneous electric field.

[358] Schwarzschild explained the fact that the earlier classical calculation yielded the correct order of magnitude for the Stark effect without any reference to the quantum hypothesis in the following way: the frequencies of the electrons in stationary orbits are of the order of magnitude of the limiting frequencies of the hydrogen series; now there existed a relation between these limiting frequencies and the quantum of action. Through that backdoor Planck's constant sneaked into Schwarzschild's classical result (Schwarzschild, 1914a, p. 24).

[359] Schwarzschild's approach to the atom as a multiply periodic system went back to Carl Gustav Jacob Jacobi (see Jacobi, 1866). Jacobi had solved the Kepler problem (of attraction of one body by the gravitational field of a second heavier body) by using the partial differential equation for the action function (the so-called Hamilton–Jacobi equation); especially, he had shown that two separate differential equations arose, each involving only one set of dynamical variables. Later his method of the separation of variables in the differential equation for the action function was extended by Paul Gustav Stäckel in his *Habilitation* thesis dealing with, as the astronomers said, 'conditionally periodic' systems of any finite degree of freedom (Stäckel, 1891). As suitable variables for the separation were the action-angle variables; they had been employed, for example, by the French astronomer Charles-Eugène Delaunay in his theory of the lunar motion (Delaunay, 1860, 1867). Henri Poincaré (1892, 1893, 1899) and Carl Wilhelm Ludwig Charlier (1902, 1907) used them extensively in their books on celestial mechanics.

his method to calculate the Stark effect in one-electron atoms and the motion of a free body rotating around its centre of gravity.[360]

While Epstein gave the first satisfactory solution of the Stark effect on the basis of the Bohr–Sommerfeld theory of atomic structure, Peter Debye, another former student and collaborator of Sommerfeld's, dealt with the influence of external homogeneous magnetic fields on the spectral lines.[361] In order to obtain his results on the Zeeman effect, which he first presented to the Göttingen Academy on 3 June 1916 (Debye, 1916a) and later, in September, also submitted to *Physikalische Zeitschrift* (Debye, 1916b), Debye used Schwarzschild's method of integrating conditionally periodic systems. In particular, he solved the Hamilton–Jacobi differential equation for the one-electron atom in the presence of a static magnetic field of strength $|H|$ in spherical coordinates, (the radius) r and (the angles) θ and ϕ, that is,

$$
\frac{\partial S}{\partial t} + \frac{m_e}{2}\left\{\left(\frac{\partial S}{\partial r}\right)^2 + \frac{1}{r^2}\left(\frac{\partial S}{\partial \theta}\right)^2 + \frac{1}{r^2\sin^2\theta}\left(\frac{\partial S}{\partial \theta}\right)^2\right\} - 2\pi\nu_L\frac{\partial S}{\partial \phi}
$$

$$
+ \frac{m_e}{2}(2\pi\nu_L)^2 r^2\sin^2\theta + \frac{e^2 Z}{r} = 0, \tag{147}
$$

where ν_L was given by the equation

$$
\nu_L = \frac{|e|\,|H|}{4\pi m_e c}, \tag{147a}
$$

and, was, therefore, equal to the classical Larmor frequency. In the approximation that the term quadratic in ν_L could be neglected, this equation yielded a solution for the action function S, in which S appeared as the sum of three integrals, R, Θ and Φ, plus a term proportional to the time; the integral R depended only on the radial variable r, the integral Θ only on the angle θ, and the integral Φ only on the azimuthal angle ϕ. These integrals had now to satisfy Sommerfeld's quantum conditions; i.e., if one extended the integrals R, Θ and Φ over a full period of the variables involved—the path of integration in R ranged from a minimum value r_{min} to a maximum value r_{max} and back to r_{min}, the integration path in Θ ranged from $\theta = \theta_0$ to $\theta = \pi - \theta_0$ with θ_0 a constant angle, and the integration path in Φ ranged from $\phi = 0$ to $\phi = 2\pi$—then each of them had to be equal to an integral multiple of Planck's constant h. Denoting their associated quantum numbers by m_1, m_2 and m_3, Debye found for the action

---

[360] In his calculation of the Stark effect Schwarzschild assumed that the atomic system exhibited a certain degree of degeneracy: he claimed, in particular, that the electrons in the presence of electric fields also described closed orbits; hence the motion was degenerate and he had to introduce special quantum conditions (similar to the ones Planck had used in late 1915) which took into account this degeneracy. In a further paper, submitted in August 1916, Epstein pointed out the essential equivalence of Schwarzschild's treatment to his own (Epstein, 1916e).

[361] Debye had already concerned himself earlier with Bohr's theory of atomic constitution and had used Bohr's model of the hydrogen molecule (Debye, 1915a).

integral the result

$$S = f(r,\theta,\phi) + \left[\frac{2\pi^2 m_e e^4 Z^2}{h^2} \cdot \frac{1}{(m_1 + m_2 + m_3)^2} + m_3 h\nu_L\right]t, \quad (148)$$

where $f(r,\theta,\phi)$ is a function of the position coordinates alone. Now the terms in the square bracket represent the energy of the system; hence the frequencies of the one-electron atom in the static homogeneous magnetic field became

$$\nu = \frac{2\pi^2 m_e e^4 Z^2}{h^3}\left[\frac{1}{(m_1 + m_2 + m_3)^2} - \frac{1}{(n_1 + n_2 + n_3)^2}\right] + (m_3 - n_3)\nu_L, \quad (149)$$

where the quantum numbers m_1, m_2, m_3 define the initial and n_1, n_2, n_3, the final state of the atom. Evidently, the first term had to be interpreted as the frequency of the free atom, and the second term as the shift due to the magnetic field. From the occurrence of the Larmor frequency ν_L in the result, Debye concluded that the normal Zeeman effect 'also occurs here with the right magnitude as a consequence of the general quantum hypothesis' and that the 'usual triplet arises if one admits for m_3 and n_3 only the smallest integral numbers' (Debye, 1916b, p. 511).

Debye also used Schwarzschild's approach to derive Sommerfeld's results concerning the relativistic fine structure (Debye, 1916c). At the same time, Sommerfeld independently concerned himself with the Hamilton–Jacobi differential equation for the one-electron inhomogeneous magnetic and electric fields and for the relativistic one-electron atom (Sommerfeld, 1916d). Like Debye, he now postulated the quantum conditions for the action function S; i.e., he required that the equations

$$\int \frac{\partial S}{\partial q_i}\, dq_i = n_i h \quad (i = 1, \ldots, f) \quad (150)$$

be satisfied for multiply periodic systems of f degrees of freedom. These were equivalent to the integrals, $\int p_i\, dq_i$, for the partial differential quotients, $\partial S/\partial q_i$, did represent the momentum variables p_i. Sommerfeld called the integrals, $\int (\partial S/\partial q_i)\, dq_i$, with paths extending over a full period of the system, the 'moduli of periodicity' of S, and formulated the quantum condition as: 'The moduli of periodicity of the Jacobi action function are integral multiples of Planck's quantum of action' (Sommerfeld, 1916d, p. 499). The solution of a quantum-theoretical multiply periodic system thus implied two steps: first, one had to introduce variables in which the action function consisted of a sum of terms, each depending only on *one* pair of canonically conjugate variables—in these variables, then, the problem was separable; second, one had to evaluate each term in the action function, especially to calculate their moduli of periodicity.

The evaluation of the angular phase integrals presented no difficulty; however, the radial phase integral possessed the general structure

$$\int \frac{\partial S}{\partial r} \, dr = \oint \sqrt{A + 2\frac{B}{r} + \frac{C}{r^2}} \, dr, \tag{151}$$

where the quantities A, B and C were independent of the radial variable.[362] Sommerfeld now carried out the integration, which went along a closed path from $r = r_{min}$ to $r = r_{max}$ and back (which was indicated by the closed circle) in the complex r-plane. There the integrand, the square-root function, possessed a cut on the real axis from r_{min} to r_{max} and two poles, one at $r = 0$ and the other at $r = \infty$. Then he replaced the original contour of integration by an equivalent one, i.e., two circles around the singularities at $r = 0$ and $r = \infty$ and applied the residue formula, obtaining[363]

$$\int \sqrt{A + 2\frac{B}{r} + \frac{C}{r^2}} \, dr = -2\pi i \left(\sqrt{C} - \frac{B}{\sqrt{A}} \right). \tag{152}$$

By adding this method of complex integration to Schwarzschild's extension of the Hamilton–Jacobi theory of conditionally periodic systems, Sommerfeld arrived at a powerful tool for dealing with atomic problems.[364] The Sommerfeld–Schwarzschild theory would dominate the calculations of atomic structure in years to come.

---

[362] The radial phase integrals of the one-electron atom (in the presence of a magnetic field) and of the relativistic one-electron atom did have the structure given in Eq. (151). In the case of the Stark effect the position coordinates were the parabolic ones, ξ and η; then the phase integrals, $\int p_\xi \, d\xi$ and $\int p_\eta \, d\eta$, could both be transformed to

$$\int \sqrt{A + \frac{2B}{x} + \frac{C}{x^2} + Dx} \, dx = \int \sqrt{A + \frac{2B}{x} + \frac{C}{x^2}} \, dx + \frac{1}{2} D \int \frac{x}{\sqrt{A + 2B/A + C/x^2}} \, dx,$$

where A, B, C and D are quantities independent of x. The evaluation of the second integral can be reduced to that of the first: if $\int \sqrt{A + B/x + C/x^2} \, dx$ becomes a function $J(x)$, then the second integral will be $(D/2)(dJ(x)/dx)$.

[363] The original path in the complex plane goes from r_{min} below the real axis to r_{max} and then from r_{max} above the real axis to r_{min}; thus it goes around the cut in the mathematically positive sense, i.e., counterclockwise. This path can be replaced by two circular paths enclosing the singular points $r = 0$ and $r = \infty$, where the paths around both singularities proceed in the mathematically negative sense. The residue theorem states that the contributions of the two integrals are $-2\pi i\sqrt{C}$ and $+2\pi i(B/\sqrt{A})$, respectively.

[364] Sommerfeld himself applied the method especially to the relativistic calculation of the Zeeman effect. He arrived at the result: '*The Zeeman effect is not influenced by the fine structure of the hydrogen lines following from the relativistic theory*' (Sommerfeld, 1916d, p. 502).

In spite of the fruitfulness of the theory of multiply periodic systems in dealing with the problems of atomic physics, a difficulty remained: the method worked only for those problems in whose dynamical description the variables associated with each degree of freedom could be separated. However, not all Hamiltonians occurring in atomic physics possessed the property of separability. Hence Einstein, in a contribution entitled '*Zum Quantensatz von Sommerfeld und Epstein*' ('On the Quantum Law of Sommerfeld and Epstein') and presented at the meeting of the German Physical Society in Berlin on 11 May 1917, proposed a more general method, which made it possible to treat nonseparable problems as well (Einstein, 1917b). In it he started from the observation that for an arbitrary system of f degrees of freedom only the sum, $\sum_i p_i\, dq_i$, where i runs over all the degrees of freedom, was an invariant quantity and might, therefore, be assigned the definite value of an integral multiple of Planck's constant. Still Einstein claimed that it was possible to derive several quantum conditions similar to the ones used by Sommerfeld, Schwarzschild, Debye and Epstein. For this purpose he considered integrals of the sum $\sum_i p_i\, dq_i$, extended over a closed path in the space of the position variable q_i. Einstein found that for any problem of atomic theory there existed a finite number of closed paths for which the integral $\oint \sum_i p_i\, dq_i$ assumed different—invariant—values; i.e., all integrals over closed paths assumed one or the other of these different values. He concluded: 'In this sense one may then prescribe a finite number of conditions, $\int \sum_i p_i\, dq_i = n_i h$, as quantum conditions' (Einstein, 1917b, p. 85). The number of quantum conditions, he argued further, did not exceed f, the number of degrees of freedom of the multiply periodic system under investigation. If it was smaller than f, the system was degenerate. A little later Paul Epstein showed that at least for degenerate systems Einstein's method did not always yield unambiguous results, i.e., the integrals $\int \sum_i p_i\, dq_i$ around closed paths depended then on the choice of p_i-variables (Epstein, 1917). In such cases additional considerations, provided by the so-called adiabatic hypothesis, had to be added in order to solve the problem. We shall discuss these developments in Section II.5.

II.5   Three Principles of Atomic Theory

Sommerfeld's work, which extended Bohr's theory of atomic structure, received enthusiastic appreciation from two experts. Thus the experimental spectroscopist Friedrich Paschen wrote to Sommerfeld: 'You will see that this paper has put spectroscopy on a new basis' (Paschen to Sommerfeld, 1 April 1916). And Niels Bohr, after receiving the reprint of Sommerfeld's paper, wrote: 'I thank you so much for your paper, which is so beautiful and interesting. I do not think that I have ever read anything which has given me so much pleasure' (Bohr to Sommerfeld, March 1916).[365] Bohr further reported in his letter: 'I myself have

---

[365] The letter from Bohr to Sommerfeld was translated into German by Harald Bohr and sent from Copenhagen to Munich in March 1916. (Paschen referred to it in his letter of 1 April 1916 to Sommerfeld, quoted above.)

worked this winter rather intensely on the quantum theory, and I have just completed a paper for publication, in which I have tried to show that it is possible to present the theory in a logically consistent manner, comprising all the diverse applications. In it I have made much use of Ehrenfest's idea concerning adiabatic transformations, an idea which seems to me to be very important and fundamental; I have also discussed a large number of phenomena, including dispersion.' His approach to dispersion, Bohr explained, was different from Debye and Sommerfeld's previous one insofar as it did not connect dispersion with the motion of electrons in stationary orbits. 'I write to you all this,' he continued, 'just to tell you how happy I was to obtain your paper before publishing my own. I immediately decided to delay the publication and to reconsider all this in the light of everything for which your paper has opened my eyes.' He expected to be able to finish the paper within a few weeks and just pointed out one result: 'It seems to me that my point of view throws some new light and supports, though only in a formal sense, your assumptions, as for instance, in your beautiful application of relativity theory.'[366] Finally, Bohr informed Sommerfeld about another piece of work: 'I also hope to send you soon a paper on the statistical applications of the quantum theory; I have worked rather intensively on it and have obtained, e.g., close agreement with the measurements on the specific heat of hydrogen' (Bohr to Sommerfeld, March 1916). However, time passed on and Bohr did not send any paper to Sommerfeld in 1916 and 1917.[367] In spring 1918 his paper was finally ready. Bohr had received several papers from Munich and he thanked Sommerfeld in a letter, dated 7 May 1918. 'I wanted to write to you since long,' he wrote, 'but I wanted at the same time to be able to report to you something about my own investigations.' Then he spoke about his past work on an extensive article summarizing the various aspects and applications of quantum theory. 'Because of numerous tasks and the great, uninterrupted progress in the theory it has been very difficult for me to complete the paper,' Bohr reported, 'and a short while ago I therefore decided to publish it in several parts' (Bohr to Sommerfeld, 7 May 1918). He announced that he had already dispatched to Sommerfeld a copy of the first part, in which he had presented certain 'new points of view' and that he would soon send the second part containing some applications.

---

[366] As Bohr mentioned earlier in his letter, the theory of fine structure agreed perfectly with the data which his Manchester colleague, Evan Jenkin Evans, had obtained recently (Evans and Croxson, 1916). It should be noted that Sommerfeld's application of relativity had also impressed Paschen. Walther Gerlach, then working in Tübingen, used to recall later repeatedly that Paschen, after seeing the complete explanation of the fine structure with the help of Sommerfeld's relativistic formula, exclaimed: 'Now I believe relativity theory!' In order to test the question even more carefully, Sommerfeld asked his doctoral student Karl Glitscher to compare the relativistic theory of the electron and Abraham's theory of the rigid electron (Abraham, 1902a, b) in connection with the fine structure of spectral lines. Glitscher succeeded in calculating the fine structure also with the help of Abraham's rigid electron and found that Paschen's data on one-electron atoms and the X-ray doublets definitely ruled out the rigid electron (Glitscher, 1917).

[367] Bohr worked hard on the revision of his article at that time, beginning it in Manchester and continuing it after his return to the University of Copenhagen as Professor of Theoretical Physics in fall 1916.

Bohr gave his series of papers on quantum theory the title 'On the Quantum Theory of Line Spectra.' He signed the introduction to the first part, bearing the subtitle 'On the General Theory,' in November 1917; it was published on 27 April 1918 (Bohr, 1918a).[368] In this introduction Bohr referred in detail to the reasons which had guided his new work. After mentioning his 1913 theory of the hydrogen atom and atomic spectra, he pointed out that this theory accounted only for singly periodic systems (i.e., periodic systems depending on one period and frequency); it neither described the differences between the hydrogen spectrum and the spectra of other elements, nor accounted for the effects of electric and magnetic fields on the hydrogen spectrum. He went on to report about the resolution of the difficulties accomplished by Sommerfeld's generalization of the theory to multiply periodic systems: the brilliant success in explaining the fine structure of spectral lines, Paul Epstein's and Karl Schwarzschild's complete description of the Stark effect, and Sommerfeld's and Peter Debye's treatment of the Zeeman effect. Finally, he stated:

> In spite of the great progress involved in these investigations many difficulties of fundamental nature remained unsolved, not only as regards the limited applicability of the methods used in calculating the frequencies of the spectrum of a given system, but especially as regards the question of the polarisation and intensity of the emitted spectral lines. These difficulties are intimately connected with the radical departure from the ordinary ideas of mechanics and electrodynamics involved in the main principles of the quantum theory, and with the fact that it has not been possible hitherto to replace these ideas by others forming an equally consistent and developed structure. Also in this respect, however, great progress has recently been obtained by the work of Einstein [1916d; 1917a] and Ehrenfest [1914, 1916]. On this state of the theory it might therefore be of interest to make an attempt to discuss the different applications from a uniform point of view, and especially to consider the underlying assumptions in their relations to ordinary mechanics and electrodynamics. Such an attempt has been made in the present paper, and it will be shown that it seems possible to throw some light on the outstanding difficulties by trying to trace the analogy between the quantum theory and the ordinary theory of radiation as closely as possible. (Bohr, 1918a, p. 4)

In his letter to Sommerfeld Bohr had written that in his work on quantum theory an important role was played by the adiabatic hypothesis, which Paul Ehrenfest had developed in several papers during the previous years. He made a prominent reference to it again in his published paper:

> As you will see [he wrote in a letter to Ehrenfest, to whom he mailed a copy of part I] the considerations are to a large extent based on your important principle of

---

[368] According to the scheme outlined in the introduction to his first paper, Bohr planned to publish his considerations in four parts: Part I should contain a brief discussion of the general principles and an application to systems with one degree of freedom, Part II a detailed treatment of the hydrogen spectrum, Part III a discussion of the questions arising in connection with the spectra of the other atoms, and Part IV a general discussion of the constitution of atoms. Of these four parts only the first three appeared in print: Part II was ready to print on 30 December 1918 (Bohr, 1918b), and Part III finally went into print on 30 November 1922 (Bohr, 1922d).

"adiabatic invariance." As far as I understand, however, I consider the problem from a point of view which differs somewhat from yours, and I have therefore not made use of the same terminology as in your original papers. In my opinion the condition of the continuous transformability of motion in the stationary states may be considered as a direct consequence of the necessary stability of these states, and in my eyes the main problem consists therefore in the justification of the application of ordinary "mechanics" in calculating the effect of a continuous transformation of the system. As it appears to me it is hardly possible to base this justification entirely on thermodynamical considerations, but it seems naturally suggested from the agreement with experiments obtained by calculating the motion in the stationary states themselves by means of ordinary mechanics. (Bohr to Ehrenfest, 18 May 1918)

In order to understand what the adiabatic hypothesis meant and how it achieved such an important role in the theory of atomic structure, let us look at its origin and the man who developed it.

Paul Ehrenfest was born on 18 January 1880 in Vienna, the youngest of four sons of a merchant who came from a small Jewish village in Moravia.[369] After attending the *Gymnasium* he enrolled, in October 1899, in the *Technische Hochschule* of Vienna to study mechanics, mineralogy, mathematics and chemistry. At the same time he attended courses at the University of Vienna as, for instance, Ludwig Boltzmann's on the kinetic theory of heat.[370] In 1901 Ehrenfest went to the University of Göttingen, attracted mainly by the young experimentalist Johannes Stark. He took many courses; besides Stark's, he attended the lectures of Felix Klein and David Hilbert, as well as those of Woldemar Voigt, Max Abraham, Walther Nernst, Karl Schwarzschild and Ernest Zermelo. He became acquainted with Walther Ritz, then a doctoral candidate, and Tatyana Alexeyevna Afanassjewa, who would become his wife in 1904. Before returning to Vienna he visited the University of Leyden, together with Ritz, to attend a course of Hendrik Lorentz'—on blackbody radiation; he also looked around Kamerlingh Onnes' laboratory. After receiving his doctorate in 1904 with a thesis, supervised by Boltzmann, on the motion of rigid bodies in fluids within the mechanics of Heinrich Hertz (Ehrenfest, 1904), Ehrenfest began to work on various theoretical problems, especially those of statistical mechanics. He also investigated the problem of heat radiation in several papers (Ehrenfest, 1905; 1906a, b, c). Following Boltzmann's death in September 1906, he spent some time at the University of Göttingen and then went with his Russian wife to St. Petersburg. Though he did not succeed in obtaining a proper position there, he lived among many young scientists, such as Abram Fedorovitch Joffé and the mathematician Alexander Friedmann. After years of intensive work he completed, assisted by Tatyana, the article on '*Begriffliche Grundlagen der statistichen*

---

[369] Details of Paul Ehrenfest's life and scientific background may be found in Martin Klein's biography of him (M. J. Klein, 1970b).

[370] At the University of Vienna, Ehrenfest's fellow student from the *Gymnasium*, Gustav Herglotz, studied mathematics. Ehrenfest became friends with Herglotz and his colleagues, Hans Hahn and Heinrich Tietze, all of whom became reputed mathematicians.

Auffassung in der Mechanik' ('The Conceptual Foundations of the Statistical Approach in Mechanics'), which Felix Klein had requested for the *Encyklopädie der mathematischen Wissenschaften* (P. and T. Ehrenfest, 1912). Early in 1912 he travelled to Germany and Austria to seek an academic position.[371] Later in May he received two letters: one from Sommerfeld offering him the position of a *Privatdozent* at the University of Munich; the other came from Lorentz, who had decided to retire from his Leyden chair and become Curator of Teyler's *Stichting* (Foundation) at Haarlem. In his letter, dated 13 May, Lorentz indicated the possibility that Ehrenfest might become his successor. This possibility materialized; in October 1912 the Ehrenfests moved to the Netherlands, and on 4 December Paul delivered his inaugural lecture on the difficulties, which he called the 'crisis' of the ether theory of light ('*Zur Krise der Lichtaether Hypothese,*' Ehrenfest, 1913a). Ehrenfest quickly established himself at Leyden: besides giving lecture courses at the University and at Kamerlingh Onnes' laboratory (on Nernst's heat theorem), he organized a weekly physics colloquium already during his first semester; he revived the *Christiaan Huygens Society* (which met once every two weeks), and he got a reading room for the students, among the first of whom were Dirk Struik and Hendrik Kramers.[372]

In Leyden Ehrenfest's active research interest turned more and more towards problems of quantum theory.[373] On 23 December 1912 he reported in a letter to Lorentz an idea which had occurred to him in connection with the quantum hypothesis. He made the observation: 'For infinitely slow compression of a volume filled with radiation and bounded by mirrored walls, the following quantity remains constant for all vibrational modes of the volume: E/ν (E the energy, and ν the frequency [of the radiation]). We may also write $\delta'(E/\nu) = 0$ (where δ' is an "adiabatic, reversible" variation).' Ehrenfest then asked the question: 'If one passes now from sinusoidal oscillations to any other periodic motions, which quantity (replacing E/ν) will then remain constant under the influence of an "adiabatic reversible" action?' (Ehrenfest to Lorentz, 23 December 1912). The answer seemed to be contained in a theorem going back to Ludwig Boltzmann and Rudolf Clausius, stating that the average kinetic energy of a system does not change under an adiabatic action.[374] Later this theorem had played a role in the theory of so-called monocyclic systems, the mechanical systems invented by Hermann von Helmholtz in order to represent thermody-

---

[371] Ehrenfest left on 19 January 1912 and visited Berlin (to see Planck), Leipzig (to see Herglotz), Munich (to see Sommerfeld), Vienna (to see Philipp Frank and Fritz Hasenöhrl), Brünn (to see Tietze), Prague (to see Einstein), and finally Czernowitz (to see Hahn).

[372] Ehrenfest modelled the reading room after Felix Klein's mathematical reading room in Göttingen. The son of the late Johannes Bosscha donated the necessary funds for the '*Lesekamer Bosscha.*'

[373] After 1906 Ehrenfest had published only one long paper on quantum theory, an analysis of the quantum aspects in the theory of heat radiation (Ehrenfest, 1911).

[374] The theorem was indicated in Boltzmann's first mechanical proof of the second law (Boltzmann, 1866, Section IV), and it returned in the papers of Clausius on the same subject (Clausius, 1871) and of C. Szily (1872). For the origin of the adiabatic principle, see Martin Klein's biography of Paul Ehrenfest (M. J. Klein, 1970b, Chapter 11, pp. 264–292).

namic behaviour. In 1902 Lord Rayleigh had pointed out in the discussion of radiation that, if one studied waves contained in a slowly contracting cavity (which was considered an 'adiabatic' process in the sense of thermodynamic changes), the ratio of energy over frequency remained constant (Rayleigh, 1902).[375] Nearly a decade later, in 1911, Ehrenfest had used Rayleigh's result to derive Wilhelm Wien's displacement law (Ehrenfest, 1911). Towards the end of 1912 he finally generalized it into a theorem applicable to any system of f degrees of freedom, which depends on r parameters, $a_1, \ldots, a_r$, and is periodic for all values of the variables and the parameters. This theorem stated: the average kinetic energy of the system increases or decreases in the same portion as the frequency under an adiabatic change (see the letters from Ehrenfest to Lorentz, 23 December 1912, and Ehrenfest to Joffé, 20 February 1913, quoted in M. J. Klein, 1970b, p. 216, Fig. 10, also pp. 261–263). Ehrenfest used his new theorem right away for the purpose of obtaining the quantization of the energy of the rotator in a paper submitted in May 1913 (Ehrenfest, 1913b). Since the motion of a one-dimensional oscillating system can be transformed adiabatically into a rotating motion, its kinetic energy must be quantized in integral multiples of $h\nu/2$, ν being the frequency of rotation (Ehrenfest, 1913b, p. 453, and footnote 2 on pp. 453–454).[376] Ehrenfest published a detailed account of his ideas later that year; on 29 November 1913, Hendrik Lorentz communicated a paper by his successor, entitled 'A Mechanical Theorem of Boltzmann and its Relation to the Theory of Energy Quanta,' to the Amsterdam Academy (Ehrenfest, 1913c). In this paper Ehrenfest stated what he called the 'adiabatic relation' as

$$\frac{\overline{T}_A}{\nu_A} = \frac{\overline{T}_B}{\nu_B}, \tag{153}$$

where $\overline{T}_A$ and $\overline{T}_B$ denote the average kinetic energies and ν_A and ν_B the frequencies of systems A and B. Systems A and B are related by adiabatic transformations; i.e., they may be obtained from one another by an infinitesimally slow change of one or several parameters.

Hardly any of the quantum physicists paid any attention to Ehrenfest's idea—except Albert Einstein. In a paper dealing with certain problems of quantum theory he became interested in the question whether Boltzmann's relation between the entropy and the number of states could also be used in the

---

[375] Lord Rayleigh used this result in particular for the derivation of the Stefan–Boltzmann law.

[376] There is an interesting relation of Ehrenfest's argument to the one raised previously at the first Solvay Conference. At the end of the discussion of Einstein's report, Lorentz raised the question as to what would happen in the case of a quantum pendulum whose length was shortened. Einstein replied: 'If one changes the length of the pendulum infinitely slowly and continuously, then the energy of the oscillation remains equal to $h\nu$, provided it has been $h\nu$ in the beginning; the energy of the oscillation changes as [the frequency] ν. The same is true of an electric oscillatory circuit possessing no resistance and of the free radiation' (Einstein in Eucken, 1914, p. 364). Starting from this conclusion Einstein, together with Otto Stern, had proposed to quantize the energy of the rotator in a way different from Ehrenfest's: they proposed to take integral multiples, $nh\nu$, or—with zero-point energy—half-integral multiples, $(n + \frac{1}{2})h\nu$ (Einstein and Stern, 1913).

case that the system under consideration depended on a parameter (say the volume) which changes in the course of time. He noted: 'This question cannot be answered without a specific hypothesis. The most natural hypothesis which presents itself here is Ehrenfest's adiabatic hypothesis, which may be stated thus: For a reversible change of λ [the quantity under consideration], each quantum-theoretically possible state goes over into another such state' (Einstein, 1914b, p. 826). With the adiabatic hypothesis, then, the entropy relation remained valid. The main interest of physicists, however, turned to an apparently different aspect of the quantum theory: the theory of atomic and molecular structure. In summer 1913, before publishing the paper in which he proclaimed the kinetic energy of a periodic system as an 'adiabatic invariant,' Ehrenfest came to know about Bohr's theory of the hydrogen atom. 'Bohr's work on the quantum theory of the Balmer formula (in the *Philosophical Magazine*) has driven me to despair,' he commented. 'If this is the way to reach the goal, I must give up doing physics' (Ehrenfest to Lorentz, 25 August 1913). In any case, he had no intention of working in the direction of Bohr's work; he rather pursued his own ideas, which seemed to lead him far away from it.[377] However, after Sommerfeld had published his papers on the extension of Bohr's theory in 1916, Ehrenfest recognized the connection between Sommerfeld's work and the 'adiabatic hypothesis,' and he wrote to Sommerfeld: 'Even though I consider it horrible that this success will help the preliminary, but still completely monstrous Bohr model on to new triumphs, I nevertheless heartily wish physics at Munich further success along this path' (Ehrenfest to Sommerfeld, May 1916; translated in M. J. Klein, 1970b, p. 286). In the same letter, Ehrenfest also pointed out the connection between Sommerfeld's work and his ideas on the adiabatic invariance. Sommerfeld replied on 30 May 1916 that he wished to know more about the adiabatic hypothesis, and mentioned that Bohr had praised it already (in Bohr to Sommerfeld, March 1916, quoted above). On receipt of this letter Ehrenfest prepared an extended paper, entitled 'On Adiabatic Changes in Connection with the Quantum Theory,' which Lorentz communicated to the Amsterdam Academy on 24 June 1916; it was published in *Proceedings of the Amsterdam Academy* and, with minor modifications, also in *Philosophical Magazine* and *Annalen der Physik* (Ehrenfest, 1916).

Ehrenfest stated the main goal of his research right away in the introduction of his paper. He wrote: 'In an increasing number of physical problems the foundations of classical mechanics (and electrodynamics) are used together with the quantum hypothesis, which is in contradiction with them. It remains of

---

[377] Ehrenfest did continue to contribute to quantum theory. Thus in 1914 he published two papers on this subject: one on the application of Boltzmann's relation of entropy to probability (Ehrenfest, 1914); the other, together with Heike Kamerlingh Onnes on a simplified deduction of Planck's radiation law from combinatorial arguments (Ehrenfest and Kamerlingh Onnes, 1914). In the former paper he dealt with the statistical weights of quantum systems, establishing a condition—the 'δG-condition'—which was related to the previous theorem on adiabatic invariants; it stated that the statistical weights are adiabatic invariants (Ehrenfest, 1914, p. 660). During 1915 Ehrenfest did not publish any paper on quantum theory.

course desirable to come here to some general point of view from which each time the limit between the "classic" and the "quantum" region may be drawn' (Ehrenfest, 1916, p. 576). In order to proceed with this plan he used the adiabatic hypothesis for quantum systems in the form presented by Einstein, namely: 'If a system is exposed to adiabatic influences then the "admissible" motions are transformed into [other] "admissible" one[s]' (Ehrenfest, 1916, p. 577). This formulation not only included his previous applications to periodic systems (of one degree of freedom), but also the extensions of the quantum hypothesis to conditionally periodic systems having several degrees of freedom—as proposed by Planck, Sommerfeld, Epstein, Schwarzschild and Debye (Ehrenfest did not refer to Ishiwara and Wilson)—provided the adiabatic hypothesis was defined in those cases, too. Thus he stated: '*For a general set of parameter values $a_1, a_2, \ldots$* [on which the conditionally periodic system depends, besides its dynamic variables] *only those motions are possible that are adiabatically related with motions possible for special values $a_{10}, a_{20}, \ldots$* (that is, which can pass into those by a reversible change)' (Ehrenfest, 1916, p. 579). With this hypothesis he then proved the existence of certain adiabatic invariants, i.e., of quantities which remain unchanged under infinitesimally slow changes. Such adiabatic invariants were the ratio $2\overline{T}/\nu$ for periodic and E/ν for harmonic motion, with $\overline{T}$ the average kinetic energy, E the total energy and ν the frequencies of the systems under investigation.[378] These adiabatic invariants had to be identified with Planck's phase-space integral or Sommerfeld's phase integral (for one degree of freedom) and, therefore, assumed—due to quantum hypothesis—values that were integral multiples of Planck's constant. Also in the case of Sommerfeld's hydrogen atom with two degrees of freedom he concluded that both phase integrals, the azimuthal and the radial ones, were adiabatic invariants. Only in the case of the relativistic hydrogen atom was Ehrenfest not certain and stated: 'In order that we might make a conclusion from the viewpoint of the adiabatic hypothesis, it would have to be investigated, which quantities are adiabatically invariant in this case' (Ehrenfest, 1916, p. 587). He also noticed a difficulty in the example 'when the vibrations in an *anisotropic* field of force are changed in an adiabatic reversible way into those of an *isotropic* field' (Ehrenfest, 1916, p. 589), i.e., when the system in the course of motion becomes degenerate.

Ehrenfest's paper on the adiabatic invariants did not impress Sommerfeld too much. His cautious presentation of all the difficulties involved in the problem—indeed his entire style—was not to Sommerfeld's liking. After Bohr's recommendation of the adiabatic hypothesis, he had expected that it would also be of great practical help in actual problems. Now all that Ehrenfest had to say in this respect was that his (Sommerfeld's) treatment of the nonrelativistic hydrogen atom could be justified. But already in the case of the relativistic hydrogen atom

---

[378] In the case of the Zeeman effect, Ehrenfest claimed that the angular momentum of the undisturbed electron plus the angular momentum of the Larmor precession represented an adiabatic invariant; in the case of the Stark effect it was the angular momentum around an axis parallel to the direction of the field that was an adiabatic invariant (Ehrenfest, 1916, p. 580, footnote 1).

and in the cases of Stark and Zeeman effects the question arose whether adiabatic invariants existed at all. However, Ehrenfest left the detailed investigation of these problems to his student Johannes Martinus Burgers.[379] Burgers, in a series of papers, concluded that the phase integrals were indeed adiabatic invariants for all mechanical systems, say of f degrees of freedom, having the following properties: first, each momentum p_k can be expressed as a function of the canonically conjugate position variable q_k, and of f integration constants of the Hamiltonian equations, as well as the parameters a (which can be changed infinitely slowly); second, each coordinate q_k describes a librational motion (i.e., all q_k vary between two fixed limits); third, there are no rational relations connecting the periods of the different degrees of freedom (Burgers, 1916a, b; 1917; 1918). In other words, Burgers proved that for separable (first requirement), conditionally periodic (second requirement), and nondegenerate (third requirement) systems, the phase integrals will remain constant under infinitely slow changes of the parameters. This result satisfied Sommerfeld completely.[380]

Besides Ehrenfest's adiabatic hypothesis, Bohr mentioned in his memoir of 1918 two recent papers of Einstein dealing with the emission and absorption of radiation (Einstein, 1916d; 1917a), as the contributions which had brought about 'great progress' in quantum theory. Einstein had attempted in these papers to provide a derivation of Planck's radiation law based on 'assumptions that were free of contradictions' ('*widerspruchsfreie Voraussetzungen*,' Einstein, 1916d, p. 318). For that purpose he used certain aspects of Bohr's theory of spectra and considered the interaction of radiation and matter. Now matter consists of atoms or molecules which, according to Bohr, possess stationary states, and their interaction with radiation results in the emission or absorption of certain frequencies and, associated with it, the transitions from one stationary state to another.

---

[379] J. M. Burgers was born on 13 January 1895 at Arnhem, the Netherlands. From 1914 to 1918 he studied at the University of Leyden; he was assistant to Ehrenfest from 1916 to 1917. After obtaining his doctorate in 1918 (the first under Ehrenfest) he became a professor at the Technical University at Delft, where he stayed until 1955. Then he went to the University of Maryland as Research Professor in the Institute for Fluid Dynamics and Applied Mathematics.

[380] After the publication of Ehrenfest's paper on the adiabatic hypothesis Sommerfeld wrote a letter to him, in which one can detect some disappointment about the results (Sommerfeld to Ehrenfest, 16 November 1916). In the same letter Sommerfeld also remarked what a pity it was that Walther Ritz had died already and could not participate in the new development. Sommerfeld thought that Ritz' imagination and optimism might have helped considerably in solving practical problems. Martin Klein, in his biography of Ehrenfest, remarked that Sommerfeld's reference to Ritz, whose approach to physical problems was so strikingly different from Ehrenfest's, 'must have had a crushing effect on Paul Ehrenfest' (M. J. Klein, 1970b, p. 291). Quoting a passage from the third (1922) edition of *Atombau und Spektrallinien*, Klein concluded that Sommerfeld's attitude towards the adiabatic hypothesis was completely negative. Perhaps it might be more appropriate, in order to illustrate this point, to quote from the first edition of Sommerfeld's *Atombau und Spektrallinien*, where he had expressed a rather positive opinion. After presenting the content of the adiabatic hypothesis, he wrote:

> The general trust in [the validity of] the principle rests, however, in the agreement between its consequences and the demands of experience. In particular, we must emphasize that, according to an investigation of J. M. Burgers [1916a], it leads for "conditionally-periodic systems" to the same quantum orbits and quantum conditions as the method of the separation of variables ... and it is therefore justified for all appropriate problems, which can be compared with experience, such as the ones which have been treated [successfully] by the separation method. (Sommerfeld, 1919, p. 433)

Concerning one type of such transmissions, which is connected with the emission of radiation, Einstein made the following fundamental assumption: 'This transition will occur without external influences. One can hardly consider it to be of any other nature than radioactive transitions' (Einstein, 1916d, p. 321).

The consistent ('*widerspruchsfreie*') derivation of Planck's radiation law had been one of Einstein's central problems, going back to 1906. After 1909 (when, in his analysis of the radiation law from the point of view of fluctuations, he had discovered two terms characterizing corpuscular and undulatory features of radiation), he had tried to explore new methods, which involved two kinds of arguments: on one hand, he had exploited thermodynamical arguments; on the other, he made use of the assumption that certain processes obeyed laws similar to that of radioactive decay. The law of radioactive decay stated that the number of atoms actually decayed per unit time is proportional to the number of all atoms N that can decay, i.e.,

$$\frac{dN}{dt} = -\lambda N, \tag{154}$$

where λ is a constant whose inverse is related to the lifetime of the radioactive atoms (λ^{-1} is the time in which the radioactive substance decays to its eth part, e being the base of natural logarithms). In order to derive Eq. (154), Egon von Schweidler had assumed that each radioactive atom possessed the probability $\lambda\,dt$ of decaying in a small time interval dt, and that this probability did not depend on N, the number of radioactive atoms (Schweidler, 1905). On the basis of this assumption he had predicted statistical deviations of the number of observed decayed atoms from the law (154), which had been confirmed experimentally.[381]

The discovery of the basically statistical behaviour of radioactive decays had impressed Einstein deeply, especially since further investigations on the fluctuation of the intensity of γ-rays emitted from radioactive substances had lent some support to his conception of light-quanta.[382] Thus he applied the statistical description to other atomic and molecular processes as well. For instance, in his thermodynamic derivation of the photochemical equivalence law in early 1912 he had assumed that the dissociation of molecules (which are hit by incident radiation) proceeds in accordance with the statistical law of radioactive decay (Einstein, 1912a). Indeed, already a year earlier he had written to his friend Michele Besso in connection with the interaction between γ-rays and matter: 'The process of absorption (ionization accompanied by a large release of energy) really does have similarity with a radioactive process' (Einstein to Besso, 10 September 1911). In spite of the fact that Einstein increasingly became involved

[381] For substances with large decay times the average deviation from Z, the number of decayed atoms according to Eq. (154), becomes $\sqrt{Z}$. Deviations of such magnitude had indeed been measured, e.g., by Edgar Meyer and Richard Regener (1908) and Hans Geiger (1908).

[382] In a paper on the nature of γ-rays, Edgar Meyer had concluded that the extension of a γ-ray perpendicular to the direction of propagation was small, thus supporting the light-quantum hypothesis (Meyer, 1910, p. 662).

in the task of generalizing relativity theory to incorporate gravitation, he did not entirely abandon thinking about quantum problems in the following years.[383] In April 1914 he moved from Zurich to Berlin to assume the special position which had been created for him at the Prussian Academy and the *Kaiser-Wilhelm-Gesellschaft*. Although, in his inaugural address to the Prussian Academy on 2 July, he dealt exclusively with the problem of relativity theory (Einstein, 1914a), and this problem took most of his working time well into the year 1916 (Einstein, 1914c, 1915b, d, e, f, g; 1916c), he did not entirely forget quantum theory. Thus he spoke on 24 July 1914 at the German Physical Society in Berlin on some quantum-theoretical questions, especially a thermodynamical derivation of Planck's law and Nernst's heat theorem (Einstein, 1914b). In approaching these problems, Einstein studied the interaction between radiation and matter, assuming that the molecules contained resonators of frequency ν, which could absorb only discrete energies, $nh\nu$, with $n = 0, 1, 2, \ldots$, and h Planck's constant. Again he described certain chemical and quantum processes in the molecules in analogy with radioactive decay processes.[384] During the following year he even entered into an experimental investigation connected with the quantum hypothesis: in experiments, carried out together with Wander Johannes de Haas at the *Physikalisch-Technische-Reichsanstalt*, Berlin, he sought to determine whether Ampère's molecular currents existed (Einstein and de Haas, 1915; Einstein, 1915a, 1916b). The proof which they obtained for the existence of such currents seemed to imply, in Einstein's opinion, the existence of zero-point energy. In 1916, after the completion of general relativity theory, Einstein returned to his problems of quantum theory. Thus he reported in May to Besso: 'I discovered a nice simplification of the thermodynamical derivation of the photochemical $h\nu$-law, just in the manner of Van't Hoff' (Einstein to Besso, 14 May 1916). And three months later he announced proudly, again to his friend Besso: 'A splendid flash came to me concerning the absorption and emission of radiation . . . . A surprisingly simple derivation of Planck's formula, I would say *the* derivation. Everything completely quantum' (Einstein to Besso, 11 August 1916).

Einstein presented his new derivation of Planck's radiation law first on 21 July 1916 at a meeting of the German Physical Society in Berlin (Einstein, 1916d) and he spoke about it again at a later meeting on 27 October. (See *Verh. d. Deutsch. Phys. Ges.* **18**, p. 367 (1916).)[385] His main goal was to obtain Planck's relation between the radiation density, ρ_ν, and the average energy of the resonator, $\overline{U}_\nu$,

---

[383] In the midst of his collaboration with Marcel Grossmann on the mathematical structure of generalized relativity theory, Einstein (towards the end of 1913) wrote: 'At the moment I am labouring on the quantum problem, though not really with great hope of succeeding' (Einstein to Besso, undated; see *Einstein–Besso Correspondance*, p. 50).

[384] Einstein noted explicitly that the concepts of physical and chemical changes of a molecule become identical. As he remarked: 'The quantum-like change of the physical state of a molecule does not seem to be fundamentally different from a chemical change of the latter' (Einstein, 1914b, p. 823).

[385] Einstein's paper containing the material he presented in October did not appear in the *Verhandlungen der Deutschen Physikalischen Gesellschaft*, but first in the *Mitteilungen der Physikalischen Gesellschaft, Zürich*, in honour of Alfred Kleiner, who had died on 3 July 1916. Kleiner, in 1905, had accepted Einstein's thesis for the doctorate.

i.e.,

$$\overline{U}_\nu = \frac{\rho_\nu c^3}{8\pi\nu^2},\tag{155}$$

from purely quantum-theoretical assumptions concerning the interaction between matter, notably atoms or molecules, and radiation, without referring to classical electrodynamics. As we have already mentioned, Einstein made use of one of the fundamental assumptions of Bohr's theory of atomic structure, i.e., that the atoms exist in stationary states having discrete energy values ϵ_n, $n = 0, 1, 2, \ldots$.[386] Then, due to Boltzmann's principle, the probability for an atom to be in the stationary state with index n was given by the relation

$$W_n = g_n \exp\left(-\frac{\epsilon_n}{kT}\right),\tag{156}$$

g_n being the statistical weight of the state n and k the Boltzmann constant. Now the atoms in a stationary state denoted by n may pass into a stationary state m of higher energy ($\epsilon_m > \epsilon_n$) by the absorption of a light-quantum of frequency ν_{nm}, gaining the energy $\epsilon_m - \epsilon_n$. Inversely, atoms passing from the higher state m to the lower state n will emit a light-quantum of frequency ν_{mn}, losing the energy $\epsilon_m - \epsilon_n$. Einstein assumed that a transition from the higher to the lower state connected with the emission of radiation can occur without any external influence. About this process he said: 'One can hardly think of it in any other way except as a radioactive reaction' (Einstein, 1916d, p. 321). The number of such transitions per unit time thus should be proportional to N_m, the number of atoms in the state m, with a proportionality constant A_m^n. He further assumed that in the presence of radiation of the characteristic frequency ν_{mn} and density ρ_ν two processes may happen: first, transitions from the lower state n to the state m through absorption of radiation, their number being equal to $B_n^m N_n \rho_\nu$, (N_n being the number of atoms in state n, and B_n^m a proportionality constant characterizing the *absorption* of radiation of frequency ν_{mn}); second, transitions from the state m to the state n, their number being given by $B_m^n N_m \rho_\nu$ (N_m being the number of atoms in state m, and B_m^n a proportionality constant characterizing the *emission* of radiation of frequency ν_{mn} in the presence of external radiation of density ρ_ν and the same frequency). In thermal equilibrium as many transitions should occur from the state n to m as vice versa. Hence the sum of the two terms, $A_m^n N_m$ and $B_m^n N_m \rho_\nu$, must be equal to the term $B_n^m N_n \rho_\nu$. By substituting for the ratio N_n/N_m at a given temperature T the expression following from Eq. (156), Einstein obtained the equation

$$A_m^n g_m = \rho_\nu \left[B_n^m g_n \exp\left(\frac{\epsilon_m - \epsilon_n}{kT}\right) - B_m^n g_m \right].\tag{157}$$

---

[386] One could perhaps speculate that Einstein already had the idea of stationary states before Bohr. (See the letter of de Hevesy to Bohr, 23 September 1913, quoted in Section II.4.)

This equation described a relation between the density of radiation of frequency ν_{mn} and the (absolute) temperature T. In order to eliminate the specific factors characterizing the particular transitions, Einstein was guided by the classical limiting case ('*den Grenzfall der klassischen Theorie*,' Einstein, 1916d, p. 320). The classical limit demanded, however, that in the case of infinite temperature also the density for a given, finite frequency, should be infinite. This condition yielded the equation

$$B_n^m g_n = B_m^n g_m.$$ (158)

With its help, Eq. (157) could be written as

$$\rho_\nu = \frac{A_m^n / B_m^n}{\exp((\epsilon_m - \epsilon_n)/kT) - 1}.$$ (159)

This equation possessed, as Einstein recognized immediately, the structure of Planck's law. Then he used Wien's displacement law to conclude that the ratio A_m^n / B_m^n must be proportional to the third power of the frequency, and the energy difference, $\epsilon_m - \epsilon_n$, proportional to the frequency ν.[387] Einstein concluded his derivation of Planck's law with the remarks:

> I admit freely, of course, that the three hypotheses concerning outgoing and incoming radiation do not at all become substantiated results by the [mere] fact that they lead to Planck's radiation formula. But the simplicity of the hypotheses, the generality with which the consideration can be carried through easily, as well as the natural connection [of the consideration employed] with the limiting case of Planck's linear oscillator (in the sense of classical electrodynamics and mechanics), persuade me to regard it as very probable that [all] this constitutes the fundamental outline of the future theoretical derivation. In favour of the theory [presented here] is also the fact that the statistical law assumed for the emission process is none other than Rutherford's law of radioactive decay, and that the result expressed by [the equation, $h\nu = \epsilon_m - \epsilon_n$] together with Eq. [(159)] is identical with the second hypothesis in Bohr's theory of [line] spectra. (Einstein, 1916d, p. 322)[388]

In the second paper, presented in October 1916, Einstein investigated a specific aspect of his previous derivation of Planck's law (Einstein, 1916e, 1917a). He asked the question as to how the momentum of the atom changes during the processes of absorption and stimulated emission ('*Einstrahlung*') and spontaneous emission ('*Ausstrahlung*') of radiation. According to the kinetic theory of matter the momentum of the molecule at temperature T must exhibit fluctua-

---

[387] Wien's displacement law demanded that A_m^n / B_m^n be const. ν^3; the constant could be found from the classical limit, i.e., the Rayleigh–Jeans' law, as $8\pi h / c^3$.

[388] In the final Section 3, Einstein added some remarks on the photochemical equivalence law, which had been at the beginning of his new considerations. (See Einstein to Besso, 14 May 1916, quoted earlier.)

tions described by the formula

$$\frac{\overline{\Delta^2}}{\tau} = 2RkT, \tag{160}$$

where Δ is the momentum transfer to the molecule during the short time interval, τ, k Boltzmann's constant and R a proportionality factor related to the friction acting against the motion of the atoms (molecules). In the case that only a few atoms exist in the cavity, one can neglect the interaction between them. Then both sides of Eq. (160) have to be explained by the interaction of atoms and radiation, including a transfer of momentum from the radiation to the atoms and vice versa. Now Einstein applied the light-quantum hypothesis, according to which any increase of the energy of the atom, say by the amount $\epsilon_m - \epsilon_n$, must be accompanied by the momentum transfer $(\epsilon_m - \epsilon_n)/c$, and every loss of the same energy (whether caused by emission without any influence or under the influence of the presence of radiation) is accompanied by a loss of momentum, $-(\epsilon_m - \epsilon_n)/c$. In calculating both $\overline{\Delta^2}$ and R with the help of this assumption, Einstein found that the fluctuation equation (160) was satisfied when the radiation density was given by Planck's law, Eq. (159). He concluded:

> The most important result seems to me to be the momenta transferred to the molecule [atom] in the processes of absorption and emission. If any of our assumptions concerning the transferred momenta were changed, then Eq. [(160)] would be violated. It hardly seems possible to reach agreement with this relation, which is demanded by the [kinetic] theory of heat, in any other way than on the basis of our assumption. (Einstein, 1917a, p. 127)

To Besso he summarized his results as follows:

> The essential thing, however, is that the statistical consideration, which leads to Planck's formula, has become uniform and as general as one can imagine, implying no other assumption concerning the special nature of mediating molecules [atoms] than the most general quantum hypothesis. Thereby the result follows that for any elementary energy transfer between radiation and matter, the amount of momentum $h\nu/c$ goes over to the molecule [atom]. Hence it follows that every such quantum process is a completely directed event. And, as a consequence, the existence of light-quanta is as good as assured. (Einstein to Besso, 6 September 1916)[389]

Ehrenfest's adiabatic hypothesis and Einstein's statistical treatment of the transitions in atoms entered as fundamental ingredients in Bohr's new memoir on the quantum theory of line spectra (Bohr, 1918a). Bohr had originally attempted in 1915 to extend and generalize the successful theory of the hydrogen atom to

---

[389] Besides the light-quanta, Einstein also showed some interest in other aspects of the quantum theory, for example, in the theory of multiply periodic systems. (See the end of Section II.4.)

systems having a more complex structure. Since by 1918 such generalizations had already been proposed, the task remained to see whether the methods of Planck, Sommerfeld, Epstein, Schwarzschild and Debye were based on consistent principles. In this investigation Bohr put special emphasis on close relations between the methods of the quantum and the classical theories. It was here that he expected to receive crucial help from the work of Ehrenfest and Einstein.

In his new approach to the quantum theory of atomic structure, Bohr retained of course the accepted results of his 1913 theory. Thus he took over the two basic assumptions: namely, (i) that *atomic systems can, and can only, exist permanently in certain series of states . . . denoted as the "stationary states" of the system,'* and (ii) *'that the radiation absorbed or emitted during a transition between two states is "unifrequentic"'* (Bohr, 1918a, p. 5). The frequency ν of this 'unifrequentic' radiation was given by the differences of the energies, W' and W'', of the stationary states involved, i.e.,

$$h\nu = W' - W'', \tag{161}$$

where h denoted Planck's constant. Bohr knew that with these assumptions *'the ordinary laws of electrodynamics cannot be applied* to these states without radical alterations.' On the other hand, however,

> In many cases, . . . the effect of that part of the electrodynamical forces which is connected with the emission of radiation will at any moment be very small in comparison with the effect of the simple electrostatic attractions or repulsions of the charged particles corresponding to Coulomb's law. Even if the theory of radiation must be completely altered, it is therefore a natural assumption that it is possible in such cases to obtain a close approximation in the description of the motion in the stationary states, by retaining only the latter forces. In the following we shall therefore, as in all the papers mentioned in the introduction, for the present *calculate the motions of the particles in the stationary states as the motions of mass-points according to ordinary mechanics* including the modifications claimed by the theory of relativity, and we shall later in the discussion of the special applications come back to the question of the degree of approximation which may be obtained in this way. (Bohr, 1918a, p. 6)

These basic principles had also been used by Sommerfeld and others in generalizing Bohr's theory; indeed, for Sommerfeld the fact that ordinary mechanics continued to remain valid had been especially important. However, in 1918 Bohr emphasized that both classical mechanics and electrodynamics broke down completely if one attempted to describe the transitions between stationary states. The classical theories did not even work in the case of temperature radiation, where they accounted only for the limiting region of low frequencies. Now Einstein had succeeded in deriving Planck's law on the basis of the above two assumptions of Bohr's theory of atomic structure 'by introducing certain supplementary assumptions about the *probability of transitions of a system between two stationary states* and about the manner in which this probability depends on the density of radiation of the corresponding space' (Bohr, 1918a, p. 6). These assumptions of Einstein's were taken over by Bohr into the quantum theory of line spectra.[390]

In his presentation of the general theory of atomic structure Bohr incorporated Ehrenfest's adiabatic hypothesis, which he preferred to call 'the principle of "mechanical transformability"' (Bohr, 1918a, p. 8).[391] He claimed that it will be 'of great importance in the discussion of the conditions to be used to fix the stationary states of an atomic system among the continuous multitude of mechanically possible motions' (Bohr, 1918a, pp. 8–9). The task of fixing the stationary states of an atomic system involved two aspects. One was the solution of the practical problem of calculating the energy of the stationary states, including their statistical weights.[392] The other argument concerned the resolution of a fundamental difficulty involved in the definition of the energy difference and, therefore, also of the frequency of transition between two states of an atomic system. 'In fact,' said Bohr, 'we have assumed that the direct transition between two such states cannot be described by ordinary mechanics, while on the other hand, we possess no means of defining an energy difference between two states if there exists no possibility of a continuous mechanical connection between them' (Bohr, 1918a, p. 9). With the help of Ehrenfest's principle, however, the difficulty could be bypassed. Especially, it enabled one to transform two stationary states of one system into two stationary states of another system, whose energy difference was arbitrarily small.

In dealing with the transformation of mechanical systems, Ehrenfest's principle encountered a limitation. Thus Bohr recognized: 'It is clear that, when by a slow transformation of a conditionally periodic system we approach a degenerate system of this kind, the time-interval which the orbit takes to pass close to any possible configuration will tend to be very long and will become infinite when the degenerate system is reached' (Bohr, 1918a, p. 22). As a result, Bohr concluded that Sommerfeld's phase integrals will not represent 'adiabatic invariants' in the case of degenerate systems.[393] Hence, if a system of f degrees of freedom is periodic, i.e., it has one characteristic period θ, then instead of the phase integrals separately only the sum,

$$I = \int_0^\theta (p_1\dot{q}_1 + p_2\dot{q}_2 + \cdots + p_f\dot{q}_f) = nh, \qquad (162)$$

---

[390] As we have mentioned earlier, Bohr had himself thought about using statistical assumptions in quantum theory; he had done so already before Einstein's derivation of Planck's law from statistical assumptions in 1916.

[391] Bohr argued that 'the above notation might in a more direct way indicate the content of the principle and the limits of its applicability' (Bohr, 1918a, p. 8, footnote 1).

[392] Ehrenfest had shown that if the statistical weights of the stationary states of one system were known, then the statistical weights of all other systems obtained from the first one by an adiabatic change were also known; the statistical weights of states, which could be mechanically transformed into each other, just became identical.

[393] In the case of a degenerate system 'the quantum conditions involve an ambiguity since in general for such systems there will exist an infinite number of different sets of coordinates which allow a separation of the variables and which lead to different motions in the stationary states, when these conditions are applied' (Bohr, 1918a, p. 20).

becomes a well-defined constant of motion. Evidently, this motion possesses only one periodic degree of freedom and the energy depends only on one quantum number n. As an example, let us consider Sommerfeld's hydrogen atom neglecting relativistic effects; although the motion extends over a plane, one degree of freedom is degenerate and the energy is a function of the sum of the azimuthal and the radial quantum numbers. In principle the motion of an electron may even possess three degrees of freedom, but only the relativistic atom under the action of an electric or magnetic field represents a nondegenerate system. Bohr knew the free one-electron atom, in which the electron moves in a planar (Kepler) orbit, could never be mechanically transformed into an atom on which there acts an external magnetic field. 'The frequency of a slow variation of the orbit [i.e., the frequency of the Larmor precession impressed upon the orbit by the magnetic field] will be seen to be proportional to the intensity of the external field,' he argued, 'and it is therefore obviously impossible to establish the external field at a rate so slow that the comparative change of its intensity during a period of variation is small' (Bohr, 1918a, p. 23). He rather expected that, 'during the establishment of the field, *the system will in general adjust itself in some unmechanical way* until a stationary state is reached in which the frequency (or frequencies) of the above-mentioned slow variation of the orbit [the Larmor precession] has a relation to the additional energy of the system due to the presence of the external field' (Bohr, 1918a, pp. 23–24).[394] It meant that the energy of the hydrogen atom under the influence of a magnetic or an electric field could not be related by ordinary mechanical elements to the energy of the free atom, in contradiction to the successful calculations of the Stark effect and the (normal) Zeeman effect by Epstein, Sommerfeld and Debye.

Bohr discovered, however, a possibility of escaping from this fundamental difficulty by applying a new argument. Especially, he discussed the relation between the theory of spectra of atomic systems and the classical theory of radiation. In the case of a periodic system of one degree of freedom the quantum-theoretical approach is based on the frequency condition, Eq. (162), and the quantum condition, i.e.,

$$I = \int_0^\theta p\dot{q}\, dt = nh, \tag{163}$$

where n denotes the quantum number and h Planck's constant. In any stationary state the displacement of the charged particle may be expressed by means of a

---

[394] This property of the nontransformability of nondegenerate into degenerate systems caused certain difficulties in determining the statistical weights of degenerate systems. Bohr, therefore, proposed to assume that while the *a priori* probability will be the same for all states of a nondegenerate system the statistical weights of a degenerate system may be found in the following way: one considered a small perturbation removing the degeneracy and then counted the number of states falling in the limit of zero perturbation into one state of the degenerate system (see Bohr, 1918a, pp. 26–27). The justification for this assumption was implied in the analogy between the classical and the quantum-theoretical description of atomic systems, which will be discussed below.

Fourier series,

$$x = \sum_{\tau} C_{\tau} \cos 2\pi (\tau \omega t + c_{\tau}), \tag{164}$$

where C_{τ} and c_{τ} are constants and the τ-sum extends over all positive integral values, $1, 2, 3$, etc. According to the classical theory of radiation this system emits a spectrum consisting of a series of lines having discrete frequencies $\tau\omega$, i.e., the ground frequency ω and its higher harmonics. According to the quantum theory, however, the frequencies are rather defined by Eq. (161). Evidently, Bohr could not expect to find a simple relation between classical and quantum frequencies. However, he noticed that in the limit of very large quantum numbers the situation was very different, especially when he considered the transition frequencies between neighboring orbits, say n' and n'' ($\tau = n' - n'' \ll n', n''$). Then the term difference on the right-hand side of Eq. (161) turned into $(I' - I'')\omega$, because the frequencies of revolution in the initial and final states were practically the same (namely, ω), and on substituting for I' and I'' the values $n'h$ and $n''h$ Bohr obtained the equation

$$\nu = \frac{\omega}{h}(I' - I'') = (n' - n'')\omega \tag{165}$$

for the quantum frequencies. 'As far as the frequencies are concerned,' he commented, 'we see therefore that in the limit where n is large there exists a close relation between the ordinary theory of radiation and the theory of spectra based on Eqs. [(161)] and [(163)]' (Bohr, 1918a, p. 15). Having achieved this connection between the frequencies, Bohr went further and claimed that 'a relation, as that just proved for the frequencies, will, in the limit of large n, hold also for the intensities of the different lines in the spectra' (Bohr, 1918a, p. 15). Then, since in classical theory the intensities of radiation having the frequency $\tau\omega$ were directly determined by (the absolute) squares of the Fourier coefficients C_{τ}, Bohr concluded: 'We must therefore expect that for large values of n these coefficients will on the quantum theory determine the *probability of spontaneous transition* from a given stationary state for which $n = n'$ to a neighbouring state for which $n - n'' = n' - \tau$' (Bohr, 1918a, p. 15)

The fact that the quantum-theoretical results passed over, in the limit of high quantum numbers, into the classical ones had been noticed by Bohr already in his first paper on atomic structure (Bohr, 1913b). In Section 3 of that paper he had justified his quantization of the kinetic energy (i.e., that it assumed values that were half-integral multiples of $h\omega$, ω being the frequency of revolution) by considering the ratio of ω and the frequency of transition between neighbouring states having high quantum numbers n and $n - 1$, respectively. He had noted then: 'According to the ordinary electrodynamics we should therefore expect the ratio between the frequency of radiation and the frequency of revolution is also very nearly equal to 1' (Bohr, 1913b, p. 13). He had even considered this relation further by studying also the frequencies associated with transitions from high quantum numbers n to states with quantum numbers $n - \tau$, where τ assumed small integral values, and found these values to be $\tau\omega$ in the case of the hydrogen atom. 'The possibility of an emission of a radiation of such a frequency,' he had remarked, 'may also be interpreted from analogy with the ordinary electrodynamics, as an electron rotating round a nucleus in an elliptical orbit will emit a radiation which according to Fourier's theorem can be resolved into homogeneous components, the frequencies of which are $\tau\omega$, if ω is the frequency of revolution of the electron' (Bohr, 1913b, p. 14).

Bohr's analogy relation between a certain high quantum number limit of the quantum-theoretical frequencies and the classical frequencies may be regarded as being connected with certain earlier observations. Thus Max Planck had stated in his *Vorlesungen über Wärmestrahlung*: 'If one were allowed to assume the quantum of action h as being infinitely small, then one would obtain the [classical] law of uniform energy distribution' (Planck, 1906, p. 178). Another way of turning quantum-theoretical relations into the corresponding classical ones had been noticed in discussing the radiation law: if one took the distribution law for large values of the ratio T/ν, where T is the absolute temperature and ν the frequency under consideration, then Planck's formula passed into the classical law (Einstein, 1905b, p. 136). Later, people required that correct quantum-theoretical results should approach classical results in the limit of high temperature T; for example, Albert Einstein and Otto Stern had argued in favour of Planck's second quantum hypothesis, according to which the one-dimensional resonator obtained the average energy

$$\overline{E}_\nu = \frac{h\nu}{\exp(h\nu/kT) - 1} + \frac{h\nu}{2} \tag{166}$$

(instead of the first hypothesis, according to which $\overline{E}_\nu$ was smaller by $h\nu/2$), for this expression became, in the limit of infinite T, equal to kT (Einstein and Stern, 1913, p. 551). The high temperature implied, of course, that each resonator had an energy $\overline{E}_\nu$ consisting of a large number of energy-quanta $h\nu$. Hence, in principle, the classical laws could be found from the quantum-theoretical ones by taking the limit of large quantum numbers.[395] Still neither Planck nor Einstein had been really motivated to look more deeply into the connection between the classical theory and a certain limit of the quantum theory. Bohr had been the first to emphasize it when he spoke, in his letter of 21 March 1913 to Rutherford, about 'the most beautiful analogy between the old electrodynamics and the considerations used in my paper [i.e., Bohr, 1913b].'[396]

---

[395] Such a conclusion may be drawn from Einstein's analysis of the blackbody radiation law in 1906: there Einstein argued that Planck's use of (classical) electrodynamics 'would indeed be plausible if in all parts of the spectrum, which can be observed, $\epsilon = (R/N)\beta\nu$ [$= h\nu$] were small compared to the average energy of a resonator' (Einstein, 1906c, p. 203).

[396] Bohr discovered this analogy—which led, as we said earlier, to a further justification of his quantum *Ansatz* for the kinetic energy of the hydrogen atom—after he had completed the manuscript of his first paper on atomic structure and had sent it (on 6 March 1913) to Rutherford. In the letter of 21 March, quoted above, he proposed 'some small alterations and additions,' among them the 'beautiful analogy.'

After 1913 Bohr slowly developed further the idea of the analogy between the classical theory and the quantum theory. He noticed that the phase-space considerations of Planck and others supported his views. The main point in these considerations was also the fact that in the limit of high quantum numbers the quantum-theoretical description became identical with the classical one; for instance, the closed curves representing the motion of a quantum system in phase space and the continuous classical distribution of phase-space points led to the same results. A similar situation occurred in dealing with aperiodic motions: thus Planck could replace the quantum-theoretical expression for the probability of a monatomic gas by the classical expression in case that only the phase cells denoted by high quantum numbers were occupied (Planck, 1916b). In his paper of 1916, which he withdrew from publication after having seen Sommerfeld's contributions to the Bavarian Academy (Sommerfeld, 1915b, c), Bohr again referred to an analogy argument concerning the frequency: he especially stated that in the case of the harmonic oscillator would the frequency condition, Eq. (161), allow in principle also for the higher harmonics, but the analogy to classical electrodynamics would exclude them; however, they would occur in the case of the anharmonic oscillator. 'But this is just what we would expect from analogy with the ordinary theories of radiation,' Bohr explained, 'since the motion of a periodic system which is not a harmonic vibrator can be resolved in harmonic terms corresponding with frequencies which are entire multiples of the frequency of revolution' (see Bohr, *Collected Works*, Vol. 3, 1976, p. 5).[397] However, a very important point that had still to be brought into the analogy considerations stemmed from Einstein's statistical treatment of blackbody radiation: Einstein had, in particular, calculated the quantum-theoretical intensity of the transitions and he had also connected it, in the limit of high temperatures, with the classical intensities obtained from the law of Rayleigh and Jeans (Einstein, 1917a, p. 124).

In the first part on the quantum theory of line spectra, Niels Bohr introduced a crucial step going beyond any previous analogy arguments. After discussing the relation between the classical coefficients C_τ of the Fourier series, Eq. (164), and the intensities of the quantum-theoretical radiation emitted from a periodic system (of one degree of freedom) in the limit of high quantum numbers, he noted:

> Now, this connection between the amplitudes of the different harmonic vibrations into which the motion can be resolved, characterised by different values of τ, and the probabilities of transition from a given stationary state to the different neighbouring stationary states, characterised by different values of $n' - n''$, may clearly be expected to be of a general nature. Although, of course, we cannot without a detailed theory of the mechanism of transition obtain an exact calculation of the latter probabilities, unless n is large, we may expect that also for small values of n the amplitude of the harmonic vibrations corresponding to a given value of τ

---

[397] Bohr repeated the same argument in an early draft of the quantum theory of line spectra (see Bohr, *Collected Works*, Vol. 3, 1976, pp. 48 52). There is also quoted some empirical evidence in favour of it derived from the observation of infrared spectra of diatomic gases (Kemble, 1916b).

will in some way give a measure for the probability of a transition between two states for which $n' - n''$ is equal to τ. Thus in general there will be a certain probability of an atomic system in a stationary state to pass spontaneously to any other state of smaller energy, but if for all motions of a given system the coefficients C in [(164)] are zero for certain values of τ, we are led to expect that no transition will be possible, for which $n' - n''$ is equal to one of these values. (Bohr, 1918a, p. 16)

Evidently, the extension of the analogy 'in some way' to small quantum numbers represented a most daring assumption. Still Bohr could refer for support to some of the earlier work on quantum theory. Especially, in the considerations on the quantum-theoretical phase space of Planck and others certain assumptions had been mentioned concerning the definition of equal *a priori* probabilities; these assumptions did indeed extend a situation that existed in the limit of high quantum numbers to the regions of low quantum numbers. Bohr, therefore, felt confident that an extension of his relations for frequencies and intensities, which were valid in the high quantum number limit, might eventually be justified also when low quantum numbers were involved.

Bohr did not encounter any difficulty in generalizing the analogy approach to conditionally periodic systems of several degrees of freedom. For this purpose the position coordinates of the charged particles in the atoms and molecules had to be expressed as a multiple Fourier sum, say for the x-coordinate of a system of f degrees of freedom,

$$x = \sum_{\tau_1, \ldots, \tau_f} C_{\tau_1, \ldots, \tau_f} \cos 2\pi \left[(\tau_1\omega_1 + \tau_2\omega_2 + \cdots + \tau_f\omega_f)t + c_{\tau_1, \ldots, \tau_f} \right], \quad (164')$$

where the coefficients $C_{\tau_1, \ldots, \tau_f}$ and the phase constants $c_{\tau_1, \ldots, \tau_f}$ depend on f integral numbers τ_k and f (mean) frequencies of oscillation, ω_k, $k = 1, \ldots, f$. (The multiple sum in Eq. (164') extends over the positive and negative integral values of the τ_i.) The absolute squares of the coefficients, $C_{\tau_1, \ldots, \tau_f}$, then determined the classical intensities of the emitted radiation, and they should be the classical analogues of the intensities associated with the quantum frequencies ν,

$$\nu = \frac{1}{h} \left[W(n'_1, n'_2, \ldots, n'_f) - W(n''_1, n''_2, \ldots, n''_f) \right], \quad (165')$$

where $W(n'_1, n'_2, \ldots, n'_f)$ and $W(n''_1, n''_2, \ldots, n''_f)$ denoted the initial and final stationary states characterized by the quantum numbers n'_k and n''_k, $k = 1, \ldots, f$, respectively. Evidently, the quantum frequencies ν, given by Eq. (165'), differed sharply for low quantum numbers from the classical ones; they approached the latter only in the limit of high quantum numbers, when the right-hand side of Eq. (165') could be replaced by $(1/h)\sum \omega_k(I'_k - I''_k)$—$I'_i$ and I''_i being the values of the phase integrals for the ith degree of freedom $(k = 1, \ldots, f)$—which sum then became equal to the classical frequency $\sum \tau_k \omega_k$ if one put $\tau_k = n'_k - n''_k$.

The reference to the analogous situation in classical theory removed many of the problems on which Bohr hit in dealing with the quantum theory of line spectra and which he treated in detail in the second part of his memoir (Bohr, 1918b). One of these was the question of how to calculate in quantum theory the energy of the stationary states of an atom under the influence of an external electric or magnetic field. We have mentioned earlier that the theory of mechanical transformation (or Ehrenfest's adiabatic hypothesis) had run into certain difficulties in that case, because the atom under the influence of an external field had to be treated with more quantum conditions than the free atom; that is, the free atom possessed fewer degrees of freedom and a new degree of freedom could not be introduced by an adiabatic transformation. Now, in classical theory the perturbation method was applied: that is, the external field perturbs the orbit of the electrons in an atom, and this perturbation gives rise to an energy correction, which is small if the intensity of the perturbing field is small. Bohr noticed that in quantum theory the situation could also be described by the perturbation method, and he stated:

> We were led to assume from the general formal relation between the quantum theory of line spectra and the ordinary theory of radiation, that it is possible to obtain information about the stationary states of the perturbed system from a direct consideration of the slow variations which the periodic orbit undergoes as a consequence of the mechanical effect of the external field on the motion. Thus, if these variations are of periodic or conditionally periodic type, we may expect that, in the presence of the external field, the values for the additional energy of the system in the stationary states are related to the small frequency or frequencies of the perturbations, in a manner analogous to the relation between energy and frequency in the stationary states of an ordinary periodic or conditionally periodic system. (Bohr, 1918b, p. 42)

This assumption then led him to justify the previous treatments of the Stark and the Zeeman effects of the hydrogen atom as well as Sommerfeld's theory of fine structure.

Finally, the detailed application of the analogy argument provided the key to another problem: the intensities and polarization of the components arising in the Zeeman and Stark effects. Bohr summarized his conclusions as follows:

> Thus we shall expect that there will exist an intimate connection between the probability of the spontaneous transition between two stationary states of the perturbed system, for which $n = n'$, $n_k = n'_k$ [$k = 1, \ldots, f - 1$] and $n = n''$, $n_k = n''_k$ respectively, and the values in these states of the coefficient $C_{\tau, t_1, \ldots, t_{f-1}}$ in the expressions for the displacements of the particles, for which $\tau = n' - n''$ and $t_k = n'_k - n''_k$. If for instance, for a certain set of values of τ and $t_1, \ldots, t_{s-1}$, the coefficient $C_{\tau, t_1 \ldots t_{f-1}}$ in the expressions for the displacements in every direction will be equal to zero for all motions of the perturbed system, we shall expect that the corresponding transitions between two stationary states will be impossible in the presence of the given external field; and if this coefficient is zero for the displace-

ment of the particles in a certain direction only, we shall expect that the corresponding transitions will give rise to the emission of a radiation which is polarised in a
plane perpendicular to this direction. (Bohr, 1918b, p. 59)

In the case of the Stark effects these arguments could be used at once and
yielded definite results, especially when connected with another analogy argument which followed from the conservation of angular momentum.

Bohr already had formulated this argument in the first part of his memoir
(Bohr, 1918a). There he had considered a periodic system possessing an axis of
symmetry, say the z-axis, and he had concluded that in transitions, in which the
angular momentum around the z-axis, $n_z(h/2\pi)$ (n_z being the quantum number
of the phase integral connected with the z-coordinate), does not change, linearly
polarized (i.e., polarized parallel to the z-axis) radiation will be emitted, while in
transitions in which the angular momentum around the z-direction changes by a
unit, i.e., $\pm h/2\pi$, circularly polarized radiation will come out. He had then
remarked:

> Now it is easily seen that the ratio between this amount of angular momentum and
> the amount of energy $h\nu$ emitted during the transition is just equal to the ratio
> between the amount of angular momentum and energy possessed by the radiation
> which according to ordinary electrodynamics would be emitted by an electron
> rotating in a circular orbit in a central field of force. In fact, if a is the radius of the
> orbit, ν the frequency of revolution and F the force of reaction due to the
> electromagnetic field of the radiation, the amount of energy and of angular
> momentum round an axis through the centre of the field perpendicular to the plane
> of the orbit, lost by the electron in unit of time as a consequence of the radiation,
> would be equal to $2\pi\nu aF$ and aF respectively. Due to the principles of conservation
> of energy and of angular momentum holding in ordinary electrodynamics, we
> should therefore expect that the ratio between the energy and the angular momen-
> tum of the emitted radiation would be $2\pi\nu$, but this is seen to be equal to the ratio
> between the energy $h\nu$ and the angular momentum $h/2\pi$ lost by the system
> considered above during a transition for which we have assumed that the radiation
> is circularly polarised. This agreement would seem not only to support the validity
> of the above considerations but also to offer a direct support . . . of the assumption
> that, *for an atomic system possessing an axis of symmetry, the total angular momentum
> round this axis is equal to an entire multiple of $h/2\pi$.* (Bohr, 1918a, pp. 34–35)

The analogy arguments thus established seemed to be in accord with the
known empirical results on hydrogen lines and they also agreed with the
theoretical results obtained independently in Munich. 'Thank you very much for
your paper [Bohr, 1918a] and your kind letter [Bohr to Sommerfeld, 7 May
1918],' wrote Sommerfeld to Bohr on 18 May 1918. 'Your paper was awaited
eagerly since long and was immediately studied by everybody. But the joy of Dr.
Rubinowicz (he is my assistant, but only until next month; after 15 June he will
return home—to Czernowitz—, where he is an assistant at the physical institute)
was a little mixed about your paper.'

Adalbert Rubinowicz had spent two years, from 1916 to 1918, at the University of Munich.[398] He had witnessed the development of Sommerfeld's extension of Bohr's theory and had then tried to contribute to it himself. Thus he chose to investigate the following question: 'Since, according to classical electrodynamics, the conservation law of [angular] momentum should be valid for the system— atom plus the electromagnetic field emitted in electronic transition—, the question arises how far this very fundamental theorem, whose importance extends far beyond classical electrodynamics, is also satisfied in the process of emission in Bohr's theory' (Rubinowicz, 1918a, p. 441). He then attempted to establish the hypothesis that 'both the atom and the propagating radiation are quantized, each by itself, and that an "ether oscillator" can absorb only one energy-quantum' (Rubinowicz, 1918a, p. 442). For this purpose he calculated the angular momentum of radiation emitted from an electron as it moves in a Kepler orbit around the nucleus according to classical theory, obtaining an expression involving the energy radiated away, ΔW, and the parameters of the orbit; then he substituted for ΔW the quantum-theoretical expression $h\nu$, with ν the frequency of the emitted radiation and h Planck's constant, and for the parameters of the initial orbit (i.e., the orbit taken by the electron before the emission of radiation) those obtained from Bohr's hydrogen calculation. Thus he obtained an inequality for the difference of the azimuthal quantum numbers, $n' - n''$, denoting the initial and the final state, respectively, i.e.,

$$|n' - n''| \leqslant 1. \tag{167}$$

This inequality implied the following selection rule ('*Auswahlprinzip*'): 'A transition of the electron from the stationary orbit [in an atom or molecule], which is connected with the emission of radiation, cannot occur in an arbitrary way, but is always connected with the condition that the azimuthal quantum number determining the angular momentum changes at most by one unit' (Rubinowicz, 1918a, p. 444). By studying the polarization of the emitted radiation he concluded further that in cases of changes (± 1) of the azimuthal quantum number the polarization will be circular, and in the case of no change at all it will be linear.

In the second part of his work, Rubinowicz applied his selection rule (167) to explain the observed fine structure of hydrogen lines (Rubinowicz, 1918b). He found, in particular, that each doublet line should consist—in contrast to Sommerfeld's original theory—only of three lines. While the new selection rule

---

[398] Adalbert (or Wojciech, according to the Polish version of his name) Rubinowicz was born on 22 February 1889 at Sadágora, Bukowina. From 1908 to 1914 he studied at the University of Czernowitz, receiving his doctorate in 1914. After his return from Munich he became a *Privatdozent* in Czernowitz in 1918, and two years later professor at the University of Lubljana, Yugoslavia. In 1922 he was appointed professor at the Polytechnic Institute of Lwów, from there he moved to the University of Lwów (1937–1941). After World War II he finally became Professor of Theoretical Physics at the University of Warsaw (1946–1960). He died in Warsaw on 13 October 1974.

seemed to work well for the data of hydrogen, in the case of the He^+-ion there existed more lines than could be accounted for by Eq. (167). Rubinowicz explained their existence, however, by assuming the 'incipient influence of the electric field—which exists in every Geissler tube—on the process of radiation,' i.e., by a Stark effect (Rubinowicz, 1918b, p. 466). He also extended his considerations to the Zeeman and Stark components, which were described by an extra quantum number, the one determining the projection of the angular momentum in the direction of the magnetic field, say n_z. In that case the selection rule also applied to n_z, yielding

$$|n_z' - n_z''| \leqslant 1; \qquad (168)$$

that is, the quantum number n_z should change only by amounts 0 or ± 1. Transitions without the change of n_z corresponded to a linearly polarized emitted radiation, the others to circularly polarized radiation. Thus Rubinowicz exluded the other lines, i.e., the ones given by transitions with $|n_z' - n_z''| > 1$, which Sommerfeld and Debye had admitted in their treatments of the Zeeman effect. The selection rule (168) also worked, as Rubinowicz showed, in the case of the Stark effect.

Sommerfeld was very satisfied with the results which Rubinowicz had obtained concerning the selection rules. He especially liked one aspect about which he wrote to Bohr in detail:

> On the occasion of the celebration of Planck's 60th birthday on 23 April [1918] I gave a talk in Berlin, in which I touched upon the question: reconciliation of quantum theory and wave theory . . . . My point of view, presented in that lecture (also Rubinowicz' point of view in his paper), is the following: The wave process occurs alone in the ether, which satisfies Maxwell's equations and acts quantum-theoretically like a linear oscillator of undetermined frequency v. The atom supplies only a certain amount of *energy* and *angular momentum* as the ingredients of the wave process; but it has nothing to do directly with the (wave-) oscillation itself. From the energy the ether takes—according to your hv-law—its frequency, and from the angular momentum its polarization.

In this way, Sommerfeld went on to report, the superfluous Zeeman components of his earlier theory (Sommerfeld, 1916d) could be avoided. He finally added: 'You arrive at the same results through your interesting comparison of classical and quantum-theoretical emission in the case of large quantum numbers. Your method has of course a wider range, but the interpretation just sketched seems to me physically more instructive' (Sommerfeld, 18 May 1918).

Indeed, for practical purposes the selection rules of Rubinowicz seemed to yield the same results as Bohr's analogy arguments. Sommerfeld tended to prefer the selection rules because of their 'physical' interpretation, although he was aware that Bohr's analogy principle ('*Analogieprinzip*') reached farther, as it provided the key for approaching a great variety of problems of atomic structure.[399] Still he found it difficult to accept it, for it seemed to have been imposed as something 'extraneous' on quantum theory ('*etwas Fremdartiges*,' Sommerfeld, 1919, p. 401).[400] Bohr, on the other hand, was confident that his 'new method of treatment will be of use in the discussion of more complex spectra' (Bohr to Richardson, 15 August 1918). He presented it on many occasions, as for instance in a lecture on 'Problems of the Atom and the Molecule,' which he delivered on 25 April 1919 at Ehrenfest's Institute in Leyden. In his lecture, entitled '*Über die Serienspektren der Elemente*' ('On the Series Spectra of Elements'), which he gave in Berlin at the German Physical Society on 27 April 1920, he discussed his analogy principle before an illustrious audience: it included O. von Baeyer, Max Born, Albert Einstein, James Franck, H. Geiser, Otto Hahn, George de Hevesy, Gustav Hertz, Paul Knipping, Walther Kossel, Rudolf Ladenburg, Alfred Landé, Wilhelm Lenz, Lise Meitner, Max Planck, Peter Pringsheim, Arnold Sommerfeld, Otto Stern, Ernst Wagner and Wilhelm Westphal.[400a]

In his talk at Berlin, Bohr tried to emphasize those points of view which seemed to be important 'for a consideration of the present status of the theory and for the possibility of its development in the near future' (Bohr, 1920, p. 425). He assigned a special role to his analogy principle, to which he referred already in his introduction by stating:

> Although it is impossible to follow in detail the process of radiation connected with the transition between two stationary states with the help of the usual electromagnetic concepts—according to which the properties of radiation emitted from an atom are directly determined by the motion of the system and its decomposition into harmonic components—, it still turns out that there exists a far-reaching *correspondence* between the various types of possible transitions between these states, on one hand, and the various harmonic components in which the motion of the system can be decomposed, on the other. This correspondence is of such a nature that the theory of spectra in question must be considered in a certain sense as the rational generalization of the ideas of the usual radiation theory. (Bohr, 1920, p. 427)

Bohr applied what he called 'correspondence' immediately to the case of the

---

[399] Sommerfeld presented Bohr's analogy principle—Bohr had referred to his method in a letter to Sommerfeld, dated 27 July 1919, as the '*allgemeines Analogieprincip der Quantentheorie*' ('the general analogy principle of quantum theory')—in the first edition of his *Atombau und Spektrallinien*. (See Sommerfeld, 1919, pp. 401–403.)

[400] Sommerfeld maintained his reservations against the analogy principle (later called the correspondence principle) also during the following years. Thus in 1920, on the occasion of the second edition of *Atombau und Spektrallinien*, he again wrote to Bohr: 'From the additions to my book you will see that I have tried to give a better appreciation of your correspondence principle than in the first edition . . . Nevertheless I must confess that I still find the fact that the origin of your principle is foreign to quantum theory embarrassing, however much I recognize that it clearly expresses the connection between quantum theory and classical theory' (Sommerfeld to Bohr, 11 November 1920).

[400a] See the photograph of the participants in *Naturwissenschaftliche Rundschau 31*, p. 492 (Keller, 1978).

hydrogen atom, obtaining the Rydberg constant (as he had shown already in his 1913 paper) by assuming the coincidence of the quantum-theoretical and the classical emission frequencies in the limit of high quantum numbers. Emphasizing the 'deeper significance of the correspondence indicated here,' Bohr argued that it consisted not only in the relation between the frequencies, but also between the intensities. He then discussed the series spectra of higher elements, and the Stark and Zeeman effects and the fine structure of the hydrogen lines, based on his principle of 'general correspondence' ('*allgemeine Korrespondenz*,' Bohr, 1920, p. 442; '*Korrespondenzprinzip*,' *ibid.*, p. 443). Thus, in his lecture at Berlin, Bohr finally raised the analogy argument to the rank of a general principle of atomic theory.

During the following year Bohr developed the theory of atomic structure on the basis of the correspondence principle (see, e.g., his report to the third Solvay Conference in April 1921: Bohr, 1923e). At about the same time he discussed the polarization of the emitted spectral lines, in reply to a note of Rubinowicz's. In a paper entitled '*Zur Polarisation der Bohrschen Strahlung*' ('On the Polarization of Bohr's Radiation'), which had been submitted in January 1921 to *Zeitschrift für Physik* and had appeared in March, Rubinowicz had treated the problem of spectral lines by invoking the assumption that 'the radiation emitted by an atom is already quantized,' and had derived from it again his selection and polarization rules (Rubinowicz, 1921). In a paper, submitted to *Zeitschrift für Physik* in June 1921, Bohr raised fundamental objections ('*prinzipielle Einwände*') against this assumption (Bohr, 1921b). He criticized in particular the view, which was in the back of Rubinowicz's argument, that atoms and radiation could be considered as one total quantum system—a view which had been taken already by Debye (1910) and William Wilson (1915)—claiming that it would not lead to a theory of atomic spectra because the coupling between atoms and radiation was still completely unknown. In comparing what he called the 'coupling viewpoint' ('*Kopplungsgesichtspunkt*') with his own 'correspondence viewpoint' ('*Korrespondenzgesichtspunkt*'), Bohr arrived at the following conclusion: the coupling viewpoint did not allow one to decide about the polarization of the emitted radiation without invoking additional assumptions, though it did provide a 'purely formal method of treating' ('*rein formale Behandlungsweise*,' Bohr, 1921b, p.8) a large variety of phenomena, such as the photoelectric effect and blackbody radiation; the correspondence point of view, though it could be applied successfully to the problems of atomic structure, did not help in solving the riddles of radiation theory, —'however, it does happen that every extension of the application of quantum theory brings the nature of this riddle into a sharper focus' (Bohr, 1921b, p. 9). Both points of view, he concluded, represented attempts 'to let the quantum theory appear as a reasonable generalization of the classical theory of radiation, although one may say that they approach this goal from opposite sides' (Bohr, 1921b, p. 9).

Thus, in the early 1920s, Niels Bohr arrived at a definite point of view about how to proceed forward in atomic theory. He wanted to make maximum use of what he called the 'more dualistic prescription' ('*mehr dualistische Behand-lungsweise*,' Bohr, 1921b, p. 5). In it the atom was regarded as a mechanical system having discrete states and emitting radiation of discrete frequencies, determined (in a nonclassical way) by the energy differences between stationary states; radiation, on the other hand, had to be described by the classical electrodynamic theory. This dualistic approach certainly contradicted the point of view which Einstein had taken in atomic and quantum theory. During his visit to Berlin, Bohr seized the opportunity he had long wished for ('*die lange erwünschte Gelegenheit*,' Bohr to Einstein, 24 June 1920) of hearing Einstein's opinions about these fundamental questions. After Bohr's visit Einstein wrote to him: 'Not often in my life has a person given me such joy by his mere presence as you did . . . . I have learned much from you, especially about how you look at scientific matters' (Einstein to Bohr, 2 May 1920). In the following years Bohr developed the theory of atomic structure on the foundation of his correspondence principle, and many joined him in this task. Bohr became the principal leader and philosopher of those who worked in atomic theory, and the final breakthrough in the discovery of a new atomic mechanics was achieved under his guidance. As for their views on atomic theory and the description of nature, Bohr and Einstein became complete antipodes. Still, after several decades, Einstein wrote: 'That this insecure and contradictory foundation [of quantum theory] was sufficient to enable a man of Bohr's instinct and tact to discover the major laws of the spectral lines and of the electron shells of atoms together with their significance for chemistry appeared to me like a miracle—and appears to me as a miracle even today. This is the highest form of musicality in the sphere of thought' (Einstein, 1949, pp. 45–47).

Chapter III

The Bohr Festival in Göttingen

On 11 November 1918 an armistice ended World War I, which had lasted more than four years. This war had affected the international relations of science more deeply than any other event during the previous hundred years. Scientific researchers active in enemy countries had been cut off from one another almost completely. While the hostilities were going on, there was no direct movement of scientific journals from one side to the other. Thus Arthur Eddington in England was supplied with copies of Albert Einstein's publications on general relativity and gravitation theory by the Dutch astrophysicist Willem de Sitter; similarly Niels Bohr from Denmark informed Arnold Sommerfeld in Germany about the progress of spectroscopy during the early war years in England. Naturally, the war had also hampered regular research and teaching programmes in the universities; the students, research scholars and younger professors had been called up to bear arms, leaving the lecture halls and laboratories empty. As a rule only the older professors had stayed at their places and had tried to carry on their work alone. For example, Ernest Rutherford had performed his investigations, leading to the discovery of the transformation of nitrogen nuclei into oxygen nuclei, with the help of the laboratory steward, William Kay; Friedrich Paschen, less fortunate in that respect, had done all the experimental work on the fine structure in the spectrum of ionized helium himself. Worst of all, the war had caused irreparable losses: Henry Gwyn Jeffreys Moseley and Fritz Hasenöhrl had been killed in action, and many other reputed scientists and promising talents had shared their fate.

In spite of all hardships, research had been carried on in atomic and nuclear physics after the outbreak of the war. The scientists who remained in their positions were assisted by foreigners from nonbelligerent countries: thus the Dane Niels Bohr had replaced Charles Galton Darwin as Reader in Physics in Manchester; in Göttingen Peter Debye, a Dutch citizen, had developed—together with his Swiss student Paul Scherrer—a most effective method for studying the crystal structures by X-ray diffraction. And Sommerfeld had been helped even by a so-called 'enemy alien,' the Russian physicist Paul Epstein, who had been interned in Munich: Epstein had applied Sommerfeld's theory of the phase integrals to the Stark effect. Some physicsts had banded together in military agencies to do research on war projects: thus Ralph Howard Fowler and Edward Arthur Milne had joined the Anti-Aircraft Experimental Section of the British Munitions Inventions Department; and Rudolf Ladenburg had played a leading role in the '*Artillerie-Prüfungskommission*' ('Artillery Testing Commission') in Berlin, to which Max Born, Alfred Landé and Ernst Madelung had also belonged. Born and Landé had used an important part of their time to investigate nonmilitary problems, like the properties of crystals, in which they employed Bohr's models of atomic and molecular structure. Of course, a few neutral countries in Europe served as oases of peace, where scientific research went on without interruption or restriction. One such oasis was Holland; there, at the University of Leyden, Paul Ehrenfest continued to guide his research students. The other was Denmark; there, at the University of Copenhagen, Niels Bohr had been installed in 1916 as Professor of Theoretical Physics and worked on atomic theory. But at most places, which had formerly been active scientifically in Europe, the universities and the laboratories waited for the war to finish.

With the end of the war the professors and students returned to the universities. However, they met many changes on their return; perhaps the most drastic change was the loss of the former internationalism of science. Though efforts were made on all sides to re-establish international relations, in many scientific societies and international conferences participants from Germany and Austria were not invited or admitted. It took many years until these restrictions were removed. But also on the national level important changes had occurred in the different countries; in most cases the continuous scientific tradition had been broken, and a new generation of students had to bridge the gap between old and new knowledge and experience as rapidly as possible. The difficulties in scientific research were enhanced by the fact that the war had exhausted the resources of most countries; the general economic shortage especially hit experimental research hard, for new equipment could not be easily acquired.[401] It was under such circumstances that Max Planck in Berlin emphasized that 'a sincere and resolute tending of the spiritual heritage was never more necessary than in the present time of the beginning of an economic impoverishment' (Planck, 1919, p. 548).

Interestingly enough, the new situation seemed to affect the field at the forefront of physics—atomic physics—much less than would have been expected; in fact, its development was favoured compared to many other fields. Of course, research in atomic physics shared the general lack of appropriate financial assistance; but this disadvantage seemed to be more than compensated by the attitude of people towards new ideas and their serious desire to concern themselves with the fundamental conceptions underlying atomic phenomena, namely, the quantum and the relativity theories. These theories were beginning to be taught in regular courses at many universities, and they attracted large numbers of students; they became the fields of new and intensive research. Some new centres of atomic physics emerged, adding their numbers to those that existed already before the war, and in certain cases the new centres surpassed the old ones in excellence within a few years.

Of the places, where atomic and quantum physics had been pioneered, several

---

[401] In some countries, especially in Germany, the consequences of the war led within a few years to a terrible inflation, in which many people lost their former prosperity.

preserved their high reputations in the 1920s. Thus the Cavendish Laboratory at Cambridge University flourished, especially after Ernest Rutherford came from Manchester to succeed the retiring Joseph John Thomson. In Munich Arnold Sommerfeld, who, since 1906, had trained many excellent disciples and introduced them to the forefront of research in theoretical physics, soon gathered a new, rapidly increasing crowd of very talented pupils. Indeed, it was after World War I that he really established what later was called the 'Sommerfeld school.'[402] Besides Cambridge and Munich an increasing number of universities became involved in research in atomic physics. Leyden and Copenhagen had entered the field during the war, and now their importance grew steadily. Paul Ehrenfest, because of his painstaking teaching methods and his own research on fundamental questions, attracted students and visitors from many countries, including Russia, the home of his wife. Ehrenfest also succeeded in obtaining for Einstein a special, distinguished visiting professorship at Leyden, beginning in 1920; besides giving his lectures, Einstein collaborated with Ehrenfest occasionally on problems of quantum theory.[403] In the same year, 1920, Niels Bohr's Institute became ready, and soon visitors arrived in Copenhagen from abroad to perform experimental and theoretical research; the early visitors included Svein Rosseland from Norway, Bohr's former Manchester colleague George de Hevesy, and James Franck from Berlin. With the help of Hendrik Kramers, his assistant, Bohr began to train research students in atomic theory. Berlin, where quantum theory had originated, and where the pioneers like Max Planck, Albert Einstein and Walther Nernst still taught and were active in research, fell somewhat behind in the 1920s. The number of quantum physicists, both experimentalists and theoreticians, was still the largest by far in Berlin; but, on one hand, the old pioneers did not concern themselves particularly with the problems of atomic structure and, on the other hand, many scientists who later made important contributions to atomic theory left Berlin to accept professorships at other universities. Thus Max Born left Berlin for Frankfurt, from where he went to Göttingen; Fritz Reiche and Rudolf Ladenburg went to Breslau. Their loss was only partially compensated by the activities at the *Physikalisch-Technische Reichsanstalt*, where Hans Geiger directed the laboratory for radioactivity with Walther Bothe among his collaborators.

As time went on, Berlin's former leading position at the frontier of quantum theory in Germany was taken over by Göttingen. The University of Göttingen, the *Georgia Augusta*, had been founded about seventy-five years earlier than the University of Berlin. In mathematics and physics, especially, it had obtained a great, worldwide reputation. In Max Born and James Franck, Göttingen had won two advocates of the most modern aspect of quantum physics: the physics

[402] It should be mentioned that up to 1914 he already had had such distinguished pupils as Peter Debye, Paul Ewald, Alfred Landé and Wilhelm Lenz.

[403] Ehrenfest tried to get Einstein completely over to Leyden by arranging for him the offer of a professorship at the University with maximum salary and minimum obligations, but Einstein declined to leave Germany at that critical time.

of atomic structure. The special collaboration of mathematicians and physicists, which had existed in Göttingen since the times of Carl Friedrich Gauss and Wilhelm Weber and had been recently revived by the mathematicians Felix Klein, David Hilbert and Hermann Minkowski, would eventually provide the mathematical foundation of atomic theory. The new Göttingen era began with a dramatic event: in June 1922 Niels Bohr delivered a series of seven lectures on the theory of atomic structure, in which he introduced his audience to the latest status and details of the subject. The *Bohr Festival*, as Bohr's visit to Göttingen and his lectures there came to be called, not only determined the future career of several young participants, but also aroused senior people like Born enough to begin to pursue active research on Bohr's theory. Thus it initiated the last phase of a development, in which the old quantum theory of Planck, Einstein, Bohr and Sommerfeld was replaced by the new scheme of quantum mechanics.

III.1   The Göttingen Tradition of Mathematics and Physics

The town of Göttingen, situated in the state of Lower Saxony, traces its origin back to a village, called Goding or Gutinga, the mention of which existed already in documents around the middle of the tenth century. The town received its charter and municipal rights from the German Emperor, Otto IV, in the early thirteenth century and played an important role in the Hanseatic League. Göttingen joined the Reformation movement in 1531, and suffered serious damage in the Thirty Years War. Its situation improved again under the Hanoverian Kings of Great Britain, especially after the university was established in 1734.[404] The University of Göttingen, which was named the *Georgia Augusta* after its sovereign, owed its creation mainly to the efforts of Gerlach Adolf Freiherr von Münchhausen (1688–1770), the Hanoverian Minister of King George II. As Curator of the University since its opening in 1737 he personally took care of its affairs. He determined its style and programme, and won for it its first professors, especially the famous Swiss anatomist, physiologist, botanist and poet Albrecht von Haller.[405] Haller not only made the new university internationally respectable, but from 1745 onwards he also edited the *Göttingischen gelehrten*

---

[404] In 1714 George Louis, Duke of Braunschweig-Lüneburg and Elector of the Holy Roman Empire, succeeded Queen Anne and became King George I of England. He was succeeded in 1727 by his son, King George II, who reigned until 1760.

[405] Albrecht von Haller was born on 16 October 1708 in Berne, Switzerland. He studied medicine at the Universities of Tübingen and Leyden, obtaining his medical doctorate in 1727. After visits to London, Oxford and Paris he went to Basle, where he became interested in mathematics, botany and philosophy. From 1729 he practiced medicine as a physician in Berne and at the same time published poems, which made his name famous in Europe (especially the long poem '*Die Alpen,*' which was contained in the first edition of his poems in 1732). In 1736 he was called to the chair of medicine, anatomy, surgery and botany at the University of Göttingen. After resigning from this position in 1753 he returned to Berne, where he assumed several official positions and completed his *Bibliotheca Medica*; besides, he wrote three philosophical novels, in which he described his views on various types of government. Von Haller died on 12 December 1777 at Berne.

Anzeigen—contributing himself several thousand articles on nearly every branch of knowledge. In 1751 Haller became the first president of the *Königliche Societät der Wissenschaften zu Göttingen* (The Royal Society of Sciences at Göttingen), an academy which was designed to promote scientific research at the university.[406]

The new University of Göttingen had, in the first place, to serve the needs of the Hanoverian State, namely, to educate physicians, theologians and administrators. In addition to the conventional chairs there were established others in more practical domains, such as economics and agriculture. An observatory was also installed, motivated in no small measure by the enormous practical applications of astronomy in British navigation. Within a few decades of the foundation of the University, chairs were established in mathematics and physics. Thus Abraham Gotthelf Kästner held the chair of mathematics from 1756 to 1800; in addition, from 1763, he directed the observatory.[407] In the latter position he had succeeded Johann Tobias Mayer, Professor of Economics and Physics from 1751 to 1762.[408] Mayer's chair in physics was later occupied by the veterinarian Johann Christian Polycarp Erxleben, whom Georg Christoph Lichtenberg succeeded in 1780.[409,410] Lichtenberg not only made Erxleben's book on experimental physics, by his additions, one of the most popular textbooks in the German language, but he also delivered excellent lectures and contributed to research on electricity.[411] He gained an even wider reputation through his literary writings; thus, beginning in 1778, he edited the *Göttinger Taschenkalender* (*Göttingen Pocket Calendar*), and, from 1780, together with the naturalist Johann Georg Adam Forster (1754–1794), the *Göttinger Magazin der Wissenschaft und Literatur*

[406]The planning of the 'Royal Society of Sciences in Göttingen' ('*Königliche Societät der Wissenschaften zu Göttingen*,' *Göttinger Gesellschaft*, or Göttingen Academy) goes back to von Münchhausen. He envisaged the Göttingen Society or Academy in a form different from the other academies, say the Prussian Academy of Sciences. The Göttingen Academy, he thought, should be a kind of university institute; its members should be selected professors of the university, who should be able to perform research in it (normally the professors had just to teach) and to discuss their results openly. Von Haller formally remained President of the Society until his death in 1777; after his return to Berne, however, the practical affairs were carried on by the philologists Johann David Michaelis from 1753 to 1770 and Christian Gottlob Heyne, who then continued until 1812.

[407]Abraham Kästner was born in Leipzig on 27 September 1719. He studied philosophy, physics and mathematics at the University of Leipzig; he also received his *Habilitation* there (in 1739) and became professor of mathematics (in 1746). Ten years later he moved to Göttingen. He died in Göttingen on 20 June 1800.

[408]Johann Tobias Mayer was born at Marbach on 17 February 1723. He obtained his knowledge of mathematics from his father. He contributed, in particular, to the theory and observation of lunar motions and established a catalogue of close to 1000 zodiacal stars. He also improved on the techniques of mapmaking. Mayer died in Göttingen on 26 February 1762.

[409]Erxleben was born in 1744 at Quedlinburg in Brandenburg. He became one of the best known veterinarians of his time. He died already in 1777 at Göttingen.

[410]Lichtenberg was born in Oberramstadt near Darmstadt on 1 July 1742, the seventh child of a pastor. He studied mathematics and science at the University of Göttingen under Kästner. In 1769 he was appointed extraordinary professor in pure and applied mathematics at Göttingen. After two longer stays in England he was appointed to the chair of physics in Göttingen in 1775; he occupied it until his death on 24 February 1799.

[411]Lichtenberg discovered, for instance, the so-called Lichtenberg figures, i.e., star-like figures originating from the dust put on an insulator when an electric discharge passes through it.

(*Göttingen Magazine of Science and Literature*).[412] Lichtenberg's collection of physical instruments became the starting point of the physics laboratory equipment at the University of Göttingen.

In the nineteenth century a great tradition in mathematics and physics became established in Göttingen, the origin of which was connected with the names of Carl Friedrich Gauss and Wilhelm Weber. Gauss, nearly a generation older than Weber, was born on 30 April 1777 at Brunswick (Braunschweig). He impressed his teachers at the elementary school with his knowledge of mathematics and continued to excel in mathematics and classical languages at the secondary school. At the age of fourteen Gauss was brought to the attention of Carl Wilhelm Ferdinand, Duke of Brunswick, who assumed the support of his continuing education. He studied at the *Collegium Carolinum* in Brunswick from 1792 to 1795. He was strongly attracted to philological studies, but he soon found a more compelling interest in mathematics. At *Collegium Carolinum* he mastered the more important works of Euler and Lagrange and Newton's *Principia*. From October 1795 to September 1798 Gauss studied at the University of Göttingen. At Göttingen his work on *Disquisitiones Arithmeticae* (Arithmetical Researches) was practically completed by 1798. In September 1798 he went to the University of Helmstedt, where there was a good mathematical library; the librarian and professor of mathematics there at that time was Johann Friedrich Pfaff (1765–1825), with whom Gauss became well acquainted. Gauss was awarded his doctor's degree *in absentia* by the University of Helmstedt in 1799 for his dissertation on '*Demonstratio nova theorematis omnem functionem algebraicam rationalem integram unius variabilis in factores reales primi vel secundi gradus revolve posse*' ('A New Proof that Every Rational Integral Function of One Variable Can Be Resolved into Real Factors of the First or Second Degree'). The Duke of Brunswick paid for the printing of Gauss' doctoral dissertation (University of Helmstedt, 1799), and 'granted him a modest pension which would enable him to continue his scientific work unhampered by poverty' (Bell, 1965, p. 232).[413] In 1807 Gauss was appointed Professor of Mathematics at the University of Göttingen as well as director of the observatory, a position which he had earned very well by his previous discoveries and publications, such as the '*Disquisitiones Arithmeticae*' (1801) and the calculation of the orbit of the planetoid Ceres, discovered on 1 January 1801 by Giuseppe Piazzi of Palermo.[414]

---

[412] Lichtenberg developed the art of short and witty formulations and was one of the masters of aphorism. With his essays on scientific and philosophical subjects he showed himself to be an advocate of enlightenment, and a fighter against superstition and intolerance. He also criticized the movement of *Sturm und Drang* in literature as well as Johann Kaspar Lavater's science of physiognomy.

[413] There is an excellent short account of the life of Carl Friedrich Gauss in *Men of Mathematics* by Eric Temple Bell. (See 'The Prince of Mathematicians' in Bell, 1965, pp. 218–269.)

[414] Before he became Professor of Mathematics at Göttingen, the list of Gauss' papers and discoveries was already very impressive. For example, in the 1790s he had studied the frequency of primes among integers; he had discovered the fact that a regular heptadecagon was inscribable in a circle, using only the compass and the straight edge, and the method of least squares; he had worked

Gauss was honoured by memberships in various academies, including the Paris Academy of Sciences (1804) and the Royal Society of London (1804). He continued to work and publish on various topics of astronomy and mathematics. Thus, in 1809, he published *Theoria motus corporum coelestium*, his principal memoir on theoretical astronomy. The main mathematical topics to which Gauss contributed were algebra, arithmetic (the fundamental theorem, i.e., every natural number can be represented as a product of primes in a unique way), complex numbers, elliptic functions, infinite series, number theory, theory of surfaces, non-Euclidean geometry, analytic functions and topology. In physics he contributed, for instance, to crystallography, optics and capillarity. Having been commissioned in 1818 to make a geodetic survey of the Kingdom of Hanover, he occupied himself for years with triangulation; he invented new theoretical methods as well as instruments for practical use. Gauss, the unquestioned 'Prince of Mathematicians,' became at the same time a pioneer in many fields of physics.[415]

Gauss and Wilhelm Weber collaborated on the fundamental problems of electricity and magnetism. Weber, since 1831 Professor of Physics at the University of Göttingen, and Gauss devised, for example, an electromagnetic telegraph arrangement extending over a distance of about 1 km between the Observatory at the *Geismartor* and the Physical Institute at the *Leinekanal* (1833). They made observations of the magnetic field of the earth at various locations—Gauss invented the magnetometer for that purpose—and stimulated scientists in many countries to follow their example in establishing a network of geomagnetic data. Their fruitful collaboration was interrupted by the consequences of the constitutional crisis of 1837: Weber joined the protest of his six colleagues at the University of Göttingen (including the historian Friedrich Christoph Dahlmann and the brothers Wilhelm and Jakob Grimm) against the suspension of the constitution by the new King Ernst August and was deprived of his professorship.[416] After remaining in Göttingen for several years as a private person, Weber

---

on the theory of elliptic functions, on linear differential equations and on the vector representation of complex numbers; he had given the proof of the fundamental theorem of algebra (for which he obtained his doctorate).

[415]'Recognition came [to Gauss] with spectacular promptness after the rediscovery of Ceres. Laplace hailed the young mathematician at once as an equal and presently as a superior. Some time later when Baron Alexander von Humboldt (1769–1859), the famous traveller and amateur of the sciences, asked Laplace who was the greatest mathematician in Germany, Laplace replied "Pfaff." "But what about Gauss?" the astonished Von Humboldt asked, as he was backing Gauss for the position of director at the Göttingen observatory. "Oh," said Laplace, "Gauss is the greatest mathematician in the world"' (Bell, 1965, p. 242).

[416]After the Napoleonic wars, the Dukedom of Braunschweig-Lüneburg had been reestablished in 1814 as the Kingdom of Hanover, whose king at the same time was King of Great Britain (George III until 1820, George IV from 1820 to 1830, and William IV from 1830 to 1837). King William's brother, Adolphus Frederic, Duke of Cambridge and Viceroy of Hanover, had appointed a commission to draw up a new constitution, which was accepted in 1833. In 1837 Ernest Augustus (Ernst August), Duke of Cumberland and fifth son of George III—and not Queen Victoria—succeeded William as King of Hanover. He held very autocratic beliefs and immediately declared the liberal constitution of 1833 as invalid.

took up the professorship of physics at the University of Leipzig in 1843, but he returned to his chair in Göttingen six years later. Weber showed that all electrical quantities could be expressed with the help of the units of length, mass and time, thus completing with Gauss (who proved the same for magnetic quantities) the establishment of a universal system of units for mechanics and electromagnetism. He worked for years on a theory of electrodynamics, starting from Gustav Theodor Fechner's idea that electric currents were caused by positive and negative electric charges moving in a conductor with the same velocity in opposite directions. A central assumption of this theory was that the forces between charges did not act instantaneously over arbitrary distances, but were transferred with a finite velocity, the velocity of light *in vacuo*.[417]

Although Weber's electrodynamic theory, on which, for instance, Gustav Kirchhoff worked and Rudolf Clausius tried to improve later, was ultimately superseded by Maxwell's, its author must be counted among the great pioneers of electrodynamics. In comparison with Weber, Johann Benedikt Listing (1808–1882), who took his place in Göttingen after 1837, gained only a modest reputation. However, due to the fact that Listing remained a professor also after Weber's return in 1849, the University of Göttingen obtained two chairs in physics. In 1881 Weber was succeeded by his student Carl Victor Eduard Riecke, who worked on magnetism and electrodynamic theory (Weber's), on the phenomena of pyroelectricity, on thermodynamic potentials and on the kinetic theory of matter.[418] Listing was succeeded, in 1883, by Woldemar Voigt, a student of Franz Neumann's in Königsberg.[419] Voigt contributed hundreds of

[417] Wilhelm Weber and Rudolf Hermann Arndt Kohlrausch (1809–1858) determined the velocity of light (*in vacuo*) by comparing the amounts of charge in a Leyden jar as obtained by a method depending on electrostatic attraction, on one hand, and by another method depending on the effects produced by discharging the jar, on the other (Weber and Kohlrausch, 1856). They concluded that the velocity of light *in vacuo* represented the velocity with which electromagnetic action is propagated.

[418] Eduard Riecke was born in Stuttgart on 1 December 1845. He studied mining at the *Polytechnikum* of his hometown, and mathematics at the University of Tübingen under Carl Neumann (from 1866 to 1868). After teaching for several months at a *Gymnasium* in Stuttgart, he continued his studies of mathematics and physics at the University of Göttingen. There Wilhelm Weber and his assistant Friedrich Kohlrausch (son of Rudolf Kohlrausch, see footnote 417) looked after him; he obtained his doctorate in May 1871 and his *Habilitation* in physics and mathematics already in June 1871. Two years later he was promoted to extraordinary professor and in 1881 to full professor of physics. He took over Weber's Institute and remained there until his death on 11 June 1915.

Riecke investigated theoretical problems even when he worked experimentally. Often he derived new consequences from available data. He was a pioneer in the electron theory of electrical conductivity (Riecke, 1898). In his later years he concerned himself primarily with the conduction of electricity in gases and with atmospheric electricity. He also wrote a very successful two-volume textbook, *Lehrbuch der Experimentalphysik*, the first edition of which appeared in 1896.

[419] Woldemar Voigt was born in Leipzig on 2 September 1850. He attended the *Nikolaischule* in Leipzig; upon graduation from there in 1868 he entered the University of Leipzig, but his studies were interrupted by service in the Franco-Prussian War. After 1871 Voigt continued his studies at the University of Königsberg. At first he was undecided between a career in physics and a career in music. Music had played an important role in his life, for Felix Mendelssohn and Robert Schumann had often visited his parents' house. At Königsberg, Voigt was greatly influenced by Franz Neumann, under whom he completed his doctorate in 1874 with a thesis on the elastic constants of rock-salt.

papers on a large variety of experimental and theoretical questions. He devoted himself in particular to crystal physics, in which his teacher Franz Neumann had been a pioneer, but he also worked on problems of elasticity, thermodynamics, and magneto- and thermo-optics. Thus, for instance, he became one of the first physicists to deal with the Zeeman effect immediately after its discovery in 1896. In a paper on oscillations in incompressible elastic media Voigt applied for the first time a set of transformation equations that later came to be called the Lorentz transformations (Voigt, 1887).

While the physics tradition of Lichtenberg and Weber was continued by Riecke and Voigt in a respectable manner, the mathematical tradition was accorded a more splendid continuity. Gauss' successor was Johann Peter Gustav Lejeune Dirichlet.[420] Dirichlet's work was not prolific but profound, and Gauss, in a letter to Alexander von Humboldt, dated 9 July 1845, called his papers jewels, adding, 'and one does not weigh jewels in a grocer's balance' ('*und Juwele wägt man nicht mit der Krämerwaage*'). So, for example, he applied the infinite series to investigate the properties of prime numbers, and he found many

---

After serving as a teacher in his former school, the *Nikolaischule* in Leipzig, he received his *Habilitation* at the University of Leipzig in 1875, and in the same year was called to Königsberg as an extraordinary professor of physics in order to take over the teaching duties of Franz Neumann. In 1883 Voigt was appointed *Ordinarius* for theoretical physics in the University of Göttingen; he and Riecke were to share a new physical institute, which became ready only in 1905. His principal scientific interest was in the understanding of crystals, but at the turn of the century he became concerned with the Zeeman effect and the electron theory. The *Dictionary of Scientific Biography* says:

> Voigt's extensive theoretical and physical researches on the nature of crystals were summarized in *Magneto- und Elektro-Optik* (1908) and *Kristallphysik* (1910). These treatises reveal the elegance of his mathematical treatments and the great orderliness that his research had brought to the understanding of crystals. The elastic, thermal, electric, and magnetic properties of crystals were ordered in magnitudes of three types: scalar, vector, and tensor. In fact, it was Voigt who in 1898 had introduced the term "tensor" into the vocabulary of mathematical physics.
> Even though Voigt devoted considerable time to his research and his students, and even though he acquired more administrative responsibility at Göttingen, he never gave up an active interest in music and musicology. He was recognized as an expert on Bach's vocal works and in 1911 published a book on Bach's church cantatas. Voigt often referred to the study of physics in musical terms. To him the region of science that represented the highest degree of orchestration and that possessed the utmost in rhythm and melody was crystal physics. It was altogether fitting that on 15 December 1919 his funeral bier was carried from his house to its final resting place to the strains of a Bach chorale. (Goldberg, *Dictionary of Scientific Biography*, Volume XIV, 1976, p. 62)

[420] Dirichlet was born on 13 February 1805 in Düren (close to Aachen). After attending the *Gymnasium* in Bonn and Cologne (where George Simon Ohm was his teacher in mathematics), he spent the years from 1822 to 1826 at Paris. There he came to know the mathematicians Sylvestre-François Lacroix, Joseph Fourier and Adrien Marie Legendre, as well as Alexander von Humboldt. Dirichlet received an honorary doctorate from the University of Bonn in spring 1827; then he travelled via Göttingen (he met Gauss) to Breslau, where he became *Privatdozent* (1827) and extraordinary professor (1828). In 1828 he went to Berlin and taught at the *Allgemeine Kriegsschule* and the University of Berlin (*Privatdozent*, 1829; extraordinary professor, 1831; *Ordinarius*, 1839). In spring 1855, after the death of Gauss, he accepted the call to the University of Göttingen. In 1831 he had been elected to the Prussian Academy; in 1854 the Paris Academy of Sciences elected him a Foreign Member; in 1855 he was awarded the *Pour le Merite* on the recommendation of Alexander von Humboldt. He died in Göttingen on 5 May 1859, shortly after the death of his wife Rebekka Mendelssohn-Bartholdy. (For more details, see the biography of Dirichlet by Biermann, 1959.)

theorems which eventually proved to be fundamental for later progress. After Dirichlet followed a man whose name was to shine most brightly in the world of mathematics: Georg Friedrich Bernhard Riemann, a student of Gauss'.[421] Riemann had obtained his doctorate in 1851 with a thesis on the foundations of the general theory of functions of a complex variable (Riemann, 1851). On 10 June 1854, he delivered his 'trial' or 'probationary' lecture ('*Probevorlesung*') on 'The Hypotheses which Lie at the Foundations of Geometry' ('*Über die Hypothesen, welche der Geometrie zu Grunde liegen*', Riemann, 1867) to become a *Privatdozent*. In this lecture Riemann proposed a new approach to geometry, which covered the ordinary Euclidean geometry as well as the non-Euclidean geometries of Nikolai Lobatchevski, Johann Bolyai and Carl Friedrich Gauss as special cases; especially, he used n-dimensional manifolds and defined a measure or metric in them by introducing an expression for the line element in terms of differential forms of second degree. At the end he discussed the relation between those geometries and the physical space, asking the question as to which geometry corresponded to physical reality and answering:

> This question can only be resolved by starting from the present, experimentally substantiated interpretation of phenomena, which was founded by Newton; this should then be gradually revised by taking into account the facts which cannot be explained at present. The investigations which, like the one carried out here, start from general concepts, may serve just one purpose: that this work [i.e., the gradual revision of the present view about the geometry of physical space] is not hindered by the use of [too] limited concepts and that progress in recognizing the relations between phenomena is not slowed down by traditional prejudices. (Riemann, 1867; reprint, 1953, p. 286)

Riemann's extraordinary early work on function theory and on the foundation of calculus and geometry had won him the highest esteem. He had been appointed *Extraordinarius* in 1857 and, less than two years later, he became the obvious candidate for the mathematics chair of Gauss and Dirichlet. As professor of mathematics, Riemann worked on various problems of mathematics and theoretical physics: for instance, on the propagation of acoustic waves, on prime numbers, and on variational calculus. Riemann fell ill in July 1862 with pleurisy and continued ill health forced him to spend long periods in the mild climate of Italy. On 20 July 1866 he died at Selasca, Lago Maggiore, before he reached his

---

[421] Bernhard Riemann was born on 17 September 1826 in Breselenz in the Kingdom of Hanover. He was educated by his father, then at the *Lyceum* in Hanover (1840–1842) and the *Gymnasium* at Lüneburg (1842–1846). In spring 1846 he entered the University of Göttingen to study philology and theology. Besides attending lectures in these fields, he also sat in courses on science and mathematics; for example, in the winter semester 1846–1847 he attended Gauss' lectures on the method of least squares. He changed his main field to mathematics and went in summer 1847 to the University of Berlin, where Jacobi, Lejeune Dirichlet and Jakob Steiner taught (thus he attended Dirichlet's lectures on analytic mechanics and higher algebra). After his return to Göttingen he attended, in particular, Wilhelm Weber's lectures on experimental physics and concerned himself with the philosophical writings of Johann Friedrich Herbart (1776–1841), a former professor of philosophy at the University of Göttingen. He obtained his doctorate in 1851 and his *Habilitation* in 1854.

fortieth birthday. The next professor of mathematics, Alfred Clebsch, also did not live to the age of forty: he came to Göttingen in 1868, but died already on 7 November 1872.[422] After Clebsch, the pioneer of invariant theory, there followed Hermann Amandus Schwarz, who worked on function theory and variational calculus.[423] When Schwarz left in 1892 to assume a professorship at the University of Berlin, Heinrich Weber succeeded him.[424] After Weber moved to Strasbourg, David Hilbert was called to Göttingen.

At the time of Hilbert's arrival, the University of Göttingen already possessed a second chair in mathematics, which had been established in 1886 and occupied by Felix Klein.[425] Klein was born on 25 April 1849 at Düsseldorf. He was educated at the University of Bonn, where he had enrolled in 1865 at the age of sixteen; he studied mathematics and physics with Julius Plücker, obtaining his doctorate in 1868. He visited Göttingen in spring 1869 and Berlin in the fall. While he found considerable encouragement from Clebsch in Göttingen, Karl Wilhelm Theodor Weierstrass (1815–1897), the almighty advocate of strict mathematical rigour at the University of Berlin, did not approve of Klein's intuitive attitude towards mathematics. Klein left Berlin in summer 1870 and, together with his friend Sophus Lie, visited Paris; there he came to know the mathematicians Jean Gaston Darboux (1842–1917) and Camille Jordan (1838–1922), who influenced his later work on geometry and algebraic equations.[426] Returning to Göttingen in the beginning of 1871, Klein obtained his *Habilitation* in fall 1872; on Clebsch's strong recommendation, he was immediately called to the chair of mathematics at Erlangen. In his inaugural lecture, entitled '*Vergleichende Betrachtungen über neuere geometrische Forschungen*' ('Comparative Considerations on More Recent Geometrical Researches') and later called simply the '*Erlanger Programm*,' he summarized his views on geometry: the various geometries were

---

[422] Alfred Clebsch was born at Königsberg on 19 January 1833. He studied mathematics and physics at the University of Königsberg (1850–1854), receiving his doctorate in 1854. Four years later he became Professor of Theoretical Mechanics at the *Technische Hochschule*, Karlsruhe, and in 1863 Professor of Mathematics in Giessen. Clebsch worked on the theory of invariants, the calculus of variations, partial differential equations, the theory of curves, Abelian functions, and the application of determinants to various fields of mathematics. In 1868, together with Carl Neumann, he founded the journal *Mathematische Annalen*.

[423] Hermann Schwarz was born in Hermsdorf, Silesia, on 25 January 1843. He obtained his doctorate in 1864 at the University of Berlin, and then became professor of mathematics at Halle (1867–1869) and Zurich (1869–1875). He died in Berlin on 30 November 1921. His name is still connected with 'Schwarz' inequality.

[424] Heinrich Weber was born on 5 March 1842 at Heidelberg. He served as professor of mathematics in Heidelberg, Zurich, Königsberg and Berlin, before he came to Göttingen. He worked on partial differential equations and on algebra. Weber died in Strasbourg on 17 May 1913.

[425] Klein had already been considered in 1873 as a possible candidate for succeeding Clebsch.
It should be mentioned that after the Austro-Prussian War of 1866 the Kingdom of Hanover was incorporated in the State of Prussia. This did not damage the prosperity and distinction of the University of Göttingen. On the contrary, a strong competition immediately developed between the Universities of Berlin and Göttingen, especially in mathematics.

[426] Camille Jordan had just published a book on algebraic equations (Jordan, 1870), based on the ingenious earlier work of Évariste Galois (1811–1832). Klein and Lie studied this book and obtained from it inspiration for their later work on group theory.

organized according to their group structure, i.e., according to the invariants of their associated mathematical groups (Klein, 1872). While Lie investigated the cases of continuous groups, Klein turned to the study of the properties of the discrete groups of linear transformations in the plane and the three-dimensional space; he did this especially in Munich, where he had accepted the chair of mathematics at the *Technische Hochschule* in 1875. Five years later he moved again, this time to Leipzig. There he followed and expanded on Riemann's earlier and not yet fully appreciated ideas connecting geometry with physical problems. As Richard Courant was to note later (in his obituary notice of Klein):

> Klein became the most ardent and successful apostle of the Riemannian spirit, leading it with irresistible force increasingly to a completely dominant position against a more critical point of view in mathematics, which [the critical point of view] carried some danger for the freedom of development. If today's mathematics is able to go on building with calm assurance on the foundation of Riemann, then the greatest and decisive credit is due to Klein. (Courant, 1925, p. 767)

In the Leipzig years Klein also worked, in hot competition with Henri Poincaré, on geometric function theory, especially on the theory of algebraic functions. After a breakdown in his productivity because of complete exhaustion in 1882, Klein turned to the theory of the icosahedron. In Göttingen he studied the theory of modular and automorphic functions; he also worked on the theory of the top together with his student Arnold Sommerfeld (Klein and Sommerfeld, 1897, 1898, 1903, 1910). But in Göttingen he mainly developed a personal style of lecturing on broad fields of mathematics and devoted his talents to directing seminars and establishing new programmes of mathematical teaching and research. After Schwarz left for Berlin, Klein became the uncrowned king of mathematics in Göttingen. He played a leading role in the reorganization of mathematical instruction, both at the high school and the university levels. He insisted not only on an education leading to research in pure mathematics, but also on keeping an open mind about the application of mathematics to science and technology. He tried to give expression to these ideas in *Mathematische Annalen*, the journal he edited, and especially in the *Encyklopädie der mathematischen Wissenschaften*, the encyclopedia he helped to develop and the contributions to which he supervised for thirty years. Thus the volumes of the encyclopedia contained all fields of pure and applied mathematics, ranging from the logical foundations of mathematics to geophysics, attesting to Klein's universal outlook.

As Klein turned his attention more and more to the applications of mathematics and to his duties in academic organization, he also devoted much energy to another important task: stimulated by a visit to the United States in 1893, where he observed the growth of enormous power from the intimate connection between social forces and vast material resources, he thought of establishing a closer relation between scientific research and technology in Germany, especially at the University of Göttingen. Against the great resistance of the defenders of pure science, on one hand, and the existing technical institutions, on the other, he received the effective help of Friedrich Althoff, then the official responsible for universities in the Prussian Ministry of Education. A private organization, called the *Göttinger Vereinigung zur Förderung der angewandten Mathematik und Mechanik* (Göttingen Association for the Advancement of Applied Mathematics and Mechanics), which Klein founded together with Henry Theodor von Böttinger and in which industrialists and Göttingen professors were represented, supported financially the plans to create new university institutes devoted to applied research. So by 1905 four new institutes joined the previously existing institutes and the observatory (directed since 1901 by Karl Schwarzschild): the *Geophysikalisches Institut* (Geophysical Institute), the *Institut für angewandte Elektrizität* (Institute for Applied Electricity), the *Institut für angewandte Mechanik* (Institute for Applied Mechanics) and the *Institut für angewandte Mathematik* (Institute for Applied Mathematics). The geophysical institute was installed and directed by Emil Wiechert.[427] Hermann Theodor Simon, a former assistant of Riecke's, became the Director of the Institute of Applied Electricity, which obtained a separate building, next to the new physics building housing the institutes of Riecke and Voigt.[428] Ludwig Prandtl, the new Professor of Applied Mechanics, moved into a wing of the old Physics Institute at the *Leinekanal*.[429] Like Prandtl, Carl Runge, the new Professor of Applied Mathematics, also came from Hanover. He had studied mathematics at the Universities of Munich and

[427] Emil Wiechert was born at Tilsit on 26 December 1861. He did fundamental research on electrons at the University of Königsberg before coming to Göttingen as extraordinary professor in geophysics. He was promoted to full professor in 1905; then he devoted himself to seismic research and the development of appropriate instruments. When he died on 19 March 1928, Gustav Heinrich Angenheister (1878–1945) became his successor.

[428] Hermann Simon was born on 1 January 1870 at Kirn (on the river Nahe). He studied physics at the Universities of Heidelberg and Berlin, obtaining his doctorate under August Kundt in 1894 (with a thesis on the dispersion of ultraviolet radiation). Then he took the position of an assistant to Eilhard Wiedemann in Erlangen and obtained his *Habilitation* in 1896. Two years later Riecke brought him to Göttingen as his assistant. At Easter 1900 he went to Frankfurt-am-Main as docent for physics and director of the physical laboratory, but by the fall of the next year he returned to Göttingen as extraordinary professor and director of the department of applied electricity, a department which Theodor Des Coudres had developed since 1895 within Riecke's Institute. In 1907, after the department had been turned into a separate institute (with a new building completed in December 1905), Simon was promoted to a full professorship. He died on 22 December 1918 and his former student, Max Gustav Hermann Reich (1874–1941), became his successor.

[429] Ludwig Prandtl was born on 4 February 1875 at Freising. He studied at the *Technische Hochschule*, Munich, and the University of Munich, obtaining his doctorate in 1900 under August Föppl. After working at the machine-tool factory MAN at Nuremberg, he was called in 1901 to a professorship at the *Technische Hochschule*, Hanover. During World War I he obtained an enlarged institute in Göttingen, the *Aerodynamische Versuchsanstalt*. In 1925 additional buildings were added and Prandtl's Institute became the *Kaiser-Wilhelm-Institut für Strömungsforschung, verbunden mit der Aerodynamischen Versuchsanstalt*. Prandtl performed fundamental research in many fields of fluid dynamics, especially with respect to its application to aerodynamical problems, such as wing theory, and trained many students and collaborators. He retired as director of the institute in 1946 and died in Göttingen on 15 August 1953.

Berlin (at the latter under Weierstrass and Kronecker) and had been appointed *Extraordinarius* for mathematics at the *Technische Hochschule*, Hanover. Besides his research on mathematical questions, he had made himself a reputation in physics through his work on spectroscopy, begun in collaboration with Heinrich Kayser. In Göttingen his official task was to teach and do research in applied mathematics; thus he belonged to the group of professors of mathematics. He obtained an institute of his own, the first one on applied mathematics in Germany; it was housed in the old physics building. Runge soon established, completely in the spirit of Klein's intentions, close relations with his colleagues in physics and applied physics. Woldemar Voigt, who, like Runge, was interested in the anomalous Zeeman effect, welcomed his new colleague and reserved a room in his institute for Runge's spectroscopic apparatus. Of course, Runge continued to collaborate, as in Hanover, with Prandtl. But he also became interested in the researches at the institutes of Simon and Wiechert and he developed friendly relations with the astronomer Karl Schwarzschild.[430] Thus Runge soon became the best advocate and executor of Klein's ideas about bringing mathematics and its applications together at Göttingen.

Before Klein withdrew from active work on pure mathematics he had made sure that this field would remain well represented at Göttingen. In spring 1895 Heinrich Weber left Göttingen to accept the chair of mathematics at the University of Strasbourg, and Klein saw to it that Hilbert, whose work and creative potential he knew very well, got the chair.[431] David Hilbert was born on 23 January 1862 at Königsberg, the son of the district judge (*Amtsrichter*) Otto Hilbert. After receiving his secondary school education from 1870 to 1880 at the *Friedrichskolleg* and the *Wilhelmsgymnasium*, where he did not particularly excel, Hilbert enrolled in the University of Königsberg to study mathematics. At Königsberg, Heinrich Weber (professor until 1883) and Ferdinand Lindemann (professor since 1883) became his teachers.[432] Lindemann directed Hilbert's attention to the theory of invariants. However, two persons exerted the decisive influence on Hilbert: they were Hermann Minkowski and Adolf Hurwitz. Minkowski, though two years younger than Hilbert, had entered the university one and a half years before him and had become well known through his brilliant work on number theory. Hilbert and Minkowski became friends. In 1884 they were joined by Hurwitz, the new *Extraordinarius* at Königsberg, a man

---

[430] In the early years of his Göttingen period Runge investigated, for instance, the radioactivity of air, and he accompanied Schwarzschild on a solar eclipse expedition to Algeria in 1905.

[431] Felix Klein, in trying to get Hilbert to Göttingen, performed a masterpiece of diplomacy. He told Adolf Hurwitz, another possible candidate, that he would not propose his name on the list of the faculty, and he put Hermann Minkowski behind Hilbert on the second place, thus increasing Minkowski's chance to become Hilbert's successor at Königsberg. He discussed his plans with Althoff, who agreed fully (see Reid, 1970, p. 46). He also defended his choice in Göttingen. 'My colleagues accused me at that time that I wanted to offer the appointment to an easy-going young colleague,' Klein recalled later. 'I answered: I am inviting the most difficult one.' (See Blumenthal, in Hilbert, 1935, p. 399.)

[432] Hilbert spent only the second semester of his studies at the University of Heidelberg with Lazarus Fuchs.

of immense knowledge in mathematics.[433] 'On innumerable, at times daily, walks,' recalled Hilbert later, 'did we in those eight years rummage through all the nooks of mathematical knowledge, and Hurwitz—with his broad and many-sided, as well as secure and well-organized knowledge—was always our guide' (Hilbert, 1920, p. 77). On obtaining his doctorate with a thesis on spherical functions in early 1885 (Lindemann was the supervisor), Hilbert travelled to Leipzig (visiting Klein) and Paris (where he met Charles Hermite). After returning to Königsberg the following year, he submitted a *Habilitation* thesis on a problem of the theory of invariants of binary forms and became a *Privatdozent*. Stimulated by a visit, during the Easter vacation of 1888, of Paul Gordan (1837–1912)—the 'king of invariants'—from Erlangan, Hilbert turned his full attention to invariant theory. Hilbert devised a general theory of algebraic forms, from which he derived a simplified version of Gordan's proof that the system of invariants associated with binary forms is finite (Hilbert, 1888; 1889a, b). In 1892 when Hurwitz accepted the call to the *Eidgenössische Technische Hochschule* in Zurich, Hilbert succeeded him as extraordinary professor; in the following year, upon Lindemann's move to Munich, Friedrich Althoff appointed him full professor. Althoff also assisted in getting Minkowski in 1894 to Königsberg as *Extraordinarius*. However, the two friends did not stay together a long time, for Hilbert left already in 1895 for Göttingen.

In Göttingen, Hilbert continued to work on the theory of number fields. At the Munich meeting of the *Deutsche Mathematiker-Vereinigung* (German Mathematical Association) in September 1893, Hilbert and Minkowski had been requested to submit in two years a report on that subject. While Minkowski retired in the course of time from active collaboration, Hilbert completed in 1897 a long memoir, entitled '*Die Theorie der algebraischen Zahlkörper*' ('The Theory of Algebraic Number Fields') in short the *Zahlbericht* (Hilbert, 1897). This monumental work absorbed all of Hilbert's energies, and he continued to publish papers on number fields until 1899. Then he turned to a new subject. Having lectured occasionally on geometry—for example, in summer 1894 at Königsberg on non-Euclidean geometry and in the winter semester 1898–1899 at Göttingen on the elements of Euclidean geometry—he published, in connection with the inauguration of the Gauss–Weber monument in Göttingen, a book on '*Die Grundlagen der Geometrie*' as a *Festschrift* (Hilbert, 1899). In his book, as in his lectures, Hilbert treated geometries in a systematic way as based on a set of few fundamental axioms. Though the axiomatic method, already introduced into geometry by Euclid, had attracted new attention in the nineteenth century—thus

---

[433] Adolf Hurwitz was born on 26 March 1859 in Hildesheim. He studied, beginning in 1877, first at the *Technische Hochschule* in Munich (with Klein), then at the University of Berlin (with Weierstrass and Kronecker), and again in Munich and at the University of Leipzig (both with Klein). He obtained his doctorate under Klein in Leipzig in 1881 and his *Habilitation* the following year at the University of Göttingen. In 1884 he was appointed *Extraordinarius* for mathematics in Königsberg. From Königsberg he went in 1892 to Zurich as *Ordinarius* at the E.T.H. He retained this position until his death in Zurich on 18 November 1919. Hurwitz contributed to many mathematical problems; he especially applied function-theoretic methods to algebraic problems.

Moritz Pasch (1843–1930) had used it in his book on modern geometry (Pasch, 1882)—Hilbert's treatment went far beyond all previous approaches: Hilbert not only gave systems of axioms on which geometries could be built, he also showed the mutual independence and consistency of the systems of axioms determining each geometry.

Die Grundlagen der Geometrie (*The Foundations of Geometry*) was a most successful book, and its success initiated the most brilliant period in its author's life. Hilbert became even better known in the world of mathematics through the International Congress of Mathematicians, held in August 1900 at Paris. On 8 August he delivered a talk on 'Mathematical Problems,' whose goal he outlined in his introduction by saying: 'Who of us would not be glad to lift the veil behind which the future lies hidden; to cast a glance at the next advances of our science and at the secrets of its developments during the future centuries? What particular goals will there be toward which the leading mathematical spirit of coming generations will strive? What new methods and new facts will the new centuries disclose in the wide and rich field of mathematical thought?' (Hilbert, 1900, p. 253). After referring to certain problems on which the interest of the mathematicians had concentrated in the past, such as Johann Bernoulli's problem of the line of quickest descent, Fermat's problem or the three-body problem in mechanics, he gave a selection of twenty-three problems to his audience.[434] Some of these problems were: No. 1, to prove Cantor's 'continuum hypothesis' that any set of real numbers can be put into one-to-one correspondence either with the set of natural numbers or with the set of all real numbers (i.e., the continuum); No. 2, to prove the internal consistency of the axioms of arithmetic; No. 4, to investigate the problem of straight lines as being the shortest distance between two points; No. 5, to investigate Lie's concept of continuous transformation groups in the case that the functions defining the groups are not differentiable; No. 6, to establish the axioms of theoretical physics; No. 13, to prove the impossibility of solving a general equation of the seventh degree by means of functions of two arguments; No. 19, to determine whether the solutions of 'regular' problems in the variational calculus are necessarily analytic.

The 23 Paris problems won Hilbert a unique reputation. Everybody recognized that he had indeed dared to 'lift the veil behind which the future lies hidden.' During the following years mathematicians rose to attack this or that problem which Hilbert had mentioned, and numerous aspiring disciples came to Göttingen to learn perhaps a solution or two from the master who had raised such important questions. While up to the end of 1900 twelve students had completed their doctorates under Hilbert (including Otto Blumenthal, the first one, in 1898, and Max Dehn in 1900), thirty-nine followed between 1901 and 1910, including Felix Bernstein and Georg Hamel in 1901, Erhard Schmidt in 1905, Ernst Hellinger in 1907, Hermann Weyl in 1908, Andreas Speiser and

---

[434] At the International Congress of Mathematicians in Paris, Hilbert actually discussed only ten problems in order to cope with the available time. However, an extract of his talk in French, containing all 23 problems, had been distributed in advance.

Alfred Haar in 1909, and Richard Courant and Erich Hecke in 1910.[435] Hilbert thus founded a school of phenomenal brilliance, which had never been equalled before in the history of mathematics.

Although leading an increasingly numerous crowd of students to important achievements, Hilbert did not neglect his own work. First, he turned to problems of the variational calculus, connected with his problem No. 19; in particular, he proved Dirichlet's principle, which had fallen into discredit in the second half of the nineteenth century (Hilbert, 1901). After 1901 he became absorbed in an even more important question. E. Holmgren, a student from Uppsala, had brought with him Ivar Fredholm's first note on the theory of integral equations (Fredholm, 1900). Fredholm had used his new method to attack Dirichlet's problem, and this stimulated Hilbert to work on the problem, too. Hilbert presented the results in six papers, contributed to the Göttingen Academy from 1904 to 1910 (Hilbert, 1904a, b; 1905, 1906a, b; 1910).[436] In his *Foundations of the Theory of Linear Integral Equations*, Hilbert not only discussed the problems considered earlier by Fredholm, such as the solution of the boundary value problem in potential theory, but went on, as was his custom, far beyond what had been done previously; he also dealt with the question of eigenoscillations occurring in the solution of partial differential equations. However, what was most important, he developed systematically a general and far-reaching theory of principal-axes transformations or orthogonal transformations of infinite bilinear forms with specific kernels. Thus he reduced the problem of solving linear integral equations or partial differential equations to a problem of the theory of invariants. Hilbert's results were extended by many students: for example, Erhardt Schmidt generalized his methods and formulated what later was called the theory of 'Hilbert spaces' (Schmidt, 1908), and Hermann Weyl dealt in his thesis with singular integral equations (Weyl, 1908). The full power of the theory would show up especially in problems of theoretical physics, to which Hilbert himself contributed.

Before Hilbert published his first paper on integral equations, there occurred a very happy event. It began in summer 1902, when Hilbert received a call to succeed Lazarus Fuchs in the chair of mathematics at the University of Berlin. After some hesitation he decided to stay on in Göttingen, and at the same time conceived a great plan. He let it be known that he would not go to Berlin if a third professorship in mathematics were created at the University of Göttingen, to which Hermann Minkowski would be invited. The coup succeeded and the almighty Althoff was persuaded. Hilbert recalled later with some modesty: 'It was again Althoff, who transplanted Minkowski on the soil most suitable for his effectiveness; with a boldness unequalled in the annals of Prussian universities, Althoff created a new chair here in Göttingen' (Hilbert, 1909b, p. 91; *Gesammelte*

---

[435] Max Dehn was also the first to solve one of Hilbert's Paris problems. The following year he solved problem No. 3 in the sense conjectured by Hilbert (Kellog, 1902).

[436] Hilbert discussed his first ideas and results on integral equation theory in summer 1901. And one student, Oliver Dimon Kellogg, wrote a thesis on this subject as early as 1902.

Abhandlungen 3, 1935, p. 355). When Minkowski arrived in fall 1902, Hilbert was completely happy. Now he could finally enjoy, after many years, the daily presence of a close friend, whom he had missed for so long. And he knew perfectly how Minkowski would stimulate the mathematics in Göttingen.

Minkowski, born on 22 June 1864 at Alexoten, Russia, came to Germany in his early youth. Already at the age of eight he entered the *Gymnasium* at Königsberg, which he completed at the age of fifteen. He enrolled at the University of Königsberg in spring 1890 and studied five semesters there (mainly with Heinrich Weber and Woldemar Voigt); then he studied three semesters at the University of Berlin (with Kummer, Kronecker, Weierstrass, Helmholtz and Kirchhoff). In 1883 he won the *Grand Prix des Sciences Mathématiques* of the Paris Academy of Sciences by solving the prize problem for 1881: this was to express integral numbers by sums of five squares of integers, with the help of a new theory of quadratic forms (Minkowski, 1884). Camille Jordan encouraged the young mathematician by writing to him. 'Work, I beg you, to become an eminent geometer.' (See Hilbert, 1909b, p. 74; *Gesammelte Abhandlungen 3*, 1935, p. 341.) The results of the prize essay also formed the basis of the doctoral thesis, which Minkowski presented in 1885 to the University of Königsberg. He continued to work on the quadratic forms of (finitely) many variables and received his *Habilitation* in 1887 at the University of Bonn. In Bonn he began an important work by creating what he called the 'geometry of numbers'; that is, he established a theory of (integral) numbers based on geometrical perception (*geometrische Anschauung*). In this theory he emphasized the concept of the convex field, i.e., of a field which contains also the line connecting the two points in it. He associated this concept with a new type of geometry, which may be described by the fact that in each triangle the sum of two sides is greater than the third ('Minkowski geometries'). Minkowski had been stimulated to develop the geometry of numbers by the letters which Charles Hermite had written in 1850 to Carl Gustav Jacob Jacobi; and nearly half a century later, after reading the first part of Minkowski's book on this subject (Minkowski, 1896), the old Hermite wrote: 'I think I see the promised land' ('*Je crois voir la terre promise*'). (See Hilbert, 1909b, p. 82; *Gesammelte Abhandlungen 3*, 1935, p. 348.)

Minkowski continued to work on his number theory after leaving Bonn, where he had become extraordinary professor in 1892; in 1894 he moved to another extraordinary professorship at the University of Königsberg (where he succeeded Hilbert as full professor in 1895). In fall 1896 he went to the E.T.H. in Zurich. When he joined Hilbert in Göttingen six years later, he turned to new problems. Physics, to which he had been attracted in Bonn through Heinrich Hertz, began to occupy him increasingly. He especially studied the theories of electrodynamics, which was a fashionable subject at that time.[437] Minkowski first gave several lectures about it in the Seminar, which he conducted together with Hilbert. At the meeting of the Göttingen Academy on 21 September 1907 he finally gave a long report on his results concerning the fundamental equations of electrodynam-

---

[437] Minkowski had concerned himself earlier with the problem of capillarity and his encyclopedia article grew out of it (Minkowski, 1907).

ics, discussing in particular their invariance with respect to Lorentz transformations (Minkowski, 1908). The importance of this invariance led him to a new concept of space and time, which he expounded in his lecture at the *Naturforscherversammlung* in Cologne on 21 September 1908, where he remarked: 'Henceforth space by itself, and time by itself, are doomed to fade away into mere shadows, and only a kind of the union of the two will preserve an independent reality' (Minkowski, 1909, p. 104). Thus the Lorentz invariance should determine all physical laws and lead, for instance, also to a change of Newton's law of gravitation. As Minkowski was about to extend this work, he was suddenly stricken one day with a violent attack of appendicitis; he did not recover and died on 12 January 1909, mourned by all his friends and colleagues in Göttingen.

Minkowski's death hit David Hilbert deeply. It interrupted all the plans he had envisaged, especially the plans about physics. Since Minkowski had come to Göttingen, Hilbert had embarked with him upon a systematic study of physical theories. Both of them had given courses of lectures and seminars on topics of theoretical physics: for example, Minkowski lectured in summer semester 1904 on the mechanics of continua and in the following winter semester Hilbert and Minkowski conducted problem sessions on advanced mechanics; in summer 1905 they held a joint seminar on electron theory; in summer semester 1906 and winter semester 1906–1907 Hilbert gave lectures on continuum mechanics; in summer 1907 Minkowski lectured on heat radiation; in winter semester 1907–1908 they conducted a joint seminar on the partial differential equations of physics. While Minkowski had also become absorbed in working on electron theory and relativity theory, Hilbert's concern was with the principles of mathematics and providing a solution to a problem which had been posed by the English mathematician Edward Waring in the eighteenth century: Waring had asserted that all positive integral numbers can be represented as a finite sum of powers of integral numbers. Hilbert's solution of Waring's Problem was ready to be presented in the joint Seminar with Minkowski in the middle of January 1909. After Minkowski's death, Hilbert presented his solution to the Göttingen Academy on 6 February 1909, dedicating it to the memory of his friend, who had done so much for number theory (Hilbert, 1909a). From then on he missed Minkowski, but carried on the Seminar with Edmund Landau, Minkowski's successor.[438] Landau, though a worthy successor with respect to number theory —his specialty was the application of analytic methods to number theory—

---

[438] Edmund Landau was born in Berlin on 14 February 1877. He studied mathematics at the Universities of Berlin and Munich from 1893 to 1899 and received his doctorate in 1899 at the University of Berlin. In 1901 he became *Privatdozent* for mathematics at the University of Berlin. Landau remained professor of mathematics at Göttingen until 1934. He died in Berlin on 19 February 1938.

In searching for a successor to Minkowski, Klein and Hilbert looked for a young mathematician, whose achievements were still ahead of him. This requirement ruled out Adolf Hurwitz, and the final candidates were Oskar Perron and Edmund Landau. The decision was made by Klein, who said: 'Oh, Perron is such a wonderful person. Everybody loves him. Landau is very disagreeable, very difficult to get along with. But we, being such a group as we are here, it is better that we have a man who is not easy' (Reid, 1970, p. 118).

showed no interest in geometry and even less in applied mathematics, not to speak of mathematical physics. His only association with physics was indirect: he immediately invited a young mathematician from Copenhagen, Harald Bohr (Niels Bohr's younger brother), who had solved a problem of number theory mentioned in one of Landau's works, to come to Göttingen to work with him. In any case, Hilbert knew that in executing his plans concerning physics, he could not count on Landau. Fortunately, an accidental fact, which was not connected with physics at all, provided a possible chance for assistance: the Wolfskehl endowment.

In 1906 the mathematician Paul Wolfskehl of Darmstadt bequeathed a sum of 100,000 Marks to the *Königliche Gesellschaft der Wissenschaften* to be awarded to the first person who, during the next one hundred years, would publish a *complete* proof of Pierre de Fermat's *Last Theorem* of 1637. In this theorem Fermat denied the existence of integers, x, y, z, and n, which satisfy the equation

$$x^n + y^n = z^n \qquad (169)$$

for $x, y, z \neq 0$ and $n > 2$.[439] Wolfskehl had also stated in his testament that the interest on the 100,000 Marks should be used by the Göttingen Academy towards the best progress of mathematical science. When the Göttingen mathematicians learned about Wolfskehl's endowment, they responded favourably. Thus Klein remarked: 'I have taken note of it with great interest; the matter has to be considered quite seriously and is very important.' While Hilbert agreed with Klein's remarks, Minkowski added: 'I knew the late Dr. Wolfskehl very well personally and, because of his great interest for mathematics and in particular number theory, I expected from him a bequest of this kind. The topic proposed

---

[439] Pierre de Fermat's (1601–1665) greatest work was the so-called 'theory of numbers' or 'the higher arithmetic.' E. T. Bell writes:

It was Fermat's custom in reading Bachet's *Diophantus* to record the result of his meditations in brief marginal notes in his copy. The margin was not suited for the writing of proofs. Thus, in commenting on the eighth problem of the Second Book of Diophantus' *Arithmetic*, which asks for the solution in rational numbers (fractions or whole numbers) of the equation $x^2 + y^2 = a^2$, Fermat comments as follows: "On the contrary, it is impossible to separate a cube into two cubes, a fourth power into two fourth powers, or, generally, any power above the second into powers of the same degree: I have discovered a truly marvellous demonstration [of this general theorem] which this margin is too narrow to contain" (Fermat, *Oeuvres*, III, p. 241). This is his famous Last Theorem, which he discovered around 1637.

To restate this in modern language: Diophantus' problem is to find whole numbers or fractions x, y, a such that $x^2 + y^2 = a^2$; Fermat asserts that *no* whole numbers or fractions exist such that $x^3 + y^3 = a^3$, or $x^4 + y^4 = a^4$, or, generally, such that $x^n + y^n = a^n$ if n is a whole number greater than 2.

Diophantus' problem has an infinity of solutions; specimens are $x = 3$, $y = 4$, $a = 5$; $x = 5$, $y = 12$, $a = 13$. Fermat himself gave a proof by his method of infinite descent for the impossibility of $x^4 + y^4 = a^4$. Since his day $x^n + y^n = a^n$ has been proved impossible in whole numbers (or fractions) for a great many numbers n (up to all primes less than $n = 14000$ if none of the numbers x, y, a is divisible by n), but this is not what is required. A proof disposing of *all* n's greater than 2 is demanded. Fermat said he possessed a "marvellous" proof.

After all that has been said, is it likely that he had deceived himself? . . . One great arithmetician, Gauss, voted against Fermat. However, the fox who could not get at the grapes declared they were sour. Others have voted for him. Fermat was a mathematician of the first rank, a man of unimpeachable honesty, and an arithmetician without a superior in history. (Bell, 1965, pp. 71–72)

by him is such that one may hope for a solution still in this century.'[440] After the Academy accepted the endowment, an announcement immediately appeared in *Physikalische Zeitschrift*, issue of 1 September 1907, informing about the existence of the Wolfskehl Prize and the problem to be solved. At the same time the Göttingen Academy had an announcement ('*Bekanntmachung*') printed, giving all the details and fixing 13 September 2007 as the date by which the problem should be solved.[441] As soon as the Wolfskehl Prize became known many mathematicians (including crackpots) submitted 'solutions'; thus, after a short time the *Wolfskehl Commission*—consisting of Ehlers, Hilbert, Klein, Minkowski and Runge—had to issue a new statement, reminding the authors that they had to solve Fermat's problem with the help of serious mathematical tools, i.e., with the methods of number theory. There were those who asked Hilbert to submit the proof of Fermat's *Last Theorem* himself to win the Wolfskehl Prize, but he laughed it off by saying that, 'Why should I kill the goose that lays the golden egg?' The 'golden egg' he had in mind was the interest on the principal, amounting to 5000 Marks per annum.

In 1908 the Wolfskehl Commission decided to use a part of the interest to invite prominent scientists as guest speakers to the University of Göttingen.[442] In March 1909 a sum of 2500 Marks was authorized as honorarium for Henri Poincaré, who was invited as the first visitor to give lectures in April 1909. Poincaré gave six lectures on selected topics of pure mathematics and mathematical physics. In his first talk on 22 April he spoke on Fredholm's integral equations, the theory which had just been developed by Hilbert and his students. The mathematicians in Göttingen wondered about the choice of this topic, for they 'were suprised that Poincaré would come and talk to *us* about integral equations' (Reid, 1970, p. 120). Still, the guest from Paris had something new to say about the subject, to which he had contributed himself. In his next two talks he considered the applications of integral equations to the theory of tides (of the sea) and to Hertzian waves, respectively. Then he spoke on Abelian integrals and Fuchsian functions, on transfinite numbers, and finally in his last lecture on 28 April, entitled '*La mécanique nouvelle*'—the only one he gave in French—he discussed the theory of relativity.[443]

In the following year Hendrik Antoon Lorentz was invited. From 24 to 29 October 1910, he delivered six lectures on 'Old and New Problems in Physics.'

---

[440] The mathematicians wrote these remarks on the letter which F. Leo, the Secretary of the Göttingen Academy, had appended to a copy of Paul Wolfskehl's last will before sending it to Klein, Hilbert, Minkowski and Runge. (The letter was dated 29 September 1906.)

[441] '*Bekanntmachung*,' see *Nach. Ges. Wiss. Göttingen*, 1908, p. 103.

[442] A part of the interest was used, for instance, to cover the expenses of printing the announcements, etc. Certain amounts were awarded for partial solutions of Fermat's problem: for example, the *Commission* decided on 17 November 1909 to make an award of 1000 Marks to A. Wieferich, who had proved a special case of Fermat's *Last Theorem*, namely, the fact that Eq. (169) has no solution if n is a prime number with $2^{n-1} - 1$ not divisible by n^2.

[443] The lectures were published as a booklet (Poincaré, 1910). Incidentally, Poincaré—in his last lecture on relativity—did not mention the name of Einstein.

Lorentz especially discussed the questions of relativity theory and radiation theory: he devoted his first lecture to Maxwell's equations and the concept of the ether; the second to Einstein's principle of relativity; the third to a possible modification of the law of gravitation (Lorentz mentioned, for instance, the observed shift of the perihelion of Mercury, which could not be described by Newton's law) and some special questions of electron theory; the fourth to the phenomena of radiant heat; and the last two lectures to the derivation of Planck's law. The lectures were subsequently edited by Max Born and published in *Physikalische Zeitschrift* (Lorentz, 1910b). In 1911 a prize of 5000 Marks was given to Ernst Zermelo for his achievement in axiomatizing set theory.[444] In May 1912 Arnold Sommerfeld was invited to give a series of lectures on recent advances in physics; in these lectures he also mentioned the discovery of X-ray interference phenomena in crystals by von Laue, Friedrich and Knipping. In April 1913, Hilbert organized what was later called the *Kinetische Gas-Kongress*, i.e., a week of lectures on various topics of the kinetic theory of matter and electricity, given by Max Planck, Peter Debye, Walther Nernst, Marian von Smoluchowski, Arnold Sommerfeld and Hendrik Antoon Lorentz. In summer 1915 Einstein visited Göttingen, and during the next three years the distinguished invitees were, respectively, Marian von Smoluchowski, Gustav Mie and Max Planck.[445] At Hilbert's request extra funds from the Prussian Ministry of Education were added to augment the income from the interest on the Wolfskehl endowment, making it possible to invite a prominent scientist for up to a semester each year.[446] As the first Wolfskehl professors, the mathematician Alfred Haar and the physicist Peter Debye were envisaged for visits during the year 1914.

In the disbursement of funds from the income on the Wolfskehl endowment the field of physics was evidently the great beneficiary, and Felix Klein and David Hilbert were undoubtedly behind that. They felt that, after Minkowski's death and his replacement by the pure mathematician Landau, physics had suffered a serious loss which had to be made up. Hilbert, especially, had lost a partner with whom he used to discuss the problems of physics. Still he wanted to

---

[444] In summer 1911 Albert A. Michelson visited Göttingen and gave several lectures. We have not found evidence that he was paid from the Wolfskehl funds. However, Max Born noted that Michelson was 'the next visiting professor (1911)' after Lorentz, implying that he was invited by the *Wolfskehl Commission* (Born, 1978, p. 146). Carl Runge was in America in 1909–1910, and probably invited Michelson then to visit Göttingen. Iris Runge noted in the biography of Carl Runge that Michelson lectured at Göttingen in 1911 for a semester (from May to July), but not that he was paid from the Wolfskehl funds (I. Runge, 1949, p. 149). The money for his visit may have been supplied by the *Verein* supporting the applied physics chairs.

[445] Smoluchowski lectured from 20 to 22 June 1916 on topics of kinetic theory (Smoluchowski, 1916); Mie discussed Einstein's theory of gravitation and his own unified theory of matter from 5 to 8 June 1917 (Mie, 1917); and Planck gave four lectures on quantum theory from 13 to 17 May 1918. (See the announcement in *Physikalische Zeitschrift 19* (1918), p. 176.)

[446] The request for augmenting the Wolfskehl funds by 5000 Marks per annum was made on 7 June 1913.

go ahead with an attack on his Paris problem No. 6, the axiomatization of physics; the solution of this problem, i.e., the systematic mathematical foundation of physical disciplines, he believed, required a mathematician. 'Physics,' he said frequently, 'is much too hard for physicists' (Reid, 1970, p. 127). Of course, he knew that the physicists had provided many ideas and made many discoveries in recent decades, ranging from Heinrich Hertz' electromagnetic waves, through the discovery of the electron, the electron theory and relativity theory, up to radiation and quantum phenomena. In order to learn about the details and digest this wealth of information he needed help and he thought of having a special assistant in physics. He asked Sommerfeld for advice, and Sommerfeld supplied him with several of his young Ph.D's. The first to come as 'Hilbert's physics tutor' was Paul Ewald, who had studied earlier in Göttingen and had obtained his doctorate in spring 1912 at the University of Munich.[447] Hilbert assigned to Ewald various topics in physics that he himself wanted to know about. As Ewald recalled later: 'I remember that one thing he assigned me was the following. There was a long-standing controversy about the number of constants of elasticity in a crystal—it went back to the founders of the field—and Hilbert wanted me to read up on it and tell him who was right.' Ewald studied the literature and reported about it to Hilbert. 'A few years later,' he added, 'the whole problem—which had held up crystal physics for more than fifty years—was solved by Max Born' (quoted in Reid, 1970, p. 129). Ewald had also to inform Hilbert about the kinetic theory of matter and the problems of radiation theory. When he went back to Munich after one year, Sommerfeld sent Alfred Landé to Göttingen. As Landé recalled: 'All kinds of subjects, the physics of solid bodies and of spectra and fluids and heat and electricity, everything that came to him I was to read and, when I found it interesting, to report it to him . . . That was really the beginning of my whole career as a scientist. Without Hilbert, I would probably never have read all those papers, certainly not have digested them. When you have to explain something to someone else, then you first must really understand it and be able to put it into words' (quoted in Reid, 1970, p. 134). In December 1914 Landé left Göttingen to serve in the Red Cross. In war times he could not be replaced, but by then Peter Debye had joined the Göttingen faculty as Professor of Physics, and with him Hilbert then began to organize seminars on the structure of matter, the most fundamental field of physics.

Hilbert began to work on problems of physics in 1912, when he published two

---

[447] Peter Paul Ewald was born in Berlin on 23 January 1888. He studied at the Universities of Cambridge (1905), Göttingen (1906–1907) and Munich (1907–1912), and received his doctorate under Sommerfeld in 1912. After spending a year with Hilbert, he returned to Munich, where he became *Privatdozent* in 1914. In 1921 he was appointed extraordinary professor at the *Technische Hochschule*, Stuttgart, and in the following year was promoted to a full professorship. He retained this position until 1937, when he went to Cambridge; then he took up a chair at Queen's University, Belfast, and in 1949 he moved to a professorship at the Polytechnic Institute of Brooklyn. He retired in 1959; later, Paul Ewald and his wife settled in Ithaca, New York, to be close to their daughter Rose and son-in-law Hans Bethe.

papers, the first on kinetic gas theory and the second on radiation theory. The main point of the paper on the 'Foundation of Kinetic Gas Theory' was that Hilbert, by applying his method of linear integral equations, systematically established the scheme of kinetic gas theory on the basis of *one* fundamental formula: the Maxwell–Boltzmann collision equation (Hilbert, 1912a). Starting from the collision equation as an axiom, he derived the principal theorems and equations of the theory, such as the second law of thermodynamics, Boltzmann's expression for the entropy of a gas, the equations of motion including friction, the equation of heat conduction and the diffusion equation.[448] In the second paper, entitled 'Foundation of Elementary Radiation Theory,' Hilbert again applied the theory of linear integral equations. He started from the basic concepts of emission and absorption (of radiation) and derived Kirchhoff's law concerning the relation between emission and absorption of radiant bodies (Hilbert, 1912b). Hilbert returned to the radiation problem in a second note, in which he formulated the four basic axioms of the theory (Hilbert, 1913).[449] Finally, he completed the foundation of elementary radiation theory in a third contribution by demonstrating the consistency of the axioms employed (Hilbert, 1914).

The work on the kinetic theory of matter and an elementary radiation theory represented only the prelude to a still more ambitious project: to establish a theory of matter based on the principle of relativity. Hilbert was stimulated principally by two developments: the attempt of Gustav Mie in Greifswald to build a complete theory of the electron and of matter by invoking systematically the field concept (Mie, 1912a, b, c), and Einstein's attempt to generalize the theory of relativity. In summer 1915 Einstein visited Göttingen to give a lecture on the status of his efforts concerning the new theory. In a letter to Arnold Sommerfeld, Einstein wrote about the reception of his ideas: 'I had the great joy of seeing in Göttingen that everything is understood to the last detail. With Hilbert I am just enraptured. An important man' (Einstein to Sommerfeld, 15 July 1915). Hilbert had an equally strong impression of Einstein, and in the following months he worked hard to complete a paper, entitled '*Die Grundlagen der Physik*' ('The Foundations of Physics'), which he presented to the Göttingen

[448] It should be mentioned that at about the same time as Hilbert worked on the foundations of kinetic theory, Paul Hertz arrived in Göttingen as a *Privatdozent*. He lectured in the winter semester 1912–1913 on kinetic gas theory. Paul Hertz was born on 29 July 1881 at Hamburg. Before coming to Göttingen he had worked on the mechanical foundations of thermodynamics (Hertz, 1910). He was promoted to an extraordinary professorship in 1921 and left Göttingen in 1933. He died in Philadelphia on 24 March 1940.

[449] The first axiom stated that in thermal equilibrium the emitted and the absorbed energy of a body must be equal. According to the second axiom, at each point of space equilibrium must exist for every frequency of radiation. The third axiom demanded that the characteristic quantities determining the radiation of a given wavelength (i.e., the velocity of light and the emission and absorption coefficients) must be uniquely defined at each point of space. The fourth axiom finally stated that in thermal equilibrium the radiation density should depend uniquely on the physical properties of matter at a given point.

Academy on 20 November 1915 (Hilbert, 1915). In this paper he obtained, independently of Einstein, the field equations of gravitation.[450]

Already in his obituary of Minkowski, Hilbert had indicated the goal he tried to achieve six years later. He had then emphasized the fact that Minkowski had used the relativity principle as a fundamental postulate ('world postulate'), from which the laws of nature might be derived: '*Minkowski's electrodynamic equations* are a necessary consequence of the world postulate; they are as certain as the latter' (Hilbert, 1909b, p. 94; reprinted in *Gesammelte Abhandlungen 3*, 1935, p. 358). Now, in 1915, he felt prepared to continue Minkowski's previous work and to establish, on the basis of general postulates, even more fundamental laws of physics than Minkowski's electrodynamic laws. He had carefully studied the most important papers, in which relativity theory and electron theory had been generalized since 1909, especially those of Einstein, Max Abraham, Gunnar Nordström and Gustav Mie. While he appreciated the ingenious ideas of these physicists deeply, he still expected to improve upon them by applying his axiomatic method. 'Following the axiomatic method,' he remarked in the introduction to *Die Grundlagen der Physik*, 'in fact from two simple axioms, I would like to propose a new system of basic equations of physics. They are of ideal beauty and, I believe, they contain the solution of the problems of Einstein and Mie *at the same time*' (Hilbert, 1915, p. 395). Starting from Minkowski's four-dimensional world, Hilbert assumed that at each world-point, x_μ ($\mu = 1,..,4$), the events were determined by the ten gravitational potentials, $g_{\mu\nu}$ ($\mu, \nu = 1,..,4$), of Einstein and the four electromagnetic potentials, A_μ ($\mu = 1,..,4$). The laws of physics, he asserted, followed from two fundamental axioms. As the first, he took what he called 'Mie's axiom of the world-function'; it stated that there existed the 'world-function' H, depending on the $g_{\mu\nu}$ and their first and second spatial derivatives and on the electromagnetic potentials A_μ and their spatial derivatives. From this function H, so to say the Hamiltonian of the universe, the laws of physics could be derived by taking the variations of the integral, $\int H \sqrt{g}\, d^4x$, where g is the determinant of the metric tensor $g_{\mu\nu}$. The second axiom, which he called 'the axiom of general invariance,' stated that H must be invariant with respect to arbitrary coordinate transformations.[451] On the basis of these two axioms Hilbert derived three important theorems and, by applying them to the

---

[450]When Hilbert presented his memoir containing the new equations of gravitation on 20 November to the Göttingen Academy, Einstein had already contributed three notes on his theory to the Prussian Academy in Berlin on 4, 11, and 18 November 1915 (Einstein, 1915d, e, f). However, in all three communications Einstein gave equations of gravitation that were different from Hilbert's. Only in his last communication of 25 November 1915 did Einstein write, five days after Hilbert, the final equations of gravitation (Einstein, 1915g). In agreement with this fact Hilbert, in an addendum to his published paper, remarked: 'It seems to me that the differential equations of gravitation so realized [by me] are in agreement with the beautiful theory of general relativity proposed by Einstein in his later [25 November 1915] memoir' (Hilbert, 1915, p. 405). (For an account of the history of Einstein's and Hilbert's work on the theory of gravitation, see Mehra, 1973a, 1974.)

[451]Hilbert remarked in a footnote that in Einstein's theory thus far—at least up to 18 November 1915—the general covariance of H had not been established (Hilbert, 1915, p. 396, footnote 2).

world-function H, he obtained 14 equations of motion for the fourteen potentials $g_{\mu\nu}$ ($g_{\mu\nu}$ being symmetrical in the two indices) and A_{μ}.[452] Ten of these equations involved variations with respect to the gravitational potentials $g_{\mu\nu}$, and Hilbert called them the equations of gravitation; the other four involved only variations with respect to the electromagnetic potentials and they represented generalized Maxwell's equations. Now the equations of gravitation coincided, as Hilbert found out later, with Einstein's equations of general relativity, while the electrodynamic equations corresponded to the ones derived earlier by Gustav Mie. Hilbert noticed, however, an important fact: because of the four conditions, which the function H had to satisfy in his theory, the fundamental equations of electrodynamics depended on those of gravitation. Hence he concluded that, '*the electrodynamic phenomena are the consequences of gravitation*' (Hilbert, 1915, p. 397). Hilbert concluded his paper by remarking that his two axioms had made it possible to formulate the new concepts of space, time and motion in the sense of Einstein's theory of general relativity, and he further expressed his conviction that,

> by means of the fundamental equations established here the innermost, hitherto hidden processes within atoms, will be explained, and in general it must be possible to achieve a reduction of all physical constants to mathematical constants. Thus altogether the possibility comes closer that physics would become in principle a science of the nature of geometry, much to the most splendid glory of the axiomatic method, which, as we see, makes use of the most powerful instruments of analysis, namely, variational calculus and the theory of invariants. (Hilbert, 1915, p. 407)

Hilbert's *Grundlagen der Physik* and Einstein's theory of general relativity were considered by Felix Klein as the most brilliant applications of his Erlangen programme.[453] On 25 January 1918 he contributed a note, entitled '*Zur Hilberts ersten Note über die Grundlagen der Physik*' ('On Hilbert's First Note Concerning the Foundations of Physics'), to the Göttingen Academy, in which he clarified in particular the meaning of Hilbert's four conditions (Klein, 1917). In two memoirs, presented to the Academy later that year, Klein dealt with energy-momentum conservation in general relativity theory and unified the different treatments of the problem by Einstein, Hilbert and Lorentz (Klein, 1918a, b) 'My goal,' wrote Klein, 'is to state the *mathematical* conceptions as clearly as possible,

---

[452] The first of these three theorems stated that for any function, which is invariant under general coordinate transformations and satisfies a variational principle (such as the world-function H), four differential equations follow which are automatically satisfied. Later these identities of Hilbert were recognized to be special cases of the identities in differential geometry, which Luigi Bianchi had discovered earlier (the so-called 'Bianchi identities').

[453] In contrast to Hilbert and Klein, Einstein did not wish to see Hilbert's theory connected with his own. Thus he wrote to Ehrenfest: 'I do not like Hilbert's formulation. It is needlessly specialized and, as far as "matter" is concerned, unnecessarily complicated. It is not honest (= Gaussian) in design, [and reflects] the pretension of a superman by a camouflage of techniques' (Einstein to Ehrenfest, 25 May 1916). Einstein especially disliked the fact that Hilbert had made use of Mie's theory of matter.

while touching upon the physical problem only briefly. I feel a certain satisfaction in this, since [my] old ideas of 1871–72 obtain decisive validity . . . . Einstein's papers, upon which I shall comment here, indeed demonstrate that at times, although not systematically, he resorts to the same conceptual freedom which I have proposed in my Erlangen programme' (Klein, 1918b, pp. 394–395 and 399). Klein's work was closely connected with the studies of Emmy Noether concerning the relations between symmetry properties and conservation laws.[454] Emmy, daughter of the Erlangen mathematician Max Noether (1844–1921), had come to Göttingen in 1916.[455] On 26 July 1918 Felix Klein communicated her paper, entitled '*Invariante Variationsprobleme*' ('Invariant Variational Problems'), to the Göttingen Academy (Noether, 1918). In this paper she investigated the consequences of the invariance of the action-integral for a dynamical system under the transformations of a continuous (Lie) group and proved what later was called 'Noether's theorem,' which stated: if the action integral or the equation of motion of a dynamical system are invariant (symmetric) under the group transformations, then certain quantities—which generate the group according to the procedure of Sophus Lie—will be conserved quantities. Thus Emmy Noether completed a line of thought which had begun with Bernhard Riemann's 'trial lecture' ('*Probevorlesung*') in 1854. In particular, she established in full generality the relation between the conservation laws of physics and the symmetry properties of the dynamical laws, which had been suggested fifty years earlier by Hermann Helmholtz in his communication, entitled 'On the Facts which Lie at the Basis of Geometry,' to Göttingen Academy (Helmholtz, 1868). In this memoir Helmholtz had considered a solid body in space and had related the constants of motion of the body to the geometrical properties of space and of the body itself; he had referred, then, to Riemann's 1854 lecture which had just been published (in 1867) and stated that he had arrived at these conclusions independently of Riemann. Later, Georg Hamel, Gustav Herglotz and Friedrich Engel tried—with the help of Sophus Lie's theory of continuous transformation groups —to formulate the ideas of Riemann and Helmholtz for special cases. Noether's theorem of 1918, however, provided the ultimate and most general proof of Riemann's conjecture. Emmy Noether had dedicated her paper to Felix Klein— Riemann's successor and advocate of his mathematical spirit—on the occasion of the fiftieth anniversary of his (Klein's) doctorate.

Hilbert regarded the results of Klein and Noether with satisfaction, for they supported his own conceptions about the foundations of physics. In a second communication to the Göttingen Academy on 23 December 1916, he completed his axiomatic treatment of the foundations of physics (Hilbert, 1917). He added,

[454] At the end of his first paper on energy momentum conservation, Klein thanked Emmy Noether for her assistance and referred to her forthcoming papers (Klein, 1918a, p. 189).

[455] Amalie Emmy Noether was born on 22 March 1882 at Erlangen. She obtained her doctorate under Paul Gordan in 1906. In 1919 she became *Privatdozent* at the University of Göttingen, and three years later she obtained the title of an extraordinary professor. Having been forced to leave Göttingen by the Nazis in 1933, she found a position at Bryn Mawr College in Pennsylvania, where she died on 14 April 1935.

in particular, an axiom concerning the structure of space and time and formulated the principle of causality.[456] With such foundations he then discussed the results of the theory of general relativity. In the following years, relativity theory was extended by Hermann Weyl and Albert Einstein in the direction indicated by Hilbert, namely, by attempting to combine the theories of gravitation and electrodynamics. In 1924 Hilbert had the basic content of his two memoirs on *Die Grundlagen der Physik* reprinted in the *Mathematische Annalen* (Hilbert, 1924). He justified this republication by the argument: 'I am quite sure that the theory I have developed here contains an enduring core, and provides a framework within which there is sufficient freedom for the future development of physics in accordance with the ideal of field-theoretical unity.' But he immediately added: 'Whether in fact the ideal of a pure field-theoretical unity is a definitive one, or if certain extensions and modifications of it will be necessary, in order to obtain in particular the theoretical foundation for the existence of the electron and the proton, as well as the establishment—free of contradictions—of the laws valid in the interior of the atom—to answer this is the task of the future' (Hilbert, 1924, p. 2).

The future development, however, led away from the field-theoretical ideas of Einstein and Hilbert. While Felix Klein died on 22 June 1925, a few weeks before Werner Heisenberg completed a paper which inaugurated the scheme of a consistent quantum theory (Heisenberg, 1925c), Hilbert lived to see how some of the questions he had raised were answered. These answers implied fundamentally new concepts; however, they also made use of important results of Hilbert's theory of linear integral equations, and thus of one of his basic tools for dealing with the foundations of physics. Hilbert himself contributed to the 'axiomatic' formulation of quantum mechanics (Hilbert, von Neumann and Nordheim, 1928). But his main concern shifted again to the foundations of mathematics, especially arithmetic. It began with a lecture, given to the Swiss Mathematical Society in Zurich on 11 September 1917, entitled 'Axiomatic Thinking' ('*Axiomatisches Denken*'), in which he summarized the successes achieved to date with the help of the axiomatic method in mathematics, physics and logic (Hilbert, 1918). The axiomatic method was then criticized thoroughly by the Dutch mathematician Luitzen Egbertus Jan Brouwer (1881–1966) and Hermann Weyl. Hilbert replied to their criticism in lectures to the Copenhagen Mathematical Society and the *Mathematisches Seminar* of the University of Hamburg in spring and summer 1922, which he published under the title '*Neubegründung der Mathematik*' ('The New Foundation of Mathematics,' Hilbert, 1922). In these lectures Hilbert turned towards an analysis of the methods of mathematical proofs. He carried on these investigations for many years, approaching but not reaching his final goal.[457] Hilbert even lived on to see, after the rise of the Nazis

---

[456] The space-time axiom stated that if the quadratic form $\sum g_{\mu\nu} x^\mu x^\nu$ ($\mu, \nu = 1, .., 4$) is transformed to sum of squares, then three of them have the positive sign and one the negative sign.

[457] In 1930 Kurt Gödel established a theorem which made it doubtful that a proof of consistency of a complete set of axioms of arithmetic may be achieved on the basis of the existing axiomatic methods (Gödel, 1931).

in 1933, the breakdown of the tradition of mathematics and physics in Göttingen when Hermann Weyl (Hilbert's successor since 1930), Richard Courant, Edmund Landau, Emmy Noether, Max Born, James Franck and many others had to leave the *Georgia Augusta*. Hilbert died on 14 February 1943 in the midst of a new, destructive world war.

III.2   The Continuity of the Tradition: Richard Courant, Max Born and James Franck in Göttingen

During the second decade of the twentieth century several of the senior Göttingen professors of mathematics and physics—men like Eduard Riecke, Felix Klein, and Woldemar Voigt—reached the age of retirement. Thus a considerable turnover in the faculty was bound to occur; the change, however, was smoothened by certain circumstances. In 1911 Felix Klein suffered a breakdown in his health because of strenuous work and too many duties. He retired from teaching in 1913 so as to be able to devote himself to purely organizational tasks.[458] Klein's professorship was filled by Constantin Carathéodory, who had obtained his doctorate in 1904 under Minkowski and who had worked—besides the variational calculus, which was the subject of his thesis—on the axiomatic foundation of thermodynamics.[459] When he was called to Berlin in 1918 as the successor of Hermann Amandus Schwarz, Erich Hecke—a former student of Hilbert's—succeeded him, but only for a year; then Hecke left to take a professorship at the newly founded University of Hamburg.[460] He was followed by another of Hilbert's former students: Richard Courant.

Courant was born on 8 January 1888 at Lublinitz, Upper Silesia. He received

---

[458] Felix Klein later on resumed lecturing for a small circle of advanced scholars in his home. There he gave, for instance, his course on the development of mathematics in the nineteenth century (Klein, 1926, 1927).

[459] Constantin Carathéodory was born of Greek parents in Berlin on 13 September 1873; his father was in the diplomatic service. He received his high school education in Brussels; there he also studied engineering at the *École Militaire de Belgique* from 1891 to 1895. After spending some years as an engineer—for example, at Aswan in Egypt from 1898 to 1900 during the construction of the (first) dam—Carathéodory decided to pursue mathematics and became a student again, first at the University of Berlin (1900–1902, especially under Hermann Amandus Schwarz) and later at the University of Göttingen. In Göttingen he obtained his doctorate in 1905 and became a *Privatdozent* immediately afterwards. In 1909 he obtained a professorship of mathematics at the *Technische Hochschule*, Hanover, and in 1910 he moved to the *Technische Hochschule* in Breslau. From Breslau he went to Göttingen in 1913 as Felix Klein's successor, and from there to Berlin in 1918. He finally moved to Munich in 1924. Both in Berlin and in Munich he occupied the chair of mathematics. Carathéodory died in Munich on 2 February 1950.

[460] Erich Hecke was born on 20 September 1887 at Buk, in the province of Posen in Prussia. He studied mathematics at the Universities of Breslau and Göttingen, obtaining his doctorate under Hilbert in 1910 with a thesis on the theory of modular functions. In 1912 he became a *Privatdozent* at Göttingen; in 1915 he was called to an extraordinary professorship at the University of Basle (and was promoted to a full professorship the following year). After the Göttingen intermezzo he occupied the chair of mathematics in Hamburg from 1919 until his death on 13 February 1947; he died in Copenhagen.

his school education at Glatz and Breslau.[461] He began to study mathematics and physics at the University of Breslau in spring 1905, one year before obtaining his *Abitur*, and he was particularly impressed by the lectures of Jakob Rosanes. Two years later he went to Zurich for a semester to attend the courses of Adolf Hurwitz, and in fall 1907 he continued his studies at the University of Göttingen, following the path of Otto Toeplitz and Ernst Hellinger, his fellow students from Breslau. Hellinger helped him to become accustomed to the higher standard of mathematics at Göttingen and introduced him to the *Lesezimmer*, the Mathematical Reading Room. Besides mathematics, Courant was deeply interested in philosophy and attended the lectures of the philosopher Edmund Husserl. Soon he became a member of the group of exceptionally talented students of Klein, Hilbert and Minkowski, which included Alfred Haar (then Hilbert's assistant), Erich Hecke (whom Courant knew from Breslau), Ernst Meissner (a student of Minkowski's) and Hermann Weyl.[462] He obtained his doctorate under Hilbert in February 1910 with a thesis on Dirichlet's principle; he had applied the latter to problems of conformal mapping (Courant, 1910).[463] After serving his year of compulsory military service from fall 1910 to 1911, he became *Privatdozent* at the University of Göttingen in 1912 and married Nelly Neumann, a fellow student from Breslau.[464] The flourishing period of Göttingen mathematics—during which, besides Courant and Toeplitz, Erich Hecke and Hermann Weyl were *Privatdozenten* in mathematics, Max Born in theoretical physics and Theodore von Kármán in mechanics—came to an end with the outbreak of World War I. Courant, who was seriously wounded in 1915, returned to Göttingen in late 1918 to resume his lecturing duties (having been awarded meanwhile the title of 'professor'). He and Nelly had been divorced in 1916. Courant married Nina Runge, daughter of Carl Runge, early in 1919. In spring 1920 he was called as

[461] Richard Courant came from a Jewish family, which had moved to Breslau from Lublinitz via Glatz in Silesia. Edith Stein, the Catholic philosopher, was a cousin of his father Siegmund Courant. For details of Courant's background and life, see Constance Reid's biography (Reid, 1976).

[462] Courant succeeded Alfred Haar as Hilbert's assistant in 1908 and retained this position until 1910. His duties included taking notes of Hilbert's lectures and working them out for the Mathematical Reading Room (*Lesezimmer*).

[463] In his thesis, Bernhard Riemann had introduced a theorem (Riemann, 1851, §18), which he later called 'Dirichlet's principle,' and applied it to obtain fundamental theorems of the theory of Abelian functions (Riemann, 1857). This theorem, which Riemann had learned in Dirichlet's lectures, stated the equivalence of a variational problem in a plane, on the one hand, and the first boundary value problem of potential theory, on the other. The variational problem consisted in finding the function that would make the double integral, $\int\int[(\partial\phi/\partial x)^2 + (\partial\phi/\partial y)^2]\,dx\,dy$, a minimum. According to Dirichlet's principle the existence of the minimum should guarantee the solution of the boundary value problem for the Laplace equation ($\partial^2\phi/\partial x^2 + \partial^2\phi/\partial y^2 = 0$). Later, Weierstrass showed that Riemann's argument was not rigorous; then many mathematicians studied the problem until Hilbert in 1900 showed a way of rescuing Dirichlet's principle, which had led to powerful applications (Hilbert, 1901). The work on the Dirichlet principle continued even after Hilbert's contributions to it and after Courant's doctoral thesis under Hilbert (Courant, 1910); Courant himself obtained important results with its help (see, e.g., Courant, 1950).

[464] Courant's *Habilitation* thesis again dealt with Dirichlet's principle. In his inaugural lecture on 23 February 1912 he spoke on a related subject, namely, 'On Existence Proofs in Mathematics.' (See Reid, 1976, p. 40.)

professor of mathematics to the University of Münster, but he returned to Göttingen in fall of the same year, finally as ordinary (i.e., full) professor.

Courant became the genuine successor to Felix Klein, especially as an organizer. He had shown organizational ability during the war, when he had approached Ferdinand Springer, the publisher, with the plan of editing a series of up-to-date monographs in mathematics and it applications.[465] The first volume of this series, which was called 'Die Grundlehren der mathematischen Wissenschaften in Einzeldarstellungen—Mit besonderer Berücksichtigung der Anwendungsgebiete' ('The Fundamental Topics of Mathematical Sciences in Monographs—With Special Reference to Applications'), appeared in 1921: it was Wilhelm Blaschke's Vorlesungen über Differentialgeometrie (Lectures on Differential Geometry, including the geometrical foundations of relativity theory). Three years later the series had already run up to volume seventeen, which was a German translation of Edmund Taylor Whittaker's Treatise on Analytical Dynamics (in the second edition), and included, as No. 12, the first volume of Die Methoden der mathematischen Physik (The Methods of Mathematical Physics), which Courant had written on the basis of the notes of Hilbert's lectures on mathematical topics that played a role in physics (Courant and Hilbert, 1924).[466] In the early years of his professorship in Göttingen, Courant also started a very ambitious project: he thought of establishing a mathematics institute—similar to those of experimental sciences—which would house all existing and future mathematical activities including the Reading Room. And he immediately introduced a new activity, the Anfängerspraktikum (a mathematical laboratory, similar to the ones existing in physics and chemistry), where the beginning students obtained the practical opportunity of solving problems to reinforce the knowledge gained in the lectures. After several years of thinking and planning, the prospects of establish-

---

[465] Ferdinand Springer, together with his cousin Julius, directed the Julius Springer Verlag, which had been founded in 1842 by their grandfather Julius Springer (1817–1877); Julius headed the publication of books on the engineering sciences, and Ferdinand those on medicine, natural sciences and art.

Since January 1913 Springer Verlag had been publishing Die Naturwissenschaften, the weekly journal for science, medicine and technology, edited by Arnold Berliner (first in collaboration with Curt Thesing, then from 1914 with August Pütter and, after 1922, alone), which later (in 1924) became the official journal of the Gesellschaft Deutscher Naturforscher und Ärzte and of the Kaiser Wilhelm-Gesellschaft zur Förderung der Wissenschaften. After 1922 the journal was supplemented by a series of annual volumes, entitled 'Ergebnisse der exakten Wissenschaften,' which was also edited by Berliner and published by the Springer Verlag. Courant got to know Ferdinand Springer through Berliner.

[466] Courant's series was continued in the following decades; thus John von Neumann's Mathematische Grundlagen der Quantenmechanik appeared as No. 38 of the series (von Neumann, 1932).

In 1925 the Göttingen physicists Max Born and James Franck started, again with the Springer Verlag, a similar series in physics, entitled 'Struktur der Materie in Einzeldarstellungen'; the first few volumes of this series were: Ernst Back and Alfred Landé's Zeemaneffekt und Multipletstruktur der Spektrallinien (Back and Landé, 1925), Max Born's Vorlesungen über Atommechanik (Born, 1925), and James Franck and Pascual Jordan's Anregung von Quantensprüngen durch Stöße (Franck and Jordan, 1926), respectively.

In 1918 Ferdinand Springer had also started a new mathematical journal, the Mathematische Zeitschrift (which, in the beginning, was edited by Leon Lichtenstein). Many of Courant's papers were published in this journal.

ing the mathematical institute improved: with the help of Harald Bohr, Courant approached the International Education Board for financial support, and by the end of 1926 a sum of $350,000 was pledged for constructing the building and equipping it (see Reid, 1976, p. 108). Three years later the new institute was completed. Among the distinguished speakers at the dedication on 2 December 1929 were such former Göttingen luminaries as Hermann Weyl and Theodore von Kármán. Hilbert, happy in the knowledge that his late colleague Felix Klein's dream had been realized, remarked: 'There will never be another institute like this! For to have another such institute, there would have to be another Courant—and there can never be another Courant' (quoted in Reid, 1976, p. 126).[467]

Courant's efforts were directed towards institutionalizing the great tradition of Göttingen mathematics, and passing on to future generations the experiences of such outstanding mathematicians as Felix Klein and David Hilbert, who—at least partially—had also withdrawn from active scientific work. He made sure that young people of outstanding talent were attracted to, and found their place in, the University of Göttingen. He also obtained for himself two paid assistantships and filled them first with Hellmuth Kneser (the son of Adolf Kneser, his former teacher at Breslau) and Carl Ludwig Siegel (who had earlier been in Göttingen, but had gone with Erich Hecke to Hamburg). Courant further won Emil Artin, who had taken his degree with Gustav Herglotz in Leipzig, as a member of the *Mathematisches Institut*.[468] Although the plan to add Hermann Weyl in 1922 among the ranks of mathematics professors in Göttingen failed, because of Weyl's refusal to leave Zurich, the prewar reputation of the University of Göttingen as a great centre of mathematical learning was soon re-established. Again students and postdoctoral fellows came to Göttingen in large numbers, both from Germany and abroad: they included Kurt Otto Friedrichs, Otto Neugebauer and Hans Lewy, who came in 1922, then Bartel Leendert van der Waerden from Amsterdam and the Russian Pavel Sergeevic Alexandroff. Later, young mathematicians came with fellowships of the International Education Board or the Rockefeller Foundation: e.g., the M.I.T. mathematician Norbert Wiener and the young John von Neumann. And, of course, from the *Praktikum* there emerged mathematicians who made great reputations in the future, such as Willy Feller (who had come without any formal education from Yugoslavia) and Franz Rellich, who came from Austria.

The great success which he had achieved in the tasks of teaching and administration also stimulated Courant's scientific work. After the war he had resumed his work with a long paper on the eigenvalues of partial differential

---

[467] In 1933, after the National Socialists came to power in Germany, Courant was removed from his professorship; in the fall of that year he went to England and finally to the United States. He accepted a professorship of mathematics at New York University, which he occupied from 1934 to 1958. At New York University he built a new institute of mathematics, comparable to the one in Göttingen. Courant died on 27 January 1972 in New Rochelle, New York.

[468] In 1922 the philosophical faculty of the University of Göttingen split into a faculty of natural sciences and a philosophical faculty; at that time Courant obtained the permission to use officially the name '*Mathematisches Institut*.'

equations, published in *Mathematische Zeitschrift* (Courant, 1920).[469] During the 1920s he invested much work in writing books, especially the *Courant–Hilbert*, in which many original contributions of his own were contained. Besides, together with his collaborators Friedrichs and Lewy, he worked on the application of the difference calculus to solve elliptic and hyperbolic differential equations. Altogether he wrote a series of good and important papers, but there was no special single field in which he was the outstanding expert. Still, with all his activities, he represented the heart of the *Mathematisches Institut*, where, in 1925, Gustav Herglotz succeeded Carl Runge (who retired in 1924 and died in 1927).[470] And, when Hilbert retired from his chair in 1930 and Hermann Weyl succeeded him, a bright future seemed to lie ahead of Göttingen mathematics. More than anyone else, Weyl—who had worked on and contributed to deep and fundamental problems, such as the theory of Riemannian sheets, the continuum hypothesis, the unification of gravitation and electromagnetism, the foundations of mathematics, and group theory and its application to physics—had to be considered as the heir to the mathematical tradition initiated by Gauss and Riemann at Göttingen.[471]

Undoubtedly, the mathematical tradition of Göttingen had continued to flourish in the post-World War I years. But its glory was equalled, if not surpassed, by the simultaneous development of physics. This might be considered surprising, for the physicists after Wilhelm Weber could not compete in reputation with their mathematical collegues. In 1900 Eduard Riecke and Woldemar Voigt were, respectively, fifty-five and fifty years old. They had sought to uphold and extend what they had learned from their teachers: thus Riecke had made an original contribution to the electron theory of metals (in the spirit of Weber), and Voigt had investigated primarily the properties of crystals (in the tradition of

---

[469] Courant developed a method of obtaining the eigenvalues of the differential equations as extremum values of certain integrals associated with the differential equations, analogous to the situation in Dirichlet's principle.

[470] Gustav Herglotz was born on 2 February 1881 at Wallern in Bohemian Forest. He studied at the Universities of Vienna (1899–1900) and Munich (1900–1902) and received his doctorate in astronomy under Hugo von Seeliger in 1902. Then he went to Göttingen and became a *Privatdozent* in mathematics and astronomy in 1904. Three years later he was promoted to an extraordinary professorship in astronomy. In 1908 Herglotz was called to the *Technische Hochschule*, Vienna, as an extraordinary professor of mathematics, and he moved to the University of Leipzig the following year as a full professor. He occupied the chair of applied mathematics at the University of Göttingen from 1925 to 1947. He died on 22 March 1953 at Göttingen.

Herglotz contributed to many problems of theoretical astronomy (e.g., the three-body problem), theoretical physics (e.g., electron theory, mechanics of continua, gravitation theory), and mathematics (power series, hypergeometric and spherical functions, differential geometry, and analytical and algebraic number theory).

[471] Hermann Weyl was born on 9 November 1885 at Elmshorn, Schleswig-Holstein in Prussia. He studied mathematics at the Universities of Munich (1905–1906) and Göttingen (1906–1908); he received his doctorate in 1908 under Hilbert with a thesis on singular integral equations. In 1910 he became a *Privatdozent* at the University of Göttingen and, just three years later, he was called to the E.T.H. in Zurich as a professor of mathematics, a position which he held until 1930—when he succeeded Hilbert in Göttingen. He left Göttingen in spring 1933 and in the fall of that year assumed a professorship at the Institute for Advanced Study, Princeton, New Jersey. He retired in 1951 and died on 8 December 1955 in Zurich.

Franz Neumann). Voigt had also performed pioneering work on the Zeeman effect. But both Riecke and Voigt seemed to be concerned too much with the past and did not play a leading role in preparing for the physics of the twentieth century. However, at their institutes, they supported men like Max Abraham, Johannes Stark and Walther Ritz, who worked at the forefront of research on electron theory and atomic physics. On the other hand, important impulses for the development of the new physics were given by the Institute of Physical Chemistry, which Walther Nernst had directed since 1894 and where he had performed the investigations leading to his *heat theorem*. After Nernst left Göttingen for Berlin in 1905, the Physico-Chemical Institute provided less stimulus, but now the mathematician Hermann Minkowski filled the gap: he contributed to the progress of physics by his fundamental work on electrodynamics and relativity theory. Then, following Minkowski's unexpected and untimely death, David Hilbert became deeply interested in the problems of physics. He not only did research on selected problems of physics, but also gave lectures on them: for example, he lectured on the molecular theory of heat in the winter semester 1912–1913 and on the theory of electrons in the summer semester 1913. Hilbert's enthusiasm for physics infected all his mathematical colleagues except Edmund Landau. Indeed, it became fashionable among mathematicians to work on problems of atomic physics and radiation theory.[472] In addition, the Wolfskehl lectures acquainted many people in Göttingen with the foremost topics of physical research. Peter Debye, one of the lecturers at the *Kinetische Gas-Kongress* in April 1913, impressed the Göttingen mathematicians and physicists so much that in summer 1914 he was invited as Professor of Physics to the University of Göttingen. To make Debye's appointment possible, Woldemar Voigt resigned from his chair, retaining the personal title of a professor. Debye immediately began to lecture on various aspects of modern physics: for example, in the winter semester 1914–1915 he gave a course on quantum theory, which he repeated regularly in the following years, and in the summer semester 1915 he lectured on the kinetic theory of dielectric phenomena. Besides, he conducted, together with Hilbert, a Seminar on 'The Structure of Matter' ('*Seminar über die Struktur der Materie*'), which was a continuation of the earlier Hilbert–Minkowski seminars. Debye himself worked, both experimentally and theoretically, on the problems of the structure of matter: in particular, he concerned himself with the interference of X-rays by crystals (Debye, 1915b) and developed, together with Paul Scherrer, a new method—which came to be called the 'Debye–Scherrer' method—for analyzing the structure of crystals (Debye and Scherrer, 1916a, b). This method introduced a definite advantage over the existing methods of research in crystal structures, such as the ones of Laue, Friedrich

---

[472] Hilbert even assigned problems of theoretical physics to his doctoral students. Thus Ludwig Föppl, son of August Föppl (the professor of technical mechanics at the *Technische Hochschule*, Munich), worked on the stability of electron configurations in atoms for his thesis (Föppl, 1912); Hans Bolza (1913) and Bernhard Baule (1914) investigated the theory of dilute gases; and Kurt Schellenberg (1915) studied electrolysis. At the same time Hermann Weyl also published his first papers on physical problems, especially the problems of radiation (Weyl, 1912a, b, c).

and Knipping, and of the Braggs, because the substances could be used as powders rather than as good crystals. Thus Debye and Scherrer were able to study the structures of many more materials than had been investigated before. After only six years in Göttingen—though six most successful years—Debye accepted in spring 1920 the call to become Professor of Experimental Physics and Director of the Physics Laboratory at the E.T.H. in Zurich.

In early 1920 a new situation developed with regard to the physics professorships at the University of Göttingen. Riecke and Voigt were no more, and Debye was about to leave for Zurich. Eduard Riecke had died already on 11 June 1915 and Woldemar Voigt on 13 December 1919. During wartime Riecke's position had not been filled; instead Robert Wichard Pohl, a *Privatdozent* from Berlin, had been appointed in 1917 as an extraordinary professor.[473] In summer 1920 Pohl, who had become known for his work on the photoelectric effect, succeeded to Riecke's chair and began a new field of investigation: the photoelectric action in solids.[474] Later he turned to the study of the processes of absorption of light by crystals (especially of alkali halogenides), of luminescence, and of photochemistry. Pohl founded his own school of devoted collaborators and followers in Göttingen (who were called the '*Pohlierten*'—the '*Pohlians*'); he also introduced a large number of students to physics through his lectures and demonstrations and successful textbooks on experimental physics. In this respect, Robert Pohl continued the tradition of Erxleben, Lichtenberg and Riecke.

Although Pohl and his associates at Göttingen worked experimentally on effects connected with quantum theory, their field was not in the centre of interest in those years. As Pohl recalled later: 'We were altogether a bit isolated with our investigations. At that time there were these wonderful results of Bohr, and all that could be explained on Bohr's theory was particularly interesting' (Pohl, AHQP Interview, 25 June 1967, p. 7).[475] He had in mind, in particular, the experiments of his Göttingen colleague James Franck, who was appointed professor of physics and one of the two successors of Peter Debye in 1920. Franck owed his appointment to the other (real) successor, Max Born. Born, who had been selected first for the Göttingen professorship, recalled later that he had hesitated initially to accept the position because he lacked the knowledge to run a big experimental laboratory and to give lectures on experimental topics in physics. While discussing matters in Berlin with the appropriate official for universities in the Prussian Ministry of Education, a Director Wende, he recognized that on the document presented to him, besides Debye's chair, two

---

[473] R. W. Pohl was born on 10 August 1884 at Hamm, Westphalia. He studied at the Universities of Heidelberg (1903–1904, under Georg Quincke) and Berlin (1904–1906), obtaining his doctorate under Warburg in 1906. He became a *Privatdozent* at the University of Berlin in 1911 and did his military service from 1916 to 1918.

[474] Before Pohl was promoted to a full professorship Max Wien, the experimental physicist at the University of Jena, had received an offer from Göttingen (in late 1919), but he had declined.

[475] Pohl's work was appreciated later on when semiconductors came to be used for technical purposes. Pohl retired in 1952 from his Göttingen chair. He lived to the age of ninety-two and died on 5 June 1976.

extraordinary professorships were listed—connected with the names of Pohl and Voigt, respectively. Now Voigt had retired in 1914 in favour of Debye, but had retained a 'personal' professorship, which was supposed to expire upon his death. However, due to a clerical error, the notation 'to be cancelled after the death of the occupant' stood after Pohl's name instead. 'I am in general not very quick to grasp a situation, but in this case I did,' Born wrote in his *Recollections*. 'So I quickly pointed out to Director Wende that there was another vacancy, and that this changed the whole situation. For if they could appoint a second experimentalist to be the head of Voigt's former department I would not hesitate to accept the post, considering the directorship of the two laboratories as purely formal. Wende laughed, saying that this was obviously an error of the copyist; but he very soon saw the possibilities. You must understand we were still living in revolutionary times. Like most of the officials in the Ministry of Education, he was a new man, keen to do new things. One of his aspirations was to stimulate science and in particular to foster the old scientific centre of Göttingen. He therefore accepted my proposal and asked me to nominate a candidate for the second Extraordinariat' (Born, 1978, p. 200).[476] And, after some thought, he decided to propose his friend James Franck for the second professorship, because 'The experiments which he, in collaboration with [Gustav] Hertz, had done to demonstrate Bohr's quantum theory of atoms I regarded as most important and fundamental. I had known Franck since my student days and loved him as a most honest, reliable and good-humoured fellow. I also knew that he and Pohl had been at the same school in Hamburg and were great pals, which guaranteed a frictionless collaboration between the two experimental departments' (Born, 1978, p. 200). In Göttingen the news about the appointment of the two new physics professors was very welcome. 'Franck plus Born are the best imaginable replacement for Debye,' Hilbert wrote to Courant. 'I am very happy about this arrangement' (see Reid, 1976, p. 82). Hilbert could indeed be very pleased with Born, who had not only made a considerable reputation as a theoretical physicist but was particularly suitable to represent the Göttingen tradition.

Max Born had descended from an academic family: his grandfather, Marcus Born, had been the first Jewish physician to be appointed as the district medical officer in Prussia and his father, Gustav Born, was professor of anatomy at the University of Breslau.[477] Max was born in Breslau, the capital of the province of Silesia, on 11 December 1882.[478] He grew up in the well-educated atmosphere of

---

[476] According to another recollection of Born (in his commentary to the Born–Einstein letters), Wende did not take the responsibility; he consulted his superior, the minister—Professor Becker—who agreed with Born's request. (See *The Born–Einstein Letters*, 1971, pp. 26–27.)

[477] Max Born's mother, Margarethe Kaufmann, came from a Breslau family active in the textile business. She loved music and knew some of the celebrated musicians of her time, including Johannes Brahms, Clara Schumann and Pablo Sarasate. Unfortunately she died already in 1886 when Born was four years old. Four years later Max and his sister Käthe had a stepmother, when their father married Bertha Lipstein from a well-to-do Jewish family originating from Russia.

[478] Although many Polish Jews came to Breslau to avoid the pogroms, the Borns had been settled in Germany much earlier. The family of Born's father originated from Görlitz, Lower Silesia, while

science and culture—among his father's friends, for instance, was Paul Ehrlich, the founder of chemotherapy. He attended primary and secondary school in his hometown.[479] In spring 1901 Born entered the University of Breslau and began to attend lectures in almost everything that came along: e.g., physics, chemistry, zoology, philosophy, logic, mathematics and astronomy.[480] In particular he was attracted to astronomy: he took several courses in it and even learned to handle the astronomical instruments at the old-fashioned Breslau Observatory which stemmed partly from Bessel's time. Since, however, he disliked the endless numerical calculations that were necessary in astronomy, he soon gave up the plan of becoming an astronomer and began to concentrate on mathematics.[481] From among the Breslau professors he would always remember Jakob Rosanes; Rosanes' lectures on linear algebra included an introduction to matrix calculus.[482] In the mathematics courses Born met and became friends with Otto Toeplitz and Ernst Hellinger. 'As a student in Breslau,' Born wrote later, 'I was much under the influence of Toeplitz, who was my senior by one year, and though my interest in algebra was not great he insisted in my learning matrix calculus properly and occasionally refreshed my knowledge when we were together again as young teachers in Göttingen' (Born, 1940, p. 617).[483] He also

---

his mother's family came from Silesia. Other Jewish scientists, who were in Breslau at the same time as Max Born—such as Oskar Minkowski (Hermann Minkowski's brother, later a famous pathologist), Otto Toeplitz and Ernst Hellinger—were mostly of Eastern European origin.

[479] At the secondary school, an average German *Gymnasium*, Born studied Latin, Greek and mathematics. His mathematics teacher, Maschke, also taught chemistry and physics. He was a clever experimenter, for as Born recalled: 'At the time Marconi's experiments on wireless communication became known, Maschke repeated them in his little lab with me and another boy as assistants' (Born, 1968, p. 16).

[480] The University of Breslau was founded in 1811 to replace the one in Frankfurt-an-der-Oder. Formerly there existed a Jesuit college in Breslau, which had a beautiful Baroque building.

[481] Physics at Breslau was represented by the old Oskar Emil Meyer (1834–1909), brother of the chemist Julius Lothar Meyer, who had propagated kinetic theory. He did not give very inspiring lectures and was soon replaced—because of illness—by the young mathematician Ernst Neumann (the grandson of the famous Franz Neumann). Chemistry did not attract Born either, for it seemed to consist just of memorizing facts.

[482] Jakob Rosanes was born at Brody, Austria Hungary, on 16 August 1842. He studied mathematics at the Universities of Berlin and Breslau, obtaining his doctorate in 1865. He became a *Privatdozent* at Breslau in 1870, extraordinary professor in 1873 and full professor in 1876. Rosanes became the rector of his university for the year 1903–1904, retired in 1911 and died at Breslau on 6 January 1922. He worked on various problems of algebraic geometry and invariant theory; he also wrote a book on chess.

[483] Otto Toeplitz was born on 1 August 1881 in Breslau. He studied mathematics at the Universities of Breslau and Berlin, and received his *Habilitation* in 1907 at the University of Göttingen. In 1911 he was appointed extraordinary professor of mathematics at the University of Kiel (promoted to full professorship in 1920). Toeplitz moved to the University of Bonn in 1928 as professor of mathematics and remained there until 1935. Then he was without a position. In 1939 he went to Palestine as scientific advisor to the administration of Hebrew University. He died in Jerusalem on 15 February 1940.
Influenced by Hilbert's treatment of integral equations, Toeplitz worked in particular on the algebra of systems of infinitely many linear equations. Later he also wrote on the history of mathematics.

got to know Clemens Schaefer, a young *Privatdozent*, who gave courses on Maxwell's electrodynamics.[484]

In those days students at German universities used to move often from one place to another, either 'attracted by a celebrated professor or a well-equipped laboratory, in other cases by the amenities and beauties of a city, by its museums, concerts, theatres, or by winter sport, by carnival and gay life in general' (Born, 1955a, p. 43). Thus Born spent the summer semester 1902 at Heidelberg and the summer semester 1903 at Zurich. In Heidelberg he met James Franck, who became one of his lifelong friends.[485] In Heidelberg, Born learned the most from Leo Königsberger, who dwelt more on the fundamental ideas than the pursuit of rigour in teaching mathematics, while at Zurich he attended Adolf Hurwitz' course on elliptic functions.[486] He used to return regularly to Breslau for the winter semesters, where he would meet Toeplitz and Hellinger again; from them he 'learned that the mecca of German mathematics was Göttingen and three prophets lived there: Felix Klein, David Hilbert and Hermann Minkowski' (Born, 1968, p. 18). So he decided to follow his friends and make the pilgrimage to Göttingen in summer 1904. Soon after his arrival there he came into close contact with Hilbert, who asked him to work out a manuscript of his lectures on function theory for the *Lesezimmer*. He was also introduced to Minkowski through a letter from his stepmother, who knew the mathematician from Königsberg. However, the relations with Felix Klein did not develop so smoothly. Born did not attend Klein's lectures regularly, but he participated in the joint Seminar of Klein and Runge on the theory of elasticity in the winter semester 1904–1905. He gave a talk on the problem of stability in elastic media, which was well received. 'Klein was very favourably impressed by it and wrote me a letter that I should do this as a prize paper for the faculty,' Born recalled later. 'This was very rare and a great honour, but I declined it' (Born, Conversations pp. 83–84).[487] Born's refusal offended the almighty Klein; he fell into

---

[484] Clemens Schaefer was born on 24 March 1878 in Remscheid, in the Ruhr region. He received his education at the Universities of Bonn (1897–1898) and Berlin (1898–1900). After taking his doctorate in 1900 at the University of Bonn with a thesis supervised by Emil Warburg, he served as an assistant to Heinrich Rubens. In 1903 he became a *Privatdozent* at the University of Breslau; seven years later he was promoted to an extraordinary and in 1917 to a full professorship. After an interlude at the University of Marburg (1920–1926), he returned to Breslau as Director of the Physics Institute (1926–1945). After World War II he went to West Germany and joined the University of Cologne. He retired in 1950 and died in Cologne on 9 July 1968.

Schaefer contributed to the problems of infrared rays, the structure of molecules and crystals, and acoustics. He wrote textbooks both on experimental and theoretical physics.

[485] Franck recalled later: 'At that time I met in the lecture hall [at Heidelberg] the mathematician Max Born, and we became quick friends. Born got a visit from his sister, and I must confess, that his sister contributed considerably to my pleasure at Heidelberg' (Franck, AHQP Interview, First Session, p. 13).

[486] Königsberger, born on 15 August 1837 at Posen, West Prussia, was professor of mathematics variously at Greifswald, Heidelberg, Dresden, Vienna, and again Heidelberg. He worked on elliptic functions and differential equations, and wrote a two-volume biography of Hermann von Helmholtz. He died in Heidelberg on 15 December 1921.

[487] Originally Born had not intended to give a talk in the Seminar himself. But it happened that, shortly before the due date, he found that he had to replace another speaker. As he recalled: 'I had

disgrace with him, although eventually he sat down to work on the problem. He submitted the essay, entitled 'Researches on the Stability of the Elastic Line in the Plane and in Space under Various Boundary Conditions,' just as Klein had wished; it received the prize of the Göttingen faculty and was accepted as a doctoral dissertation (Born, 1906).[488]

The difficult personal situation with Klein drove Born into physics.[489] After four months of military service in winter 1906–1907—Born was soon released, for he suffered from asthma, which had plagued him since his early youth—he left in April 1907 for England and the University of Cambridge in order to learn more about physics. There were two reasons why he chose to go to Cambridge: first, it was regarded in those days as a great centre of experimental research; second, Joseph John Thomson, the discoverer of the electron lived and taught there, and electron physics had always aroused a great interest in Göttingen.[490] Born was accepted as an advanced student of Gonville and Caius College. He attended the lectures of Joseph Larmor on electrodynamics and J. J. Thomson's experimental demonstrations; while he found Larmor's lectures old-fashioned and his Irish accent all but incomprehensible, he was very impressed by Thomson. He even took part in an experimental course at the Cavendish Laboratory.

---

only a few days to prepare myself. And I saw that there was such a lot of literature—impossible to read in a few days—so I thought I would better do it myself . . . . I had learned quite recently the theory of variations, so I worked out a new method to determine, with the help of mathematical criteria of minima, the stability of such lines [i.e., of the so-called *elastic* lines in continuous media]' (Born, Conversations, p. 83).

[488] Born had earlier obtained from Hilbert a subject for his doctoral thesis; it was the determination of the roots of Bessel functions, but he made no progress on it. Born's thesis on the problem of physics was accepted by the faculty, hence he was not listed officially as having obtained his doctorate under Hilbert. But Hilbert did examine him in mathematics, and Born remembered the following story about it. He had asked Hilbert in advance about the topic on which he would examine him. 'In what area do you feel yourself most poorly prepared,' Hilbert asked. 'Ideal theory,' replied Born. Hilbert said nothing more, and Born assumed that he would be asked no questions in that area. However, when the day of the examination arrived, all of Hilbert's questions were on the theory of ideals. 'Ja, ja,' Hilbert said afterwards, 'I was just interested to find out what you know about things about which you know nothing' (Reid, 1970, p. 105).

Since Born wanted to avoid being examined by Felix Klein, he chose astronomy as one of his minors. He had attended Karl Schwarzschild's Seminar on the atmosphere of planets, in which he also learned kinetic theory.

[489] In Göttingen Born attended several courses on physics. He liked Woldemar Voigt's lectures on optics, and recalled: 'We made all the elementary optical experiments on diffraction, interference, etc., with our own hands. To every experiment belonged a sheet, where the experiment was explained, and one had to fill out lines about what one was doing and then to write down the measurements and so on. And it was awfully well prepared and I learned a lot there' (Born, Conversations, pp. 81–82).

Voigt's course, together with Schwarzschild's lectures on optical instruments provided Born with a thorough knowledge of optics, which he used later when he wrote a book on that subject. On the other hand, Born did not like Johannes Stark's course on radioactivity. As he explained later: 'It was all so dogmatic, no proofs which I considered to be necessary, and I left it after a few weeks. And this is the reason I never have done nuclear physics at all. One has to learn it young, otherwise one cannot do it' (Born, Conversations, p. 8).

[490] The first to be interested in electrons had been Wilhelm Weber. Later Wiechert, Riecke, Voigt, Abraham and Runge had written papers dealing with electron theory. The subject of electron theory had also been treated in the Hilbert–Minkowski Seminar in summer 1905, which Born had attended.

Thomson, at that time, also lectured on his model of atomic structure (Thomson, 1904a), and Born would later choose it as the subject of his trial lecture (*Probevorlesung*) for the *Habilitation* (Born, 1909d). In England, Born made another acquisition: he bought the two volumes of J. Willard Gibbs' scientific papers, which had just appeared (Gibbs, 1906). 'I bought these books by Gibbs just by chance in a shop,' he recalled, 'and I read them and I was fascinated by them, and I studied them every day in Cambridge' (Born, Conversations, p. 7). Gibbs' works helped to turn him into a physicist; when he returned to Breslau in summer 1907, he went straight to the physics institute rather than to the mathematicians.[491] There he met Rudolf Ladenburg, who had just returned from Cambridge himself. Born found that, 'He was really a physicist, and there [at the physics institute] I learned first the real problems of physics, like the optical coherence of waves and such things that I had never heard before, because Voigt was quite formal' (Born, Conversations, p. 2). In Breslau, besides Clemens Schaefer and Rudolf Ladenburg, Erich Waetzmann was another *Privatdozent* (he became one in 1907), and in 1908 Fritz Reiche joined as an assistant after completing his doctorate under Planck. Lummer and Pringsheim continued their work on heat radiation also in Breslau, and they gave lectures on it. Born, who learned about this subject for the first time, decided to get involved himself. He went to Lummer's Institute, where the professor gave him 'what was called then a black body—a tube of porcelain material with a heating arrangement and a table with a gas fire and a cooling mantle of water around' (Born, Conversations, p. 7). But, not being 'very gifted for this type of work', Born just caused a flooding, upon which he gave up the experiments. Instead, together with Ladenburg, Waetzmann and Reiche, he started reading papers on new and interesting subjects. 'And there one day,' Born recalled later, 'we came across the relativity paper of Einstein [1905d]. Reiche knew it already from Planck, and he told us we must read it; and there I became excited, I found this marvellous' (Born, Conversations, p. 85).[492] Born also studied Einstein's other papers on relativity and tried to connect their content with what he had learned in Minkowski's Seminar in 1905. Since many questions arose from this study, he wrote to Minkowski in Göttingen asking him for help. Instead of answering these questions, however, Minkowski invited Born to return to Göttingen and become his assistant and collaborator in relativity theory. This response made Born extremely happy. He met Minkowski again at the *Naturforscherversammlung* in Cologne in September 1908 and listened to his lecture on 'Space and Time'

---

[491] In 1907 Born published his second paper on variational principles in thermodynamics together with Erich Oettinger of Karlsruhe (Born and Oettinger, 1907). In this paper Born cited the papers of Lorentz and Gibbs and generally demonstrated his knowledge of the literature. (Born's first paper on variational principles was his thesis: Born, 1906.)

[492] In fact Ladenburg became acquainted with Einstein personally in 1908; he had been sent by Planck to Berne to invite Einstein as a speaker for the next (i.e., 81st) *Naturforscherversammlung* in Salzburg. 'His [Ladenburg's] account was the first I heard of Einstein the man,' Born recalled. 'Ladenburg was enthusiastic and made us curious of the great unknown' (*The Born–Einstein Letters*, 1971, p. 1).

(Minkowski, 1909). There he decided to accept his offer. After arriving in Göttingen in early December 1908 he saw Minkowski nearly every day and discussed with him the problems of electron theory. He soon worked on a paper, which he completed during the Christmas vacation (Born, 1909a).[493] The unexpected death of Minkowski on 12 January 1909 put an end to a promising collaboration, but it did not ruin Born's career. In summer 1909—with the help of Hilbert, Runge and Voigt—he became a *Privatdozent* in physics at the University of Göttingen, and in September he spoke at the 81st *Naturforscherversammlung* in Salzburg on the dynamics of the electron with Einstein in the audience (Born, 1909c).[494] As Minkowski's last collaborator to work on relativity theory he was charged with editing his unpublished papers on that subject.[495] But he wanted to do more, namely, to extend Minkowski's work. Therefore, in his *Habilitation* thesis, he discussed the concept of a rigid body in relativistic kinematics and dynamics, a concept which the founders of relativity theory had always used in some sense without giving it a proper definition (Born, 1909b).[496] Although Born's definition of the rigid body in relativity theory encountered certain objections and a controversy arose in the following year—in which Max von Laue and Fritz Noether (Emmy Noether's brother) took part—Gustav Herglotz used it successfully in developing his relativistic mechanics of deformable bodies (Herglotz, 1911).[497]

Born's scientific work at Göttingen between 1909 and 1911 grew more or less from the ideas which Minkowski had expressed in his long memoir on the electrodynamics of moving bodies. Thus he established for himself a reputation as an expert on relativity and electron theory, and as such he was invited by Albert A. Michelson—who spent the summer semester 1911 in Göttingen—to visit Chicago in 1912 to give a series of lectures on relativity theory. However, in spite of the good beginning, Born made only slow progress in the problems he treated. One reason may have been the fact that he employed somewhat clumsy mathematical tools; for instance, following Minkowski, he tried to formulate problems with the help of matrices, which were soon replaced by the more

---

[493] In his first paper on relativity theory, Born formulated an action principle. He had spoken on this problem already in the Hilbert–Minkowski Seminar in 1905.

[494] For his *Habilitation* thesis, Born submitted an extended version of his earlier work on the relativistic electron (Born, 1909b).

[495] On Hilbert's advice, Minkowski's widow requested Born to examine and organize Minkowski's manuscripts on physical problems. On the basis of some surviving notes of Minkowski and his previous conversations with him, Born constructed a paper, which was subsequently published in *Annalen der Physik* (Minkowski, 1910). It is of interest to note that another paper based on Minkowski's surviving manuscripts appeared in 1915; Sommerfeld then edited Minkowski's lecture on relativity theory (Minkowski, 1915).

[496] Born noticed that the definition of the rigid body presented no difficulties, except when the motion was accelerated. However, when the accelerated body moved in a straight line, a rigid body could still be defined as the one which obeyed the simple dynamical laws of a mass point.

[497] Born tried to formulate a relativistic mechanics of the continua himself. In a paper submitted in May 1911 to *Physikalische Zeitschrift* he introduced the first step (Born, 1911). However, before making further progress he came to know that Herglotz had already done the job.

elegant concepts of 'four-' and 'six-vectors' (Sommerfeld, 1910b, c).[498] In Born's furture work also elegance would often be sacrificed, especially when he sought to arrive at the most general results. Of course, the difficulties that faced the extension of relativity theory after 1910 were extremely serious, and Born lacked the ingenious vision which enabled Einstein to go forward nonetheless to create the theory of general relativity. He rather preferred to take over the physical ideas of others and to formulate them with the help of the rigorous mathematical methods he had learned. When he later saw Einstein's completed theory of general relativity, he gave up working on relativity and electron theory for nearly twenty years.[499]

Unlike relativity theory, the other great theoretical conception of the early twentieth century—the quantum theory—had not yet made a great impact in Göttingen.[500] Born, who had learned about Planck's theory in Breslau, lectured in the summer semester 1911 on heat radiation. But this did not stimulate him to become personally involved in quantum theory. An impulse had to come from outside, and it did in the following manner. At Göttingen, Born lived, until he married in August 1913, in a boarding house on *Nikolausberger Weg*, and there (in 1912) he met Theodore von Kármán, the assistant of Ludwig Prandtl. They soon became friends and started to discuss all new problems in physics and mechanics. And one day they came across Einstein's paper on the specific heat of solids (Einstein, 1906g). As von Kármán recalled:

> I brought Einstein's paper to Born and we both studied it. Unfortunately experiments with different substances soon showed that Einstein's formula [with just one characteristic frequency] was limited. It worked only for higher temperatures, but not for the lowest temperatures. We wondered why. And we finally agreed that the discrepancy was due to the fact that Einstein's approach while basically correct was too simple. (Kármán and Edson, 1967, p. 67)[501]

---

[498] Born continued to use matrix methods later on—as, for example, in his paper dealing with energy momentum conservation in Mie's electrodynamics (Born, 1914a).

[499] In 1916 Born published a pedagogical review article on Einstein's theory of general relativity in the *Physikalische Zeitschrift* (Born, 1916a), which prompted Einstein to write to him: 'This morning I received the corrected proofs of your paper for the *Physikalische Zeitschrift*, which I read with a certain embarrassment but at the same time with a feeling of happiness at being completely understood by one of the best of my colleagues. But, quite apart from the material contents, it was the spirit of positive benevolence radiating from the paper which delighted me—it is a sentiment which all too rarely flourishes in its pure form under the cold light of the scholar's lamp' (Einstein to Born, 27 February 1916). There started a friendship between Einstein and Born, which lasted until Einstein's death.
    Later Born occasionally lectured on general relativity theory to the general public (see Born, 1920d). After 1933 Born worked again on electron theory and published papers on a nonlinear generalization of Maxwell's electrodynamics. (See, e.g., Born, 1934; Born and Infeld, 1934.)

[500] During the summer semester 1907 Minkowski gave a course on heat radiation, but we do not know whether he also presented Planck's theory in detail. Max Abraham, Planck's former student, had announced a course on heat radiation for the winter semester 1910–1911, but before giving it he went to Milan.

[501] The Göttingen people learned about the status of the problem of specific heats from Walther Nernst, who used to visit Göttingen frequently. (See, e.g., Courant, AHQP Interview, p. 5.)

In order to improve on Einstein's theory, Born and von Kármán thought about the structure of crystals, which they imagined as a three-dimensional lattice of coupled atoms. Now crystal lattices were a rather well-known scientific topic in Göttingen, especially around Woldemar Voigt, who however imagined crystals usually as continuous media with certain symmetry properties. 'But he mentioned that it [the crystal] could be reduced to a crystal lattice,' Born recalled, 'and when I started my work with Kármán, we took it for granted that there are real lattices' (Born, Conversations, p. 89). The two friends felt that they were on safe ground with this assumption, for Erwin Madelung had already used molecular lattices two years earlier in order to explain the observed discrete infrared radiation orginating from diatomic crystals (Madelung, 1909; 1910a, b). By investigating systematically the motion of atoms in three-dimensional crystal lattices, they obtained the characteristic spectrum of the crystals and arrived at a satisfactory explanation of the deviation of the empirical data from Einstein's formula of 1906 (Born and von Kármán, 1912). In arriving at the results, Born made use of his great mathematical erudition; he applied especially the theory of infinite sets of algebraic equations developed by his friends Ernst Hellinger (1907) and Otto Toeplitz (1911). Before Born and von Kármán published their formula for the specific heats in their second paper on crystal lattices (Born and von Kármán, 1913a), they learned about a different theory, which Peter Debye had presented somewhat earlier and which led to similar conclusions (Debye, 1912b). However, they felt, in spite of Debye's priority, that their description was more appropriate to the physical situation.[502] Crystal lattices and their dynamics became Born's main field of interest in the following years.[503] As he remarked later: 'From a personal point of view the occupation with vibrations in crystal lattices was of great practical importance for me; it opened up for me my own field of research, the kinetic theory of solids, which—though it does not penetrate into the depths of the ultimate principles—provided a large number of problems —not exhausted even today—for special investigations' (Born, 1956, p. 97).

Born's new field of interest certainly lay close to that of Hilbert and others in Göttingen. After all, in April 1913 Hilbert organized the *Kinetische Gas-Kongress*, in which, for instance, Peter Debye discussed the problems of the specific heats

---

[502] At about the same time the information about the X-ray interference patterns of von Laue, Friedrich and Knipping became known in Göttingen, for Sommerfeld discussed these matters during his visit there in summer 1912. Interestingly enough von Laue's results on X-ray interference were not referred to in the papers of Born and von Kármán. As Born remarked later: 'I remember that I at the same time as von Laue was thinking about using X-ray diffraction to prove the lattice structure. I worked it out theoretically and I was so busy with that and then suddenly appeared von Laue's papers . . . . So you see I took these things seriously' (Born, Conversations, pp. 89–90).

[503] The time around 1912 was a very happy period in Born's life. Born, von Kármán, Albrecht Renner (a medical student) and Hans Bolza (a student of Hilbert's) rented a house in *Dahlmannstrasse* and hired a nurse for housekeeping and preparing meals. The house was called '*ElBoKaReBo*,' the abbreviated version of Bo[rn]Ka[rman]Re[nner]Bo[lza]. Paul Ewald and the graduate student Ella Philipson regularly came to the meals. The bachelors' idyll came to an end when Paul Ewald married Ella Philipson, and Max Born (on 2 August 1913) married Hedwig Ehrenburg, the daughter of the Leipzig law professor Viktor Ehrenburg.

of solids. Born started with an especially ambitious problem, namely, to calculate the properties of diamond from lattice theory (Born, 1914b). The crystal of diamond had been analyzed before by William Henry Bragg and his son (Bragg and Bragg, 1913).[504] To Born it represented a particularly simple structure because each atom in the lattice possessed only four nearest neighbours; hence he had to consider just two elastic constants if he took into account only the nearest-neighbour interaction. He solved the 24 equations of motion for the eight fundamental mass points in the diamond lattice and thus calculated the frequency spectrum and the specific heat of the substance.[505] Born also succeeded in determining the elastic constants of diamond. His theory represented a milestone in the kinetic theory of solids, for he had arrived, by straightforward calculation, at a complete and satisfactory description of the properties of the crystal.[506] Born was confident that with his work he was on the right track of developing 'a systematic theory of crystal lattices, which explained all simple properties of the crystals without invoking new hypotheses' (Born, 1915a, p. 391). And, being really ambitious, he decided to write a book on the theory, which he called 'The Dynamics of Crystal Lattices,' and which he completed the following year (*Dynamik der Kristallgitter*, Born, 1915c).[507]

When Born wrote the book on the dynamics of crystal lattices, he already had left Göttingen and occupied, since early 1915, an extraordinary professorship at the University of Berlin.[508] He started to lecture on electrodynamics, but after

[504] Paul Ewald attended the Birmingham Meeting of the British Association in September 1913, where William Henry Bragg presented the experimental results on diamond.

[505] Born found that for low temperatures the specific heat of diamond approached a Debye curve, while for higher temperatures it came close to an Einstein curve.

[506] In contrast to Born's treatment, Debye's—and, to a lesser extent, Einstein's—description of, say, the energy of the crystal could only be considered as being 'phenomenological.' In discussing the theory of the solid state, Born often seemed to be in competition with his Göttingen colleague Peter Debye, for he frequently pointed out where Debye's phenomenological approach failed.

[507] The physical idea underlying Born's crystal lattices was that point-like material objects—ions, atoms or electrons—occupy the lattice points in elementary cells. He assumed that between these material objects there existed forces, which could be derived from a potential and decrease very rapidly with the distance. Due to the short range of these forces, Born was able to replace—for the purpose of calculating most of the crystal properties (except the ones connected with the surface)— the finite lattice by an infinite one, which simplified the evaluation considerably. By generalizing the methods used in the specific case of diamond to arbitrarily complicated lattices, Born arrived at many important results. For example, he solved the old riddle whether, in the most general lattice, there are 21 or 15 elastic constants, in favour of the higher number. (He showed that the six Cauchy relations, which allowed the reduction to 15 constants only, came about by the explicit use of the continuum theory.) Born further studied the optical properties of crystals (Born, 1915a). Especially, he solved the equation of mass points on which were impressed periodic oscillations of wavelength λ (much larger than the lattice parameter δ), by expanding the amplitudes in powers of the small parameter $2\pi\delta/\lambda$. He found in the first order the Fresnel formula describing the refraction of waves in terms of the elastic constants of the crystal; by expanding to second order and restricting himself to a crystal with only one optical axis he obtained a description of double refraction.

[508] In early 1914 a second chair of theoretical physics was established at the University of Berlin in order to relieve Max Planck of his teaching burdens. It was intended to invite Max von Laue to take this chair, but the plan failed to materialize for two reasons. On one hand, World War I broke out in August 1914 and most students left the universities to do military service; on the other, von Laue was

some time he was called up for military service. Being one of the few younger physicists in Germany who did not take part in actual fighting, he worked first with a small group at the Döberitz Camp, close to Berlin, on the problem of wireless communication between airplanes and the ground.[509] After a short training course as an aircraft wireless operator, Born joined the *Artillerie-Prüfungskommision* (Artillery Testing Commission), a new department founded by his friend Rudolf Ladenburg. The task of this department consisted in developing various methods of 'scientific ranging,' optical, acoustical, seismometric, electromagnetic, etc.[510] In Ladenburg's department he found many physicists, among them his former Göttingen colleague Erwin Madelung and Alfred Landé, who had been his student and later Hilbert's 'physics tutor.' Born helped Ladenburg in hiring more scientists, for 'It was my main interest in the war to save people from being killed' (Born, Conversations, p. 93).[511] Although military service restricted the activities of the *Extraordinarius* Max Born, he still enjoyed considerable freedom in his office. Since the military authorities possessed little knowledge of physics and mathematics, they could not control his work. Hence Born and Landé started, besides their work on sound ranging, an intense collaboration on some problems of crystal dynamics. Born was also allowed to attend regularly the seminars and colloquia at the University of Berlin as well as the meetings of the German Physical Society.[512] In Berlin, physics at that time was very exciting—Albert Einstein, Max Planck, Walther Nernst, Fritz Haber and Heinrich Rubens being the leading personalities. Almost all of these men

---

not available, for at that time the *Akademie für Sozial- und Handelswissenschaften* in Frankfurt-am-Main was turned into a regular university having five faculties, including one for science, and von Laue was appointed professor of physics. Hence the plans in Berlin were changed; an extraordinary professorship was established and Born was called to it.

[509] Born's experience of military service was not all that extensive. He had not been able to complete his obligatory year of military service after his doctorate because of a serious attack of asthma. It should be mentioned that in Germany during World War I there did not exist any special programs, at least in the beginning, in which scientists could do technical work for the military. This was different in Great Britain. As Born remarked: 'In Britain, I think, they were much more patriotic' (Born, Conversations, p. 95).

[510] Ladenburg, whom Born had first met in 1908 at Breslau, had seen action with a cavalry regiment during the war, but had been wounded and sent back to Berlin. There he developed the idea of sound-ranging and was allowed by the military administration to establish a department and recruit appropriate people.

[511] Born did not always succeed in this endeavour. Later on he recalled with unhappiness the case of one of the best students in mathematics from Göttingen, Herbert Herkner, who served in an infantry regiment: 'I tried and tried to get him back and then at last I got permission and sent a telegramme to his regiment. And the night before it arrived he was killed' (Born, Conversations, pp. 93–94). He wrote an obituary of Herkner—a rare mark of honour for a student—which appeared in *Naturwissenschaften* (Born, 1918a).

[512] Evidently, Born's lecturing duties during the war were not heavy; the students were simply not around. So, after the course on electrodynamics in summer 1915 was interrupted, he announced courses on the kinetic theory of matter during the following three semesters (of which he probably gave only one); in the summer semester 1917 he announced a course on thermodynamics; and in the following three semesters (until the winter semester 1918–1919, when he probably delivered the lectures) a course on the dynamics of crystal lattices.

were deeply involved in the quantum theory and from them Born learned about its problems. Above all, he became friends with Einstein. They met frequently in Einstein's apartment in *Haberlandstrasse*—which was very close to Born's office in *Spichernstrasse*—to discuss physics and politics and to play music together.

In Berlin Born continued to work on the kinetic theory of matter, especially on problems of the solid state. But he also extended his interest to the theory of liquids and gases. So, for instance, he answered Debye's question of whether he could explain the double refraction in fluids; he discovered that an optical activity would arise in case the constituents of the molecules formed asymmetric tetraeders (Born, 1915b). The theory of fluids and the dispersion of light by molecules also occupied him in the following years.[513] Thus he extended the dispersion theory of Drude and Lorentz by incorporating fluctuation phenomena —such as those treated by Marian von Smoluchowski in his theory of critical opalescence (Smoluchowski, 1908)—to describe the scattering properties of aniso-tropic molecules (Born, 1917). In the same problem, he also considered Bohr's model of atomic constitution; in particular, he derived the most general formula for the electric moment **P** of the diatomic molecule,

$$\mathbf{P} = \sum_j \frac{\mathbf{A}_j(\mathbf{A}_j \cdot \mathbf{E})}{\omega_j^2 - \omega^2} , \tag{170}$$

where **E** is the vector of the electric field of the incident radiation having frequency ω, and the $\mathbf{A}_j$ are fixed vectors depending on the specific molecule and its eigenfrequencies ω_j (Born, 1918b).[514]

Born became increasingly better acquainted with Bohr's theory of atomic structure from lectures and discussions. For instance, Sommerfeld visited Berlin twice in 1918 to speak about his own researches at the German Physical Society: first, on 26 April, at a meeting of the Society in honour of Planck's sixtieth birthday; second, at a meeting of the Society on 26 July (when he also acted as President of the Society, having been elected on 31 May 1918), he spoke on X-ray spectra. He proposed then to remove the theoretical difficulties connected with the orbits occupied by several electrons by assuming that each electron moved on a separate ellipse; that is, he introduced what he called the '*Ellipsen-verein*,' a system of ellipses with the atomic nucleus at one focus. Born and Landé exploited this idea of Sommerfeld's systematically in their calculations of crystal properties (Born and Landé, 1918a). Especially, they determined Φ, the potential

---

[513] Born treated, for example, the properties of anisotropic fluids in several publications, using ideas similar to the ones applied by Paul Langevin and Pierre Weiss in ferromagnetism and Peter Debye in his theory of polar molecules (Born, 1916b; 1918b).

[514] In order to obtain Eq. (170), in which $(\mathbf{A}_j \cdot \mathbf{E})$ denotes the scalar product of the vectors $\mathbf{A}_j$ and **E**, Born averaged over the phases of the electrons in Bohr orbits. Equation (170) yielded Rayleigh's scattering formula for low frequencies ω, but yielded deviations from it in the case of higher frequencies.

energy of ionic crystals, obtaining the result

$$\Phi = \frac{\overset{(-1)}{\Phi}}{r} + \frac{\overset{(-5)}{\Phi}}{r^5}, \tag{171}$$

which should describe the attractive binding of the oppositely charged nearest neighbours ($\overset{(-1)}{\Phi} < 0$) as well as the repulsive action of the more distant ions having the same charge (where r denotes the distance). The repulsive potential term $\overset{(-5)}{\Phi}$ (which is larger than zero) depended on the quantum numbers of the Bohr orbits. While Born and Landé first assumed that all the electron orbits associated with a particular ion were in the same plane, they revised this assumption in the following investigations (Born and Landé, 1918b, c). After comparing the theoretical results with the data on the compressibility of crystals, they concluded:

> The plane electron orbits do not suffice; the atoms are evidently spatial objects. This conclusion appears to us to be as important as the results and researches on the X-ray spectra; in spite of the successes obtained there with planar systems of [electron] rings, we must insist on an extension of the theory in the direction mentioned. (Born and Landé, 1918c, p. 216)[515]

Born and Landé continued their investigations and sought to determine details of the electronic configuration in ionic crystals. For instance, they used the new potential energy expression

$$\Phi = \frac{\overset{(-1)}{\Phi}}{r} + \frac{\overset{(-9)}{\Phi}}{r^9}, \tag{171a}$$

obtained from the spatial electron orbits (where $\overset{(-9)}{\Phi}$ describes the repulsion due to interaction with next-to-nearest neighbours), in order to derive far-reaching

---

[515] In calculating the compressibility Born and Landé first made a crucial error, which Born recalled later as follows:

> We calculated the mutual energy of these [electron] rings—the sums of the energy of a pair—and we worked it out. And we forgot that one has to write one-half of that [in order not to count every pair twice]. These ring models fitted beautifully and gave us the correct values of the compressibilities. So we thought this was wonderful, and we gave it to Einstein for publication in the Berlin Academy. And the next morning I came to my military department and there was Landé sitting quite depressed and he said, "You must destroy the paper—it is quite wrong." And he told me that he had found this mistake . . . . I ran to Einstein, and he laughed—I have never heard him laugh so much. And he said: "This is so marvellous that you have made a mistake; I thought you never make one. You must not destroy it. Of course, we shall not submit it at once—I will given it back to you. But I expect to have it in a week or two again in an improved form." (Born, Conversations, pp. 99–100)

Born followed Einstein's advice and sat down again to work with Landé. Finally they arrived at the solution: in the case of spatially distributed orbits the second repulsive term in Eq. (171) had to be replaced by a term $\overset{(-9)}{\Phi} / r^9$.

conclusions. Thus Born calculated with it the energy liberated in the course of the formation of alkali-halogenides, finding a reasonable description of the existing data (Born, 1919a).[516] He also discovered a method for determining the so-called 'electron affinity' of halogen atoms, i.e., the amount of energy gained when an electron is added to a halogen atom to form a halogen ion (Born, 1919b). Since no data existed to check his result, Born invented an indirect method: he related the electron affinity to the energy necessary to split alkali-halogenides into alkali and halogen atoms and compared the latter with the experimental results, again with satisfactory agreement.[517] The success of these calculations persuaded Born that the forces in ionic crystals were of purely electrical origin. Thus, starting from the kinetic theory of solids, he had been able to establish a quantitative connection between physical and chemical properties. In an essay for the *Naturwissenschaften*, entitled '*Die Brücke zwischen Chemie und Physik*' ('The Bridge between Chemistry and Physics'), Born expressed the great hope of the physicist to unify the physical and chemical forces by reducing both to the interaction between the elementary constituents of matter, i.e., the electrons and the atomic nuclei. He said:

> The physics of today already possesses pictures of atoms, which certainly approach reality to some extent, and with these numerous mechanical, electrical, magnetic and optical properties of substances can be explained. Now one should not be stopped by chemical properties, but attempt also to reduce them to atomic forces to the extent they are known. In this respect Nernst's theorem provides valuable preliminary preparation by reducing the intricate complex of chemical forces to a number of simple constants [especially the chemical constants, see Section I.6]. It is now the task of the molecular physicist to calculate these constants—which can be determined by the physical chemists through calorimetric and other measurements —from the properties of atoms; with this begins a new and inconceivably big era of thermochemical research. (Born, 1920b, pp. 373–374)

Born, who considered himself a molecular physicist, was prepared to contribute further to this important field of research.

In spring 1919 Born left Berlin and went to Frankfurt-am-Main to take up the position of an ordinary (i.e., full) professor of physics at the University of Frankfurt, while his predecessor there, Max von Laue, moved to Berlin.[518] For a time Born had hesitated to leave Berlin and his friends and colleagues there, but Einstein encouraged him to go to Frankfurt, telling him to 'accept uncondition-

---

[516] Born noted that in the case of lithium-halogenides the data seemed to be explained best by a power term $\Phi^{(-5)}/r^5$, but in other cases inverse powers of the distance between r^{-7} and r^{-11} resulted.

[517] Fritz Haber, who had helped Born in getting the right data, developed a method of representing Born's procedure (Haber, 1919). It became known to physico-chemists as the *Born–Haber cycle*.

[518] The offer to von Laue to come to Berlin had been made already in early 1918. But it took more than a year before the second professorship of theoretical physics, envisaged already in 1914, was established at the University of Berlin.

ally' and arguing that 'one should not refuse such an ideal post, where one is completely independent' (Einstein to Mrs. Born, 8 February 1918). At Frankfurt Born began to lecture in summer semester 1919, choosing, at first, courses on such subjects as mechanics and quantum theory. He quickly adapted himself to the new place, finding a suitable home to live in with his family; his friend Ernst Hellinger, then professor of mathematics at the University of Frankfurt, lived with them. The institute, which he took over, was small but comparatively well-equipped. He found able assistants in Otto Stern, *Privatdozent* at Frankfurt since 1914, and Elisabeth Bormann. Alfred Landé, his wartime collaborator, also came over to Frankfurt as *Privatdozent*. For Born it was no problem to organize appropriate lecture courses for the students, who had begun to flock to the universities—including Frankfurt—after World War I.[519] But running an institute after the war was not an easy task, especially to obtain equipment for experimental work. Germany had lost the war and, apart from all the losses that had been incurred, enormous and not-yet-fixed reparations had to be made to the Western Allies. The Weimar Republic had therefore to fight against tremendous economic and financial problems and could not support universities on the same scale as the former, prosperous *Kaiserreich*. Industry was likewise hindered by the reparations and the fact that the Rhein-Ruhr region, the most important industrial region remaining in Germany, continued to be occupied by French troops for years. Rampant inflation devalued German currency, and Born—like any other institute director—was faced with the problem that working within a fixed budget did not cover the steadily rising expenses of research. At that difficult time, however, some assistance came from various sources, especially from the *Notgemeinschaft der Deutschen Wissenschaft* (Emergency Association of German Science), which was founded on 30 October 1920 and which distributed additional funds to the universities.[520] Born also obtained financial help from Einstein, who 'tried to squeeze some funds out from the *Kaiser Wilhelm-Institut*

---

[519] In Berlin, Born had learned a lot from Max Planck. Thus he also started to give systematic courses in Frankfurt: in summer 1919 he lectured on the mechanics of particles, the following winter semester on the mechanics of continua, and in the winter semester 1920–1921 on electrodynamics. Otto Stern supplemented Born's courses by lecturing in summer 1919 on kinetic theory of gases, in the winter semester 1919–1920 on thermodynamics, and in winter 1920–1921 he took over the course on mechanics of continua (which followed Born's course on particle mechanics in summer 1920). The aim was to present the main courses on theoretical physics (mechanics, thermodynamics and electrodynamics) every year. Besides the main courses, special courses were also given: e.g., on atomic and quantum theory. Finally, Born conducted, together with Stern and Landé, a Seminar on modern problems of theoretical physics.

[520] Shortly after the war the former Prussian Minister of Education, Friedrich Schmidt-Ott, discussed with Max Planck and Adolf von Harnack, then President of the *Kaiser Wilhelm-Gesellschaft*, the possibilities of helping to promote scientific research in Germany and Austria. The Prussian Academy became involved, and on 19 April 1920 Max Planck—on behalf of the Academy—requested Schmidt-Ott to head the *Notgemeinschaft der Deutschen Wissenschaft*. The aim of this foundation was to examine, together with the academies in Germany and Austria, the situation in scientific research and to support scientists with funds and instrumentation. Committees, consisting of small numbers of experts, were established in each field of research. The *Notgemeinschaft* received its funds mainly from the Government of the Weimar Republic and the German states.

[of Physics]' (Einstein to Mrs. Born, 1 September 1919), and who also shared other monies with his friend. In addition, Born capitalized on the public interest that existed at the time in Einstein's relativity theory: he gave lectures for the general public on it and charged entrance fees, which he put into his institute.[521] Finally, he was supported by some private individuals: thus the rich Frankfurt jeweller, G. Oppenheim, who had previously helped to establish the physics chair at the university of Frankfurt, provided Born a diamond for his experiments on elasticity; and the American banker, Henry Goldman, donated considerable sums in dollars which, not being subject to inflation at that time, were the more valuable.

The major part of Born's budget went for experimental research. For instance, Stern developed in Frankfurt the method of atomic beams, produced from metals heated in a vacuum, for the investigation of atomic properties. In his first important work using this method he measured the thermal velocity distribution of atoms (Stern, 1920a, b, c, d). And soon he became involved in looking for an experimental verification of spatial quantization (Stern, 1921). The experiment, which he later performed together with Walther Gerlach—who had joined the University of Frankfurt in 1920 as a *Privatdozent*—yielded early in 1922 the final result that a beam of silver atoms was split by an inhomogeneous magnetic field into two parts (Gerlach and Stern, 1922a). Born also performed certain experiments himself: at the *Naturforscherversammlung* in Bad Nauheim in September 1920 he presented the results of a measurement of the mean free path of silver atoms in air (Born, 1920c). But his main efforts were on the theoretical side: he continued to study the kinetic theory of matter, to investigate the electron affinity of atoms (with Elisabeth Bormann) and the dispersion of light by diatomic molecules (with Gerlach). Thus he accomplished within a year a respectable amount of work, and then he received the call to Göttingen. Again he hesitated initially to accept the invitation. After all, he felt well in Frankfurt, where he had succeeded in establishing a good institute and in acquiring some good friends. Would he also find in Göttingen some wealthy people, as in Frankfurt, who supported his activities? And would he find such capable collaborators as Otto Stern and Alfred Landé, whom he could not take along?[522] The decision was made as soon as Born succeeded in getting the additional professorship from the Ministry. With James Franck, he felt certain, he would be able to continue the Göttingen tradition of physics in a manner worthy of the place.[523]

---

[521] A book grew out of Born's public lectures, entitled '*Die Relativitätstheorie Einsteins und ihre physikalische Grundlagen*,' which sold very well (Born, 1920d).

[522] For some time Born tried to have Otto Stern appointed as his successor in Frankfurt, but he did not succeed. Stern left Frankfurt in late 1921 to take up a professorship at the University of Rostock, and Landé was called in 1922 as an extraordinary professor to the University of Tübingen. Born's successor in the Frankfurt chair was Erwin Madelung.

[523] Born's coming to Göttingen was delayed by the problem of finding a suitable accommodation for his family. This was not an easy enterprise in those days. It took until spring 1921 when the problem was solved and the Borns moved to Göttingen.

Born appeared to be the ideal man for Göttingen. He had grown up at the University under the influence of David Hilbert and Hermann Minkowski. He had executed a large part of Hilbert's programme of 'axiomatizing' physics: for instance, he had established the kinetic theory of solids on the basis of just a few assumptions concerning atoms and the forces between them; he had started to do the same with the theory of molecules, and was beginning to 'axiomatize' chemistry. Moreover, he was very experienced in giving carefully organized lectures on all aspects of theoretical physics; in Frankfurt he had also shown considerable talent in running the experimental institute and getting the necessary funds. Born knew that the same was expected of him now at Göttingen. Thus he wrote to Einstein, still from Frankfurt: 'Franck has now settled in Göttingen. He must have enough freedom there, and so I am busily collecting money for him. So far I have got 68,000 Marks. It is not at all easy to inspire laymen with some interest in our work. I must have more money. Wien got a whole million for re-equipping his Institute in Munich. I believe that what Wien has, Franck should also get' (Born to Einstein, 12 February 1921). In collecting money Born addressed himself in particular to the industrialist Carl Still of Recklinghausen (in the Ruhr region), to whom Richard Courant had introduced him.[524]

Unlike Born, James Franck had never been in Göttingen before. He was born on 26 August 1882 at Hamburg, the son of a Jewish banker.[525] After attending the *Wilhelmsgymnasium* there, he entered the University of Heidelberg in 1901 and began to study geology, mathematics and physics, but mainly chemistry.[526] Though he loved the surroundings of Heidelberg, he did not find the professors at the University very inspiring; in the following year he went to Berlin.[527] In Berlin, however, he quickly became interested in physics under the influence of Emil Warburg. He began to be involved in measuring the mobility of ions and received his doctorate in 1906. Following a few months as an assistant in Frankfurt-am-Main, he returned to Berlin and became Heinrich Rubens' assistant. Franck continued to remain interested in the problem he had treated in his thesis; he also followed the work of British physicists, like Joseph John Thomson, in this field. He collaborated with Robert Wichard Pohl on ion mobilities and the

[524]Carl Still, the son of a Westphalian farmer, had started off as a mechanic and built up a large firm, which built coke ovens and similar installations. He was profoundly interested in science and frequently invited the physicists and mathematicians of Göttingen to his countryseat in Rogätz on the Elbe.

[525]James Franck's ancestors could be traced back to the eighteenth century in the region near Hamburg. The banking firm, J. Franck and Company, was founded by James' grandfather Jacob Franck, Jr. His son James Franck married Ingrid Josephson of Göteborg.

[526]At school James Franck was not very interested in classical languages. Instead he became attracted to physics at the time when Röntgen discovered X-rays. He liked mathematics, especially geometry, and did chemical experiments at home. At school he also learned some English and French.

[527]Franck found the lectures on inorganic chemistry very old-fashioned. For example, he never heard anything about Svante Arrhenius' work on electrochemistry. The professor of physics was the old Georg Hermann Quincke, who did not give inspiring lectures either.

determination of the velocity of X-rays, with Wilhelm Westphal on the charge of ions, with Robert Williams Wood—who came as a guest of Rubens to Berlin in fall 1910—on the fluorescence of iodine and mercury vapours, with Peter Pringsheim on the electrical and optical properties of the chlorine flame, and with Lise Meitner on radioactive ions. In 1911 he received his *Habilitation* at the University of Berlin. In the same year Gustav Hertz obtained his doctorate and became Franck's successor as assistant to Rubens. Franck and Hertz then started a series of investigations on the ionization potentials (of the atoms and molecules) of various elements in connection with the quantum hypothesis. Within three years they arrived at their results on the collision of electrons with mercury atoms (Franck and Hertz, 1914a, b), which were considered right away by Bohr as a confirmation of his model of atomic constitution. Interestingly enough, Franck and Hertz did not know Bohr's theory at all and interpreted their results not as an excitation potential, but as the ionization potential of the atoms. Only a couple of years later did Franck turn to Bohr's interpretation.

Meanwhile the war had broken out, and Franck voluntarily joined the army; he became an officer. Having fallen seriously ill, he returned to Berlin in 1917 and joined the *Kaiser Wilhelm-Institut für Physikalische Chemie*, whose director was Fritz Haber.[528] Franck became head of the Physics Division and worked to re-establish research in the post-war years, one of his collaborators being Fritz Reiche. In April 1920 he met Niels Bohr, who on the 27th of that month gave a lecture on the series spectra of elements at the German Physical Society. Franck was very impressed by Bohr's personality and became a great admirer of the Danish physicist. Bohr reciprocated his feelings and wrote some months later:

My stay in Berlin, though regrettably short, has been a particularly beautiful and stimulating event for me; and one of the greatest joys I experienced there was meeting you and getting to know you. I have always followed your important investigations with the greatest interest, and it was for me extremely interesting to see your experimental setup and to hear your opinions about the results. I have lately been especially concerned with your new experiments on helium, and, while

---

[528] Fritz Haber was born on 9 December 1868 in Breslau. He studied chemistry and physics at the University of Berlin (1886, under August Wilhelm Hofmann and Hermann von Helmholtz), the University of Heidelberg (1886–1889, under Robert Bunsen) and at the *Technische Hochschule* in Berlin (1889–1891), where he received his doctorate in 1891 with a thesis in organic chemistry under the supervision of Karl Liebermann (1852–1914). The following year he joined the *Eidgenösische Technische Hochschule* (E.T.H.) in Zurich and investigated problems of chemical technology. Then he returned to Germany, worked for a while in his father's chemical company, and went to the *Technische Hochschule* in Karlsruhe in 1894. There he became *Privatdozent* in 1896, *Extraordinarius* in 1898 and *Ordinarius* in 1906. In Karlsruhe, his research dealt first with the decomposition of hydrocarbons by heat, then with the oxydation-reduction process and later, after 1905, with the synthesis of ammonia from nitrogen and hydrogen. Finally, together with Carl Bosch of the *Badische Anilin und Soda-Fabrik*, he developed the method for the large scale production of ammonia. In 1911 Haber went to Berlin as Director of the newly founded *Kaiser Wilhelm-Institut für Physikalische Chemie und Elektrochemie*. During World War I he became involved in chemical warfare. After the war he started to rebuild his institute, and it became an internationally known centre of research. Haber received the Nobel Prize in Chemistry for 1918. He resigned the directorship of the institute in April 1933 and left Germany in early summer. He died in Basle on 29 January 1934.

contemplating the mechanism of the collision between atoms and electrons, I hit upon a problem, about which I would be very grateful to hear your views. (Bohr to Franck, 18 October 1920)

With the last remarks Bohr referred to the results of Franck and Paul Knipping on helium excitation; they had observed there a possible excitation potential of 21.9 V (Franck and Knipping, 1920, pp. 326–327). Bohr now wanted to know about the probability of such collisions, which according to his theoretical estimate should be very small, and he wrote to Franck:

> In connection with my interest in the field of experimental physics that you have opened, I thought of the possibility that you might perhaps one day come to Copenhagen to give your support for some time to our work in the new Institute for Theoretical Physics, which is being established here and which is being especially equipped for experimental research on spectroscopic problems.

Bohr wished to have Franck's advice for the installation of experimental apparatus and he invited him to come for some time to Copenhagen. He had just obtained funds from private sources to pay for the stay of a distinguished foreign physicist (*'einen bedeutenden ausländischen Physiker'*) for a short time. He wrote to Franck:

> Therefore I hasten to ask you whether you have the desire and the possibility of giving us the honour and joy of coming to Copenhagen for a few months in the beginning of the new year ... I cannot quite express the anticipation with which we would all look forward to your visit, and how much I, in particular, would rejoice to have the opportunity of discussing with you personally the problems in which both of us are so interested. (Bohr to Franck, 18 October 1920)

Franck accepted the invitation with pleasure; after 1921 he returned repeatedly to Copenhagen and established a close relationship with the physicists there— especially, of course, with Niels Bohr.

In Göttingen the three new professors of physics, Robert Pohl—who directed what was called the First Experimental Institute—, James Franck—who directed the Second Experimental Institute—, and Max Born—who directed the Institute for Theoretical Physics—, soon arranged their respective work and duties. Franck left to Pohl the teaching of physics to beginners—the big physics course —and restricted himself to giving lectures on specialized topics. 'I had the laboratory course as the main thing, with the *Praktikum*,' he recalled later. 'And we used a good deal of the instruments of Voigt to make an advanced laboratory course where more complicated things were measured' (Franck, AHQP Interview, 13 July 1962, p. 4). On the other hand, Born, who continued to work in Göttingen on some of the experiments he had started in Frankfurt, soon gave them up and left experimental research to Franck and Pohl. Instead he organized systematic three-year cycles of courses on theoretical physics, consisting of six series of lectures distributed over six semesters: they were on (1) mechanics of

particles and rigid bodies; (2) mechanics of continuous media; (3) thermodynamics; (4) electricity and magnetism; (5) optics; (6) elements of statistical mechanics, atomic structure and quantum theory. Each course consisted of four hours of
lectures per week and a tutorial. Born supplemented these main lectures by
continuing series of lectures on special topics: thus, in the summer semester 1921
and the following winter semester he gave a series of lectures on the kinetic
theory of solids, and in the winter semester 1922 he dealt with magneto- and
electro-optics.[529] Later, when Born's assistants became *Privatdozenten*, they
helped him with the main courses of lectures; then the cycle of main courses was
repeated after three semesters. Born, Franck and Pohl, together with Max Reich,
the professor of applied electricity, had a joint colloquium, in which Ludwig
Prandtl, Emil Wiechert and the astronomer Johannes Hartmann participated
regularly, and where David Hilbert and Richard Courant showed up occasionally. The style of the colloquium was very informal. As Born recalled: 'It was
customary to interrupt the speaker and to criticize ruthlessly. We had the most
lively and amusing debates, and we encouraged even young students to take part,
by establishing the principle that silly questions were not only permitted but even
welcomed' (Born, 1978, p. 211).[530] The different personalities and styles of the
three physics professors contributed importantly to the atmosphere. Maria
Goeppert-Mayer characterized them by relating the following incident: 'I was
once in the experimental laboratory. I came down the steps. Franck, Born, and
Pohl were standing talking to each other. They were all friends of my parents.
Pohl just nodded; Born said hello; and Franck stretched out his hand' (Goeppert-Mayer, AHQP Interview, p. 5). The students were more or less identified
with their professors, and Pohl referred to them '*die Pohlierten, die Franckierten,
die Bornierten*,' respectively. Naturally, due to their excellent personal relations,
Born and Franck worked together more closely. So they not only organized the
institute parties together (Franck and Born belonged historically to the same
institute), but also jointly conducted the *Physikalisches Proseminar*, in which
selected topics, mainly of quantum physics, were discussed.[531]

In research, Franck and Born continued their earlier interests. The investigation of the collision between electrons and atoms or molecules remained (in the
early 1920s) the principal field of research of James Franck and his collaborators, who included Walter Grotrian, Günther Cario and Patrick Maynard Stuart
Blackett, a guest from England. With some of his associates, Born carried on

[529] 'Born gave very thorough, but rather difficult lectures. Born's lectures were difficult to
understand,' recalled Friedrich Hund. 'He presented much more than we do nowadays, hence his
influence on the physicists was smaller. At that time things had fallen apart: the experimentalists did
not really learn theoretical physics, and the theoretical physicists did not learn proper experimental
physics . . . . Of course, the situation was better for the advanced students' (Hund, AHQP Interview,
First Session, p. 14).

[530] The colloquium took place in the small auditorium (*Kleiner Hörsaal*), and 60 to 70 people used
to attend it.

[531] For example, the Ramsauer effect was discussed in the *Proseminar*. The latter also replaced the
Seminar on the Structure of Matter (*Struktur der Materie*) which Hilbert tried to reestablish with
Born.

investigations on crystal dynamics; Sommerfeld had also asked Born to write a comprehensive article on that subject for the *Encyklopädie der mathematischen Wissenschaften* (Born, 1923b). During the writing of the encyclopedia article, Born came across a number of problems suitable for doctoral theses, and a number of students—among them Carl Hermann and Gustav Heckmann—obtained their degrees by working on them. Even in his later Göttingen years, Born returned occasionally to crystal dynamics.[532] Indeed, the kinetic theory of solids, the problems of which he had first attacked in 1912 and which he had continued to develop in the following decade, remained dear to him until the end of his career. While Franck showed little interest in questions of the solid state, he appreciated the investigation of molecular problems, which was the other field of Born's research in the early 1920s. A fruitful collaboration developed between Franck and Born, and they even wrote a joint paper on molecular physics (Born and Franck, 1925a, b). And, finally, the field in which Born became involved during the early twenties—Bohr's theory of atomic structure—united the two friends in a common cause and determined the success of the Born–Franck era of Göttingen physics.[533]

III.3   Progress in Atomic Models from 1913 to 1921

In his three papers of 1912 on the constitution of atoms and molecules Niels Bohr had constructed, based on the conception of the nuclear atom, detailed models of atoms and molecules (Bohr, 1913b, c, e). The hydrogen atom had the simplest structure; in it an electron moved in a circular (or an elliptical) orbit around the nucleus (carrying one unit of positive charge). But immediately after discussing the theory of the hydrogen atom, Bohr turned to atoms containing several electrons. Again he pictured these atoms as systems consisting of a positively charged nucleus surrounded by electrons moving in circular orbits (Bohr, 1913b, §5). 'We shall assume,' Bohr remarked in his second paper, 'that the electrons are arranged . . . in coaxial rings rotating round the nucleus' (Bohr, 1913c, p. 477). The dimensions of these rings were determined by the quantum condition for the angular momentum, which Bohr borrowed from John William Nicholson. To this quantum condition he added 'as condition of stability, that

---

[532] Together with Maria Goeppert-Mayer, Born wrote another review article on crystal dynamics for the *Handbuch der Physik* (Born and Goeppert-Mayer, 1933).

[533] The great era of Göttingen physics ended in 1933. In May of that year Born was deprived of his Göttingen chair and left Germany. Franck also resigned, but he stayed on until the fall, when he went to Niels Bohr's Institute in Copenhagen. In 1935 he went as a professor of physics to Johns Hopkins University, Baltimore, from where he went to the University of Chicago in 1938. He retired in 1947. In 1964 he returned for a visit to Göttingen, accompanied by his second wife, Hertha Sponer; there he died on 21 May 1964. Born, on the other hand, settled in Great Britain. He stayed for a period in Cambridge as Stokes Lecturer (1933–1936), and from there helped German refugee scientists in getting positions. In 1936 Charles Galton Darwin, who was about to become Master of Christ Church College, Cambridge, invited Born to become his successor as Tait Professor of Natural Philosophy in the University of Edinburgh. Born retired in 1953 and settled in Bad Pyrmont close to Göttingen. He died there on 5 January 1970.

the total energy of the system in the configuration in question is less than in any neighbouring configuration satisfying the same condition of the angular momentum of the electrons' (Bohr, 1913c, p. 477). Bohr also knew that, 'Corresponding to different distributions of the electrons in the rings, however, there will, in general, be more than one configuration which will satisfy the condition of the angular momentum together with the condition of stability' (Bohr, 1913c, pp. 477–478). The problem became, then, to determine in the case of each atom or molecule the configuration of the electrons under these conditions. An important question was, of course, to study the stability of a given electron ring against displacements of individual electrons perpendicular to the plane of the ring.[534] Bohr applied classical dynamics to this problem and arrived at the following conclusion: the number of electrons (n) which can rotate in a single ring around the nucleus of charge $Z|e|$ increases very slowly with the increasing atomic number Z; 'for $Z = 20$ the maximum value is $n = 10$; for $Z = 40$, $n = 13$; for $Z = 60$, $n = 15$.' He further noticed that, 'a ring of n electrons cannot rotate in a single ring round a nucleus of charge $n|e|$ unless $n < 8$' (Bohr, 1913c, p. 482.)[535] Finally he concluded: 'The calculation indicates that only in the case of systems containing a great number of electrons will the planes of the rings separate; in the case of systems containing a moderate number of electrons, all the rings will be situated in a single plane through the nucleus' (Bohr, 1913c, p. 483). Bohr also considered, under the same dynamical assumptions, the mutual influence of different rings, and observed: 'Unless the ratio of the radii of the rings is nearly unity the effect of outer rings on the dimensions of inner rings is very small, and that the corresponding effect of inner rings on outer [rings] is to neutralize approximately the effect of a part of the charge on the nucleus corresponding to the number of electrons on the [inner] ring' (Bohr, 1913c, pp. 485–486). Based on these results he began to construct models of atoms and molecules.

In the case of neutral atoms the number of electrons equal the atomic number Z. The first system to be considered beyond the hydrogen atom was the helium atom consisting of a nucleus bearing positive charge $2|e|$ and two electrons. Bohr found, in particular, that in the state of the lowest energy (the ground state) the two electrons are bound more tightly than the electrons in the hydrogen atom; both move on the same orbit having the quantum number 1 (which, in a circular orbit, is just the quantum number of angular momentum). He denoted this configuration by the symbol 2(2), stating that both of the electrons assume the orbit of the lowest quantum number. The ionization energy of the helium atom

---

[534] In his first paper on atomic constitution, Bohr had noticed a difficulty with respect to the stability consideration. A ring of electrons rotating round a positive nucleus was unstable against the displacements of electrons in the plane of the ring; in order to overcome this instability Bohr turned to the hypothesis of quantized angular momentum (Bohr, 1913b, §5). He had concluded that 'the ordinary principles of mechanics cannot be used in the discussion of the problem in question' (Bohr, 1913c, p. 480).

[535] These results were derived for circular orbits, but Bohr argued that also in the case of elliptical orbits the conclusions remained valid, provided the ratios of the semiaxes of neighbouring rings differed appreciably from unity.

then turned out to be 27 eV, having the correct order of magnitude compared to the observed value. Similarly he concluded that the ground state of lithium had the configuration 3(2, 1), with two electrons in the innermost ring (of quantum number 1) and one outer electron in the next ring (of quantum number 2), and the configuration of the ground state of beryllium was 4(2, 2). For atoms with a larger number of electrons the considerations quickly became very cumbersome. However, Bohr pointed to a fact which helped him considerably, namely: 'Unless the charge on the nucleus is very great, rings of electrons will only join together if they contain equal numbers of electrons, and that accordingly the numbers of electrons on inner rings will be only 2, 4, 8 . . . ' (Bohr, 1913c, p. 495).[536] Evidently, this conclusion, which arose from stability considerations *plus* assumptions concerning the construction of neutral atoms from charged nuclei by adding one electron after another, satisfied Bohr, for it was 'strongly supported by the fact that the chemical properties of the elements of low atomic weight vary with a period of 8' as well as 'the fact that the valency of an element of low atomic weight always is odd or even according as the number of the element in the period series is odd or even' (Bohr, 1913c, p. 495).[537] Due to the stability arguments concerning perturbations normal to the plane of the orbit, Bohr further held that occasionally more electrons might be assembled on outer rings than on the neighbouring inner ones; thus he claimed that the configurations 5(2, 3) and 6(2, 4)—rather than the configurations 5(2, 2, 1) and 6(2, 2, 2)—represented the ground states of the neutral boron and carbon atoms, respectively. Finally, he put the number of electrons in the outer rings of atoms 'arbitrarily' equal to the chemical valency of the corresponding elements and arrived at the following arrangements of electrons in light elements[538]: hydrogen 1(1); helium 2(2); lithium 3(2, 1); beryllium 4(2, 2); boron 5(2, 3); carbon 6(2, 4); nitrogen 9(4, 4, 1); neon 10(8, 2); sodium 11(8, 2, 1), etc. This scheme exhibited, he observed, 'a marked periodicity with a period 8' (Bohr, 1913c, p. 497). In agreement with the chemical properties of homologous elements, the outer electrons of higher homologues were less tightly bound.[539] For atoms of higher atomic weight, Bohr argued, the above periodicity with the number 8 had to be replaced by one with the number 18, the latter then representing the number of electrons in the innermost ring.

---

[536] Bohr knew that the states of lithium and beryllium having all electrons in the lowest orbit—i.e., the states 3(3) and 4(4), respectively—possessed smaller energies than the states 3(2, 1) and 4(2, 2), respectively, assumed above.

[537] The explanation was evident, since the inner rings contained only an even number of electrons, i.e., 2, 4 or 8. Further, Bohr concluded that already in the case of neon the inner ring contained eight electrons, arguing that the separation of eight electrons into two rings of four (which was also possible) lacked stability against the perturbations perpendicular to the plane.

[538] Bohr justified the assumption concerning the number of outer electrons with the help of the empirical data on the atomic volumes of substances. Early in the history of the periodic system of elements, it had been noticed that elements having valency 1 assumed the largest and those of valency 4 the smallest atomic volume.

[539] In Bohr's model the screening of the nuclear charge was more complete; hence, for instance, the valence electron of sodium could be removed more easily than that of lithium, etc.

When Bohr looked for appropriate models of the ground state of simple molecules, he noticed a remarkable difference with respect to atoms: while one highly charged atomic nucleus was able to bind a small number of electrons, two nuclei could not do so, due to lack of stability. Hence he had to assume that the molecules were formed by the interaction of systems, each containing a single nucleus surrounded by a number of bound electrons (Bohr, 1913e). In the case of diatomic molecules, for example, the electrons then moved in circular orbits in the planes of symmetry; i.e., 'in a stable configuration the greater part of the electrons must be arranged around each nucleus approximately as if the other nucleus were absent; and that only a few of the outer electrons will be arranged differently rotating in a ring round the line connecting the nuclei' (Bohr, 1913e, p. 862). As the simplest system of such kind, Bohr considered the hydrogen molecule: the two electrons should move in a ring of radius a ($= 0.95a_0$, where a_0 is the radius of the hydrogen atom in its ground state), the nuclei being at distances b ($= (1/\sqrt{3})a$) from the plane of the orbit (Bohr, 1913e, p. 863). Hence the binding energy, i.e., the negative difference between the energy of the hydrogen molecule and the energy of two free hydrogen atoms, amounted to 6×10^4 cal/gram-molecule, which did have the right order of magnitude.[540] In a similar way, Bohr constructed models for different molecules, including HCl, for instance; again he assumed that the chemical bond was provided by a ring containing two electrons. A more complicated situation should describe the H_2O molecule; here two outer rings, each containing three electrons (two from the oxygen atom and one from the hydrogen atom), establish the binding.

Initially, Bohr's models of atoms and molecules seemed to account well for the observed mechanical, optical and chemical properties, and they became increasingly the objects of scientific investigations. For example, Peter Debye treated the dispersion of light by the hydrogen molecule (using Bohr's model) and found that it agreed with the observations (Debye, 1915a). Debye's student Paul Scherrer, on the other hand, discussed in his thesis the rotation of the plane of polarized light passing through hydrogen in the presence of a magnetic field (the Faraday effect); again Bohr's model described the data correctly (Scherrer, 1916b). Friedrich Krüger calculated the specific heat of hydrogen with the help of the same model (Krüger, 1916a, b), claiming that his results agreed with Eucken's measurements (Eucken, 1912).[541] However, certain difficulties also showed up; thus Miss H. J. van Leeuwen, a student of Hendrik Lorentz, pointed out that some of the electron motions (which Debye had considered in his treatment of dispersion by Bohr's hydrogen molecules) were not stable (Leeuwen, 1915). On the other hand, Adalbert Rubinowicz claimed that such an instability

---

[540] In his paper, Bohr referred to a measurement of Irving Langmuir, who had obtained the result of 13×10^4 cal (Langmuir, 1912). After seeing Bohr's paper in late 1913, Langmuir informed him about a new measurement which gave a value between 7.6×10^4 and 8×10^4 cal (Langmuir, 1914).

[541] Paul Epstein, independently, was able to fit the data on specific heat; however, he noticed certain difficulties with the correct quantization condition (Epstein, 1916f).

would not spoil Debye's resulting dispersion formula (Rubinowicz, 1917b).[542] Surprisingly, the situation appeared to be far less satisfactory in the case of the other two-electron system: the helium atom. Debye, who evaluated the dispersion of light from Bohr's model of helium, arrived at the conclusion that it did not account for the empirical facts and remarked: 'The model of helium has yet to be discovered' (Debye, 1915a, p. 26).

Several years later Alfred Landé proposed a new helium model. He assumed that the two electrons moved in opposite directions on different orbits, whose planes made an angle different from zero with each other (Landé, 1919c). This system, he claimed, not only possessed a slightly smaller energy than Bohr's one-electron-ring atom—Landé calculated $-6.16Rh$ to $-6.20Rh$ as against Bohr's value of $-6.125Rh$, R being Rydberg's constant and h Planck's constant —but it also offered the possibility of achieving better agreement with the dispersion data. In suggesting a spatial structure for the helium atom, Landé referred to his previous experience with Born based on the consideration of ionic crystals; there the authors found that the mechanical properties could be described provided one gave up the coplanar orbits and turned to a spatial arrangement of the planes of electron motion (Born and Landé, 1918b, c). Landé now investigated the idea of 'spatial atomic structure' in many papers submitted during the following years (Landé, 1919a, b, e, f, g; 1920b, c, d, e, f; 1921a). A spatial structure of the atoms seemed to relate well to the observations in chemistry and crystal physics. Thus the periodicity of chemical properties, especially the occurrence of the period eight, had led Gilbert Newton Lewis to emphasize the importance of octets of electrons in the structure of atoms and molecules (Lewis, 1916). He visualized these octets to occupy the corners of a cube, thus suggesting the so-called 'cubical atom.'[543] Three years later Irving Langmuir extended Lewis' theory and based it on the set of following postulates: (i) the electrons in atoms are either stationary or rotate, revolve or oscillate about certain definite positions; (ii) the electrons in any given atom are distributed on a series of concentric, (nearly) spherical shells, all of equal thickness, whose mean radii form an arithmetic series 1, 2, 3, . . . ; (iii) each shell is divided into cellular

---

[542] Rubinowicz took into account the motions of the nuclei in the hydrogen molecule as well; he concluded that they should not affect the dispersion significantly. (The same should be in the cases of dispersion by oxygen and nitrogen molecules, which Sommerfeld treated (Sommerfeld, 1918a).)

[543] G. N. Lewis was born on 23 October 1875 at Weymouth, Massachusetts. He studied at the University of Nebraska (1890–1893) and at Harvard University (1893–1899), obtaining his doctorate in chemistry in 1899. Then from 1900 to 1901 he continued his studies at Leipzig (with Wilhelm Ostwald) and Göttingen (with Walther Nernst). From 1901 to 1904 he served as an instructor at Harvard. From 1904 to 1905 he was in charge of weights and measures at the Phillipine Islands. Then he was appointed at M.I.T.: first as laboratory assistant (1905–1907), then as assistant professor (1907–1908), associate professor (1908–1911) and full professor of chemistry (1911–1912). In 1912 he joined the University of California at Berkeley and occupied the chemistry chair there until his retirement in 1945. Lewis died on 24 March 1946 at Berkeley.

Earlier evidence concerning a cubical arrangement of electrons in atoms had been given by A. L. Parson (1915).

spaces or cells, occupying equal areas in their respective shells and distributed over the surface of the shell in accordance with the symmetry required by postulate (i)—thus the first shell contains two, the second eight, the third eighteen, and the fourth thirty-two cells; (iv) each cell in the first shell can contain only one electron, but all other cells may contain one or two electrons, with the occupation of the cells proceeding from the inner to the outer shells; (v) two electrons in the same cell do not attract or repel one another by strong forces (however, there may be a magnetic attraction which counteracts the electrostatic repulsion); (vi) if there are only a few electrons in the outside shell, attraction dominates, while electrostatic repulsion dominates if the shell is nearly filled (Langmuir, 1919a, b).[544] With these assumptions or postulates Langmuir explained many properties of the atoms: for example, the fact that noble gases possess 2 (helium), 10 (neon), 18 (argon), 36 (krypton), 54 (xenon), and 86 electrons (radon), respectively, and that in chemical binding pairs and octets of electrons play an essential role. In spite of this success he was perfectly aware of one great difficulty of his theory, about which he remarked as follows:

> How can these results be reconciled with Bohr's theory and with our usual conception of atoms? It is too early to answer. Bohr's stationary states and the cellular structure postulated above have many points of similarity. It seems that the electron must be regarded as a complex structure which undergoes a series of discontinuous changes while it is being bound by the nucleus or kernel of an atom. There seems to be strong evidence that an electron can exert magnetic attractions on other electrons in the atom even when not revolving about the nucleus of an atom. (Langmuir, 1919a, p. 302)

In 1919 it was indeed much too early to answer such questions about the structure of the electron. However, the work of Born and Landé on ionic crystals, from which a similar cubic picture of the electron arrangement in atoms seemed to follow, encouraged Langmuir to go ahead with the construction of atomic and molecular models. In particular, he concerned himself with the problems of the helium atom and the hydrogen molecule (Langmuir, 1921a, b). While these efforts did not lead quickly to a convincing success, the general idea of the spatial structure of atoms received some support from spectroscopy, notably X-ray spectroscopy. The latter field received enormous attention from physicists during the second decade of the twentieth century, and the resulting data allowed

---

[544] I. Langmuir was born on 31 January 1881 in Brooklyn, New York. He studied metallurgical engineering at the Columbia School of Mines (1898–1903) and then chemistry at the University of Göttingen (1903–1906), receiving his doctorate under Walther Nernst. Then he joined Stevens Institute of Technology, Hoboken, New Jersey, as an instructor in chemistry (1906–1909), and finally the General Electric Company's Research Laboratory in Schenectady, New York, where he remained active until his death: first as assistant director (1909–1932), then as associate director (1932–1950), and finally as consultant (1950–1957). He died in Falmouth, Massachusetts, on 17 August 1957.

Langmuir worked on molecular physics and chemistry, especially gaseous discharges; he developed important new methods in science (for biology) and technology (e.g., a method for producing artificial snow). He received the 1932 Nobel Prize in Chemistry for his pioneering contributions to surface chemistry.

one to approach the problem of the detailed structure of atoms in a previously unexpected manner.

X-ray spectroscopy arose from the observation of Charles Glover Barkla and his associate C. A. Sadler, who had discovered that homogeneous X-radiation emerges when X-rays of continuous frequencies are scattered by matter (Barkla and Sadler, 1908a). Barkla and Sadler had then determined the homogeneity by measuring the penetration property. However, the discovery of the interference patterns from crystals by von Laue, Friedrich and Knipping in 1912 enabled one to characterize X-rays by their wavelengths. William Henry Bragg and his son William Lawrence Bragg developed a practical method of determining the wavelength by reflection of the X-rays from the surface of a known crystal lattice. Henry Gwyn Jeffreys Moseley then succeeded in late 1913—by applying a modified version of the Bragg method—in obtaining the strongest X-ray emission lines of the atoms of many elements. The theoretical description of his results led to a confirmation of Bohr's atomic theory. At about the same time the techniques of observation were greatly refined. For instance, Maurice de Broglie in Paris got rid of the disturbing effects of spurious spectral lines (appearing as a result of crystal imperfections) by having the crystal rotate about a vertical axis contained in the reflecting surface (M. de Broglie, 1914).[545] Manne Siegbahn, then at the University of Lund, adopted de Broglie's technique and began to explore systematically, together with his students, the X-ray spectra of a large number of elements.[546] Siegbahn devised improvements in every detail of the apparatus; he thereby increased the accuracy of his measurements, achieving within ten years a resolution of wavelengths surpassing Moseley's by a factor of 1000. As early as 1914 his student Ivar Malmer published the first results on X-ray spectra of fourteen elements from yttrium to cerium (Malmer, 1914). Two years later—Malmer had meanwhile completed his doctoral dissertation—

[545] Louis César Victor Maurice, Duc de Broglie, was born on 27 April 1875 in Paris. He entered the *École Navale* in 1893 and served there until 1902. Then he established a small laboratory in his home, where he first carried out measurements of the electric charge of fine suspended particles and later turned to investigate X-ray spectra and X-ray absorption. After World War I, during which he held a commission in the French Navy (he contributed to the underwater detection of submarines and the development of radio), he continued to work on X-ray spectroscopy, shifting his interest later on to nuclear and elementary particle physics. From 1942 to 1946 he was professor at the *Collège de France*, and for fifteen years he was a member of the *Conseil Scientifique de l'Énergie Atomique* (1945–1960), including ten as President. Maurice de Broglie was elected a member of Paris *Académie des Sciences* (1924) of the *Académie Française*, and a Foreign Member of the Royal Society. He died on 14 July 1960 in Neuilly-sur-Seine.

[546] Karl Manne Georg Siegbahn was born on 3 December 1886 at Örebro, Sweden. He studied at the University of Lund (1906–1911), where he received his doctorate with a thesis on measurements of the magnetic field. He served as an assistant to Johannes (Janne) Robert Rydberg from 1907 to 1911. In 1911 he became a lecturer at the University of Lund, in 1915 deputy professor of phsyics, and finally, in 1920, he succeeded Rydberg as professor of physics. Three years later he moved to the University of Uppsala, and in 1937 he was appointed research professor of experimental physics at the Royal Swedish Academy of Sciences and director of the Nobel Institute of Physics (1939–1964). While his early work was devoted to problems of electricity and magnetism, after 1912 he shifted his interest to X-ray spectroscopy and later to nuclear spectroscopy. Manne Siegbahn received the Nobel Prize in Physics in 1924. He died in Stockholm on 25 September 1977.

Siegbahn published their joint results in an extended review article, containing especially the K- and L-series lines of many elements (Siegbahn, 1916b). At the same time he also reported the preliminary results on the M-series of heavy metals (Siegbahn, 1916a).

Though he concentrated later on the emission line spectra, Siegbahn had begun by studying another type of X-ray data, namely, the absorption spectra of elements. They had been obtained first several years earlier by Barkla and Sadler while investigating the absorption of homogeneous radiation by various substances (Barkla and Sadler, 1909). Barkla and Sadler had also introduced the concept of the 'mass absorption coefficient,' defining it as the ratio of the usual absorption coefficient α (i.e., the constant multiplying d, the thickness of the absorbing substance, in the expression for the intensity, $I = I_0 \exp(-\alpha d)$) and the density D of the absorber. They had found that α/D showed identical behaviour for nine types of homogeneous X-rays (i.e., for nine specific frequencies of the primary X-radiations) if absorbed by different materials, provided the specific frequencies did not coincide with the characteristic frequencies of these materials. Three years later Barkla had confirmed and, with a collaborator, extended the previous findings (Barkla and Collier, 1912). Manne Siegbahn, who was interested in the relation between the absorption coefficients and the wavelengths, took Barkla's α/D-values for various substances in the region of strong absorption (i.e., where the incident X-rays had nearly the characteristic wavelengths λ of the absorbing material) and plotted their logarithm versus the logarithm of λ, obtaining a straight line (Siegbahn, 1914). Hence he concluded the validity of the relation

$$\frac{\alpha}{D} = A \cdot \lambda^x, \tag{172}$$

A and x being constants. He further found that Eq. (172) remained valid for wavelengths larger than the characteristic ones. Shortly afterwards Walther Kossel of Munich submitted a paper to *Verhandlungen der Deutschen Physikalischen Gesellschaft*, in which he demonstrated—based on the same data as Siegbahn's—that the products of the mass absorption coefficients with the squares of the atomic numbers of the corresponding elements and the inverse of the characteristic wavelengths λ, respectively, assumed constant values for a large number of substances, provided one used for λ the wavelengths of characteristic K-radiation, λ_K (Kossel, 1914a). Kossel also noticed that for $\lambda > \lambda_K$ the absorption coefficients (of the elements Fe, Ni, Cu, Zn) satisfied the equation

$$\frac{\alpha}{D} = \frac{\text{const.}}{Z^2} \left(\frac{\lambda}{\lambda_K} \right)^{2.7}, \tag{173a}$$

and, for $\lambda < \lambda_K$, the equation

$$\frac{\alpha}{D} = 7.4 \frac{\text{const.}}{Z^2} \left(\frac{\lambda}{\lambda_K} \right)^{2.7}. \tag{173b}$$

In the region of the characteristic wavelength, i.e., for $0.85 \ \lambda_K < \lambda < \lambda_K$, the above equations were not exactly satisfied but, after a rapid increase, the mass absorption coefficient reached a maximum value for wavelength λ below λ_K.[547] Kossel interpreted this result as being consistent with Stokes' rule and discussed the consequences in detail in a second paper, submitted about a month later again to the *Verhandlungen* (Kossel, 1914b). 'It seems to me that there occurs a peculiar case of Stokes' rule,' he remarked in the introduction and explained: 'The full excitation of fluorescence does not take place until the hardness of the incident radiation surpasses that of the *weaker*, short-wavelength emission lines [i.e., of the K_α-doublet].' And he continued: 'I now want to show that this connection can be understood and—as far as available data permits—even be investigated quantitatively, if one pictures a view of the processes based on Bohr's atomic model. There is no doubt that a formal description would represent the connection also; however, we shall pursue in the following the more visualizable method and shall proceed as if Bohr's model had been proved to be completely right and known at least in the case of one nucleus' (Kossel, 1914b, p. 953).

At the time, when he discussed the data on X-ray absorption spectra, Kossel was an assistant at the *Technische Hochschule* in Munich.[548] He had been familiar with Munich's scientific atmosphere for several years, having visited the place quite regularly since 1911 and given a number of talks at the weekly Wednesday Colloquium of the Munich physicists.[549] Thus he had participated in the progress of physics achieved in Sommerfeld's and Röntgen's institutes during these years. In particular, he had been aware of the important discovery of the diffraction of X-rays by crystals by Max von Laue, Walter Friedrich and Paul Knipping from the very beginning. At an early stage he also became accustomed to Bohr's ideas on the structure of atoms, which began to play a role in the

---

[547] Although Eqs. (173a, b) were consistent with Siegbahn's result, Eq. (172), Kossel did not know about it. However, he referred to Siegbahn's paper in an addendum to the proofs (Kossel, 1914a, p. 909).

[548] Walther Ludwig Julius Kossel was born on 4 January 1888, the son of the physiologist Albrecht Kossel (who in 1910 won the Nobel Prize in Chemistry). He studied physics at the University of Heidelberg from 1906 to 1911 (spending the winter semester 1907–1908 in Berlin), obtaining his doctorate in 1911 with a thesis on the properties of secondary cathode rays produced in gases. He served as an assistant to his teacher Lenard from 1910 to 1913. Then he became an assistant to Jonathan Zenneck, professor of physics at the *Technische Hochschule*, Munich. In 1921 Kossel accepted the call to the chair of theoretical physics at the University of Kiel, which had been occupied before by Planck (as extraordinary professor, 1885–1889) and Lenard (1898–1907). In 1932 he moved to the University of Danzig as professor of experimental physics (1932–1945). After World War II, in 1947, he took up a chair at the University of Tübingen. Kossel died on 22 May 1956 in Heidelberg.

[549] Kossel discussed the results of his thesis in the Wednesday Colloquium in Munich on 13 and 20 December 1911; in November 1912 he addressed the Colloquium on Lenard's recent work on phosphorescence; in June 1913 he spoke on radioactive elements and the periodic system. After his appointment in Munich he twice discussed in the Colloquium the phenomena connected with the ionization of gases by electron impact, especially the results of the work of Lenard and of Franck and Hertz. (For details, see the article by John L. Heilbron on the Kossel–Sommerfeld theory: Heilbron, 1967.)

discussions of the Munich physicists soon after Bohr's first publications appeared in *Philosophical Magazine*.[550] Thus Kossel increasingly became part of the Munich scientific circle. Lenard approved of this and wrote to Sommerfeld on 25 September 1913: 'Our common student Kossel has, as far as I know, the opportunity of joining Professor [Jonathan] Zenneck as assistant. I very much wish it for his sake . . . then he would go to an entirely new territory for the third time, and I believe that anyone who desires to become a good physicist should enjoy this [the changes] whenever possible. In addition, he would then be close by in your neighbourhood, which he, as I know, values very highly.'

Although the main advance in X-ray physics, namely, the discovery of the interference patterns, had occurred at Sommerfeld's Institute, Röntgen's associates quickly became interested in the new development. They also concerned themselves at an early stage with research on the discrete X-ray spectra emitted by atoms. Especially Ernst Wagner, one of Röntgen's assistants, began to study carefully the X-ray absorption edges.[551] In a detailed examination of the absorption spectra from several metals, including silver, gold, copper, nickel and iron—and using Maurice de Broglie's method of the rotating crystal for wavelength analysis—he arrived at the following observations: first, the K-absorption edge was slightly harder than all registered emission lines (i.e., K_α-, K_β-, and K_γ-lines); second, two L-edges existed (Wagner, 1915). He had come upon the first observation earlier and explained it by referring to Albert Einstein's 1905 theory of fluorescence: he had shown, in particular, that the frequency difference between the K-aborsption edge and the K_α-emission line was equal to ν_{L_α}, the frequency of the first line in the L-spectrum (Wagner, 1914). The second observation represented a new discovery; Wagner showed further that both L-edges possessed a shorter wavelength than the known L_α-emission line.[552] In an article, entitled '*Über Röntgenspektroskopie*' ('On X-ray Spectroscopy') and published in *Physikalische Zeitschrift*, Wagner reviewed the techniques and results of the field, thereby establishing himself as one of the leading experts in the field. In interpreting the results, he systematically employed Bohr's theory of atomic structure together with Sommerfeld's relativistic extension of the theory (Wagner, 1917).

In this atmosphere at Munich, in which the physics of X-rays played a prominent role, Walther Kossel developed his theory of X-ray spectra (Kossel, 1914a, b). The fundamental question he set out to answer concerned the origin of

---

[550] Kossel even shared one of the Wednesday Colloquia (on 15 July 1914) with Niels Bohr who presented his atomic theory there. (See Heilbron, 1967, p. 455.)

[551] Ernst Wagner was born on 14 August 1876 at Hildburghausen. He first studied medicine and then physics at the Universities of Würzburg, Berlin and Munich, obtaining his doctorate under Röntgen in 1903. He became *Privatdozent* in spring 1909 and extraordinary professor in 1915 at the University of Munich. He left Munich only after he received the call to the chair of experimental physics at the University of Würzburg in November 1922; however, just six years later, on 1 November 1928, he died there. Wagner excelled in careful and critical high precision experiments; besides X-rays, he worked on the investigation of the properties of canal rays and on the absolute measurements of high pressures.

[552] The doublet character of the L-edge was confirmed by Maurice de Broglie (1916).

these spectra, a question which had not been treated adequately earlier by either Bohr or Moseley. Kossel found that the observation that the X-ray edges corresponded to higher wavelengths than the emitted lines provided the key to this problem. Thus he came to the following picture describing the origin of the X-ray fluorescence spectra:

> The fluorescence will be initiated by the complete removal of an electron, either by a photoelectric effect, caused by the incident X-rays [these will be absorbed especially when they possess the frequency of the edges], or by the collision with a moving electron, thereby producing a secondary cathode ray . . . The atom is now (singly) positively charged; however, the elementary quantum [of electricity] is not missing at the periphery of the atom, say at the valence position, as happens usually, but from the innermost ring. To this state there corresponds a very high potential energy of the system [i.e., the atom]. One may ask whether this state endures until, attracted by the single charge [of the atom], an electron comes from the outside and takes the place of the missing one. The emission [of radiation] that occurs then must have the frequency corresponding to the total energy of removing [the inner electron], and its wavelength must coincide with that of the excitation limit. Such an emission, if it exists, is in any case weak in comparison with the ones observed thus far. The most intense line [observed] possesses a far longer wavelength; it corresponds to a return [of the electron] from a finite distance [within the atom]. Since an emission having still longer wavelength has not been observed, we associate it with a process corresponding to the emission of the smallest amount of energy: we assume that the missing electron in the first ring (counted from the interior [of the atom]) will be replaced by one from the second ring. The energy thus liberated shows up in the emission of the K_α-line. Now an electron is missing in the second ring. The same state can also be produced by letting a light wave of suitable frequency fall upon the second ring or have electrons impinge upon it, whose kinetic energy surpasses the work of removing [an electron] from it. In this case, the excitation limit must be at longer wavelengths than for electrons of the inner ring, which corresponds to K-radiation; the excitation limit lying on the long-wavelength side closest to the K-radiation is observed as the region of "L-radiation." We identify the state obtained by the passing of one electron [from the second] to the first ring with the one obtained by L-excitation. Analogously to the first case, the electron might also now be replaced by the one from the next ring, which is bound more weakly. To this process, which involves the smallest energy transfer, we attribute the [emission of the] strongest line of longest wavelength, l_α. In a similar way the process may be continued further; the potential energy will decrease steadily and finally the atom just misses one of its outermost electrons, corresponding to a visible line and the energy of removing a valence electron. If finally the initial electron, which had been freed originally by high-frequency [radiation], returns [to the atom], then it takes the available position at the surface, and the frequency [connected with it] is of the order of magnitude of visible light. (Kossel, 1914b, pp. 953–955)

With the above mechanism of the emission lines, Kossel immediately explained the frequency relation

$$\nu_K = \nu_{K_\alpha} + \nu_L, \tag{174}$$

relating ν_K, the frequency of the K-absorption edge, to ν_{K_α}, the frequency of the emitted K_α-line, and ν_L, the limiting frequency of the L-series. From ν_L he determined the frequency of the L_α-line via Moseley's formula, Eq. (121), and from the result he verified the relation

$$\nu_{K_\beta} = \nu_{K_\alpha} + \nu_{L_\alpha} \tag{175}$$

for the frequency of the K_β-line. In all cases, Eq. (175) was satisfied to better than 1% (Kossel, 1914b, p. 959, see table). Hence Kossel's mechanism seemed to describe the physical situation correctly. However, Kossel still was not satisfied; in particular, he had to rule out the possibility that an electron from the first Bohr orbit of an atom may go over (via absorption of X-radiation) into a fully occupied second orbit (the final state of the L-series emission process). In a third paper on the X-ray spectra, submitted again to *Verhandlungen der Deutschen Physikalischen Gesellschaft* in August 1916, he summarized the empirical results of Ernst Wagner (1914), Ivar Malmer (1915) and William Henry Bragg (1915), and extended his theory to account for all data (Kossel, 1916b). In particular, he claimed that the duplicity of the K- and L-absorption edges was intimately connected with the associated emission lines; thus ν_{K_α} and $\nu_{K_{\alpha'}}$, the two frequencies of the K_α-line doublet, satisfied the equations

$$\nu_{K_\alpha} = \nu_K - \nu_L \tag{176a}$$

and

$$\nu_{K_{\alpha'}} = \nu_K - \nu_{L'}, \tag{176b}$$

where ν_K is the K-absorption edge and ν_L and $\nu_{L'}$ the two L-absorption edges of Wagner. A difficulty seemed to arise in explaining the existence of an L'-term in the atom, which had to exist in Bohr's atomic theory. However, in summer 1916 Kossel was able to refer to Sommerfeld's new theory of relativistic fine structure, which had been published meanwhile (Sommerfeld, 1915c; 1916b, c).

Sommerfeld, who had followed the activities in X-ray research with the greatest interest—after all, X-ray spectra had been among his favourite subjects for many years—explained the duplicity of the stationary states of an atom by the relativistic motion of the electrons in Kepler orbits around the nucleus. Electron orbits distinguished by different eccentricities had different energies: for example, the $K_{\alpha'}$-line should correspond to the transition from the lower elliptical L-state (with principal quantum number 2 and azimuthal quantum number 1), while the initial state of the K_α-line was the circular L-orbit (with principal and azimuthal quantum numbers 2). With the help of the relativistic formula, Sommerfeld evaluated the difference of the energies of the two L-states to be

$$h\,\Delta\nu = \frac{Rh\alpha^2}{16}(Z - \sigma)^4, \tag{177}$$

where R, h and α denote Rydberg's constant, Planck's constant and the fine-structure constant, respectively, and $(Z - \sigma)$ represents the effective charge acting on an electron in the two-quantum (L-) orbit (i.e., the orbit with the principal quantum number $n = 2$) (Sommerfeld, 1916c, §2). Sommerfeld found that Eq. (177) was satisfied by the available data, provided he chose for σ the value of 3.5. The nonintegrality of the constant σ, which represented the screening effect, did not bother him much for the moment—from the conceptual point of view he would have preferred, of course, an integral number—but he emphasized another difficulty, namely, 'that we cannot deduce any L-lines from the K-lines according to the combination principle' (Sommerfeld, 1916c, p. 155). That is, Kossel's Eq. (175) appeared to be violated. Moreover, Sommerfeld had to introduce two extra quantum levels, Λ and Λ', in order to account for all observed L-lines, and these could not be derived from his theory.[553] In spite of these difficulties, Sommerfeld's theory of X-ray spectra was a great success and confirmed the view—propagated by Bohr, Moseley and especially Kossel—that the X-ray spectra were due to processes which took place in the deep interior of atoms. Thus Sommerfeld concluded his discussion of these matters by saying:

It is also remarkable that so far no indication of the "periodic" system of elements has shown up in the region of the X-ray spectra. Evidently only the outer parts of the atom, in which the optical and chemical processes occur, have a periodic nature, however, the inner parts, in which those electron motions take place that give rise to the emission and absorption of X-rays, are completely uniform and are determined by the linear progression of the atomic number of the element. (Sommerfeld, 1916c, pp. 166–167)

In the following years Sommerfeld, Kossel and others turned their attention to the above-mentioned difficulties of the theory of X-ray spectra, i.e., to the problem of the validity of the combination principle, the exact distribution of electron orbits of the atom and the explanation of the extra Λ- and Λ'-terms. As for the violation of the combination principle Kossel argued that the relativistic terms calculated by Sommerfeld represented ideal or 'virtual' energy terms, which did not coincide with the real or 'physical' energy terms entering into the spectroscopic combination principle (Kossel, 1916b, pp. 357–358). Sommerfeld agreed in principle and emphasized that the disturbance of the inner orbits by the outlying electrons might cause the defect concerning the combination principle in his theory (Sommerfeld, 1916c, §8; 1916e, §5). The question concerning the number of electrons in inner shells, on the other hand, was treated by Peter Debye in his paper on '*Der erste Elektronenring*' ('The First Electron Ring,' Debye, 1917). He especially sought to derive Sommerfeld's formula for the frequency ν of the K_α-line (i.e., the higher-frequency component of the doublet),

---

[553] The question concerning the existence of the levels Λ and Λ' arose from the data of several authors, especially from the four frequencies of the L-series (M. de Broglie, 1916; Wagner, 1917, p. 465).

i.e., the equation

$$\frac{\nu}{R} = \frac{(Z - 1.6)^2}{1} - \frac{(Z - 3.5)^2}{4} \, , \qquad (178)$$

which corresponded to the transition of an electron from an L-orbit (a two-quantum orbit) to a K-orbit (a one-quantum orbit, i.e., an orbit having the principal quantum number $n = 1$). For the purpose of deriving Eq. (178) Debye modified the usual picture of the origin of spectral lines in Bohr's theory in the following way:

> In the interior of the atom there exists, to begin with, a first electron ring next to the nucleus, on which three electrons rotate at equal angular distances around the nucleus of charge $Z|e|$, each associated with one quantum of action. From this ring one electron can be removed and be brought on a circular orbit associated with two quanta; the two remaining electrons then come closer to the nucleus and describe, at an angular distance of 180° from each other, a new circular orbit around the nucleus. The transition of the three electrons from the second state to the first state creates the K_α-line. The statement thus formulated is characterized by the fact that now not one, but three, electrons participate simultaneously in the creation of the K_α-line. (Debye, 1917, p. 277)

Debye calculated the difference between the energy states according to this picture and related it via the Bohr–Einstein condition to the frequency ν_{K_α}. From the agreement with the observed frequency Debye concluded: '*All atoms* (except those of the initial elements [in the periodic system]) contain a first ring around the nucleus containing three electrons' (Debye, 1917, p. 284).

In spite of the satisfactory explanation of the K_α-frequency, Debye's theory produced new difficulties. The problem was not connected so much with the violation of the usual periodicity of the elements; after all Bohr, already in 1913, had pointed to the fact that in the innermost shell of heavier atoms the usual periodicity with two and eight electrons might be broken, and more electrons could appear on a shell. A more serious difficulty, however, arose when one wanted to extend Debye's treatment to the other lines of the K-series as well as to the other series. Lars Vegard of the University of Oslo became involved in this question.[554] He established a theory of the K- and L-series, which fitted the data reasonably well, provided he assumed seven electrons in an electron ring described by the principal quantum number $n = 2$ (or a two-quantum ring) and eight in the next ring, ten in the four-quantum ring, etc. (Vegard, 1917a, b). Thus uranium, for instance, possessed the electronic structure 92 (3, 7, 8, 10, 8, 10, 8, 14, 10, 8, 6). During the following years he continued to derive frequency formulae

---

[554] Lars Vegard was born in 1880 at Vegaarshei, Norway. He studied at the Universities of Oslo (1899–1905), Cambridge and Leeds (1908–1910) and Oslo (1910–1911), Würzburg (1911–1912) and again Oslo (1912–1913), receiving his doctorate in 1913. He then became a lecturer at the University of Oslo, and in 1918 professor of physics there; he retired in 1950.

from the electron-ring theory by employing the 'hypothesis of the increasing quantum numbers of normal electron rings' (Vegard, 1919a, b). However, his theory encountered two objections. First, it failed to explain why the frequency of the X-ray lines occurring in absorption and emission processes, respectively, did not coincide. Second, Fritz Reiche and Adolf Smekal—who repeated Debye's K_α-calculation by taking into account the mutual electric perturbations of the electrons in the K-ring, the electric perturbations created by the excited electron (participating in the emission of the K_α-line, which, according to Debye, was not in an L-ring), and the electric perturbations caused by the electrons in the L-ring—arrived at no satisfactory description of the observed X-ray data under the assumption that the K- and L-rings were coplanar (Reiche and Smekal, 1918a, b). They concluded: 'If one, therefore, sticks to Debye's view, *then one cannot assume the K- and L-rings to be in one plane.* But if one consequently assumes that the K-ring and the L-ring make an angle with each other, then one must complete the theory by taking into account the magnetic interactions' (Reiche and Smekal, 1918a, p. 304). Adolf Smekal proposed still another model for the emission of the K-lines of the X-ray spectra, in which the electric perturbations exerted by the neighbouring electron rings on each other were very small: he assumed that the K_α-line might be emitted by the transition of an electron, which is initially on a (supercharged) L-ring (rather than on a different two-quantum orbit, as proposed by Debye), to the final orbit on the (one-quantum) K-ring (Smekal, 1918).[555] In his treatment Smekal retained the validity of the combination principle, while Sommerfeld had previously raised certain doubts about it (see Sommerfeld, 1916c, e).

Sommerfeld had meanwhile pursued his theory of X-ray spectra further by adopting a new picture of the electron distribution in orbits: the so-called '*Ellipsenverein*' or grouping of ellipses (Sommerfeld, 1918b).[556] Several years earlier John William Nicholson had proposed a similar idea; in particular, he had claimed that a concentric ring-structure in the atom was not stable and had argued that either the electrons had to move in different planes (Nicholson, 1914a) or each electron described a properly oriented elliptical orbit (Nicholson, 1914c). He had argued that in this way the mutual influence of the electrons on one another would not disturb the Coulomb field of the atomic nucleus, and the

---

[555] Adolf Gustav Stephan Smekal was born on 12 September 1895 in Vienna. He studied at the *Technische Hochschule*, Vienna (1912–1913), and at the University of Graz (1913–1917), where he received his doctorate in 1917. He continued his studies from 1917 to 1919 at the University of Berlin under Planck, Einstein, Rubens and Warburg. In spring 1919 he became an assistant at the *Technische Hochschule* in Vienna; in the following year he moved to the University of Vienna to become a *Privatdozent*. He was promoted to an extraordinary professorship in 1927; a year later he accepted the call to the University of Halle as a full professor and director of the institute of theoretical physics. After World War II Smekal joined the *Technische Hochschule* in Darmstadt, but in 1949 he was appointed to the chair of experimental physics at the University of Graz. He died in Graz on 7 March 1959.

[556] In his paper, submitted in early April 1918 to *Physikalische Zeitschrift*, Sommerfeld noted that he had obtained the idea of the '*Ellipsenverein*' from Franz Pauer. (See Sommerfeld, 1918b, p. 298, footnote 1.)

same argument now played a role in Sommerfeld's treatment.[557] He proposed:

> One has to assume that the n electrons are not distributed on the same ellipse; rather, each electron describes its own ellipse, congruent to the others, around the nucleus as the focus, and each ellipse is rotated by an angle $2\pi/n$ towards the neighbouring ellipse. We wish to call the configuration so obtained an "*Ellipsenverein*" [a "grouping" or "association of ellipses"]. If the actual positions of the n electrons are connected by straight lines, there results a regular n-sided polygon, whose area increases or decreases if the n electrons approach their aphelion or perihelion, respectively, which they do by following exactly the same rhythm. From the arrangement of the n-sided polygon it is immediately evident that the repulsions, which one electron receives from the other $n-1$ electrons, add up to a resultant force which passes through the nucleus and may be represented by a diminution of the nuclear charge according to the rule $Z - s_n$ [where s_n is a number depending on the number of electrons n]. With this organization of the electrons the orbits still retain exactly the Keplerian form [i.e., they remain ellipses]. The case of the multiply-occupied circular orbit has only the advantage of one peculiarity in comparison with the general case of the *Ellipsenverien*: in it all n orbits appear to lie on top of each other, forming a common circular orbit. (Sommerfeld, 1918b, pp. 297–298)

With the help of the 'grouping of ellipses' ('*Ellipsenverein*') Sommerfeld began to devise the occupation of the levels in atoms of elements in the periodic system of elements. Though he assumed that the light atoms contained 2, 8, 8, 18 electrons on the K-, L-, M- and N-rings, respectively, in agreement with the chemical properties, he did not exclude the possibility that the internal rings of heavy atoms might contain (as Bohr had suggested) more electrons.[558] For the purpose of calculating the energy states of the atoms he returned to the arrangement of electrons in concentric rings, but he found that the stability condition required the planes of the rings to be perpendicular to each other. However, the correction caused by this spatial arrangement of the electron rings turned out to be small. Thus he concluded:

> Hence, insofar as the correction term is negligible, observation cannot tell us anything about the mutual inclination between the planes of the orbits in the atom. The experimental accuracy must still be increased considerably in order to be able to compute from the observations the angle between the orbits by using the formula for the corrections to the interaction term. (Sommerfeld, 1918b, p. 300)[559]

---

[557] Sommerfeld obtained the reference to Nicholson's paper from the Munich physicist Richard Swinne. (See Sommerfeld, 1918b, p. 298, footnote 1.)

[558] In this context Sommerfeld referred to Debye's theory of the K_α-line, in which three electrons were on the K-ring, and to a paper of Jan Kroo of Cracow, who had represented the K_α-doublet by a model, in which the normal state was represented by three electrons on the K-ring and eight electrons on the L-ring (Kroo, 1918).

[559] In the calculation, the individual terms describing the electron–electron interaction were relatively large, but they all cancelled up to a small residue.

In a further paper Sommerfeld used his new model (of the *Ellipsenverein* or grouping of ellipses) to account for the observed fine structure of the K_β-line of different electrons (Sommerfeld, 1918c). He showed, in particular, that the L-ring (which consisted partly of a circular orbit and partly of an *Ellipsenverein*) induced a splitting of the K_β-line into a doublet, whose separation was proportional to $\alpha^2(Z - \text{const.})^3$, where α denotes the fine-structure constant, Z the atomic number, and the constant a nonintegral number expressing the screening effect of the inner electrons on the nuclear charge. Since the doublet separation of the K_α-line turned out to be proportional to $\alpha^2(Z - \text{const.})^4$, which was much larger for large atomic numbers, Sommerfeld suggested to look for the K_β-doublet rather 'among the light atoms'; this suggestion seemed to be confirmed by the most recent observations of Manne Siegbahn (Sommerfeld, 1918c, p. 372).

Both Sommerfeld and Kossel tried to extend their theoretical description of short-wavelength spectra along two directions. On one hand, they dealt with the optical spectra of many-electron atoms emitting lines having complicated structures; on the other, they deepened the theory of X-ray spectra in the light of more and newer data. The optical spectra of atoms were connected with their peripheral structure, which also was assumed to be responsible for their chemical properties. Walther Kossel again took the first step towards exploiting Bohr's theory systematically for the purpose of explaining the chemical properties of atoms. In a long memoir, entitled '*Über die Molekülbildung als Frage des Atombaus*' ('On Molecular Formation Considered as a Problem of Atomic Structure'), which was received by *Annalen der Physik* on 27 December 1915 and published in the issue of March 1916, he reviewed in detail the situation concerning the binding of atoms to form the known chemical compounds. With respect to the structure of atoms he started out with what he called the 'simplest representation which can be made,' illustrating it (the representation) by studying the cases of atoms of successive elements in the periodic system. He remarked:

> The organization of the inner electrons should always remain similar for successive elements [in the periodic system]; it should only be changed continuously in its dimensions—due to the continuous increase of the nuclear charge—in such a way that it corresponds to the increase of the characteristic [X-ray] frequencies. The next electron, which appears in the heavier element [in comparison to the preceding lighter element], should always be added at the periphery; one should arrange it in such a manner that the observed periodicity [of chemical and physical properties] i.e., its [the electron's] accessibility from outside, results. This leads to the conclusion that the electrons, which are added further, should be put into concentric rings or shells, on each of which (because of the stability conditions, which we do not specify here) only a certain number of electrons—namely, eight in our case—should be arranged. As soon as one ring or shell is completed, a new one has to be started for the next element [in the periodic system]; the number of electrons, which are most easily accessible and lie at the outermost periphery [of the atom], increases again from element to element and, therefore, in the formation of each new shell the chemical periodicity is repeated. (Kossel, 1916a, p. 237)

The shell structure thus established then fully accounted for the so-called heteropolar or ionic compounds: they consist of oppositely charged ions, arising from atoms which tend to attract or get rid of an electron, such that saturated shells having two or eight electrons will form the surface of the remaining ions.[560] The success in explaining the chemical properties of atoms on the basis of their electronic structure encouraged Kossel and Sommerfeld a few years later to state an important rule concerning the nature of optical spectra. Since Kossel, while discussing molecular structure, had discovered that the ion of an element behaved chemically similar to the element preceding it in the periodic table, he now studied with Sommerfeld the same situation in the case of spectra. Empirically, two kinds of spectra could be obtained from chemical elements: the 'arc spectra' corresponding to the optical spectra of atoms and the 'spark spectra' corresponding to the spectra of ions.

We now argue [Kossel and Sommerfeld wrote in a joint paper submitted in April 1919 to *Verhandlungen der Deutschen Physikalischen Gesellschaft*] *that the spark spectrum of each element has the same character as the arc spectrum of the preceding element in the periodic system*; i.e., it consists of doublet, triplet or so-called irregular lines, depending on whether the arc spectrum of the preceding element is made up of doublets (as in alkali elements), of triplets (as in alkaline-earth elements), or of lines without any series law ('*serienlose*', as in noble gas elements). We can thus state a *displacement law* which relates, similar to the radioactive displacement law, one element to the neighbouring element in the periodic system. (Kossel and Sommerfeld, 1919, pp. 244–245)

The so-called 'spectroscopic displacement law' of Kossel and Sommerfeld could be tested in numerous examples and was found to hold in all cases without exception.

On 2 September 1919 Arnold Sommerfeld signed the foreword of his book *Atombau und Spektrallinien*, in which he discussed in detail the relation between atomic structure and discrete spectra, which had emerged from Bohr's theory and his own extension of it. This book marked the end of the first phase of the development of the theory of atomic structure (Sommerfeld, 1919). It became a most successful book, one of the great scientific classics of the twentieth century, and went through numerous editions—the second edition was brought out the following year. Later in September 1919 Sommerfeld met Niels Bohr at Lund, Sweden, where Manne Siegbahn had organized a conference. At the conference they learned about the most recent progress in X-ray spectroscopy, including the measurements of Karl Wilhelm Stenström on the M-spectra of thorium and uranium (Stenström, 1919). Stenström had found, in particular, three edges in the absorption spectra, confirming a prediction of Sommerfeld's relativistic theory: in it the K-level should be a single term, the L-level a doublet and the M-level a

---

[560] In the case of nonpolar bindings the molecular structure could not be established yet. Kossel proposed to postpone the discussion until more dispersion data might assist the theoretician in constructing adequate models.

triplet. Sommerfeld immediately analyzed the new data and worked out a complete and satisfactory description, about which he wrote to Bohr, and Bohr replied: 'I was extremely interested in everything you wrote in your letter. The regularities in the X-ray spectra, which you mention, are very beautiful, and there is no doubt that you have hit the heart of the problem' (Bohr to Sommerfeld, 19 November 1919). Sommerfeld summarized his results on the fine structure of X-ray spectra in a paper, which he submitted to *Zeitschrift für Physik* in January 1920; it was published in February (Sommerfeld, 1920b). In the same issue, Walther Kossel, to whom Sommerfeld had reported about Stenström's measurements, gave a review of the status of the theory. In particular, he discussed again the problem of the deviations from the combination principle and concluded: 'To sum up, we do not consider it to be justified to assume real deviations from the combination relations, which would require a special foundation, if not a departure in principle from Bohr's fundamental assumptions' (Kossel, 1920, p. 133). Kossel claimed that both optical and X-ray spectra obeyed the same principles and that there existed a smooth transition from the smaller to the larger frequencies. 'Further, it also appears to be the most important task to establish a network of combination relations, which is as complete as possible and embraces all terms and lines, in order to understand how the selection rules work. In this regard, the great increase in accuracy, which has been recently obtained by Mr. Siegbahn through his precision method, will become very valuable, for one has frequently to decide between components of the fine structure' (Kossel, 1920, p. 134).

Shortly thereafter many experimental and theoretical investigations on the structure of X-ray spectra were carried out. Thus Elis Hjalmar, Siegbahn's student, obtained further high precision data on the *K*-series (Hjalmar, 1920a, 1921) and on the *L*-series (Hjalmar, 1920b), and Gustav Hertz determined the accurate position of the absorption edges of *L*-spectra of the elements from cesium to neodyne (Hertz, 1920). Hertz' main result was to question the existence of the Λ- and Λ'-states, which had been proposed by Sommerfeld (1916c). Adolf Smekal also concluded the absence of these states from a theory of the fine structure of X-ray spectra, which he proposed in a series of papers starting in 1920 (Smekal, 1920, 1921a, b, c). In his theory, which explained all existing data, Smekal employed the spatial structure of the atoms by associating three quantum numbers with each electron configuration. Thus, apart from the radial and azimuthal quantum numbers (describing a planar motion), he also used a 'spatial' quantum number. Sommerfeld himself tried to incorporate the new experimental data of Hjalmar and Hertz in a paper, which was received by *Zeitschrift für Physik* on 23 February 1921 (Sommerfeld, 1921a).[561] In an addendum to this paper he referred to a letter of Niels Bohr, which appeared in *Nature*, issue of 24 March 1921 (Bohr, 1921a). In it, Bohr proposed a new arrangement of electrons in atoms. Sommerfeld, who received a copy of this

---

[561] In this paper, Sommerfeld also gave up the two states Λ and Λ', which he had introduced *ad hoc* in 1916. (See Sommerfeld, 1921a, pp. 3–4, footnote 2.)

letter from Bohr prior to publications, wrote back to Bohr on 7 March:

> Sincere thanks for [a copy of] your letter to *Nature*. It evidently constitutes the
> greatest progress in [the theory of] atomic structure since 1913. Like you in 1916, I
> had also decided to withdraw a part of the manuscript which I had submitted to
> *Zeitschrift für Physik*. In it I had pondered about the question, how the quantum
> sums, which increase systematically in the interior of the atom, decrease again at
> the surface. This question you have now definitely settled.

Bohr's new theory of atomic structure led to a complete description of the
periodic system of elements, which would be confirmed not only by the existing
spectroscopic data but also by the discovery of a new element: hafnium.

III.4 Bohr's Wolfskehl Lectures and the Theory of the Periodic System of Elements

Since its establishment in 1913, Bohr's theory of atomic structure had received
greater attention than any other attempt in this direction. This did not mean,
however, that this theory was generally accepted by physicists and that rival
views were not discussed at all. Thus, for instance, Joseph John Thomson, in two
series of lectures at the Royal Institution in 1918 and 1919, respectively, discussed the problems of atomic structure and origin of spectral lines; he invoked a
model of atoms differing widely from Bohr's and resting on the assumption that
the forces exerted on the electrons in atoms were not only of the Coulomb type
but also included repulsive terms (Thomson, 1918, 1919).[562] It also did not mean
that no opposition arose against Bohr's conceptions. For example, Johannes
Stark, in an article entitled '*Zur Kritik der Bohrschen Theorie der Lichtemission*'
('On a Criticism of Bohr's Theory of Light Emission'), expounded a strong
criticism; he not only emphasized the unclear points in the fundamental assumptions, but also concluded that the theory failed in many respects to describe the
data correctly (Stark, 1920). Arnold Sommerfeld defended Bohr's theory against
Stark's accusations (Sommerfeld, 1921c). While he agreed with Stark that certain
points of Bohr's theory, such as the question concerning the coupling of electrons
to radiation or the dependence of the emitted radiation on the final state
involved in the transition, required further clarification, he pointed out that 'one
should not forget that Bohr's theory is only eight years old' (Sommerfeld, 1921c,
p. 419). Continued investigations, he was certain, would resolve many if not all

---

[562] In his 1918 lectures, Thomson referred explicitly to a repulsive term proportional to the third
inverse power of the distance between the electron and the centre of atom (i.e., nucleus), which he
had used several years before (Thomson, 1913b). With this assumption the electron would possess a
stable position, i.e., it would be in equilibrium at the position of zero force, and the radiation would
come about from oscillations around the stable position. In his 1919 lectures Thomson chose the full
potential (i.e., Coulomb attraction plus additional repulsive forces) as being given by $(1/r^2) \cdot$
$(\sin(2c/r)/(2c/r))$, where r is the radial distance and c is the velocity of light *in vacuo*; this potential
led to several stable positions.

difficulties. As for the presumed failures of Bohr's theory to account for the empirical facts, Sommerfeld contradicted Stark in all cases. For example, he argued that the most recent measurements of the doublet splitting of hydrogen lines by Ernst Gehrcke and Ernst Gustav Lau, which seemed to yield results smaller than the ones predicted by his relativity calculations (Gehrcke and Lau, 1920), could not, because of their insufficient accuracy, be used against the Bohr–Sommerfeld theory. On the other hand, Sommerfeld claimed that all Stark effect data agreed with the theory, while he declared it to be premature to conclude any failure of the theory in describing the anomalous Zeeman effects. Altogether, he strongly opposed Stark's opinion that the success of Bohr's theory was restricted to explaining the Balmer spectrum, the frequencies of the Stark components and the normal Zeeman effect, and remarked:

> Even if we exclude, as Mr. Stark does, the wide field of X-ray spectroscopy, which has been organized and clarified only by applying Bohr's theory, and even if we forget about the questions of fine structure, which have been discussed above sufficiently, we must still count in the field of visible spectroscopy such an amount of successes in favour of Bohr's theory that one will not hesitate to call the latter the greatest progress of all times in the understanding of the atom. (Sommerfeld, 1921c, p. 429)

The great progress to which Sommerfeld referred included not only the well-known explanation of the spectra of ionized helium and the qualitative interpretation of the structure of the spectra of non-hydrogen-like atoms (i.e., the occurrence of the principal and subordinate series, etc.), but also the recent advances in the theory of band spectra. Band spectra had already been discussed in connection with quantum theory before Bohr proposed his models of atomic structures, especially by Niels Bjerrum, as an example of quantizing the rotator system (Bjerrum, 1912). Later, Karl Schwarzschild had considered the band spectra as arising from the transitions between different stationary states: a band-line with frequency ν,

$$\nu = \nu_0 + \frac{h}{8\pi^2 A}(m^2 - m'^2), \tag{179}$$

should be emitted when the electrons in the molecule make a transition liberating the energy $h\nu_0$ and the molecular rotation changes the rotational quantum number from m to m' (Schwarzschild, 1916). By inserting into Eq. (179) the data from the so-called cyanide-bands (emitted by molecular nitrogen), he had found, however, that the moment of inertia, A, thus derived was much larger than the one obtained from kinetic gas theory. Torsten Heurlinger of Lund had then investigated the situation more closely; in particular, he had observed that the nitrogen-bands could be separated into several systems of bands, the frequencies of the components being described by the formula

$$\nu = \nu_0 + c_1 m + c_2 m^2, \tag{180}$$

where m is an integer. He had then developed, from the point of view of Bohr's theory of atomic and molecular constitution, a description of the band spectra of diatomic molecules (like N_2) by associating with each state a series of quantum numbers $m, n_1, n_2, \ldots$ and deriving the frequencies as differences of terms, $\psi(m, n_1, n_2, \ldots)$ and $\psi(m', n_1', n_2', \ldots)$, where m and m' are the rotational quantum numbers of both terms and $n_1, n_2, \ldots$ and $n_1', n_2', \ldots$ the other quantum numbers of the initial and final states, respectively (Heurlinger, 1918, 1919, 1920).[563] Now, on returning to Munich after the end of the war, Sommerfeld's former student Wilhelm Lenz entered the field.[564] Lenz considered the molecular models more carefully; especially, he took into account the alteration of the moment of inertia of the molecule due to a change in the radial quantum number of the electron ring, and he was indeed able to justify Heurlinger's Eq. (180) including the term proportional to m^2 (Lenz, 1919).[565] However, in determining the quantum numbers of the ring from the band spectra and their Zeeman effects in the cases of the nitrogen and the hydrogen molecules, Lenz noticed a difficulty: the data required the quantum number n of the ring to be zero, in contradiction to Bohr's models, which required $n = 2$ in the case of hydrogen and a still higher value for nitrogen.[566] When Lenz left Munich in 1920, he had already trained Adolf Kratzer, his successor at Sommerfeld's Institute.[567] Kratzer extended the theory of diatomic molecules by including anharmonic forces between the nuclei, which changed the oscillation frequencies (Kratzer, 1920a). He would devote his efforts in future basically to the investigation of band spectra and become one of the leading experts in this field.[568]

The theory of band spectra represented only one of the subjects discussed at Sommerfeld's Institute in Munich. An increasing number of new students, including Wolfgang Pauli, Gregor Wentzel, Werner Heisenberg, Otto Laporte

---

[563] By adding the selection rules, $m' = m$, $m \pm 1$, Heurlinger found that for the term differences, $\psi(m, n_1, \ldots) - \psi(m \pm 1, n_1, \ldots)$, Eq. (180) could indeed be derived.

[564] W. Lenz was born on 8 February 1888 at Frankfurt-am-Main. He studied physics at the Universities of Göttingen (1906–1908) and Munich (1908–1911), obtaining his doctorate in 1911. He then stayed on in Munich (after 1914 as a *Privatdozent*). In 1920 he was called to the University of Rostock as an extraordinary professor; the following year he moved to the University of Hamburg as ordinary (i.e., full) professor of theoretical physics. He retired in 1956 and died in Hamburg on 30 April 1957.

[565] Without this change of the radial quantum number, only the term linear in m can be explained. (See Lenz, 1919, p. 634, Eq. (4).)

[566] In another paper on the theory of band spectra (Lenz, 1920b), Lenz discussed the spectra of iodine, which had been obtained recently by Robert Williams Wood (1918).

[567] A. Katzer was born on 16 October 1893 at Günzburg. He studied physics at the *Technische Hochschule*, Munich (1912–1914), and—after two years service in the army—at the University of Munich from 1916 to 1920, when he received his doctorate under Sommerfeld. Then he was sent to Göttingen as Hilbert's assistant for physics (1920–1921). In 1921, after his return to Munich, he became a *Privatdozent*. A year later he accepted the invitation to the chair of theoretical physics at the University of Münster.

[568] In a second paper in 1920, Kratzer studied the influence of nuclear masses (the isotope effect) on band spectra (Kratzer, 1920b). His detailed analysis of the cyanide-bands resulted in the introduction of half-integral quantum numbers to account for the rotation (Kratzer, 1922).

and, a little later, Walter Heitler, Karl Bechert and Albrecht Unsöld, helped Sommerfeld in exploring various parts of atomic theory: Pauli obtained his doctorate with a thesis on the hydrogen molecule-ion, Wentzel worked on X-ray spectra, Heisenberg on the anomalous Zeeman effect and Laporte on complex spectra. The flourishing Sommerfeld school sent its representatives to other places—where, with Sommerfeld's effective assistance, they obtained professorships and began to develop atomic theory themselves: thus Peter Debye in Zurich, Wilhelm Lenz in Hamburg, Adolf Kratzer in Münster and Erwin Fues in Stuttgart. Besides Munich, atomic theory also played an important role in Berlin and at those universities where people from Berlin went: thus in Breslau, Rudolf Ladenburg and Fritz Reiche pursued it. The publications of the results emerging from these numerous and intensive investigations assumed a prominent place in the German scientific journals. The *Zeitschrift für Physik*, a new journal, played a major role in reporting about the developments in atomic theory; it was established in 1920 to publish the growing number of contributions that was submitted to the German Physical Society for inclusion in its *Verhandlungen*.[569] Soon not only the physicists from Germany, Austria and Switzerland, but also from the Netherlands, the Scandinavian countries and the Soviet Union submitted their papers to *Zeitschrift für Physik*, making it the leading journal on atomic theory. Next to it the *Physikalische Zeitschrift* still contained a considerable number of papers on atomic theory, while the number of papers submitted to the more conservative *Annalen der Physik* in this field declined.[570]

While innumerable publications made the theory of atomic structure a really fashionable topic at the forefront of physics, the man, who had invented the theory, remained comparatively silent. Since his return to Copenhagen in fall 1916 as a professor of theoretical physics, Niels Bohr had published only two papers dealing with the fundamental aspects of the theory of line spectra (Bohr, 1918a, b) and one paper suggesting the existence of a triatomic hydrogen molecule (Bohr, 1919). Besides these papers, he had given several lectures reviewing the progress of his theory at the Physical Society of Copenhagen, whose chairman he was from 1916 to 1919; and he had lectured abroad, for example, at the University of Leyden on 25 April 1919 (on 'Problems of the Atom and the

---

[569] The number of papers submitted to the *Verhandlungen* had increased steadily since 1899. After World War I the production costs for the journal rose tremendously due to inflation and the leading scientists in the German Physical Society decided—against some opposition, e.g., from Philipp Lenard—to reduce the content of the *Verhandlungen* by restricting its pages to reports on the activities of the Society, obituary notices, etc. The extended original publications were supposed to appear in *Zeitschrift für Physik*, a journal edited by the Society (with Karl Scheel as the responsible editor) and printed by *Friedrich Vieweg und Sohn*, Braunschweig. The first issues of *Zeitschrift für Physik* appeared in February 1920; it soon became the leading physics journal, overtaking the *Annalen der Physik*. The period between the submission of a paper and its publication was much shorter in the *Zeitschrift*, and papers on modern topics, especially atomic theory, were more likely to be submitted to it.

[570] It should be mentioned, however, that many detailed experimental investigations on spectroscopy and a number of extended theoretical studies (e.g., doctoral theses) continued to appear in *Annalen der Physik*.

Molecule') and in Berlin at the German Physical Society on 27 April 1920 ('*Über die Serienspektren der Elemente*'). To an outsider the years between 1918 and the end of 1920 appeared to be a quiet period in Niels Bohr's scientific productivity. He seemed to be deeply involved in establishing his own institute, the Institute of Theoretical Physics at Blegdamsvej in Copenhagen.[571] Bohr's Institute, which had been planned with the help of the experimentalist Hans Marius Hansen and funded mainly by the Carlsberg Foundation, was dedicated on 3 March 1921.[572] Bohr had a few collaborators, the most important among them being Hendrik Anthony Kramers.

Hendrik Kramers was born on 17 December 1894 at Rotterdam. He began to study theoretical physics in 1912 under Paul Ehrenfest at the University of Leyden. After passing a predoctoral examination in 1916, he wanted to continue his studies of mathematics and physics at a foreign university. He went to Copenhagen and there wrote a letter to Bohr, in which he introduced himself and requested: 'Of course I should like very much to come in acquaintance with *you* in the first place, and also with your brother Harald. Therefore I should be very glad if you permit me, to visit you in one of these days. Perhaps you'll be so good to write me a card or telephone to my hotel *when* I may come to see you' (Kramers to Bohr, 25 August 1916).[573] After meeting Bohr, Kramers expressed the wish to become his assistant. Bohr, who was not quite sure about what to do, asked his brother Harald, who answered that 'if the young Dutchman was really so keen, he might as well be given a chance' (Rosenfeld and Rüdinger in Rozental, 1967, p. 69).[574] So Kramers stayed on in Copenhagen, and the first problem which Bohr asked him to solve was to calculate the Fourier coefficients

---

[571] When Bohr assumed his professorship at the University of Copenhagen in September 1916, he had only a single room next to the physics library of the old Polytechnic Institute. A schoolfriend of his, Aage Berlème, started a private initiative to collect money for buying the land for a new physics institute already in 1917. With the help of further private and official contributions the construction of the building began in 1919.

[572] Although the building was completed at that time, the equipment of the laboratory for spectroscopic investigations still had to be installed. It would take many years and absorb the energies of Bohr and his collaborators. Nevertheless, the experimental investigations had already been started; for example, James Franck, who had spent several weeks at Copenhagen before March 1921, had set up an apparatus for the observation of collisions between electrons and atoms. George de Hevesy, on the other hand, had done experiments during 1920–1921 at the physical chemistry laboratory of the University of Copenhagen.

[573] In visiting Copenhagen and Niels Bohr, Kramers may have been advised by Paul Ehrenfest, although Ehrenfest in 1916 was not yet in favour of Bohr's atomic theory. Oskar Klein recalled later that Kramers had told him the following in connection with his visit to Copenhagen: 'There was a student meeting at Copenhagen . . . . It was an aunt of his who invited him to go to that meeting and he went there . . . . He told Ehrenfest about it, and Ehrenfest just said that he ought to visit Bohr. Kramers very much liked to see people, so he went up to Bohr right at the beginning' (Klein, Conversations, p. 33).

[574] Oskar Klein recalled later that Kramers had told him the following story about becoming Bohr's assistant: 'He [Kramers] went to the student meeting after the [first] talk with Bohr, and then he spent all his money. He had no money to go back to Holland, so he went back to Bohr and asked him if he could borrow some money from him. Then they got into further conversation, and Bohr asked him—there might, of course, have been some days in between—to be his assistant' (Klein, Conversations, p. 33).

of the electron orbits in the hydrogen atom when perturbed by a static (external) electric field. Simultaneously, both Bohr and Kramers began to investigate the structure of the helium atom. At that time Bohr was just on the way to developing his correspondence point of view. As Oskar Klein, one of the first students of Bohr and Kramers, recalled:

> Bohr had, of course, the beginnings of it [the correspondence principle] in his first paper [on atomic constitution (Bohr, 1913b)] already, but it hadn't occurred to him that one should be able to approximate that way the probability for transitions and especially also the absence of certain transitions . . . Then Bohr saw that that was immediately clear from what he called at that time the analogy principle. . . . I think Kramers told me that it had occurred to Bohr while on a trip he had made walking in Jütland. When he came back from that he told Kramers about these things, and then Kramers immediately tried to calculate Fourier coefficients. (O. Klein, Conversations, p. 37; also AHQP Interview)

The work on the intensities of spectral lines occupied Kramers for a couple of years. He finally submitted a long paper on it in spring 1919 as his doctoral thesis to the University of Leyden (Kramers, 1919).[575]

In his second communication on atomic structure to the Bavarian Academy, in which he dealt with the fine structure of spectral lines, Sommerfeld had already published certain considerations on their intensities. There he had advanced the following argument: 'We shall assume that in all cases the circular orbit possesses the greater probability, and that an elliptical orbit is the less probable, the larger its eccentricity' (Sommerfeld, 1915c, p. 473).[576] In later papers he had formulated this idea more quantitatively as an 'intensity rule' (Sommerfeld, 1917). Thus I, the relative intensity of the fine-structure components of hydrogen emitted in transitions between elliptical orbits, characterized by the pairs of azimuthal and radial quantum numbers, (m, m') and (n, n'), respectively, should be given by the equation

$$I = \frac{n}{n + n'} \cdot \frac{m}{m + m'} , \tag{181}$$

provided the transitions were not forbidden. Statistical arguments by Karl Herzfeld and Arnold Sommerfeld had appeared to support Eq. (181): while Herzfeld had shown, in particular, that the outer orbits were rendered less probable through the action of neighbouring atoms on the atom under consideration (Herzfeld, 1916), Sommerfeld had derived Eq. (181) essentially by studying the degeneracy of the elliptical electron motions (Sommerfeld, 1917).[577] Still,

[575] Kramers went to Leyden together with Niels Bohr in April 1919 to receive his doctorate.

[576] This assumption was not only consistent with the absence of the pendulum orbit—because of infinite eccentricity it was highly improbable—but also explained the observed intensity ratios of the relativistic doublet components fairly well.

[577] In general, the electrons in an atom could move in three dimensions. Each elliptical motion in the plane, therefore, corresponded to a degenerate motion; Sommerfeld determined the degree of degeneracy for each orbit and used it to derive an equation for the intensity similar to Eq. (181).

many difficulties had remained in Sommerfeld's approach; Paul Epstein, for instance, had noted in connection with the intensities of the Stark components of the hydrogen lines that, 'In our attempts we were not successful in establishing a general point of view concerning the probability for various orbits; we can just report on negative experiences' (Epstein, 1916c, p. 517), and Sommerfeld had concluded that 'dynamical emergency measures' were necessary to explain Stark's observations (Sommerfeld, 1917, p. 109). For this reason Bohr's approach seemed to be superior, because it made use of the analogy to the (classical) electrodynamic theory of the emission of spectral lines (Bohr, 1918a). In his thesis Kramers applied the correspondence arguments systematically to the problem of the intensity of hydrogen lines, both to the fine structure and to the Stark components (Kramers, 1919). For this purpose he calculated the coefficients of the Fourier series, C_{τ_1,τ_2}, into which the position coordinates, x, y, and z, of the electron's motion in the atom could be expanded—e.g., $z = \sum_{\tau_1,\tau_2} C_{\tau_1,\tau_2} \exp\{2\pi i(\tau_1 w_1 + \tau_2 w_2)\}$, with w_1 and w_2 the angular coordinates and τ_1 and τ_2 assuming the values $1, 2, 3$, etc.—both in the relativistic case as well as the nonrelativistic case, where the atom was under the influence of an external field. Kramers then expressed these Fourier coefficients in terms of Bessel functions, J_{τ_1} and J_{τ_2}, depending on the arguments $\tau_1\sigma_1$ and $\tau_2\sigma_2$, where σ_1 and σ_2 denoted combinations of the action variables conjugate to w_1 and w_2. In classical electrodynamics the absolute squares of the Fourier coefficients, $|C_{\tau_1,\tau_2}|^2$, described the intensities of the radiation emitted by an electron in a given orbit. Kramers now proposed to take the squares of the average Fourier coefficients, $|\overline{C}_{\tau_1,\tau_2}|^2$, with $\overline{C}_{\tau_1,\tau_2}$ representing the values of the Fourier coefficients of orbits averaged between the initial and the final orbits, as describing the quantum-theoretical intensities. Thus he computed, for instance, the intensities of the hydrogen line components in the Stark effect and compared them to the data, concluding: 'On the whole it will be seen, that it is possible on Bohr's theory to account in a convincing way for the intensities of the Stark components' (Kramers, 1919, p. 341).

Kramers also dealt with the intensities of the fine-structure components, which had been described successfully before by Sommerfeld. He had in mind not only to test Sommerfeld's results from the point of view of the correspondence principle, but also to explore certain limitations of Sommerfeld's theory of fine structure in connection with an earlier proposal of Niels Bohr. Bohr had suggested originally that the fine structure of spectral lines might originate from the influence of small electric forces perturbing the Coulomb attraction of the nucleus on the electron (Bohr, 1914b). After Sommerfeld put forward his theory of relativistic fine structure, Bohr did change his point of view; nevertheless he remained convinced that perturbing electric fields in atoms might have some influence on the fine structure. Therefore he asked Kramers to consider the problem in detail, and Kramers treated the effect of electric fields both in his thesis and, more completely, in a separate paper submitted to *Zeitschrift für Physik*, which was received on 1 October 1920 (Kramers, 1920b). He used, in particular, Bohr's perturbation method; that is, he calculated the influence of small electric fields on the fine structure of the hydrogen lines with the help of the perturbation theory of periodic systems (Bohr, 1918a). Thus he obtained the following results. First, for small electric fields the corrections to the relativistic energy terms were proportional to the square of the electric field strength; as a consequence, each fine-structure component splits into polarized Stark components.[578] Second, if the electric field strength were increased, the electron would leave the plane of the unperturbed orbit and a third quantum number would have to be introduced in order to describe the energy terms; as a consequence, a complicated structure arose, mixing the Stark components and the relativistic doublets. Third, if the electric field strength was further increased, the Stark effect would dominate the line structure; again the electrons would move in a plane, a plane perpendicular to the electric field which did not contain the nucleus, and the energy terms could be described by two quantum numbers, the principal quantum number and a quantum number characterizing the Stark effect.[579] These results, Kramers thought, threw some light on the understanding of the series spectra of elements of higher atomic numbers. According to Sommerfeld these spectra could be represented by the spectra emitted by a one-electron atom, whose Coulomb field is perturbed by an additional central field (Sommerfeld, 1915c, p. 478). Now Kramers argued:

> However, one has to remember that the comparison between the perturbing action of the inner electrons exerted on the motion of the external electron in an atom, on one hand, and the action of a central field [of force], on the other, can only be viewed as an incomplete analogy; rather one has to expect that, in investigating the influence of an electric field on the atom, the action of the internal electrons will in various respects be more complicated than the action of a simple central field. (Kramers, 1920b, p. 223)

In the following years Kramers became Bohr's closest collaborator. With his mastery of mathematical formalism and his skill in solving tricky problems—such as displayed in his doctoral thesis—he was able to carry out the laborious calculations necessary to test Bohr's ideas on atomic structure. He not only performed the detailed evaluation of the energy states of the helium atom (Kramers, 1923a), but also worked on band spectra (Kramers 1923b; Kramers and Pauli, 1923) and on continuous X-ray spectra (Kramers, 1923d). As Bohr's official assistant since 1919, Kramers exhibited many other talents: especially, he was fluent in several languages (in 1917, a year after his arrival in Copenhagen, he began to give talks in Danish) and he was willing to take care of lecture

---

[578] Each of the H_α-components, for example, splits into three components, which are due to transitions in which the radial quantum number changes by 0 and ± 2 units, respectively.

[579] Kramers noted (already in his thesis) that an electric field having the strength of 300 V/cm would give rise to the Stark effect of the hydrogen, which is of the same order as the relativistic fine-structure effect (Kramers, 1919, pp. 376–377).

courses and to advise students. The early students included Oskar Klein. Klein had met Kramers first in fall 1917 at Svante Arrhenius' Institute in Stockholm, where Kramers gave a talk on atomic theory and the adiabatic principle. Kramers impressed Klein so much that he decided to continue his studies under Niels Bohr. Thus Klein arrived at Copenhagen in May 1918 and stayed there, with some interruptions, until fall 1922.[580] 'I learned a great deal from Kramers,' he recalled later. 'From Bohr I largely heard generalities as I did not have much occasion to see him at this time. Kramers was my teacher in these new things, the Hamilton–Jacobi equations, the correspondence principle and the beginnings of the Copenhagen philosophy' (O. Klein, Conversations, p. 3).[581] Kramers helped and collaborated with most newcomers to Bohr's Institute in the early 1920s including, for instance, Wolfgang Pauli and Werner Heisenberg. He always came to know Bohr's ideas first, and in many cases he had to formulate them in proper mathematics—as happened, for example, in dealing with the fundamental considerations of a new radiation theory (Bohr, Kramers and Slater, 1924). From the latter he would derive a theory of dispersion of light by atoms, his most important contribution to quantum physics (Kramers, 1924a, b).[582]

Bohr found his collaboration with Kramers on atomic theory most satisfactory, and he frequently quoted Kramers' papers (Kramers, 1919, 1920b), especially in his talk at the German Physical Society in Berlin on 27 April 1920 (Bohr, 1920, p. 447). In Berlin he gave a review of the origin of line spectra, in which he discussed the results of the erstwhile models of atomic structure in the light of the correspondence principle. There he met a large number of German physicists, who were interested in atomic and quantum theory, including Born, Franck, Kossel, Ladenburg and Landé, and he established many fruitful contacts. A few months later Ladenburg, after giving a lecture on atomic constitution at the Bunsen Congress in Halle, asked Bohr a question concerning the occupation of the outer shells in noble gases.

> May I now ask [he wrote] what you think of Kossel's idea that the halogens F, Cl, Br, I in their simple compounds (say with the alkalis) by capture of a single

---

[580] Oskar Benjamin Klein was born in Stockholm on 15 September 1894. From 1912 he studied at the University of Stockholm and the Nobel Institute for Physical Chemistry with Svante Arrhenius, obtaining his *Licentiat* in 1918. After obtaining his doctorate in 1922 from the University of Stockholm, Klein served as *Docent* there; in 1922 he went to the University of Michigan at Ann Arbor (instructor, 1923–1924; assistant professor, 1924–1925). From 1926 to 1933 he was a lecturer at the University of Copenhagen; then he accepted a call to the University of Stockholm as professor and director of the Institute of Mechanics. Klein retired in 1968, and died on 5 February 1977 at Stockholm. Klein worked on problems of quantum theory (e.g., the crossed-field problem), non-relativistic and relativistic quantum mechanics, meson theories and general relativity.

[581] When Kramers stayed on for several months in the Netherlands after obtaining his doctorate in spring 1919, Klein came into closer contact with Bohr. At that time Bohr used to dictate to him his papers and letters. Bohr also set Klein to work on certain problems of quantum theory, the first one being a treatment of the ionized hydrogen molecule.

[582] Kramers stayed on in Copenhagen until spring 1926, when he accepted the chair of theoretical physics at the University of Utrecht. In 1934 he was appointed Paul Ehrenfest's successor in Leyden, a chair which he occupied until his death on 24 April 1952 at Oegstgeest.

electron, just as the alkalis by losing an electron, assume the particularly stable configuration of the noble gases with eight electrons in the outer shell . . . ? Born's calculation of the electron affinity of the halogens and sulphur appears to me a strong support for this conception . . . Or do you still prefer your old view of two electrons in the outer shell of noble gases? (Ladenburg to Bohr, 18 June 1920)

Bohr replied to Ladenburg on 16 July 1920, confessing that he did 'not consider any conception sufficiently assured to make it possible to take a definite standpoint' and that the considerations in his earlier papers should 'only be regarded as a tentative orientation.' He saw difficulties in explaining the observed chemical properties from physical arguments, claiming that the question 'depends not only on the geometrical character of the configuration but above all on its stability properties.' And, as a consequence of the latter, one had to give up his original ring configurations and be 'forced to reckon with far more complicated motions of the electrons in atoms.' A couple of weeks later, Bohr wrote more explicitly: 'I value the work of Kossel and others [concerning the relation of chemical properties to electronic constitution] very highly; however, I feel myself strongly cautioned by the conviction that ultimately the whole question of explaining the chemical stability cannot be regarded as primarily a geometrical problem but must rather be conceived a dynamical one' (Bohr to Ladenburg, 29 September 1920). Then, in October 1920, Alfred Landé gave a lecture at Bohr's invitation to the Physical Society of Copenhagen, in which he discussed his ideas concerning the spatial symmetry of atoms. Landé's geometrical arguments did not convince Bohr, but they persuaded him to become more deeply involved in the problem of atomic constitution. In a lecture delivered on 15 December 1920 —again before the Physical Society of Copenhagen—Bohr referred to the ideas of Born and Landé on the spatial structure of atoms, especially to Landé's recent theory of the structure of diamond (Landé, 1921a), in which Landé had assumed that the electrons in the normal state of carbon moved on very eccentric orbits. Bohr then remarked:

After getting acquainted with this work through a lecture of Landé . . . , it has occurred to me that it might be possible from simple points of view to give a rational explanation of this at a first glance strange assumption. In fact, in the following I shall try to show that it seems possible, by pursuing these viewpoints further, to throw light on a great number of the problems mentioned in the foregoing and to create a hitherto lacking basis for an understanding of the peculiar stability of the elements, of which their specific properties so strongly bear witness. (Bohr, 1977, p. 58)

He then discussed the consequences of the assumption of elliptical orbits for the ground-state electrons in the case of helium, sodium and carbon, as well as a few other elements. He expressed the hope to work out, in the following months, a detailed paper on the new theory of the constitution of atoms. However, hindered as he was by many administrative duties and poor health during most of the year 1921, he did not succeed. Instead, he submitted two letters to *Nature*, in which he sketched just a few points of his theory.[583]

The stimulation to write a letter to *Nature* came from another letter which the British physicist Norman R.Campbell had published in that journal; Campbell's letter, dated 16 November 1920, had appeared in *Nature's* issue of 25 November 1920. In it Campbell had compared the Bohr–Sommerfeld theory of atomic structure and the Lewis–Langmuir (or the Born–Landé) theory of molecular structure and had argued that 'they are not really inconsistent' (Campbell, 1920, p. 408). Bohr, however, felt that Campbell had not described the situation correctly, and he wrote: 'Dr. Campbell puts forward the interesting suggestion that the apparent inconsistency under consideration may not be real, but rather appear as a consequence of the formal character of the principles of quantum theory, which might involve that the pictures of atomic constitution used in explanation of different phenomena may have a totally different aspect, and nevertheless refer to the same reality' (Bohr, 1921a, p. 104). Campbell had claimed, for instance, that the electrons in the Bohr–Sommerfeld theory do not really move on the orbits and, in justification, had cited the correspondence principle. Bohr was completely against such an interpretation.

> On the contrary [he wrote] if we admit the soundness of the quantum theory of spectra, the principle of correspondence would seem to afford perhaps the strongest inducement to seek an interpretation of the other physical and chemical properties of the elements [i.e., those described formerly by the Lewis–Langmuir–Born–Landé theory] on the same line as the interpretation of their series spectra. (Bohr, 1921a, p. 104)

He therefore proposed a new view of atomic constitution based entirely on the correspondence principle. In fact, he claimed that the correspondence principle allowed one to establish the definite arrangement of electrons in given atoms in certain groups or shells—a problem which he had been unable to solve satisfactorily by the previous methods. He wrote:

> Thus by means of a closer examination of the progress of the binding process this principle offers a simple argument for concluding that these electrons are arranged in groups in a way which reflects the periods exhibited by the chemical properties of the elements within the sequence of increasing numbers. In fact, if we consider the binding of a large number of electrons by a nucleus of higher positive charge, this argument suggests that after the first two electrons are bound in one-quantum orbits, the next eight electrons will be bound in two-quanta orbits, the next eighteen in three-quanta orbits, the next thirty-two in four-quanta orbits. (Bohr, 1921a, p. 105)

---

[583] Due to ill health Bohr did not attend the third Solvay Conference in Brussels in April 1921. Instead Paul Ehrenfest presented Bohr's report, entitled '*L'application de la théorie des quanta aux problèmes atomiques*' ('The Application of Quantum Theory to Atomic Problems,', Bohr, 1923e). In this talk Bohr did not go into any details of atomic structure, but concentrated on general points of view.

An essential feature of the new view of atomic structure was the fact, which had
been suggested already by Landé in connection with his theory of diamond, that
'for each group the electrons within certain sub-groups will penetrate during their
revolution into regions that are closer to the nucleus than the mean distances of
the electrons belonging to groups of fewer-quanta orbits' (Bohr, 1921a, p. 105).
This led to a coupling between the electrons of different shells, which would
establish 'the necessary condition for the stability of atomic configurations'
(Bohr, 1921a, p. 105).

Bohr's letter to *Nature*, dated 14 February 1921 and published in the issue of
24 March 1921, of which he sent copies to various interested physicists, received
an enthusiastic response from many sides. Thus Landé wrote to Bohr that his
new ideas were of the greatest significance for the future of atomic theory. 'It
seems to me,' he said, 'that until the appearance of your detailed account, it
makes no sense at all to work theoretically on atomic theory' (Landé to Bohr, 21
February 1921). Sommerfeld, who agreed with Landé about the importance of
Bohr's proposal, immediately wanted to make use of it: he asked his student
Gregor Wentzel to work out its consequences in the theory of X-ray spectra.
Kasimir Fajans wrote from Munich: 'The announcement [in] your letter to
"Nature" of the solution of the riddle of the periodic system has filled us
chemists with particularly great expectations' (Fajans to Bohr, 25 June 1921). All
these people waited eagerly for the detailed account of the theory which Bohr
had promised in his letter to *Nature*. 'I am especially happy,' Walther Kossel
expressed their hope, 'that we may now expect your complete exposition of this
subject very soon' (Kossel to Bohr, 15 August 1921). Instead of the expected
memoir, Bohr wrote a second letter to *Nature* on 16 September 1921. It appeared
in *Nature's* issue of 13 October (Bohr, 1921d), and in it he clarified some points
of his previous letter; in particular, he modified certain results by assigning
slightly different quantum numbers to the electrons in the outer shells. However,
Bohr gave the first complete presentation of the theory of the periodic system in
his Wolfskehl lectures, which he delivered in June 1922 at Göttingen.[584]

Since its beginning in 1913, Bohr's theory of the atomic constitution had been
followed in Göttingen with considerable interest. Especially, the mathematicians
David Hilbert and Richard Courant had immediately taken a positive attitude

---

[584] Bohr had first talked publicly on matters connected with his theory of the periodic system on
15 December 1920 at the Physical Society of Copenhagen. (A handwritten manuscript of this talk
and fragments of some notes concerning it are in the *Bohr Archives*; they were published under the
title 'Some Considerations of Atomic Structure' in Bohr's *Collected Works*, Volume 4, 1977, pp.
43–69). The next opportunity arose in connection with the third Solvay Conference in April 1921.
Bohr had planned to attend the Conference and to give a report on 'The Application of Quantum
Theory to Atomic Problems.' However, being unable to go to Brussels, he sent the manuscript of the
first part of his talk to Paul Ehrenfest, who presented it at the Conference (Bohr, 1923e). Bohr had
promised to send the second part dealing with the details of atomic models, but he never did so; the
notes dealing with it were published in his *Collected Works*, Volume 4 (Bohr, 1977, pp. 91–174).
Evidently, these notes constituted a preliminary version of what Bohr later discussed at Göttingen in
June 1922.

towards it.[585] Then Peter Debye, Woldemar Voigt's successor in Göttingen, had worked on some aspects of Bohr's theory, especially on a model of the hydrogen molecule (Debye, 1915a), the Zeeman effect (Debye, 1916a, b), and X-ray spectra (Debye, 1917), and so had his student Paul Scherrer (Scherrer, 1916b).[586] When Debye left Göttingen, Max Born and James Franck worked actively on Bohr's theory. Hilbert also became involved in it and got his student Hellmuth Kneser to treat certain problems connected with this theory (Kneser, 1921). In order to become better acquainted with the most up-to-date progress of the subject, Hilbert got an idea: he proposed to continue the programme of the Wolfskehl lectures and to invite Bohr to speak on the theory of atomic structure.[587] On 10 November 1920 the Wolfskehl Commission, consisting of Felix Klein, David Hilbert, Carl Runge, Emil Wiechert, Ludwig Prandtl, Edmund Landau, Johannes Hartmann, Richard Courant and Robert Pohl, sent a letter to Niels Bohr proposing a date in the following year. Bohr accepted, but because of ill health he had to postpone the visit to June 1922.[588]

In the period from 12 to 22 June 1922 Bohr delivered seven lectures on the theory of atomic structure. He had carefully prepared these lectures, knowing that many scientists would come to Göttingen to listen to his latest results and ideas.[589] The Wolfskehl lectures, as a rule, were announced publicly and interested people were invited from all German universities. So the audience did not consist only of the scientists from Göttingen: the mathematicians Richard Courant, David Hilbert and Carl Runge, the physicists James Franck, Max Born and Robert Pohl and their students—for example, Gustav Heckmann, Carl Hermann, Erich Hückel, Friedrich Hund, Pascual Jordan and Rudolph Minkowski. Arnold Sommerfeld from Munich brought with him Werner Heisenberg; Wilhelm Lenz and his assistant Wolfgang Pauli travelled from Hamburg to Göttingen; Alfred Landé, Walther Gerlach and Erwin Madelung came from

---

[585] The attitudes of Hilbert and Courant towards Bohr's theory of atomic constitution have been described by Constance Reid in her biographies of the two men. (See Reid, 1970, p. 135; 1976, p. 45.)

[586] In this context, it should be mentioned that the old Eduard Riecke, in his last paper, discussed Bohr's theory of hydrogen and helium (Riecke, 1915).

[587] As we have mentioned earlier, the funds available from the annual interest on Wolfskehl endowment had been augmented by extra funds from the Prussian Ministry of Education; this was to enable the University of Göttingen to invite a prominent scientist as guest lecturer for up to a semester. This arrangement was renewed after World War I.

[588] When Sommerfeld heard from James Franck about Bohr's ill health, he wrote anxiously to Copenhagen:

Dear Bohr: Ten years ago when Hilbert was overworked after completing [his theory of] integral equations and had to go to a sanatorium, I wrote to him: "The mathematical kingdom, which you have established, is already worth converting for (*eine Messe wert*)." ['Paris is well worth a Mass.' (Henry IV of France)] I want to tell you the same, for I have heard from Franck that you have gone on strike with your capacity to do work. Your mathematical-physical kingdom will last even longer and contain more citizens than Hilbert's integral-equations-empire. Do not think at all that your present work-stoppage is something serious or special. I find it only natural that you have to pay the human tribute for your most recent discoveries, which certainly have demanded an immense concentration of thought. And I would pay it *with pleasure*, if only such deep insights were given to me. (Sommerfeld to Bohr, 25 April 1921)

[589] For example, Bohr had invited Adalbert Rubinowicz to come to Copenhagen in spring 1922 to help him prepare these lectures. (See Bohr to Rubinowicz, 3 January 1922.)

Frankfurt; and Paul Ehrenfest came from Leyden. Bohr himself was accompanied by Oskar Klein and Wilhelm Oseen.[590] Altogether about one hundred people attended the lectures and participated in the events of what soon came to be called the 'Bohr Festival' ('*Bohr Festspiele*'). Thirty-five years later Friedrich Hund recalled Bohr's lectures as follows:

> Bohr did not speak very clearly, and we junior people were not allowed to sit in the front rows reserved for important guests; thus we strained to hear with our ears turned forward [towards the speaker], fighting our hunger for supper. Of course, we had read a little in Sommerfeld's *Atombau und Spektrallinien*, also Debye in 1920 had given a course of lectures on quantum theory (in an unheated lecture hall); but what Bohr presented sounded quite different, and we felt that it was something very important. The glamour that surrounded this event [Bohr's lectures] cannot be communicated in words today; for us it was as brilliant as the Händel-Festival of those days in Göttingen. (Hund, 1961, p. 1)[591]

Rudolph Minkowski, assisted by Erich Hückel, was asked to prepare notes of the lectures (see Minkowski, AHQP Interview, p. 1). They prepared a typed manuscript, which gave a detailed account of what Niels Bohr presented at Göttingen.

In his first lecture on Monday, 12 June 1922, Bohr introduced his audience to the fundamental ideas underlying his theory of atomic structure. 'The present state of physics,' he began, 'is characterized by the fact that we not only are convinced about the reality of atoms but also believe that we possess a detailed knowledge of their building stones. I shall not discuss here the development of our conceptions of these building stones, nor Rutherford's discovery of atomic nuclei, which initiated a new epoch in physics' (Bohr, 1977, p. 343).[592] He then sketched the picture of the nuclear atom consisting of a positively charged nucleus having Z positive elementary charges $|e|$, around which a number of electrons (with negative charge e) revolve; all building stones, the electrons and the nucleus, could be regarded as point masses. In describing the properties of atoms, Bohr said, 'it is not possible to make any progress with classical electrodynamics' (Bohr, 1977, p. 344). It failed to account both for the observed stability of the atom and the emitted radiation; only the quantum theory would explain the basic features of the atom. Thus Bohr established an atomic theory resting on two fundamental assumptions: the first one being the existence of stationary states, i.e., of certain definite mechanical motions in the atom, from which the electrons cannot emit radiation; the second one stating that a change of the

---

[590] The three Scandinavians stayed together in a *Pension* in Göttingen. (See Klein, AHQP Interview, p. 22.)

[591] Hund explained further: 'They [the lectures] took place, I believe, in lecture hall No. 15 of the *Auditorienhaus*. There were, as far as I remember, perhaps 150 seats, and these were certainly occupied . . . I still recall that we were excluded from the first several rows, which were reserved for the members of the German Physical Society . . . ' (Hund, AHQP Interview, 25 June 1963, p. 1).

[592] The English translation of the Minkowski–Hückel notes of Bohr's lectures is included in Bohr's *Collected Works*, Volume 4, 1977, pp. 341–419.

energy in the atom happens only through a transition from one stationary point to another, and that it is connected with the emission (or absorption) of radiation of a uniform frequency, given—up to a factor $1/h$ (h being Planck's constant)—by the energy difference between the states. Bohr demonstrated that the application of these assumptions led immediately to a satisfactory description of the spectra of the hydrogen atom and the helium ion.

In his second lecture on Tuesday, 13 June, Bohr addressed himself to the peculiar situation existing in the quantum theory of atomic structure. He said:

> So far, only concepts developed in the classical theories, such as those of the electron and electric and magnetic forces, are available to us for describing the natural phenomena; however, we assume at the same time that the picture of the classical theories is invalid. Now the question arises if there is any possibility at all of uniting the classical concepts with the quantum theory without contradiction. [So far the question had not been really settled.] However, physicists hope that the ideas of both theories possess a certain reality. (Bohr, 1977, p. 351)

The contradictory situation led, of course, to 'formidable difficulties' in formulating the quantum principles, and he proposed to use the most cautious procedure, namely, to apply the principles in practical cases and to compare the results with the observations. He then proceeded to develop the basis of the mechanical treatment of atomic systems and the radiation emitted (or absorbed) by them. The atoms, he emphasized, had to be considered as multiply periodic systems, whose motion could be expanded as a Fourier series of harmonic terms involving higher harmonics of the fundamental frequencies ω_i, the time derivatives of the angle variables w_i ($i = 1, \ldots, f$ for a system of f degrees of freedom). In the stationary states of the system the canonically conjugate variables J_i were determined by $n_i h$, the integral multiples of Planck's constant, which provided the condition of stability.[593] Bohr then expressed the action variables by phase integrals and discussed their relation to the frequencies. Thus he showed that in the limit of high quantum numbers the frequencies approached their classical values: i.e., for large n' and n'' and small $|n' - n''|$, the relation

$$\nu = \frac{1}{h} \int \omega \, \delta J = (n' - n'')\omega \tag{182}$$

followed. As an example, Bohr calculated the fine structure of a one-electron atom, together with the selection rules. In all these considerations, he pointed out, the influence of the radiation on the stationary orbits was neglected. If this

[593] Bohr did not hide the difficulty of understanding the stability of atoms against collisions. 'Ordinary mechanics,' he explained, 'is incapable of describing what happens in a collision. The reason for this fact is easily perceived. In mechanics, the state of a system determines directly only the states that are infinitesimally adjacent in time; here, however, the final state is determined beforehand and can be defined only by a consideration of the entire motion' (Bohr, 1977, p. 353).

were taken into account, the energy of the states as well as the frequencies were not accurately determined; hence the question arose whether the frequency condition remained valid. Bohr found the resolution of this apparent puzzle to lie in the fact that both the energies and the frequency occurring in a quantum system were determined 'to just the same degree of accuracy' (Bohr, 1977, p. 361). He also claimed that quantum theory must fail to describe the radiation in wireless telegraphy 'because of the immense magnitude of the radiative forces involved' (Bohr, 1977, p. 362).

In his third lecture on Wednesday, 14 June 1922, Bohr turned to several applications of the principles of atomic theory, especially of the correspondence and the adiabatic principles. After clarifying the range of validity of the adiabatic principle, Bohr treated the problems of atoms in static magnetic and electric fields, deriving the known results, including Kramers' results on the influence of an electric field on the fine structure of atoms (Kramers, 1920b).

With this background preparation Bohr attacked, beginning with his fourth lecture on Monday, 19 June 1922, his main problem: the construction of a theory of the periodic system of elements. That one could do so was not yet generally accepted. Of course, scientists did possess important experimental data concerning the properties of many-electron atoms, their spectra and their chemical behavior. Already more than half a century earlier the chemists Lothar Meyer and Dmitri Ivanovich Mendeleev had proposed the ordering of the elements in a periodic system. Later, new elements had been discovered and the periodic system was improved and completed. Towards the end of the second decade of the twentieth century the periodic system contained nearly ninety elements, from hydrogen to uranium, which were distributed according to increasing atomic numbers into seven periods, each period—except the last one—ending with a chemically inactive noble gas: the first by helium, the second by neon, the third by argon, the fourth by krypton, the fifth by xenon and the sixth by radon (which was usually called niton in those days). After Bohr's theory of the atomic constitution became known, people had tried to explain the periodic system on its basis; thus, for instance, Walther Kossel had discussed the chemical properties of atoms as exhibited in the constitution of molecules (Kossel, 1916a) and Lars Vegard had examined X-ray spectra in detail (Vegard, 1917a, b).[594] Special questions, such as the occurrence of more than eight elements in the so-called long periods had been considered by Rudolf Ladenburg (1920). Bohr had followed these attempts with great interest and, after he had developed the correspondence principle, he felt ready to enter into a discussion of the problem himself. The basis of his practical method consisted in what he called the 'Building-up Principle' (or 'construction principle,' the 'Aufbauprinzip'; see, e.g., Born, 1925, p. 211), that is, the idea that the structure of a given atom can be

---

[594] In these treatments the periodicity of certain physical properties, such as that of the atomic volume (observed already by Lothar Meyer, 1870) and of the electrical conductivity (Benedicks, 1916), provided valuable help.

obtained by considering the successive bindings of electrons to the nucleus.[595] With this method Bohr approached in the early 1920s the explanation of the periodic system, and he presented the results in detail at Göttingen. Of the existing formulations of the periodic system he preferred the one which had been given earlier by his fellow countryman Julius Thomsen because it seemed to be 'particularly well adapted' for his purpose (Bohr, 1977, p. 388).[596] Thomsen had ordered the elements in horizontal groups and vertical periods and had predicted on its basis the group of noble gases (Thomsen, 1895). Bohr showed, at the beginning of his fifth lecture on 20 June 1922, a table representing Thomsen's scheme (see Figure 1). It was this scheme which he tried to justify.

The hydrogen atom and its properties had been satisfactorily described on the basis of the quantum theory of atomic structure. In all other cases, however, the situation appeared to be completely unclear. The empirical data from atomic spectra provided, of course, the most important hints concerning the atomic structure. It had been known since long that the spectra of all elements could be represented by a Rydberg–Ritz formula of the type

$$ \nu = R \left\{ \frac{1}{\left[n'' + \phi_{k''}(n'') \right]^2} - \frac{1}{\left[n' + \phi_{k'}(n') \right]^2} \right\}, \qquad (183) $$

where the denominator, $[n + \phi_k(n)]^2$, replaces the n^2 of the Balmer formula; $\phi_k(n)$ changes only slightly as a function of the principal quantum number n and approaches a constant value α_k in the limit of high quantum numbers n. Now, in the case that one electron moves at a great distance from the other electrons, Eq. (183) could indeed be interpretated in Sommerfeld's way: i.e., definite numbers n and k describe the principal and azimuthal quantum numbers of the outer

---

[595] The importance of this principle in Bohr's work was recognized, for instance, by Hendrik Lorentz, who—in the discussion of Bohr's report at the third Solvay Conference—drew attention to the fact that Bohr had not so much discussed 'how the atom is constituted, but how it can be formed' (Lorentz, *Rapports et Discussions du Conseil de Physique tenu à Bruxelles de 1ᵉʳ au 6 Avril 1921*, p. 257). Bohr agreed with Lorentz' characterization of his method and stated explicitly in his talk at the Physical Society of Copenhagen on 21 October 1921: 'We attack the problem of atomic constitution by asking the question: "How may an atom be formed by the successive capture and binding of the electrons one by one in the field of force surrounding the nucleus?"' (Bohr, 1921e; English translation, *Collected Works*, Volume 4, 1977, p. 277). Later, he formulated the method of obtaining a picture of the atomic constitution by '*following the process of the building-up of atoms through successive capture of the electrons*' by means of what he called the '*postulate of the invariance and permanence of quantum numbers*' (Bohr, 1923c, p. 256). This postulate required that, when a new electron is captured, the quantum numbers of the electrons already bound do not change, except the ones which describe the orientation relative to the new electron.

[596] Hans Peter Jörgen Julius Thomsen (1826–1909) was professor of chemistry at the Copenhagen Polytechnic (1847–1856), at the Military High School (1856–1866) and at the University of Copenhagen (1866–1891). He worked on such chemical problems as the law of mass action of Guldberg and Waage and the properties of acids. Bohr had learned about Thomsen's periodic system during his university studies.

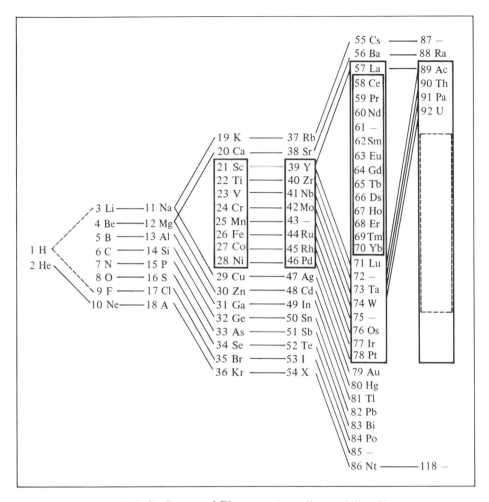

Figure 1   Periodic System of Elements According to Julius Thomsen.

electron. In this manner he explained, to a certain extent, the alkali spectra; he claimed that their doublet structure arose from the fact that the atomic core, which acted on the series electron, possessed axial symmetry and therefore gave rise to a perturbation term depending on the two allowed relative orientations of the core versus the orbit of the series electron. However, in general, in atoms 'most of the [electron] orbits do not lie far outside the other electrons during their entire course, so that the deviation from the Keplerian motion cannot be regarded as caused simply by a field of central symmetry' (Bohr, 1977, p. 375). Helium, the second element in the periodic system, represented a special case, for its atom contained two electrons. Empirically, this atom had been found to emit two types of spectra, one consisting of singlet and the other of doublet lines. Formerly these two types of spectra had been attributed to two different elements, ortho-helium and parhelium, but due to the theory of the nuclear atom they had rather to be associated with two very different types of states of a two-electron atom. In order to describe the situation adequately, Bohr proposed to investigate the mutual perturbation of the two electrons on each other, which exceeded by far the relativistic effects. He claimed that, while the motion of the first electron might be regarded as purely periodic, it fixed the eccentricity of the orbit of the inner electron for one type of helium states. As a result of his detailed calculations, carried out together with Hendrik Kramers, Bohr asserted the following: ortho-helium corresponded to a coplanar system, with the outer electron moving on a circle and the inner one on an ellipse, whose perihelion rotated with the same angular velocity as the outer electron. On the other hand, the two electrons in parhelium should both move on circles. Due to correspondence arguments, the latter states could not be stable, but Bohr argued that such a behaviour was supported by recent experiments on atomic collisions. Although this electron-ring model gave a too high ionization potential in the existing calculations (e.g., in the calculations of Edwin Kemble, 1921, who had treated two circular orbits, whose planes made an angle of $60°$ with each other), Bohr hoped that an improvement of the calculation methods, in which the mutual perturbation of the two electron orbits on each other was taken into account more accurately, would yield a better description of the data on the ionization potential of helium.

The helium calculations had been very cumbersome; however, Bohr claimed that for the higher elements the situation became better and better with each step. So in his fifth and sixth lectures—on Tuesday, 24 June, and Wednesday, 25 June—he treated all elements from the second to the seventh period. After referring briefly to the previous work of Joseph John Thomson and Walther Kossel, he began with the second period, containing the elements from lithium to neon. In the case of the ground state of lithium, Bohr concluded that the two inner electrons were bound as in the helium ground state, i.e., in two circular 1_1-orbits (where 1_1 corresponds to both the principal and the axial quantum number being equal to one), but that the outer electron assumed a 2_1-orbit. Since the latter comes very close to the 1_1-orbits of the inner electrons, the s-term of lithium should deviate strongly from the hydrogen ground state; but for higher states, in which the outer electron is further away from the helium-like core, the similarity of hydrogen and lithium states must increase. Bohr derived the doublet structure of the lithium lines from the fact that the core possessed the angular momentum 1 (in units of $h/2\pi$): it would add to the angular momentum of the outer electron to yield the total angular momentum 2 in the case of the ground state, but two possible angular momenta, namely, 2 and 3, in the case of the p-state. Thus the singlet nature of the s-term and the doublet nature of the p-term seemed to follow—although Bohr admitted that 'this argument is quite uncertain' (Bohr, 1977, p. 391). In connection with the explanation of the lithium doublets Bohr also referred to the model which Werner Heisenberg had used in the discussion of the anomalous Zeeman effect of doublet spectra (Heisenberg, 1922a). He called Heisenberg's paper a 'very promising' attempt, but found it at the same time difficult to 'justify Heisenberg's assumptions,' which involved half-integral quantum numbers (Bohr, 1977, p. 391).

Proceeding to the beryllium atom, Bohr pointed out that it did not possess any simple spectra consisting of singlet lines. He assumed that the two inner electrons in beryllium were bound in a manner similar to that of lithium (or helium) atom, only just a bit tighter, and that the two outer electrons moved in elliptical 2_1-orbits. These orbits were possible because the first outer electron was supposed to revolve very slowly, thereby shielding off the inner system only poorly.[597] In the next atom, boron, there existed three outer electrons, whose description seemed to present considerable difficulty in contrast to the carbon atom having four outer electrons. Bohr claimed that all these electrons 'are bound in 2-orbits which are oriented relative to one another in such a way that their normals are directed towards the corner of a regular tetrahedron with its midpoint in the nucleus' (Bohr, 1977, p. 392).[598] He then discussed the constitution of the succeeding elements by referring to the next stable atom, neon, of whose eight electrons four were supposed to move in 2_1-orbits and the other four in 2_2-orbits.[599] The structure of neon was also supposed to be extremely symmetric, i.e., 'the orbits of these eight outer electrons are arranged in two configurations of tetrahedral symmetry, since then at certain instants two orbital planes would coincide' (Bohr, 1977, p. 394). The reason for the new 2_2-orbits had to be found in the fact that more than four electrons could not be bound—due to considerations of stability similar to the ones of Joseph John Thomson—in 2_1-orbits. Since the electrons were bound in the circular 2_2-orbits more firmly than the ones in the 2_1-orbits, the electronegative properties of nitrogen, oxygen and fluorine followed immediately.

The same construction principles, which were applied in the second period, also worked in the third period. From the spectrum of sodium one derived—as in the case of the lithium spectrum—that the s-term deviated strongly from the corresponding hydrogen term, while in the higher terms the difference became smaller and smaller. Erwin Schrödinger, starting from Landé's theory of electron shells (Landé, 1920d, e), had attempted to explain the alkali s-terms by assuming that the orbit of the outer electron dives into the region of the innermost electron group. Thus its orbit could be thought to consist in a good approximation 'of

---

[597] According to Bohr, the motion of the two electrons did not occur as in Sommerfeld's *Ellipsenverein*; rather, while one electron is far away from the nucleus, the other comes very close to it, closer than the inner electrons.

[598] Bohr stressed the difference of his model of carbon with Alfred Landé's. Landé had replaced the inner system (the helium-like core) by a point charge concentrated in the nucleus, while in Bohr's model the outer electrons penetrated into the inner system.

[599] Evidently, nitrogen corresponded to a triply-ionized, oxygen to a doubly-ionized and fluorine to a singly-ionized neon atom.

pieces of elliptical orbits, which occur when the effective nuclear charge is 1 (at the periphery) and 9 (in the interior), respectively' (Schrödinger, 1921, p. 348).[600] By carrying out a detailed calculation, Schrödinger had arrived at the conclusion that the orbit of the external electron in sodium had to be a 2_1-orbit, because orbits with larger principal quantum number than 2 would not be able to penetrate into the interior of the atom.

Schrödinger, who knew about Bohr's letter to *Nature* before its publication—probably because Bohr had sent a copy of it to Ladenburg in Breslau, where Schrödinger was at the time—had sent a reprint of his paper on the diving electron orbits to Copenhagen. In a letter he had pointed out that it seemed 'to fit very well' with Bohr's new views on atomic structure, especially 'on the necessary interaction between the electrons of different "shells"' (Schrödinger to Bohr, 7 February 1921). And Bohr replied: 'Your paper in *Zeitschrift für Physik* naturally interested me very much. By the way, I made the same consideration myself long time ago and carried out the corresponding calculations' (Bohr to Schrödinger, 15 June 1921). And he had added that he had presented certain results concerning the lithium atom in his lecture at the Physical Society of Copenhagen on 15 December 1920, but had not been able to publish anything. Then, a year later, in his fifth Wolfskehl lecture, Bohr discussed the problem again and arrived at a different conclusion from Schrödinger's with respect to the orbit of the outer electron in sodium. He said: 'However, as with the binding of the third and seventh electron, we must assume that, because of the peculiar stability of the electron configuration already present, the eleventh electron will now be bound in a new type of orbit, namely, in a 3_1-orbit. Considerations of the firmness of the binding of this last electron lead us to the same conclusion' (Bohr, 1977, p. 394)[601] Bohr, like Schrödinger, viewed the orbit of the outer electron in the sodium atom to consist of two parts: the part lying in the region of the inner

[600] Erwin Schrödinger was born on 12 August 1887 in Vienna. From 1906 he studied physics at the University of Vienna under Fritz Hasenöhrl and Egon von Schweidler, obtaining his doctorate under the latter in 1910. From 1911 to 1914 he became assistant to Franz Exner (1849–1926), the Director of the Physics Institute, who worked on spectral analysis, colorimetry, atmospheric electricity and Brownian motion. He got his *Habilitation* in 1914, then served in World War I for the next four years. In 1918 he returned to Vienna as *Privatdozent* (1918–1920). In 1920 he went to Jena as assistant of Max Wien, but shortly afterwards he became *Extraordinarius* at the *Technische Hochschule*, Stuttgart. In the same year he received a call to the University of Breslau as full professor, from where he moved to the University of Zurich in 1921. Six years later Schrödinger succeeded Max Planck in the chair of theoretical physics at the University of Berlin. With Hitler's coming to power, he left Germany in 1933 and accepted a position at the University of Oxford; then in 1936 he went to the University of Graz. He left Graz in 1938 to join the Institute for Advanced Studies in Dublin (1939–1956). In 1957 a special chair was created for him at the University of Vienna. He died on 4 January 1961 in Vienna.

Schrödinger worked on many topics of theoretical (and some of experimental) physics: electricity and magnetism, dispersion of light, X-ray diffraction, problems of specific heat, Brownian motion, colour theory, atomic theory and general relativity. His greatest work was the invention of wave mechanics, the alternate scheme of quantum mechanics.

[601] This assignment was confirmed by A. T. van Urk, who continued Schrödinger's calculations (Urk, 1923).

electrons, which differs only little from a 2_1-orbit, and the outer part, which must at least be a 3_1-orbit. 'By considering the polarizability of the atomic core, as in the case of lithium,' he said, 'we can estimate its value for a definite principal quantum number and compare it with the polarizability of neon as calculated from its dielectric constant. The best argument between these polarizabilities is obtained when the principal quantum number 3 is assigned to the largest p-term' (Bohr, 1977, p. 394). The whole picture of diving or penetrating orbits ('*Tauchbahnen*'), consisting of an external part and an internal part, led to the term structure, $R/(n + \alpha_k)^2$, for the alkali atoms, with α_k denoting a number nearly independent of the principal quantum number.[602] The sum, $n + \alpha_k$, represented what Schrödinger called the 'apparent quantum sum' ('*scheinbare Quantensumme*') and its difference with the principal quantum number the 'quantum defect' ('*Quantendefekt*,' Schrödinger, 1921, p. 349). With these considerations the theory of non-hydrogen-like spectra reached a new level of generality.

Bohr then quickly worked out the constitution of the elements in the third period. Thus he explained the spectrum of magnesium as being caused by a magnesium core—a system similar to the sodium atom having eleven electrons—plus the twelfth (i.e., the series) electron; he assumed the latter as being bound also in a 3_1-orbit.[603] Going on to aluminum, he concluded from the spectroscopic information that 'the term corresponding to the firmest binding is not an s-term, but a p-term, to which we must assign a 3_2-orbit' (Bohr, 1977, p. 399). With respect to the next element, silicon, Bohr claimed that he could not yet decide whether the four outer electrons were on the 3_1-orbits or whether, due to the stronger mutual perturbation of the electrons (which move in closer orbits because of the higher nuclear charge), there were two separate groups consisting of two electrons each (in 3_1- and 3_2-orbits). The decision was even more uncertain in the case of the elements phosphorus, sulphur and chlorine, while in the case of argon the eight external electrons fell—as in the case of neon—into four 3_1-orbits and four 3_2-orbits.[604]

The fourth period presented a new problem to the theory, for it contained in its middle the family of ferrous metals, in which the elements following each other in the periodic system of elements differed only very little in their chemical properties. Bohr concluded from absorption data that the nineteenth electron (in the potassium atom) must be bound in a 4_1-state—it had to be an s-state, and the 3_1-state was excluded because it did not lie outside the argon-like core.[605] In the

---

[602] The penetrating orbits ('*durchdringen*', Schrödinger, 1921, p. 348) were later called 'orbits of the second kind' ('*Bahnen zweiter Art*,' Bohr, 1923c, p. 257) or 'diving orbits' ('*Tauchbahnen*,' see, e.g., Born and Heisenberg, 1924b, p. 391).

[603] The fact that two types of spectra, singlet and triplet, existed, reminded Bohr of the situation in helium; he argued that such a situation would arise whenever a new electron was added in the same type of orbit as the preceding one.

[604] Bohr found that the circular 3_3-orbits gave diameters of the atom that were too large compared to the data of the kinetic theory.

[605] The effective or apparent quantum number of the normal state was equal to 2 in the first approximation.

calcium atom the twentieth electron had also to be bound in a 4_1-state; however, as the nuclear charge increased, the 3_3-orbits became more stable than the 4_1-orbits because, as Bohr argued, they were drawn into the interior of the atom. The possibility to occupy either the 4_1- or the 3_3-orbits then explained the great length of the fourth period. 'In the elements following after scandium,' he concluded, 'the still empty 3-quantic orbits are gradually occupied' (Bohr, 1977, p. 401). As a consequence, for example, the spectra of scandium differ greatly from that of aluminum, in agreement with observation. The noble element krypton at the end of the fourth period should have four outer electrons in 4_1- and four electrons in 4_2-orbits. There remained a difficulty to be explained: why must the fully developed 3-quantic orbits contain 18 electrons? Bohr suggested this number (18) might perhaps be achieved by a perturbation of the 3_1- and 3_2-orbits due to the presence of the 3_3-orbits; this perturbation might lead to an occupation of all 3-quantic orbits, with each—the 3_1-, 3_2- and 3_3-orbit—being filled with up to six electrons. Bohr said: 'It is possible to form an idea, albeit rather uncertain, of why there appear just three groups of six electrons each. By means of a stereographic projection, it is easy to see that only in such an arrangement can the coincidence of orbits be avoided' (Bohr, 1977, p. 402). He proposed to test this idea by analyzing the X-ray spectra in detail. Due to the particular situation in the fourth period, the Kossel–Sommerfeld law was also violated: so, for example, the arc and the spark spectra of manganese were very similar. Bohr noticed that his conception had some similarity with an earlier proposal of Rudolf Ladenburg, who had described the constitution of the elements from scandium to nickel by introducing intermediate shells (Ladenburg, 1920). 'However,' he added, 'we do not assume a definite intermediate shell as something lying between the outermost and inner orbits; rather, we have found a real reason for the peculiarity of the fourth period. When we employ the quantum theory to explain the periodic system at all, we must be prepared beforehand that something new must happen here; for, from the theory of the hydrogen atom, it is impossible to exclude the occurrence of 3_3-orbits' (Bohr, 1977, p. 402). In any case, the picture of the structure of elements from scandium to nickel, with the nonclosed subshells, did account for the observed colours of their compounds and, to some extent, for the paramagnetic properties of their ions.[606]

In the fifth period, ranging from rubidium to xenon, the situation was very similar to the fourth period: the 37th electron in rubidium had to be bound in a 5_1-orbit, and the same for the 38th electron in strontium. Further, there existed the group of elements between yttrium and palladium, in which the 4_3-orbits had to be filled successively. In xenon, Bohr assumed a symmetric configuration of four electrons in 5_1-orbits and four electrons in 5_2-orbits. A more complicated

---

[606] Closed shells are usually achieved in the compounds of other elements, hence their emitted and absorbed radiation frequencies are very high. Optical frequencies can occur only with shells that are not closed. The phenomena of paramagnetism were not yet completely understood on the basis of atomic theory; nevertheless, Bohr argued that nonclosed shells indicated paramagnetism.

situation occurred in the sixth period. The period started out, as usual, with an alkaline and alkaline-earth element, i.e., the 55th and 56th electrons could be bound in 6_1-orbits; the 57th electron in lanthanum was bound in the 5_3-orbit. But then the group of rare earths exhibited a novel situation: the occurrence of 4_4-orbits, which were successively occupied by one electron (Ce), two electrons (Pr), up to eight electrons (Yb). Then came the platinum metals (from lutetium to platinum); in them, according to Bohr, increasingly the 5_3-orbits had to be occupied, while roughly one electron remained in the outermost 6_1-orbit. In the following elements, from gold to niton (the radioactive noble gas, now called radon), the 6_1- and 6_2-shells were filled up until six electrons occupied each orbit type. Bohr concluded that, 'contrary to the customary assumption, the family of rare earths, which begins with *praesodymium*, is completed with *lutetium*.' Hence he argued that 'the not yet discovered element with atomic number 72 must have chemical properties similar to those of zirconium and not to those of the rare earths' (Bohr, 1977, p. 405).[607]

The discussion of the problem of the lowest state of many-electron atoms became even more involved in the case of the elements of the seventh and last period. On one hand, this period did not exist as a complete one in nature; the next noble gas would have the atomic number 118, while in nature the elements occurred only until uranium with atomic number 92; on the other, the calculations in the seventh period became more delicate due to the existence of an increased number of types of competing orbits. In order to facilitate his task, Bohr referred—much more than he had done earlier—to empirical facts in treating the electronic structure of atoms. Thus, from the spectrum of radium, he concluded that the 87th and the 88th electrons must be bound in 7_1-orbits; and, from the absence of a family of elements similar to the rare earths in the sixth period, he concluded that the 5_4-orbits corresponded to a looser binding of the 89th and further electrons than in orbits of the types 6_1, 6_2, and 6_3. He argued that in this way one could, in principle, proceed with the building-up of further atoms having arbitrarily many electrons. At the end of the sixth lecture, Bohr told his audience: 'I hope that I have succeeded in showing that matters become simpler and simpler the farther we proceed in the periodic system. Formerly, I believed that the difficulties would become increasingly greater the more electrons there are in the atoms. However, if we do not demand too much, it actually seems that we can encounter simpler problems and fewer newer ones.' However, he immediately added a word of caution concerning the reliability of his results. 'I hardly need to emphasize,' he said, 'how incomplete and uncertain everything still is' (Bohr, 1977, p. 406).

In his seventh and last lecture on Thursday, 22 June 1922, Bohr turned to a test of his views concerning the structure of atoms. He discussed, in particular, the X-ray spectra and showed that 'our assumptions suffice to explain the stability conditions also for the X-ray spectra.' He considered this fact to be

---

[607] He noticed that this fact had already been assumed by Julius Thomsen, who had associated zirconium with element Number 72 in his system (see Figure 1).

'perhaps the strongest support for the correctness of our views' (Bohr, 1977, p. 407). Before presenting his own interpretation of the data, Bohr quickly went through the development of the understanding of X-ray spectra since Moseley's pioneering work in 1913. In this review he referred to the work of Kossel, to Sommerfeld's interpretation of fine structure and the following investigations of Sommerfeld and Vegard, as well as to the more recent studies of Smekal, Coster and Wentzel, who had analyzed the complexity of the levels involved in the emission of X-ray spectra. So Dirk Coster had concerned himself since 1920 with the X-ray data obtained in Lund, especially the L- and M-series of heavy elements (Coster, 1921a, b, c).[608] Coster had succeeded in organizing all K-, L- and M-levels, finding that there existed one K-, three L- and five M-levels in each atom (Coster, 1921c), and his organization supported Bohr's new theory of atomic constitution (Coster, 1922a, b). In his scheme, Coster had introduced the division of the levels into two types, calling them a and b, respectively. Gregor Wentzel had suggested a slightly different organization by making use of a third quantum number—besides the azimuthal and radial quantum numbers—which he named the 'fundamental quantum number' ('$Grundquantenzahl$') and denoted by the letter m (Wentzel, 1921a, b).[609] Together with his teacher Sommerfeld, Wentzel had then separated the X-ray lines into two groups, the 'regular' and the 'irregular' doublets (Sommerfeld and Wentzel, 1921). In a given X-ray spectrum the regular doublets were always followed by irregular ones; the regular doublets were characterized by a constant wavelength difference, $\Delta\lambda$, and the irregular ones by constant differences of the square roots of the frequencies, $\Delta\sqrt{\nu}$.[610] Bohr essentially agreed with the results of the recent investigations of X-ray spectra. Thus he approved of the third quantum number involved in the description, but he did not wish to associate it with the motion of the electron per se. 'The third quantum number,' he claimed, 'has to do only with the mutual orientation of the orbits' (Bohr, 1977, p. 410). Then, for example, the fact that the K-level of an atom was single immediately followed, since an electron orbit could

---

[608] Dirk Coster was born in Amsterdam on 5 October 1889. He studied at the University of Leyden (1913–1916), at the *Technicum*, Delft (1916–1919), at the University of Copenhagen (1919–1920), and at the University of Lund. In 1922 he obtained his doctorate under Ehrenfest at Leyden. Then he returned to Copenhagen for the following year, and in 1923 became Conservator at Teyler's Foundation in Haarlem. A year later he was appointed to a professorship in experimental physics at the University of Groningen. Coster died in Groningen on 12 February 1950.

[609] Gregor Wentzel was born on 17 February 1898 at Düsseldorf. He studied physics at the Universities of Freiburg (1916–1919; except 1917–1918, when he did military service), Greifswald (1919–1920) and Munich (1920–1921). He obtained his doctorate in 1921 under Sommerfeld with a thesis on the X-ray spectra. In 1922 he became *Privatdozent* at the University of Munich, and four years later (in 1926) extraordinary professor at the University of Leipzig. In 1928 he succeeded Schrödinger in the chair of theoretical physics at the University of Zurich. In 1948 he accepted a professorship at the University of Chicago. He retired in 1968 and spent his last years in Ascona, Switzerland, where he died on 12 August 1978.

Wentzel, who concentrated on the analysis of X-ray spectra in his early career, later made important contributions to wave mechanics, quantum electrodynamics and meson theory.

[610] Such irregular doublets had already been shown to exist by Gustav Hertz in the case of the L-absorption edges (Hertz, 1920). Wentzel had discussed them in detail in his thesis (Wentzel, 1921a).

be oriented to itself only in one way.[611] As for the weak X-ray lines, which Wentzel had attributed to the ionized atoms (Wentzel, 1922), Bohr—like Coster —preferred to associate them with the normal atom: they should occur as a result of jumps from a complete shell to an incomplete one.

With the survey of X-ray spectroscopy Bohr ended his Wolfskehl lectures, in which he had attempted to guide his audience through the latest status of his theory of atomic structure. In spite of the optimism evident in his lectures, Bohr did not hide the uncertainty in various specific conclusions, nor did he forget about the important role played by empirical facts in the construction of electron orbits. As he emphasized: 'A large amount of experimental material has contributed to the shaping of our view. It is a matter of taste, I believe, whether to put the main emphasis on the general considerations or on the bare empirical facts' (Bohr, 1977, p. 397). Bohr was thus perfectly aware that his fundamental theoretical principles were more or less equivalent to 'bare empirical facts.' Thus he assumed an attitude towards the problems of atomic theory which differed considerably from that of others. As Heisenberg recalled later about his first impression of Bohr's approach:

A new feature for me was that one could speak and think about these problems in a very different way from the way which Sommerfeld thought and spoke about them. One of the things which impressed me most was the kind of intuition Bohr had—Bohr knew the whole periodic system. At the same time one could easily [see] from the way he talked about it that he had not proved anything mathematically, that he just knew that this was more or less the connection. When he talked about closed shells at this time, of course, he didn't know why they were closed. [Heisenberg realized that Bohr was aware of this situation.] He took things extremely seriously and at the same time he saw that he could not really prove things. The whole picture was vague, and it was just this vagueness in connection with the enormous force of Bohr's imagination that was attractive for a young man who wanted to do some work for himself. A young man learned that he could do a lot of work himself; that it's not finished at all. At the same time it was almost clear that the whole picture could not be much different from what Bohr said; but still, all the details had to be filled out and [some] essential features were still missing, and we [had to] find out about them. (Heisenberg, Conversations, pp. 121–122; also AHQP Interview)

Many questions occurred to people in the audience concerning Bohr's way of presenting atomic theory. As Heisenberg recalled: 'It occurred to me that it was very interesting to raise criticism just to listen to what Bohr would say to this criticism. I [wanted] to see whether or not his answers would be more or less a kind of excuse or whether they would really hit some essential point' (Heisenberg, Conversations, p. 122). In particular, Heisenberg asked a question related to the quadratic Stark effect. In his third lecture, Bohr had cited Kramers' detailed calculations on the multiplet structure of hydrogen lines in weak electric fields (Kramers, 1920b). At the end of his lecture he had remarked that, while the

---

[611] Similarly he concluded that the 'uppermost level' of an atom must also be a single term.

quantum theory of atomic structure still involved many difficulties, Kramers' results should remain valid. Heisenberg, however, did not believe it. He had already thought about this problem and found that the quadratic Stark effect was almost the same phenomenon as the dispersion of light in the limit of infinitely low frequency; he had concluded, therefore, that one could just take the dispersion formula of the hydrogen atom and consider the limit. However, then one would always get resonance at the critical frequency, but this frequency did not show up at all in Kramers' result. Therefore, Heisenberg rose at the end of Bohr's lecture and told him about the above reasoning and that he did not believe in what Bohr had said about Kramers' result concerning the quadratic Stark effect. He noticed that Bohr was rather shaken by his remarks and gave an evasive answer. But Bohr remained worried and, after the lecture, he invited Heisenberg to go for a walk on the Hainberg in Göttingen. In this long walk of about three hours Heisenberg learned about how Bohr thought about the status of the entire theory. 'It was my first conversation with Bohr,' he recalled. 'For the first time, I saw that one of the founders of quantum theory was deeply worried by its difficulties' (Heisenberg, Conversations, p. 30).

Enough time was reserved for the discussions of Bohr's lectures and many people, especially Sommerfeld and Pauli, actively participated in them. Sommerfeld, for instance, was still a bit skeptical about the correspondence principle. And Bohr did not know the answer to certain questions. So, when he was asked about the detailed validity of Einstein's equations involving the absorption and emission coefficients, he replied: 'Yes, this is a question which does not depend at all on the application of quantum theory, but rather on the foundations of quantum theory; and about that one knows nothing and can say nothing' (recalled by Jordan, AHQP Interview, 17 June 1963, p. 17). But in general, people were greatly impressed by Bohr: they had learned about his latest ideas and how he had arrived at his explanation of the periodic system. Everybody returned home with a deepened understanding of the problems of atomic structure. As for Bohr, he summarized the impression of his visit in a letter to James Franck by saying: 'My entire stay in Göttingen was a wonderful and instructive experience for me, and I cannot say how happy I was for all the friendship shown me by everybody' (Bohr to Franck, 15 July 1922). To which Franck replied that, 'we considered every minute of your stay as a great gift' (Franck to Bohr, 29 July 1922). The success of the Wolfskehl lectures would soon be felt by Bohr's hosts in Göttingen and by Bohr himself.

III.5   Immediate Impact and Triumph of Bohr's Theory of the Periodic System of Elements

Although Bohr's theory of the periodic system was not the first theoretical approach to this problem, it left the deepest impression on the physicists who worked on atomic structure as well as on those who took some interest in the progress of atomic theory. Thus Albert Einstein wrote to Bohr from his trip to

Japan: 'Your recent investigations on the atom have accompanied me on my travel and have further increased my admiration for your mind' (Einstein to Bohr, 11 January 1923).[612] Many years later Einstein commented in some detail upon his impression of Bohr's work; in his *Autobiographical Notes* he discussed his own endeavors to achieve an appropriate theoretical description of the quantum phenomena which had occupied him since the publication of Planck's radiation theory, and remarked:

> All my attempts, however, to adapt the theoretical foundation of physics to this [new type of] knowledge failed completely. It was as if the ground had been pulled out from under me, with no firm foundation to be seen anywhere, upon which one could build. That this insecure and contradictory foundation was sufficient to enable a man of Bohr's unique instinct and tact to discover the major laws of spectral lines and of the electron-shells of the atoms together with their significance for chemistry appeared to me like a miracle—and appears to me as a miracle even today. This is the highest form of musicality in the sphere of thought. (Einstein, 1949, pp. 45–47)

The obvious success of Bohr's recent considerations also persuaded Arnold Sommerfeld to change his mind towards the correspondence principle, which Bohr employed as his guiding principle in those days. While in the first two editions of *Atombau und Spektrallinien* he had taken a rather skeptical attitude, for the correspondence principle seemed to arise from ideas foreign to quantum theory, he presented matters differently in the third edition which came out in spring 1922. This change was immediately noticed by Niels Bohr, who thanked Sommerfeld for the copy of the book he had received and remarked:

> At the same time I would like to express my gratitude for the friendly manner in which you have treated the work of my collaborators and myself. In the past years I have often felt very lonely in scientific matters, because my attempts to develop the principles of quantum theory systematically to the best of my ability have been received with very little sympathy. For me the question is not just didactic trifles, but a serious effort to obtain such an inner relationship [between the principles] that one might hope to create a secure foundation for further development. I realize how little the matters have been clarified and how clumsy I am in expressing my thoughts in an easily accessible form. My joy has therefore been so much greater, for I believe I noticed a change in the point of view contained in the latest edition of your book. (Bohr to Sommerfeld, 30 April 1922)

Sommerfeld did more than change his attitude in words alone; he also applied the correspondence principle in two papers, written jointly with his student Werner Heisenberg and submitted to *Zeitschrift für Physik* in August 1922. In the first paper Sommerfeld and Heisenberg discussed the relation between the

---

[612] The investigations of Bohr, to which Einstein referred above, were the ones contained in the book entitled '*Drei Aufsätze über Spektren und Atombau*,' which appeared in summer 1922 (Bohr, 1922e). These three essays included Bohr's lecture on the structure of atoms and their physical and chemical properties, given at the Copenhagen Physical Society on 18 October 1921 (Bohr, 1921e).

relativistic X-ray doublets and the sharpness of spectral lines (Sommerfeld and Heisenberg, 1922a). Bohr had argued earlier that any quantum-theoretical calculation, yielding a difference of the energy terms smaller than the radiation losses due to classical electrodynamics, was fundamentally wrong (Bohr, 1918b, p. 67). Hence the question had to be asked whether the higher-order relativistic corrections, which appeared to contribute essentially to the L-doublet states of heavy elements, still made sense. Sommerfeld and Heisenberg now calculated the classical energy losses and connected them, via a correspondence argument, to the observed linewidth, and concluded from their result that the higher-order relativistic corrections considered so far were 'perfectly consistent' with Bohr's view concerning the radiation losses (Sommerfeld and Heisenberg, 1922a, p. 393). They made an even deeper application of the corresondence principle in their discussion of the intensity of Zeeman components (Sommerfeld and Heisenberg, 1922b). In this paper they completely followed Bohr's treatment of the magnetic effect on spectral lines, including the decomposition of the position coordinates of the moving electron in a Fourier series and the use of perturbation methods; thus they arrived at a satisfactory description of the relative intensities of doublet and triplet spectra and their Zeeman effects. Sommerfeld's enthusiasm for the correspondence principle faded in the following years as a result of the new fundamental difficulties that emerged in the theory of atomic structure. These difficulties showed up in attempts to explain various phenomena: e.g., complex multiplet structure, the anomalous Zeeman effects, the energy states of the helium atom, and the Compton effect. In all these cases Sommerfeld would assume points of view differing strongly from Bohr's.

Unlike Munich, where Bohr's guidance in atomic theory no longer remained so dominant, things developed differently in Göttingen. People there, especially the mathematician Hilbert and the physicists Franck and Born, had a very great respect for Bohr. Since Niels Bohr first visited Göttingen with his brother Harald in July 1914, Hilbert had esteemed him highly as a physicist. Hilbert's regard for Bohr had even grown higher in the following years, when he himself became deeply involved in the problems of foundations of physics. Now, in the early 1920s Bohr seemed to have in hand the key to the problem of the structure of matter, which interested Hilbert particularly. As a consequence of this interest— and of Bohr's Wolfskehl lectures—Hilbert announced that he was to give a course of lectures on the mathematical foundations of quantum theory in the winter semester 1922–1923. In these lectures he discussed Bohr's theory of atomic structure, using variational methods as the main mathematical tool.[613] An even closer association with Bohr had James Franck, who continued in Göttingen his experiments on the collision between electrons and atoms, which had provided the first direct proof of Bohr's postulates of atomic theory. Franck had become acquainted with Bohr in spring 1920 in Berlin, and they had become

---

[613] A detailed manuscript on Hilbert's lectures on the mathematical foundations of quantum theory (*Mathematische Grundlagen der Quantentheorie*) is part of Hilbert's archive (*Nachlass*) at the Library of the University of Göttingen.

close friends. Max Born later recalled the special relationship of Bohr and Frank as follows:

> Franck was an ardent admirer of Bohr and believed in him as the highest authority in physics. I sometimes found this rather exasperating. It happened more than once that we had discussed a problem thoroughly and come to a conclusion. When I asked him after a while: "Have you started to do that experiment?", he would reply: "Well, no; I have written first to Bohr and he has not answered yet." This was at times rather discouraging for me, and even retarded our work to some degree. (Born, 1978, p. 211)

Franck's work in the early 1920s concentrated on checking and proving various aspects of Bohr's theory of atomic structure, its fundamental principles as much as detailed applications. So he investigated, for example, the excitation levels of atoms of several elements, including helium and mercury, or the collisions of the second kind, whose existence had been predicted by Oskar Klein and Svein Rosseland. This work of James Franck completed and extended the fundamental investigations on electron collisions with atoms, started a decade earlier in collaboration with Gustav Hertz, for which both would share the 1925 Nobel Prize in Physics.

Max Born came to be interested in Bohr's theory in a less direct manner. His first acquaintance with the models of atomic structure had arisen from the discussion of the properties of ionic crystals; these investigations had led him to discuss questions of the constitution of molecules and their formation from atoms. It was characteristic of Born's scientific work that he worked steadfastly for many years on the same subjects. Thus he did not give up in Göttingen his concern with problems of crystal dynamics and of molecules; however, now he began to take as the basis of his investigations Bohr's models of atomic and molecular constitution. For example, with his assistant Erich Hückel, he extended the quantum theory to molecules consisting of more than two atoms (Born and Hückel, 1923).[614] They assumed the atoms to be bound in the molecules by potentials, depending on the distance between the atomic centres; these centres performed oscillations against each other, having a small amplitude; by expanding the Hamiltonian in terms of powers of the small parameter associated with these oscillations, and by using appropriate quantum conditions,

---

[614] Erich Armand Arthur Joseph Hückel was born at Charlottenburg on 9 August 1896. He studied physics at the University of Göttingen (1914–1916, 1919–1921), obtaining his doctorate under Peter Debye; his thesis dealt with the structure of anisotropic fluids. After serving for a year as assistant (to Hilbert and Born) in Göttingen, he joined his teacher Debye at the E.T.H. in Zurich, and became *Privatdozent* in 1925. From 1928 to 1929 he was a Rockefeller Fellow in London and Copenhagen; the following year he received a grant from the *Notgemeinschaft* at the University of Leipzig. In 1930 he obtained a position as a lecturer at the Technical University of Stuttgart, and seven years later he became *Extraordinarius* at the University of Marburg. Erich Hückel performed pioneering work on the theory of strong electrolytes (with Peter Debye at Zurich) and on the theory of carbon binding in organic substances. He retired in 1962 and died in Marburg on 16 February 1980.

they obtained an expression for the energy of the multiatomic molecules.[615] A year later Born returned to the theory of molecules in a joint paper with Werner Heisenberg (Born and Heisenberg, 1924a). Even more systematically than in the work of Born and Hückel, they treated the following problem: 'What are the motions of a molecule considered as a mechanical system, and how far does the usual theory of band spectra represent these motions?' (Born and Heisenberg, 1924a, p. 2). That is, they looked on molecules as mechanical systems, consisting of nuclei and electrons in dynamical equilibrium, and approached the problem of calculating the quantum-theoretical energy of a molecule by expanding its Hamiltonian in terms of a small parameter λ (whose square denoted the ratio of the electron mass to the nuclear mass). In solving the equations of motion, Born and Heisenberg restricted themselves to solutions in which the nuclei did not move farther from their equilibrium positions than by distances of the order of λ times the equilibrium distances between two nuclei. Finally they brought in quantum theory by quantizing the appropriate momenta. With these conditions, and by taking into account the additional constraints of molecular constitution, they arrived at the following general results: the mechanical problem could be treated with the help of perturbation theory, the unperturbed system being represented by the motions of electrons around fixed nuclei; the first-order perturbation energy of the molecules vanished, and the second-order term (in λ^2) arose from the oscillations and rotations of the nuclei; the third-order energy term again disappeared, and the fourth-order term depended quadratically on the quantum numbers of the oscillational motion and in a more complicated manner on the quantum numbers of the rotation of the molecules (i.e., essentially the rotational motion of the nuclei) and of the orbits of the electrons in the molecules.

The investigations of Born, Hückel and Heisenberg on molecular structure were highly theoretical, with almost no practical application. Their style was entirely determined by the methods of Born. As in his work on electron theory or on crystal lattices, Born did not invent any new physical ideas in molecular theory, but made use of existing concepts. His main contribution consisted in carrying out detailed and involved calculations on the basis of given molecular models. These models were provided by Niels Bohr, and it was by means of the conceptual framework he had erected that Bohr guided the theoretical investigations of Born and his associates after 1922. Bohr's influence was also felt in another respect, namely, through his favourite mathematical method in the quantum theory of atomic structure. Since 1913, in discussing atomic systems which could not be described by a periodic motion with a single frequency, he had advocated the use of perturbation theory. In Göttingen the perturbation theory was then developed further in connection with atomic models, and it became the dominant method of approach there, the more so since Born had

---

[615] For molecules consisting of more than two atoms their energy formula became very complicated; in those cases the axis of rotation could not be considered as fixed (as in the case of diatomic molecules), hence an involved interaction between oscillational and rotational motion emerged.

been acquainted with it through his work on crystal dynamics, performed earlier in collaboration with E. Brody.[616] In 1921 Born and Brody had entered into a systematic investigation of the thermodynamic properties of solids, with the goal of achieving a complete description from first principles based on their constitution (in terms of atoms and ions). They had succeeded, for example, in obtaining the energy W of a crystal lattice, whose mass points were coupled by a force involving a small anharmonic term of order λ, as a power series in the parameter λ (Born and Brody, 1921b).[617] They had then used the result of the second-order calculation, i.e.,

$$W = h \sum_k \nu_k n_k + \frac{1}{2} \sum_{k,l} \nu_{kl} n_k n_l, \tag{184}$$

where ν_k and ν_{kl} denoted the characteristic oscillation frequencies in the crystal and n_k and n_l the integral quantum numbers associated with them, to account for the observed deviation of the specific heats from the value $3R$. Thus Born and Brody had derived the equation

$$c_v = 3R(1 - 6\sigma RT), \tag{185}$$

R being the universal gas constant and σ a numerical factor (which could be smaller or larger than zero), which indeed seemed to describe perfectly the behaviour of the specific heats of solids (at constant volume) as function of the (absolute) temperature T.[618]

Although Born applied perturbation theory in many other problems of the solid state and obtained valuable results, a number of which were reported in his encyclopedia article (Born, 1923b), this method proved to be even more important for the specific problems of atomic and molecular constitution. So Born and his students used it in particular to discuss the helium atom, the hydrogen molecule-ion and the hydrogen molecule. The rigorous treatment of these problems appeared to be very promising, and one expected quick success in confirming Bohr's theory of the atomic constitution, which achieved a major triumph towards the end of 1922.

---

[616] E. Brody, a Hungarian Jew, obtained his doctorate with a thesis on the chemical constant of monatomic gases; he submitted a paper on the content of his thesis to *Zeitschrift für Physik* in June 1921 (Brody, 1921a). Brody went to Germany after World War I and was hired by Born in Frankfurt. Born wrote to Einstein in a letter: 'I am working on my article for the Encyclopedia [on crystal dynamics], with Dr. Brody as my private assistant. He is a very clever man. Unfortunately he knows very little German, and is rather hard of hearing' (Born to Einstein, 12 February 1921).

[617] In solving the equation of motion, Born and Brody applied Poincaré's theory of integral invariants in many-body systems, in which Brody had become interested earlier (Brody, 1921b).

[618] In two later papers, Born and Brody corrected a number of errors without, however, changing the final result, Eq. (185). (See Born and Brody, 1922a, b.) Brody stayed on in Göttingen until the end of 1922; then he left to take up a professorship in Temesvar, Rumania. In 1925 he took a position with *Vereinigte Glühlampen- und Elektrizitäts- A.G.* near Budapest. Born learned later on that his former collaborator 'survived all the horrors of the Nazi occupation and the war' (Born, 1978, p. 214).

This triumph had to do with finding one of the missing elements in the periodic system. Ever since the periodic system had been first devised, attempts had been made to fill the gaps in it. While a large number of elements had been discovered in the last decades of the nineteenth and the early years of the twentieth century, some places in the periodic system had still remained unoccupied. Thus the table which Bohr presented at the Wolfskehl lectures in Göttingen (see Figure 1 in Section III.4) did not show the elements with atomic numbers 43, 61, 72, 75, 85 and 87. The properties of all but one unknown element seemed to be evident.[619] Only the nature of the element number 72, which was situated near the rare-earth elements, presented certain difficulties. A serious search for it had begun in 1907 when Georges Urbain in France and Carl Auer von Welsbach in Austria, after investigating for three years what had been called 'ytterbium probes' at that time, had finally isolated two different substances, which were later baptized ytterbium and lutecium. Both chemists had, however, assumed a third element to exist in the probes and, again after several years of laborious experiments, Urbain had announced the discovery of a new substance; he had given the name 'celtium' to this substance and placed it next to lutecium in the periodic system of elements (Urbain, 1911).[620] Several years later a difficulty had arisen with the element celtium: Henry Gwyn Jeffreys Moseley, whom Urbain had visited in early summer 1914 and asked to check the X-ray spectrum of the substance—which he had assumed to contain the element No. 72—had been unable to obtain the characteristic X-ray lines for the element. Moseley had reported his unsuccessful search in a letter to Rutherford by stating: 'Celtium has proved most disappointing. I can find no X-ray spectrum in it other than those of Lutecium and Neoytterbium [as ytterbium was called in those days], and so number 72 is still vacant' (Moseley to Rutherford, 5 June 1914). Moseley had then argued that the visible spectrum, which Urbain had associated with celtium, had to be ascribed to a mixture of lutecium and ytterbium and that Urbain's substance corresponded to a mixture of lutecium with a few percent neodymium. In spite of that failure, Urbain had continued to believe in the existence of celtium during the following years. The next advance came in 1920, when Maurice de Broglie examined the K-spectra of the very sample from which Urbain had earlier concluded the existence of celtium: his results seemed to indicate that the element number 72 was indeed present (M. de Broglie, 1920). Two years later, in May 1922, Alexandre Dauvillier, the chief assistant at Maurice de Broglie's laboratory, reported about the measurement of the L-

---

[619]The elements, numbers 43 and 75, had to be similar to manganese, the element number 61 had to be a rare earth, number 85 should represent a halogen and number 87 an alkali element.

[620]G. Urbain, born at Paris on 12 April 1872, received his education at the École Municipale de Physique et Chimie (1891–1894), and obtained his doctorate in 1899 from the University of Paris. From 1894 he served as an assistant and later as professor of chemistry at the École Municipale. In 1906 he became assistant lecturer, in 1908 professor of chemistry and in 1928 Director of the Chemistry Institute of the University of Paris. Urbain, a member of the Paris Academy of Sciences since 1921, worked mainly on rare-earth elements and on complex inorganic salts. He died in Paris on 5 November 1938.

spectrum of a sample containing the known elements ytterbium and lutecium; in addition to the lines of the latter elements, he observed two weak lines which he attributed to the element number 72 (Dauvillier, 1922a).[621] These lines, situated at 1561.8 and 1319.4 X-ray units, seemed to fit well into Moseley's diagram for the L_{α_1}- and L_{β_2}-lines for the element in question; hence Urbain concluded: 'Celtium has conclusively won its place among the chemical elements' (Urbain, 1922, p. 1349). Rutherford, who had continued to take interest in the celtium question since 1914, was informed by the French scientists about their findings and he reported their main results in a letter to *Nature*, which appeared in the issue of 17 June 1922 (Rutherford, 1922). In this letter he basically translated Urbain's paper (giving an account of the historical development of the celtium problem from 1911 to Dauvillier's X-ray analysis of 1922) and added finally the comment:

> Now that the missing element of number 72 has been identified, there remain only three vacant places of ordinal numbers—43, 61, 75—between hydrogen and bismuth in the Moseley classification of elements. With the rapidly increasing perfection of X-ray spectra and the use of powerful installations, it is to be anticipated that the missing elements should soon be identified if they exist in the earth. The law of the X-ray spectra, as found by Moseley, is an infallible guide in fixing the number of an element, even if present in only small proportion in the material under examination. (Rutherford, 1922, p. 781)

Bohr saw Rutherford's letter in *Nature* when he was back in Copenhagen in late June 1922. Its content worried him very much; after all, he had just recently stated (in his sixth Wolfskehl lecture in Göttingen on 21 June 1922) that the element with atomic number 72 should be similar to zirconium (and not a rare-earth element as concluded from Dauvillier's observations), if his conceptions of atomic structure were correct. Now, in a letter to James Franck, he admitted that his statement had been wrong and that the element number 72, 'as shown by Urbain and Dauvillier, contrary to expectation, has turned out to be a rare earth element after all' (Bohr to Franck, 15 July 1922). Rutherford's advocacy of the French results also stimulated him to write an addendum (*Nachschrift*) to the three essays on '*Spektren und Atombau*' (i.e., the German edition, which was about to appear), stating: 'After these essays were printed I learned from an article of Sir Ernest Rutherford in *Nature* about an investigation of Dauvillier on the X-ray spectra of several rare earths. It seems to follow from

---

[621] Alexandre Henri George Dauvillier was born on 5 May 1892 at Saint-Lubin-de-Joncherets, Eure-et-Loire. He was educated at the University of Paris, where he obtained his doctorate in 1920. Then he joined Maurice de Broglie's laboratory. In 1925 he became lecturer at the *École Supérieure d'Électricité* (until 1940), and in 1931 also at the *Faculté de Médecine de Paris* (until 1942). Then he was appointed director of research at the *Centre National de Recherche* and professor of cosmic physics at the *Collège de France* (in 1944). From 1935 onwards, Dauvillier directed the laboratory of cosmic physics at the Paris Observatory in Meudon. Besides X-ray spectra, he worked on problems of physical chemistry, geophysics and astrophysics. Dauvillier died on 23 December 1979 at Bagnères-de-Bigorre.

it that the element with the atomic number 72 has to be identified with the element celtium, whose existence had been speculated earlier by Urbain' (Bohr, 1922e, p. 147).[622] In order to account for the new situation, Bohr argued further, the classification given earlier must be changed in such a way as to include the elements, numbers 71 and 72, among the rare earths. This could be explained theoretically in the following way: the interaction of the electrons moving on orbits having the principal quantum numbers five and six (i.e., five- and six-quantum orbits) with the electrons on four-quantum orbits might be responsible for the fact that not all four-quantum orbits were occupied, in contrast to what seemed to follow from the earlier ideas (of J. J. Thomson and Niels Bohr) on the structure of electron groups in atoms. However, Bohr was not really satisfied with this explanation, and he wrote to Dirk Coster, whom he regarded as an expert on questions of X-ray spectra. Coster had just sent to Bohr a copy of his doctoral dissertation, for which Bohr thanked him in a letter, dated 3 July 1922, and added in a postscript:

> On my return from Göttingen I found a note by Rutherford in "Nature" of June 17 which you probably have seen; it mentions some notes by Dauvillier and Urbain concerning the identification of the element with atomic number 72 as celtium. I should be extremely grateful to you if you would write to me about your opinion of the reliability of Dauvillier's identification of the X-ray lines in question. The problem is of the greatest interest, since, as you know, the conceptions of atomic structure seem to demand essentially different chemical properties of an element with atomic number 72 than those exhibited by the rare earths. The question is apparently rather clear, but one must, of course, be always prepared for complications. They may arise from the circumstance that we have to do with a simultaneous development of two inner electron groups. (Bohr to Coster, 3 July 1922)

Coster replied to Bohr in some detail in a letter, dated 15 July 1922. 'Dauvillier's paper was not very convincing to me,' he remarked. 'You know my opinion about this author. His papers are a mixture of very good and very bad things and it is always impossible to check his statements.' However, already the next day Coster received the first-hand judgment of an expert about the reliability of the X-ray lines of celtium: Manne Siegbahn, after a visit to Paris, where he had seen Dauvillier's photographic plates, came to The Hague and met Coster on 16 July 1922. Coster wrote to Bohr immediately about Siegbahn's conclusions:

> He [i.e., Siegbahn] told me that he does not at all trust Dauvillier's work on the element 72. He could *not at all* see the lines L_{α_1} and L_{α_2} of [the element] 72 (the only lines which Dauvillier claims to have found for that element). When he told this to Dauvillier, the latter said: "Oh yes, this is perfectly possible, and the reason is that there is no clear weather today!!!" Mr. Siegbahn has authorized me to write to you

---

[622] It appears that Niels Bohr had noticed the papers of Dauvillier and Urbain earlier, but had not attributed much importance to them; this becomes evident from the development leading to the discovery of hafnium, as reported here.

the following as his opinion: "If Dauvillier wishes to claim that his probe contained about 0.01 (a hundredth of a percent) of the element 72, then one has to admit that the photographs, which he took, do not at all contradict this—but they also justify it *very little*." (Coster to Bohr, 16 July 1922)

Coster drafted a letter to *Nature*, in which he argued against Dauvillier's identification of the element number 72; but Bohr, to whom he sent a copy for approval, suggested not to publish it, for it seemed to be too polemical for the British journal and not based on new observations. Instead he informed Rutherford privately about the doubts of Coster, Siegbahn and himself concerning the existence of celtium. Now Bohr was convinced that one 'can hardly ascribe much significance to Dauvillier's result' (Bohr to Coster, 5 August 1922). And he decided to examine, with Coster, whom he had invited to work in Copenhagen for the following year, the question of X-ray spectra and its relation to atomic structure thoroughly.

Coster started to work with Bohr on the renewed analysis of X-ray data immediately after his arrival in Copenhagen in September 1922. Already by the end of October they had completed a paper, entitled '*Röntgenspektren und periodisches System der Elemente*,' which was received by *Zeitschrift für Physik* on 2 November 1922 and published in its issue No. 6 of volume 12 in January 1923 (Bohr and Coster, 1923). In this paper Bohr and Coster discussed all the X-ray data available at that time.[623] They especially included the results obtained by Manne Siegbahn and his students as well as the observations of American scientists, notably William Duane and his associates at Harvard University.[624] In a series of papers these authors had reported their measurements of the X-ray diffraction frequencies of many chemical elements; their results showed, for example, that the square root of the K-absorption frequencies, if plotted against the atomic number, did not fall on a straight line (Duane and Hu, 1919a, b, Duane and Shimizu, 1919a, b), and that the critical K-absorption frequency of a given frequency always exceeded the shortest wavelength of the emitted radiation by fractions of a percent (Duane and Hu, 1919a). Coster had already analyzed these data previously: he had represented each X-ray frequency as the difference of two spectral terms, and he had organized the terms—which, if multiplied by

---

[623] Bohr and Coster emphasized that their compilation of the data was much larger and more reliable than earlier ones, given by Sommerfeld in the third edition of *Atombau und Spektrallinien* and by Dauvillier (Dauvillier, 1922b). (See Bohr and Coster, 1923, pp. 348–349, footnote 2.)

[624] W. Duane was born at Philadelphia on 17 February 1872. He studied at the University of Pennsylvania (1888–1892), Harvard University (1892–1895), the Universities of Göttingen (1895, as Tyndall Fellow) and Berlin (1895–1897), where he took his doctorate under Max Planck. From 1898 to 1907 he was a professor at the University of Colorado; then he spent five years at the radium laboratory of Pierre and Marie Curie in Paris. In 1913 he returned to the United States and became an assistant professor of physics (1913–1917) and a professor of biophysics (1917–1934) at Harvard University. He died on 7 March 1935 in Devon, Pennsylvania.

Duane performed important investigations in the fields of radioactivity and X-ray spectroscopy. Thus he discovered with Franklin L. Hunt the so-called Duane–Hunt law, relating the minimum wavelength of X-rays to the threshold voltage of the cathode rays that excite them (Duane and Hunt, 1915).

Planck's constant, represented the energy levels of atoms (especially the lowest ones)—in a way that was consistent with Bohr's recent views on atomic structure (Coster, 1922a, b). The repetition of the analysis by Bohr and Coster in fall 1922 aimed at a further sharpening of Coster's classification of energy levels, which should enable one to decide clearly about such questions as the electron shell occupation in the element with atomic number 72. Bohr and Coster began their work by ordering the lowest levels of the atoms—as obtained from the X-ray spectra, i.e., the K-level, the three L-levels, the five M-levels, the seven N-levels, the five O-levels and the three P-levels—according to rising energy: K; L_I, L_{II}, L_{III}; $M_I, M_{II}, M_{III}, M_{IV}, M_V$, etc.[625] Then they associated with each term three quantum numbers, $n(k_1, k_2)$, where n was the principal quantum number (i.e., $n = 1, 2, 3$, etc., for the K-, L-, M-, etc., levels), k_2 was equal to the azimuthal quantum number, and k_1 represented a further quantum number, whose value was either equal to k_2 or $k_2 + 1$. For example, the K-level was described by $1(1, 1)$, the three L-levels by $2(1, 1)$ for L_I, $2(2, 1)$ for L_{II} and $2(2, 2)$ for L_{III}. In the transitions giving rise to X-ray lines the selection rules,

$$\Delta n = 1, 2, 3 \ldots, \qquad \Delta k = 1, \qquad \Delta k_2 = 0, 1, \qquad (186)$$

were obeyed.

Bohr and Coster noted that their new organization corresponded formally to the one proposed earlier by Gregor Wentzel—Wentzel's quantum numbers m and n had just to be identified with k_1 and k_2 respectively (Wentzel, 1921a)—but they suggested a different physical interpretation based on the idea of penetrating or diving orbits. According to their view an electron orbit consisted in general of an outer part and an inner part, which penetrates into the interior shells. The different nature of the optical and the X-ray spectra could then be explained as follows:

> Thus, the *typical periodicity of the chemical properties and of the optical spectra* depends on the circumstance that, for the *outermost electron orbits*, the effective quantum numbers, in contrast to the principal quantum numbers, vary only little as one goes from an element to the homologous element in the next period in the system of elements. On the other hand, the *striking lack of periodicity of the essential features of X-ray spectra* depends on the circumstance that we are here primarily concerned with the conditions of the *innermost electrons* in the atom, which move in groups that are already completely formed and that repeat themselves unchanged in all subsequent elements. (Bohr and Coster, 1923, p. 357; Bohr, *Collected Works*, Volume 4, 1977, p. 534)

Therefore, in calculating the energy terms one found a remarkable difference if an inner or an outer orbit was considered. For an outer electron it sufficed to

---

[625] Bohr and Coster were aware of the fact that certain difficulties arose with respect to the uniqueness of the term scheme of a given element. Thus J. Bergengren had observed different K-edges for different allotropical forms of phosphorus (Bergengren, 1920), and Hugo Fricke had noticed a fine structure of the K-edge for many elements (Fricke, 1920).

take into account a gross screening of the nuclear charge Z (in units of the absolute value of the electron's charge) through the 'screening factor' $\alpha_{n,k}$; for an inner electron, however, one had also to pay attention to the effect of the outer electrons, which create an additional screening. Finally, Bohr and Coster included the relativity correction and obtained a formula for the most general energy term W, i.e.,

$$W = Rh\frac{(Z - \gamma)^2}{n^2} + Rh\frac{(Z - \delta)^4}{n^4}\left(\frac{2\pi e}{hc}\right)^2\left(\frac{n}{k} - \frac{3}{4}\right). \qquad (187)$$

In Eq. (187), γ represented the 'total screening factor,' i.e., the total screening effect on the nuclear charge of the inner and outer electrons, while $N - \delta$ represented the effective nuclear charge for the relativity corrections. The screening effect was not supposed to be the same for the nonrelativistic term of the energy and the relativity correction term; in fact, from the data, δ appeared to be much smaller than γ.

From the above formula it was easy to obtain a dependence of the energy term on three quantum numbers, because the screening effects in Eq. (187) actually depended on three quantum numbers. Bohr and Coster did not go into the details of the theoretical problem, but rather confined themselves to studying the dependence of the empirical term data (for given sets of three quantum numbers, $n(k_1, k_2)$) on the atomic number: i.e., they plotted the values of the quantity T/R (T being, up to the factor h^{-1}, the energy level W) against Z, the atomic number of the elements. Thus they obtained basically regular curves, which increased monotonically with increasing Z. In these plots they rediscovered what Sommerfeld and Wentzel had called 'screening doublets' ('*Abschirmungsdublette*') as parallel curves (e.g., L_I and L_{II}), and the 'relativity doublets' ('*Relativitätsdublette*') represented by curves diverging rapidly for increasing Z (e.g., L_{II} and L_{III}). Bohr and Coster argued that the 'screening doublets' corresponded to transitions between states possessing different screening constants, $\alpha_{n',k'}$ and $\alpha_{n'',k''}$ (which arose from the effect of the inner electrons on the states considered), while the 'relativity doublets' corresponded to transitions between states having different azimuthal quantum numbers, k_2' and k_2''. However, they were not able to derive the existence of the two types of doublets from a deeper theoretical reasoning.[626] The most important observation made by Bohr and Coster was that the curves showed discontinuities—i.e., jumps in the slopes of the curves showing T/R against Z, which consisted of connected pieces of straight lines—exactly at those places where, in accordance with Bohr's views, an inner electron shell in atoms became filled up. For example, such discontinuities

---

[626] Bohr and Coster drew attention to another organization of the X-ray levels. So they called 'normal' levels those in which both quantum numbers k_1 and k_2 coincided; in them the energy calculation could be carried out by assuming the electron shells in the atoms as more or less independent. In 'anomalous' levels, i.e., levels with different k_1 and k_2, on the other hand, the interaction of neighbouring electron shells contributed materially (Bohr and Coster, 1923, pp. 364–365).

clearly showed up in the beginning and at the end of the group of rare-earth elements, especially in the N-levels. That is, between the atomic numbers 56 and 58 the slope of the curves changed suddenly from a larger to a smaller value until all four-quantum orbits were occupied; then the slope again became steeper. Unfortunately, a clear decision could not yet be made from the X-ray data as to the atomic number at which the steeper slope started. Bohr and Coster noted: 'Because of the incompleteness of the measurements for the elements in the vicinity of [$Z =$] 72, which form the end of the family of rare earths, there exists here some uncertainty about the course of the curves of the levels N_I, N_{II}, N_{III}, N_{IV} and N_V' (Bohr and Coster, 1923, p. 370; Bohr, *Collected Works*, Volume 4, 1977, p. 545). So the question concerning the constitution of the element with atomic number 72 and its chemical properties remained unanswered. However, it was soon to be decided in a more direct way by the discovery of the element hafnium.

In November 1922 the announcement was made that the Nobel Prize in Physics for the year 1922 was awarded to Niels Bohr. On 11 December 1922, Bohr delivered his Nobel address 'On the Structure of Atoms' (Bohr, 1923b). George de Hevesy recalled later: 'A few minutes before Bohr started his Nobel lecture, Coster announced by telephone the presence of the Hafnium β_1 and β_2 lines on his photograpic plate' (Hevesy, 1951, p. 683). Bohr happily mentioned the news of the discovery of the new element in Copenhagen in the last part of his lecture. After referring briefly to the story of the element Number 72, including Dauvillier and Urbain's claim that it belonged to the rare earths, he remarked:

> In these circumstances Dr. Coster and Professor Hevesy, who are both for the time working in Copenhagen, took up a short time ago the problem of testing a preparation of zircon-bearing minerals by X-ray spectroscopic analysis. These investigations have been able to establish the existence in the minerals investigated of appreciable quantities of an element with atomic number 72, the chemical properties of which show a great similarity to those of zirconium and a decided difference from those of the rare-earths. (Bohr, 1923b, p. 42)[627]

Bohr's announcement of the discovery of the element 72, having the properties of zirconium, ended a feverish search in which Dirk Coster and George de Hevesy were the main actors. It had begun in the last week of May 1922, shortly after the publication of Dauvillier's and Urbain's notes in *Comptes Rendus* (Dauvillier, 1922a; Urbain, 1922). Coster had just come over from Lund, and Bohr had consulted him and de Hevesy, who was an expert on the chemical literature on rare-earth elements. 'I agreed with Coster in that the findings of Urbain and Dauvillier in no way disprove the correctness of Bohr's prediction as

---

[627] According to Oskar Klein, Bohr spoke on this occasion without having his prepared manuscript with him. Klein reported: 'At the obligatory lecture, for which he had chosen to talk about the constitution of atoms, he discovered that he had forgotten his notes and slides at the hotel, so he had to begin without them, while they were fetched. This, however, was rather an advantage, because it forced him to improvise, as he did in private conversation' (Klein in Rozental, 1967, p. 84).

to the nature of the element,' Hevesy recalled later. He continued:

> After our talk with Niels Bohr we did not pursue the subject and none of us thought
> at that date to initiate a search for the element 72. It was a coincidence of fortunate
> events which led to the discovery of the element, the character of which was
> predicted by Bohr, in his institute. Among these events Coster's stay in Copenhagen
> after the completion of his studies with Manne Siegbahn at Lund was the most
> important one. In collaboration with Coster we intended to use radiolead as anode
> of an X-ray tube and to investigate the possible effect of intense and prolonged
> irradiation on the decay rate of radium D, which should reflect itself in a
> corresponding change in the rate of decay of radium E. While, in search of a
> suitable high-voltage aggregate, being much interested in the application of X-ray
> spectroscopy to mineral chemistry, I suggested that Coster should look for the
> element 72. In the beginning he was very reluctant to embark on this investigation,
> maintaining the view that, if element 72 was present in a mineral, its concentration
> could be expected to be only very low and, since the sensitivity of X-ray spectros-
> copy in those days was rather weak, it seemed almost hopeless to discover the
> missing element by means of this method. Finally, however, Coster yielded to my
> argument that the aim of our work should be to enable me to learn the technique of
> X-ray spectroscopy and we could just as well at the same time try to find Bohr's
> element 72. (Hevesy, 1951, pp. 682–683)[628]

After suitable preparations of the apparatus and of the probes to be investigated, Coster began to take X-ray data by the end of November 1922. Working
with a concentrate of Norwegian zirconium minerals, he first found the L_{α_1}- and
L_{α_2}-lines of a new substance, which one could associate with the atomic number
72. Still the situation was a bit complicated, for 'by a queer coincidence the
second-order zirconium K-spectrum coincided with the hafnium K_α-line'
(Hevesy, AHQP Interview, 25 May 1962, p. 14). In principle, it was possible to
avoid the complication; the second-order zirconium spectrum could be
suppressed by keeping the voltage of the X-ray tube below a certain value. 'But
we had a very primitive apparatus,' Hevesy recalled, 'and our voltage was going
up and down, so it was quite difficult' (Hevesy, AHQP Interview, 25 May 1962,
p. 15). However, the uncertainty did not last for long, and 'the final proof of the
discovery of hafnium was brought when Coster succeeded in locating the β_1 and
β_2 lines and soon was able to determine the whole X-ray spectrum of that
element' (Hevesy, 1951, p. 683). In obtaining these spectra, Coster used probes in
which Hevesy had enriched the content of the new element.[629] After measuring
some more lines of the X-ray spectra, Coster informed Bohr on 2 January 1923:

---

[628] Hevesy became interested in the element 72 from discussions with Friedrich Adolf Paneth
(1887–1958), then professor of chemistry at the University of Hamburg. In 1922 the two of them
wrote a textbook on radioactivity, in which they expressed doubts about the existence of Urbain's
celtium (Hevesy and Paneth, 1923, p. 109). After the publications of Urbain and Dauvillier, Paneth
urged Hevesy strongly to investigate the problem himself. (See Kragh, 1979, p. 184.)

[629] The original probes contained of the order of 1% of element 72, but Hevesy worked hard to
purify the material. Shortly after the discovery of hafnium, he found a method to separate it from
zirconium by fractional crystallization of ammonium-zirconium fluoride (in which also some
zirconium was replaced by hafnium) on platinum plates. Due to the large difference in the solubility
of the hafnium- and zirconium-ammonium hexafluorides, the method worked extremely well.

'I am as sure of the existence of hafnium as I was of any other well known element I investigated.' The same day they signed a letter to *Nature* in which they announced their results and said: 'For the new element we propose the name Hafnium (Hafniae = Copenhagen)' (Coster and Hevesy, 1923, p. 79).[630]

Rutherford, who saw Coster and Hevesy's letter to *Nature* before publication, immediately agreed with it. He first congratulated Hevesy, remarking that the discovery was 'an admirable example of cooperation of theory and experiment' (Rutherford to Hevesy, 8 January 1923). At the same time he wrote to Bohr:

> I have seen the paper [of Coster and Hevesy] to *Nature* and I am exceedingly pleased that you have been able to verify your conclusions so rapidly. Urbain sent his letters over to me, obviously with a wish that I would make a note of it in *Nature*, but I had forgotten at that time that your theory fixed the element as an analogue of zirconium. I cannot understand how Dauvillier made the mistake he obviously has. I am sure this confirmation will put you in good spirits for tackling the work of another year. (Rutherford to Bohr, 8 January 1923)

Georges Urbain, on the other hand, was less happy. As Hevesy recalled: 'We had a big fight with Urbain. Urbain said, "Well, I found this element long ago." But he had bad luck. Namely, after hafnium was prepared [in a purer sample], not a single one of perhaps 26 optical lines agreed with those of Urbain' (Hevesy, AHQP Interview, 25 May 1962, p. 15). The optical spectra, referred to by Hevesy, were measured by Hans Marius Hansen and Sven Werner and appeared in the issue of *Nature* of 10 May 1923 (Hansen and Werner, 1923).

The discovery of hafnium, with the properties predicted by his theory, constituted a major triumph for Niels Bohr. So it was no exaggeration when the spectroscopist Heinrich Matthias Konen of Bonn remarked that Bohr's 'achievements in spectroscopy can only be compared with Darwin's contributions to biology' (Konen to Bohr, 16 December 1922). At the turn of the year from 1922 to 1923, the physicists looked forward with enormous enthusiasm towards detailed solutions of the outstanding problems, such as the helium problem and the problem of the anomalous Zeeman effects. However, within less than a year, the investigation of these problems revealed an almost complete failure of Bohr's atomic theory.

---

[630] People in Copenhagen at first disagreed about the name of the new element. Coster and Kramers agreed on hafnium, while Hevesy and Bohr preferred 'danium' (for Denmark). In the proof of the note to *Nature*, Coster and Hevesy requested to change the name hafnium to danium; but due to some misunderstanding it was not corrected in the actual publication. This situation caused some confusion in the public reports on the discovery of element 72. (See Kragh, 1979, p. 186.)